W0049924

The Papovaviridae

Volume 1
THE POLYOMAVIRUSES

THE VIRUSES

Series Editors
HEINZ FRAENKEL-CONRAT, *University of California*
Berkeley, California

ROBERT R. WAGNER, *University of Virginia School of Medicine*
Charlottesville, Virginia

THE VIRUSES: Catalogue, Characterization, and Classification
Heinz Fraenkel-Conrat

THE ADENOVIRUSES
Edited by Harold S. Ginsberg

THE HERPESVIRUSES
Volumes 1–3 • Edited by Bernard Roizman
Volume 4 • Edited by Bernard Roizman and Carlos Lopez

THE PAPOVAVIRIDAE
Volume 1 • Edited by Norman P. Salzman

THE PARVOVIRUSES
Edited by Kenneth I. Berns

THE PLANT VIRUSES
Volume 1 • Edited by R. I. B. Francki
Volume 2 • Edited by M. H. V. Van Regenmortel and Heinz Fraenkel-Conrat

THE REOVIRIDAE
Edited by Wolfgang K. Joklik

THE TOGAVIRIDAE AND FLAVIVIRIDAE
Edited by Sondra Schlesinger and Milton J. Schlesinger

The Papovaviridae

Volume 1
THE POLYOMAVIRUSES

Edited by
NORMAN P. SALZMAN
National Institutes of Health
Bethesda, Maryland

PLENUM PRESS • NEW YORK AND LONDON

Library of Congress Cataloging in Publication Data

The Papovaviridae.

(The Viruses)
Includes bibliographies and index.
Contents: v. 1. The polyomaviruses.
1. Papovaviruses—Collected works. I. Salzman, Norman P. II. Series. [DNLM: 1. Papovaviridae. QW 165.5.P2 P218)

QR406.P36 1986 576'.6468 86-15160

ISBN-13:978-1-4612-9303-3 e-ISBN-13:978-1-4613-2221-4

DOI: 10.1007/978-1-4613-2221-4

© 1986 Plenum Press, New York
Softcover reprint of the hardcover 1st edition 1986

A Division of Plenum Publishing Corporation
233 Spring Street, New York, N.Y. 10013

All rights reserved

No part of this book may be reproduced, stored in a retrieval system, or transmitted in any form or by any means, electronic, mechanical, photocopying, microfilming, recording, or otherwise, without written permission from the Publisher

Contributors

Margaret K. Bradley, Department of Pathology, Harvard Medical School, and the Dana Farber Cancer Institute, Boston, Massachusetts 02115

John N. Brady, Laboratory of Molecular Virology, National Cancer Institute, National Institutes of Health, Bethesda, Maryland 20892

Charles E. Buckler , Laboratory of Molecular Microbiology, National Institute of Allergy and Infectious Diseases, National Institutes of Health, Bethesda, Maryland 20892

Melvin L. DePamphilis, Department of Biological Chemistry, Harvard Medical School, Boston, Massachusetts 02115

Richard J. Frisque, Department of Molecular and Cell Biology, Pennsylvania State University, University Park, Pennsylvania 16802

Roger Monier, Laboratoire d'Oncologie Moléculaire, Institut Gustave-Roussy, Pavillon de Recherche, 49805 Villejuif Cédex, France

Venkatachala Natarajan, Laboratory of Biology of Viruses, National Institute of Allergy and Infectious Diseases, National Institutes of Health, Bethesda, Maryland 20892

Norman P. Salzman, Laboratory of Biology of Viruses, National Institute of Allergy and Infectious Diseases, National Institutes of Health, Bethesda, Maryland 20892

Gerald B. Selzer, Laboratory of Biology of Viruses, National Institute of Allergy and Infectious Diseases, National Institutes of Health, Bethesda, Maryland 20892

Kenneth K. Takemoto, Laboratory of Molecular Microbiology, National Institute of Allergy and Infectious Diseases, National Institutes of Health, Bethesda, Maryland 20892

Duard L. Walker, Department of Medical Microbiology, University of Wisconsin Medical School, Madison, Wisconsin 53706

Kunito Yoshiike, Department of Enteroviruses, National Institute of Health, Shinagawa-ku, Tokyo 141, Japan

Preface

It has been more than twenty years since the isolation of polyoma virus and SV40, and the reports that they could produce tumors in animals and transformation of cells in culture. What was startling was that these biologic properties are associated with viruses that contain genetic information that is able to code for only five or six proteins.

Since that time, investigations with these viruses have been in four principal areas. One major area of study has been on cells transformed by viruses that show altered growth properties and specify new viral and cellular proteins. Transformation studies have focused on the tumor (T) antigens that are specified by the virus and are required to initiate and to maintain the transformed state. Current studies on transformation are summarized in Chapter 4. The second broad area of investigation concerns replication of viruses during a lytic cycle of infection. T-antigens that are the hallmark of transformed cells are also expressed in cells that are lytically infected and are required for viral DNA replication and also function to alter rates of transcription of the early and late viral genes. Except for T-antigen, virus replication depends on the cellular enzymatic machinery and so the description of viral macromolecular synthesis has provided valuable insights into the cellular biosynthetic pathways. These studies are described in Chapters 1–3. The studies that have medical relevance concern JC and BK viruses and there is evidence of widespread exposure of human populations to these agents. These viruses also have provided powerful tools in molecular studies and work with them is described in Chapters 5 and 6. A fourth area of investigation concerns the use of SV40 as a vector in recombinant DNA studies. These DNA recombinant procedures have had an impact on almost all molecular studies and their role in current work is apparent in every chapter of this book.

Most major areas of modern molecular biologic investigation have a counterpart in ongoing studies with the polyomaviruses. It has been the interplay between new molecular approaches and the intriguing biologic questions posed by this group of viruses that has made studies with them such a vigorous and exciting field in which to work.

Norman P. Salzman

Contents

Chapter 3

Replication of SV40 and Polyoma Virus Chromosomes

Melvin L. DePamphilis and Margaret K. Bradley

Chapter 4

Transformation by SV40 and Polyoma

Roger Monier

Chapter 5

Studies with BK Virus and Monkey Lymphotropic Papovavirus

Kunito Yoshiike and Kenneth K. Takemoto

Chapter 6

The Biology and Molecular Biology of JC Virus

Duard L. Walker and Richard J. Frisque

Appendix

Annotated Nucleotide Sequences and Restriction Site Lists for Selected Papovavirus Strains

Charles E. Buckler and Norman P. Salzman

CHAPTER 1

The Papovaviruses
General Properties of Polyoma and SV40

JOHN N. BRADY AND NORMAN P. SALZMAN

I. EARLY STUDIES WITH POLYOMA VIRUS

In the early 1950s, Gross (1953) studied mouse leukemias that developed spontaneously in AK mice. While carrying out experiments on cell-free transmission of mouse leukemia, he noted that while most C3H mice injected with filtrates developed leukemias, some also developed tumors of the parotid gland. This suggested the possibility that filtrates from the leukemic mice contained two distinct viruses, one that produced leukemias and a second that was the agent responsible for parotid gland carcinomas. He obtained evidence for the presence of two viruses by high-speed centrifugation of leukemic extracts; the leukemic agent pelleted while the supernatant fluid produced parotid tumors in newborn mice.

Drs. Steward and Eddy had studied the same system described by Gross and found that the parotid tumor agent was able to produce tumors at multiple sites and proposed the name *polyoma virus* (Greek *poly* many, *oma* tumor). An important finding was that they were able to propagate the virus in cells grown in culture (Stewart *et al.*, 1957). Mouse embryo cultures supported rapid proliferation of the virus and yielded high-titer virus preparations that produced parotid tumors on injection. Purified DNA could be prepared from polyoma virus-infected mouse embryo cell cultures and when added to normal mouse embryo cell cultures, it produced cytopathic effects and infectious virus was recovered from the tissue culture fluid (di Mayorca *et al.*, 1959). Weil (1961) subsequently

JOHN N. BRADY • Laboratory of Molecular Virology, National Cancer Institute, National Institutes of Health, Bethesda, Maryland 20892. NORMAN P. SALZMAN • Laboratory of Biology of Viruses, National Institute of Allergy and Infectious Diseases, National Institutes of Health, Bethesda, Maryland 20892.

extended these studies and determined conditions for quantitation of infectious viral DNA.

While mouse cultures are lysed after being infected by polyoma virus, infected hamster cells do not produce detectable virus but are transformed and carry the virus in a latent form (Dawe and Law, 1959; Negroni *et al.*, 1959; Habel and Atanasiu, 1959; Sachs and Winocour, 1959; Vogt and Dulbecco, 1960, 1962; Eddy, 1960).

Studies on cell transformation by polyoma virus have continued as a major area of investigation and current studies in this exciting research area are described in this volume in the chapter on transformation.

II. EARLY STUDIES WITH SV40

In the early 1950s, an inactivated poliovirus vaccine was developed by Dr. Jonas Salk, and Dr. Albin Sabin had produced a live attenuated poliovirus vaccine. Preparation of both depended on growing the virus in primary rhesus monkey kidney cell cultures. A number of viruses had previously been described that were latent within the monkey kidney, and would on occasion produce cytopathologic effects when the kidneys were dispersed and then grown in culture. In 1960, Sweet and Hilleman described a new simian virus that was present in culture fluids of rhesus monkey kidney cultures. It did not produce cytopathic effects in these cells, but was able to in kidney cell cultures of the African green monkey, *Cercopithecus aethrops.* They also reported that both the Salk and Sabin vaccines contained large quantities of this newly discovered virus. Dr. Sarah Stewart, working in the Division of Biologic Standards at the NIH, was investigating whether there was oncogenic potential associated with latent viruses present in normal monkey kidney cell cultures. In collaborative studies with Dr. Bernice Eddy, extracts from normal rhesus kidney cell cultures were injected into newborn hamsters, of which a high percentage subsequently developed tumors. Drs. Eddy and Stewart were able to show that the oncogenic potential of normal rhesus kidney cell extracts was not due to the presence of polyoma virus. Rather, the agent that produced tumors in hamsters was SV40, the same agent that Sweet and Hilleman had identified as a virus latent in rhesus kidneys. These early observations that polyoma virus and SV40 were able to produce tumors when injected into animals and transform cells in culture stimulated work on all aspects of their biologic properties. Twenty years later, we have accumulated detailed information about these viruses. New aspects of these studies continue and they are described throughout this volume.

III. VIRUS CLASSIFICATION

The Papovaviridae family contains two genera, the papilloma viruses and the polyomaviruses. The first is the subject of a separate volume in this series (see *The Papovaviridae*, Volume 2: *The Papillomaviruses*, edited by N. P. Salzman and P. Howley, Plenum Press, 1987).

In addition to polyoma and SV40, other members of the polyomavirus family are (Melnick *et al.*, 1974):

BK, JCV	Human
K	Mouse
RKV	Rabbit
HaPV	Hamster
STMV	Stump-tailed macaque
Lymphotropic, LPV	Green monkey, human (?)
SA12	Baboon

IV. VIRION STRUCTURE

SV40 and polyoma virions are icosahedral in shape, having a diameter of approximately 45 nm and a sedimentation coefficient of 240 S. The outer virion protein coat is made up of 72 capsomeres. Analysis of electron micrographs originally suggested that the papovavirus capsid consists of 12 five-coordinated (pentons) and 60 six-coordinated (hexons) morphologic units (capsomeres) arranged on a $T = 7d$ icosahedral surface lattice (Finch, 1974; Finch and Crawford, 1975; Klug, 1965). It was originally presumed that the capsids of the $T = 7d$ papovaviruses should be constructed from 420 identical protein subunits, quasi-equivalently bonded into the 12 pentameric and 60 hexameric capsomeres. More recently, Rayment *et al.* (1982) have used X-ray diffraction to produce a 22.5-Å-resolution electron density map of the polyoma capsid. These studies revealed an unexpected substructure in the hexavalent capsomere. The hexavalent morphologic unit is actually a pentamer, demonstrating that the specificity of bonding is not conserved among the protein subunits of the icosahedrally symmetric capsid. The similarities between polyoma virus, SV40, and other members of the papovavirus family suggest that all may have capsids built of pentameric capsomeres.

The molecular weight of both the polyoma and SV40 virion is estimated to be approximately 27×10^6 (Tooze, 1981), with one copy of the viral DNA constituting 12–13% of the total mass of the virion. The DNA molecules extracted from virions are covalently closed, double-stranded circles in a superhelical form (see Section V). SDS–polyacrylamide gel electrophoresis of purified SV40 virions reveals the presence of three virus-encoded proteins: VP1 (45K), VP2 (42K), and VP3 (30K). In polyoma, the molecular weights of VP1, VP2, and VP3 are 47K, 35K, and 23K, respectively (Soeda *et al.*, 1980). The major structural protein, VP1, makes up approximately 75% of the total virion protein. VP2 and VP3 each contribute approximately 5%. In addition to the virus encoded proteins, four cellular histones (Fearson and Crawford, 1972; Fey and Hirt 1975; Roblin *et al.*, 1971)—H2A, H2B, H3, and H4—are associated with the viral DNA in an internal virion chromatin complex. The structural integrity of polyoma and SV40 virions is apparently maintained mainly by calcium ions and disulfide bridges (Hare and Chan, 1968; Friedmann, 1971; Brady *et al.*, 1977, 1980; Ng and Bina, 1981).

V. STRUCTURE OF THE VIRAL DNA

One unique feature of the papovaviruses is that the viral genomes are double-stranded, covalently closed DNA molecules. The unusual properties of polyoma DNA were described in 1963 by Dulbecco and Vogt (1963) and Weil and Vinograd (1963). At this same time, the presence of viral genomes that contained covalently closed circular molecules was described for SV40, rabbit papilloma, and human papilloma (Crawford and Black, 1964; Crawford, 1964, 1965). Some of the physical states of the closed circular DNA are illustrated in Fig. 1. Vinograd and his associates established that the topological properties of form I DNA are described by three features of the molecules. The *topological winding number* (α) is defined as the number of complete revolutions made by one strand about the duplex axis when the axis is constrained to lie in a plane.

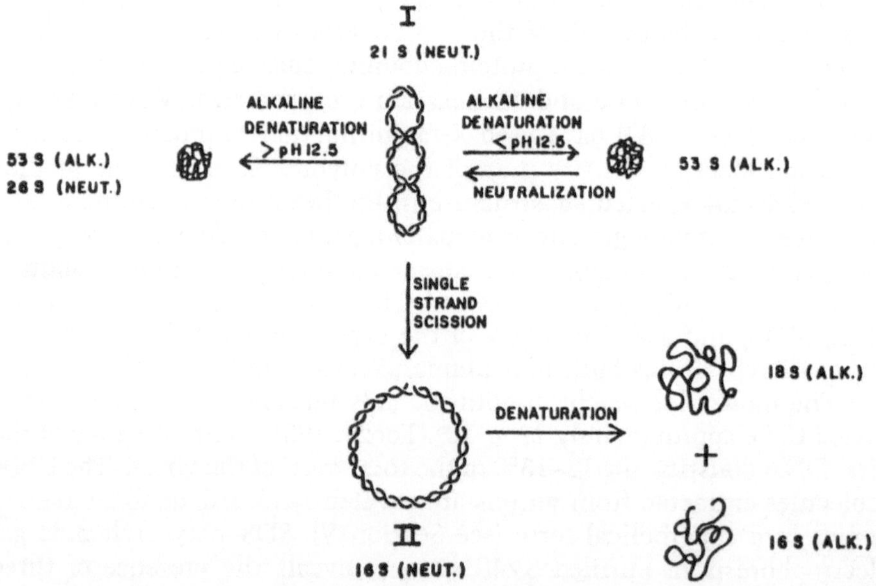

FIGURE 1. Covalently closed DNA that is extracted from virions sediments at 21 S in a neutral sucrose gradient. Cleavage of one or more phosphodiester bonds converts it to a relaxed form that sediments at 16 S in neutral sucrose gradients and when treated with alkali the nicked 16 S DNA is converted to a single-strand circle and a linear single strand that sediment at 18 S and 16 S, respectively. Form I DNA denatured at pH 12.0–12.6 sediments at 53 S and on neutralization, resediments at 21 S. Above pH 12.6, form I sediments at 53 S in alkali, and when neutralized, it sediments at 26 S. Reprinted from Kelley and Nathans (1977).

For a given closed circular duplex molecule, α is a fixed quantity which cannot be changed without interrupting the continuity of one of the two strands. The *duplex winding number* (β) is the number of complete revolutions made by one strand about the duplex axis in the unconstrained molecule, a number that depends on several physical parameters. A difference in the values of α and β will be reflected in tertiary (superhelical) turns in the molecule. Thus, the *superhelix winding number* (τ), defined as the number of revolutions that the duplex makes about the superhelix axis, is given simply by the equation: $\tau = \alpha - \beta$ (Vinograd and Lebowitz, 1966). The number of superhelical turns per 10 base pairs is called the superhelix density, σ.

When DNA is partially denatured by alkali, it loses superhelical turns and assumes a configuration equivalent to DNA II. A similar effect is observed in the presence of critical concentrations of ethidium bromide. When the two strands in DNA I are denatured by alkali at a pH no higher than 12.6, a compact structure that sediments rapidly (53 S) in alkali is obtained. If the alkali-denatured DNA preparations are neutralized, the process of denaturation is reversed and 21 S DNA I is re-formed. At pH values above 12.6, denaturation is no longer a reversible process (Westphal, 1970; Salzman *et al.*, 1973). Similar behavior is observed for the replicative form of ϕX174 (Rush and Warner, 1970). When DNA II is treated with alkali, it gives rise to a linear single strand of DNA (16 S) and a single-stranded circle (18 S).

VI. LATE VIRAL PROTEINS

The late region of polyoma and SV40 code for the viral proteins VP1, VP2, and VP3 and, in SV40, the agnoprotein (Figs. 2 and 3). The major capsid protein, VP1, is encoded in the 3′ end of the late region and is translated from a 16 S mRNA. Minor capsid proteins VP2 and VP3 are encoded at the 5′ end of the late region and are translated from a 19 S mRNA in SV40 and 18 S and 19 S mRNA in polyoma (Tooze, 1981; Soeda, 1980). The sequences that code for the late SV40 agnoprotein are located immediately upstream of the major late RNA initiation site, within the leader sequences.

The major virus-encoded protein, VP1, has both structural and biologic functions in the virion particle (see below). In polyoma, VP1 is composed of six species (designated A–F) (Fig. 4) that can be separated and identified by isoelectric focusing (Bolen *et al.*, 1981; Anders and Consigli, 1983). The pI's of the VP1 species fall between pH 5.75 and 6.75. The differences in isoelectric points are due to the posttranslational modification of the protein by host-cell machinery (Bolen *et al.*, 1981). O'Farrell and Goodman (1976) reported that SV40 virions also possess six distinct VP1 species. The pattern of six species, even though the pI's vary considerably between polyoma and SV40, may thus prove to be a common feature of the papovaviruses.

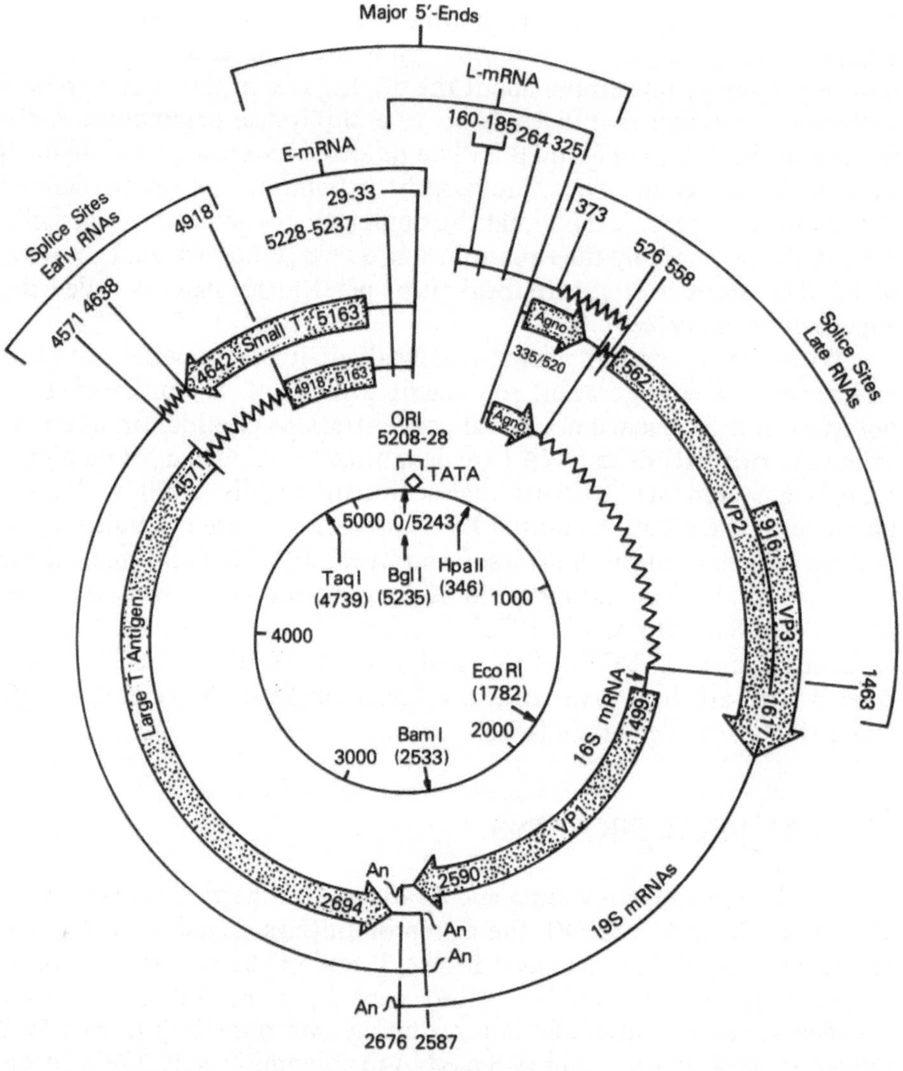

FIGURE 2. Genetic organization of the SV40 genome. The major structural features of DNA transcription, RNA processing, and RNA translation are indicated and their locations are given according to the nucleotide numbering system of Buchman *et al.* (1980). Cleavage sites of five single-site restriction endonucleases are shown for reference. Indicated on the map are the origin of replication (ORI), 5′ and 3′ ends of early (small t- and large T-antigens) and late (16 S and 19 S) mRNAs, and the early gene TATA box. The protein coding regions of mRNAs are designated by open arrows. Within these, the first number locates the A in the AUG codon, and the second number identifies the nucleotide that immediately precedes the termination codon. Intervening sequences which are spliced out of mRNAs are indicated by wavy lines. Early and late transcription proceed on the viral minus and plus strands in counterclockwise and clockwise directions. Only the principal *in vivo* 5′ ends and splice junctions are shown. Adapted from DePamphilis and Wassarman (1983).

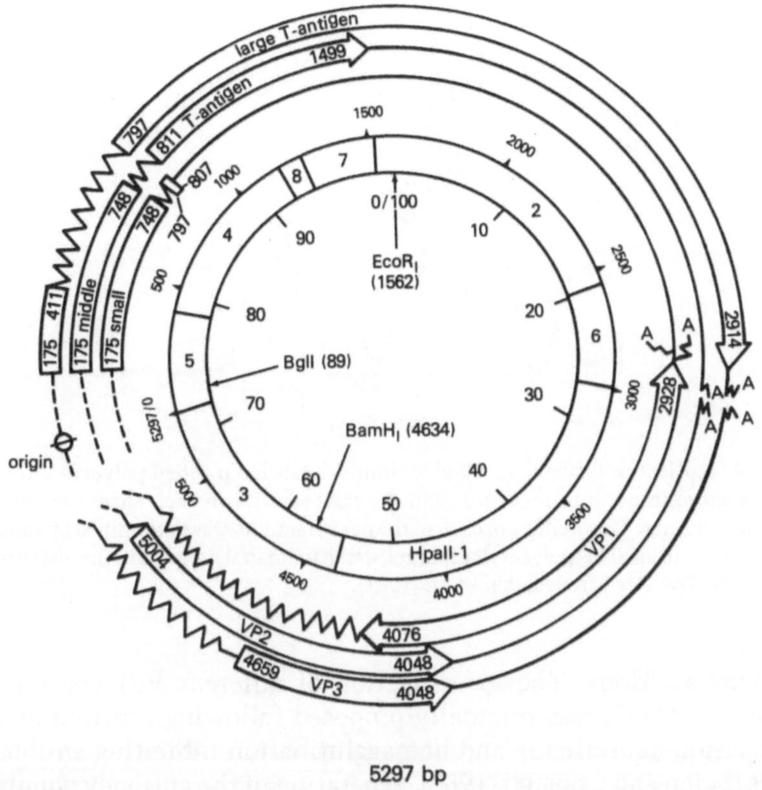

FIGURE 3. Landmarks on the DNA of polyoma virus (A2 strain). The *Hpa*II physical map is shown and has been divided into 100 units, with the single *Eco*RI cleavage site at position 0/100; all other numbers refer to corresponding nucleotide numbers in the primary DNA sequence (see Appendix, this volume). The location on the genome of the viral origin of replication and the six known virus-coded proteins (the three "early" proteins, small, middle, and large T-antigens, and the "late" capsid proteins) are indicated. Within the coding regions, the first number locates the A in the AUG codon, and the second number identifies the nucleotide that immediately precedes the termination codon. The unique coding region for middle T-antigen lies within nucleotides 811–1499, using a reading frame different from that which encodes large T-antigen in the same area of the genome. Additional details of the polyoma A2 strain are contained in a review by Griffin and Dilworth (1983). Adapted from Griffin and Dilworth (1983).

Polyoma VP1 species A was found to be tightly associated with the virion chromatin core (Bolen *et al.*, 1981). Based on comparable studies with SV40 virion chromatin complexes (Brady *et al.*, 1980, 1981), it has been suggested that this species maintains the chromatin core in a compact form and increases the potential accessibility of enzymes, including RNA polymerase, to the viral DNA. Polyoma VP1 isoelectric species D, E, and F are phosphorylated and are of particular interest since they apparently function as viral attachment proteins for the mouse cells (species E) and guinea pig erythrocytes (species D and F) which can be agglutinated

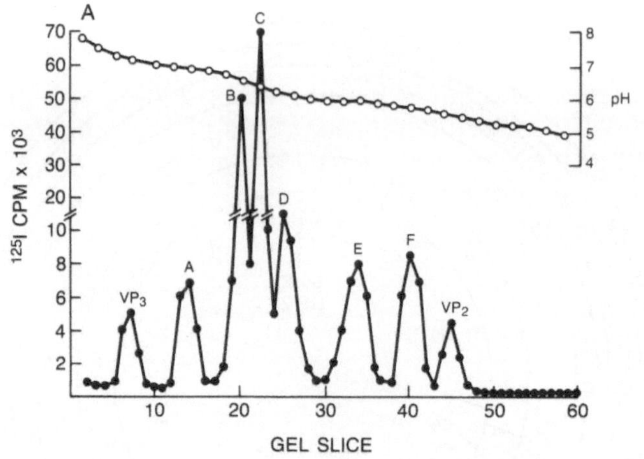

FIGURE 4. Isoelectric focusing of ^{125}I-chloramine-T-labeled purified polyoma virions. The gels were sliced into 2-mm segments, and the radioactivity in each slice was quantitated in a gamma counter. The identity of each of the peak fractions was determined by subsequent electrophoresis in an SDS–polyacrylamide gel, using iodinated proteins in the virion proteins as standards. Reprinted from Bolen *et al.* (1981).

by polyoma virions. The identification of different VP1 species as attachment proteins was originally proposed following separations of polyoma virion-neutralizing and hemagglutination-inhibiting antibody activities (Bolen and Consigli, 1980). Separation of the antibody populations was performed using a polyoma capsid affinity column. Antibodies that eluted from this column did not react with capsids and possessed only virus-neutralizing activity. Antibodies that bound to the polyoma capsid matrix column possessed only hemagglutination-inhibiting activity. VP1 species D and F are common to both polyoma virions and empty capsid shells. Antibodies that specifically inhibit virion or capsid adsorption to and hemagglutination of guinea pig erythrocytes specifically recognize VP1 species D and F. Virion VP1 species E, which is present exclusively in polyoma virions, reacted specifically with the polyoma-neutralizing antibody E and appears to represent the virion protein that is required for specific adsorption to cellular receptors of mouse kidney and mouse embryo cells, suggesting that it is the virion-associated receptor (see below). Recently, Anders and Consigli (1983a) have chemically cleaved polyoma VP1 and have localized the reactive sites of the neutralizing and hemagglutination-inhibiting antibodies to peptides derived from the C-terminal portion of VP1.

The role of SV40 and polyoma proteins VP2 and VP3 in the virion has not been rigorously analyzed. An attractive possibility is that VP2 and VP3 could form the 12 capsomeres (pentons) that lie at the vertices of the icosahedral shell. Following EGTA–DTT dissociation of polyoma virions *in vitro*, it was demonstrated that the 5 S capsomere subunits were enriched in VP2 and VP3 (Brady *et al.*, 1978). While it is tempting to speculate that the 5 S capsomeres are the virion pentons, rigorous proof requires further investigation. Based on recent analysis of the polyoma and SV40 virion chromatin complexes, VP2 and VP3 do not appear to be primarily associated with the viral DNA complexes, as described originally (Huang *et al.*, 1972; Christiansen *et al.*, 1977; Brady *et al.*, 1978, 1980).

The late region of SV40 encodes in its leader region a 61-amino-acid highly basic polypeptide, termed the agnoprotein (Jay *et al.*, 1981). Several viable deletion mutants in the agnogene coding sequence have been characterized, clearly indicating that this protein is nonessential in tissue culture cells. Nevertheless, most of these mutants grow with decreased efficiency and produce plaques of altered morphology. There is a curious lack of correlation between the size and position of the late leader deletions and the relative growth rate of the mutants. There is evidence from both genetic and biochemical studies to suggest that the agnoprotein plays a role late in the lytic cycle, perhaps in the assembly of virions (Ng *et al.*, 1985b). In addition, based on its DNA-binding properties (Jay *et al.*, 1981) as well as some of the characteristics of the deletion mutants, there is speculation that the agnoprotein may also directly affect the synthesis and/or processing of late mRNA (Alwine, 1982; Hay *et al.*, 1982) (see Chapter 2). Due to the complexity and overlapping of SV40 control sequences and the ill-defined nature of the late transcriptional signals, it has been difficult to ascribe the defects in agnogene mutants to either the loss of the agnoprotein and/or alterations in either the template or control signals for late mRNA synthesis which result from the various mutations in the agnoprotein.

Recently, a 2-bp insertion mutant in2379 was created in the coding region of this protein (Nomura *et al.*, 1986) which prevents the synthesis of the agnoprotein, and, in contrast to the more extensive mutants previously described in this region, might be expected to have a lesser effect on the template for late viral transcripts. The 2-bp insertion mutant grew significantly less well than larger deletions in the agnoprotein region, and interestingly gave rise to frequent second-site alterations in this region. All of the second-size mutations grew more efficiently than the original 2-bp insertion mutant.

VII. EARLY EVENTS IN PAPOVAVIRUS INFECTION

The uptake of virus by host cells has been studied most extensively with polyoma virus. Three distinct polyoma particles can be isolated from an infected mouse cell. Purified polyoma virions, which are the infectious agent, contain a full complement of viral structural proteins (VP1, VP2, and VP3), cellular histones (H2A, H2B, H3, and H4), and encapsidated viral DNA. All six isoelectric species of VP1 are found in highly purified virion preparations (Table I). Polyoma capsids, on the other hand, contain viral structural proteins VP1, VP2, and VP3, but lack cellular histones and viral DNA. The major species of VP1 protein found in polyoma capsids are isoelectric species B, D, and F. The third class of particles, polyoma pseudovirions, contain a full complement of viral structural protein VP1 (all six isoelectric species), VP2, and VP3, cellular histones, and fragments of mouse host-cell DNA. Early studies of polyoma virus adsorption, which presented conflicting pathways of virus uncoating and transport, were performed before there was an awareness of the capsid and pseudovirion populations (Bourgaux, 1964; Mattern *et al.*, 1966).

Using purified polyoma virions, pseudovirions, and capsids, a detailed analysis of the early events of polyoma infection has been performed by Consigli and his colleagues. By electron microscopy and autoradiography, MacKay and Consigli (1976) demonstrated that viral coat proteins and DNA arrive simultaneously in the host cell nuclei, as early as 15 min postinfection. Interestingly, when virions, pseudovirions, and capsids were used to infect mouse cells, only the virions and pseudovirions appeared to enter the nucleus. Polyoma capsids were found to be specifically located in cytoplasmic lysosomes, apparently targeted for degradation by proteases and nucleases. These early studies suggested that a nuclear transport factor(s) was present in polyoma virions and pseudovirions, but ab-

TABLE I. Isoelectric Focusing of Polyoma Virion and Capsid Structural Proteins

Protein	Species	pI	% of total[a]	$^{32}P/^{3}H$[b]	$^{14}C/^{3}H$[c]
Virions					
VP1	A	6.75	8.6	0.2	
	B	6.50	18.0		0.05
	C	6.30	37.8		0.25
	D	6.10	13.2	0.40	
	E	5.90	6.6	0.60	
	F	5.75	5.8	0.80	
VP2		5.50	4.8	1.00	
VP3		7.00	5.2	1.10	
Capsids					
VP1	A	6.60	34.6		0.20
	B	6.50	18.5		
	D	6.10	20.9	0.20	0.40
	F	5.75	16.0	0.40	0.30
VP2		5.50	5.4	0.80	
VP3		6.50	4.6	1.00	

[a] The area of the peak for each of the proteins and protein species is expressed as the percentage of the total area occupied by the protein or protein species present in the sample. Each value is the average of a number of separate determinations of pooled ^{3}H-amino acid-labeled samples.
[b] Ratio of ^{32}P-labeled proteins to ^{3}H-amino acid-labeled proteins.
[c] Ratio of ^{14}C-acetate-labeled proteins to ^{3}H-amino acid-labeled proteins.

sent in polyoma capsids (see below). Subsequently, using radiolabeled virus, it was demonstrated the following uncoating of polyoma virus in the nucleus of the host cell, a specific uncoating intermediate (190 S) consisting of the viral DNA protein complex in association with a structure of host origin could be isolated (Winston *et al.*, 1980). SDS–PAGE of the 190 S uncoating intermediate suggested the presence of nuclear matrix proteins. The role of the uncoating intermediate in polyoma viral DNA replication and early transcription remains to be established.

The nuclear transport recognition factor referred to above is apparently a biologic function that resides in the virus encoded protein VP1. More specifically, virion VP1 species E appears to represent the virion protein that is required for successful virus adsorption to specific cell receptors, which results in infection of mouse kidney cells. Thus, VP1 species E apparently fulfills the role of the polyoma virion-associated attachment protein or nuclear recognition transport factor mentioned above. Two lines of evidence support this contention. First, VP1 species E is found only in virions, never in capsids. Since, as discussed above, capsids do not compete with virions for specific cell receptors, it seems logical that capsids would lack the corresponding protein present in virions (Bolen and Consigli, 1979). Second, antibodies whose only known activity is to inhibit virion adsorption to mouse kidney cells, specifically recognize distinct antigenic determinants only on VP1 species E in immune precipitation experiments (Bolen *et al.*, 1981).

Consigli and colleagues have pursued the isolation of the mouse cell receptor by preparing immunologic reagents that could be used to both identify and isolate mouse kidney cell receptor proteins. They prepared a polyclonal antibody to an SDS–PAGE-isolated cross-linked virion–mouse kidney cell receptor complex and prepared an anti-idiotype (polyconal) antibody to the Fab fragments of a monoclonal antibody that reacted specifically with VP1 isoelectric species D, E, and F. The anti-idiotype serum was found to react positively with the monoclonal antibody but not with polyoma proteins. In addition, the anti-idiotype antibody was capable of attaching to the surface of mouse kidney cells, as well as competing with polyoma virus for cell receptors and preventing infection as determined by indirect immunofluorescence. The anti-idiotype antibody has recently been used in Western blots to identify mouse cell receptor proteins. Reactivity with seven mouse kidney cell membrane proteins with molecular weights of 12, 23, 28, 30, 38, 52 and 110K was observed (Griffith and Consigli, 1984; Marriot and Consigli, 1985; Consigli *et al.*, 1986).

VIII. PAPOVAVIRUS HOST-RANGE RESTRICTION

Polyoma virus and SV40 interact with host cells in two distinct ways. A productive infection results when the virus replicates in the cell nucleus, forming progeny virus and eventually causing death of the host cell. The second type of interaction, termed nonproductive, results when the infected cell is transformed and acquires the properties of a malignant cell. The species of cells that is infected is the primary determinant of whether a productive or a nonproductive infection occurs. For example, with SV40, infection of monkey kidney cells leads to a productive infection. In contrast, infection of mouse or hamster cells leads to a nonproductive pathway and eventual transformation of the host cell. Several elements, including the enhancer elements, DNA replication origin, and early gene functions, apparently contribute to the host range of the papovaviruses.

The enhancer element of both SV40 and polyoma is located between the early and late coding sequences (Tooze, 1981; see Chapter 2). This transcriptional regulatory sequence is absolutely required for early gene expression *in vivo* (Benoist and Chambon, 1981; Gruss *et al.*, 1981; Fromm and Berg, 1982), and thus is important for manifestation of the viral lytic cycle and transformation of nonpermissive cells. At the transcriptional level, the enhancer elements of both SV40 and polyoma are somewhat promiscuous and function at reasonable levels in most cells except undifferentiated cells (Gluzman and Shenk, 1983; de Villiers *et al.*, 1981, 1982; Khoury and Gruss, 1983; Levinson, *et al.*, 1982; Laimins *et al.*, 1982) (see below). For example, when the SV40 early promoter and enhancer are linked to the bacterial chloramphenicol acetyltransferase (CAT) gene and transfected into monkey kidney cells and mouse cells, the efficiency of enhancer-dependent gene expression is about 5-fold greater in the monkey kidney cell line (Laimins *et al.*, 1982). The decreased efficiency of SV40 early transcription in mouse cells, and thus T-antigen expression, is not, however, the sole determinant of the restricted host cell specificity of SV40. Graessmann *et al.*, (1981) have demonstrated in microinjection experiments that while elevation of DNA template number allows production of late viral protein VP1, the block to SV40 DNA replication, and thus the full lytic cycle, is not observed in mouse cells. Similarly, using an inducible metallothionein promoter linked to the SV40 early gene, Gluzman *et al.* (personal communication) have demonstrated that an increase in the level of T-antigen by > 10-fold in mouse cells does not result in efficient SV40 DNA replication. These results suggest that while the level of early gene expression controlled by these enhancer elements may be maximal in the cell that serves as the lytic host for the virus, the enhancer is not the principal determinant of host range in all cases. It seems more likely that replication-specific proteins that interact with the SV40 replication origin in monkey kidney cells and with the polyoma replication origin in mouse cells permit full expression of the lytic cycle.

There are situations in which a papovavirus enhancer element appears to specifically determine the ability of a virus to be expressed in a particular host cell. One of these mentioned above is the expression of polyoma virus EC mutants in embryonal carcinoma cells. Although the wild-type virus is unable both to express its early gene product efficiently and to replicate in undifferentiated cells, mutants in the enhancer region (Py-EC mutants) overcome both of these restrictions (Katinka *et al.*, 1980; Fujimura *et al.*, 1981; Sekikawa and Levine, 1981; Fujimura and Linney, 1982). The polyoma virus enhancer element contributes both to the expression of early viral RNA (which encodes the replication-required protein, T-antigen), and to a separate but essential requirement for viral DNA replication (de Villiers *et al.*, 1984; Veldman *et al.*, 1985). Whether the specific role of the enhancer in DNA replication is linked to its ability to stimulate transcription and/or its function in chromatin organization is unclear.

The human papovavirus, JCV, is closely associated with the degenerative neurologic disease progressive multifocal leukoencephalopathy (PML). The virus replicates efficiently only in human glial cells in tissue culture and has been repeatedly isolated from demyelinated plaques found in the brains of patients with this disease. Recent studies have shown that the JCV enhancer element functions efficiently only in glial cells (Kenney *et al.*, 1984). Thus, the tissue-specific transcriptional efficiency of the JCV enhancer element directly correlates with the tissue specificity of the disease. Whether or not there are additional restrictions (such as a replication-associated protein) that limit the growth of JCV to human glial cells is a question presently under investigation.

Major *et al.* (1985) have obtained an SV40-transformed human glial cell line that constitutively produces SV40 T-antigen, and continues to express a number of cell markers specific for glial cells. JCV replicates efficiently in these cells and its replication appears to be dependent on the production of JCV T-antigen from the transfected genome rather than on the endogenous SV40 T-antigen. These results differ from those of Li and Kelly (1985) who have shown that replication of JCV DNA *in vitro* can be supported by SV40 T-antigen with 20% of the efficiency of SV40 DNA replication *in vitro*. However, the present *in vitro* replication assay scores for only a subset of the determinants required for efficient *in vivo* replication. For example, there is no benefit seen *in vitro* after addition of transcriptional control sequences (such as the 21-bp repeats of SV40) to the minimal SV40 origin of replication (Li and Kelly, personal communication). *In vivo,* addition of these sequences stimulates replication at least five-fold (Bergsma *et al.*, 1982; Peden *et al.*, personal communication). The transformed human glial cell line should be important in determining the role of the enhancer element in the host-range restriction and disease spectra of JCV.

In addition to tissue culture systems and viral infections of animal models, recent technologic advances have allowed investigators to introduce foreign or chimeric constructs into the genomes of mice. Although experiments involving transgenic mice are limited, these studies have already demonstrated the importance of enhancer elements in the tissue-specific expression of SV40. Recent experiments suggested that the expression of the SV40 early genes, as evidenced by T-antigen production or tumor formation, occurs specifically in the choroid plexus of the cerebral ventricles (Brinster *et al.*, 1984). Another study in which the SV40 T-antigen has been linked to regulatory elements of the insulin gene showed that expression of the recombinant transgene was restricted to the β cells of the pancreas (Hanahan, 1985).

The host-range transformation-defective (hr-t) mutants of polyoma virus were selected originally for their ability to grow on polyoma virus-transformed 3T3 cells and poorly on normal 3T3 cells (Benjamin, 1970). Analysis of the growth of hr-t mutants of different cell types has led to the hypothesis that cellular permissive factors are required for efficient virus production and that the function of the hr-t gene is to regulate or induce the permissive factors (Staneloni *et al.*, 1977). Recently, Garcea and Benjamin (1983) have demonstrated that polyoma virus host-range mutants are blocked in virion assembly. In normal 3T3 cells, the hr-t mutants synthesize approximately 30–40% as much viral DNA and 80–100% as much capsid proteins as the wild-type virus. Yet, the hr-t mutants produce only 1–2% as much infectious virus. The hr-t mutant apparently fails to undergo efficient transition from 95 S replicating minichromosome to 240 S virion structures (see Section X). The block in assembly of viral particles is accompanied by the failure to induce a series of posttranscriptionally modified forms of VP1 (see above). Based on these results, it was proposed that the late viral protein VP1 is a target for the hr-t gene-controlled modification and that the modified forms are essential for encapsidation of the viral minichromosomes. The lack of VP1 modification, and thus virus production, was observed in the nonpermissive 3T3 cells but not in baby mouse kidney cells.

IX. CHROMATIN STRUCTURE

The minichromosome of SV40 has been studied extensively in an attempt to understand higher-order chromatin structure and its effect on transcription and replication. The digestion products of SV40 minichromosomes by micrococcal nuclease are similar to those of cellular chromatin (Bellard *et al.*, 1976; Ponder *et al.*, 1978; Noll, 1974). The average repeat length of DNA oligomers from SV40 minichromosomes is 187 ± 11 bp, which differs only slightly from that observed for the host monkey kidney cell, 186 ± 6 bp (Shelton *et al.*, 1978). Electron microscopy of isolated minichromosomes indicates that approximately 21 ± 2 nucleosomes are present. There is still a controversy as to the distribution of nucleosomes on the SV40 DNA. Some reports suggest that the location of nucleosomes is random (Polisky and McCarthy, 1975; Cremisi *et al.*, 1976; Shelton *et al.*, 1980) while other reports indicate that the nucleosomes are phased in a nonrandom manner (Das *et al.*, 1979; Ponder and Crawford, 1977; Persico-Dilauro *et al.*, 1977). It is clear from several studies, however, that certain regions of the SV40 DNA lack a normal nucleosome conformation in at least a percentage of the viral minichromosomes.

Analysis of the nuclease digestion products of SV40 chromatin isolated early and late in the lytic infection cycle has demonstrated a region of preferential sensitivity near the origin of replication (Scott and Wigmore, 1978; Waldeck, 1978; Varshavsky *et al.*, 1979; Gerard *et al.*, 1982). Electron microscopy has also suggested that a segment of viral DNA, approximately 350 to 400 bp in length, and located between the SV40 origin of replication and late coding sequences, is nucleosome free (Jakobovits *et al.*, 1982; Saragosti *et al.*, 1980). Herbomel *et al.* (1981) have reported that polyoma minichromosomes also contain a DNase-sensitive region on the late side of the origin of replication. Since this region of DNA contains all the known replication and transcriptional regulatory control sequences, including the origin of replication, T-antigen binding sites, early TATA box, 21-bp repeats, 72-bp enhancer repeats, and early and late RNA initiation sites, there has been considerable interest in defining the DNase-hypersensitive sites and elements responsible for maintenance or induction of the open chromatin conformation.

Cereghini *et al.* (1983) and Saragosti *et al.* (1980) have examined the DNase-hypersensitive sites in SV40 chromatin complexes. Nuclei isolated from SV40-infected cells were treated briefly with nuclease and redigested with restriction enzymes producing one or two cuts in the viral DNA. Southern blot analysis was then performed using DNA probes located next to the cleavage site. The highest frequency of DNase I cuts was observed to occur between the SV40 *Bgl*I site (m.p. 5243/0) and the *Hpa*II site at m.p. 346. The entire region was not uniformly sensitive since discrete bands were observed over the background of nondefined fragments. One DNase I-hypersensitive site was localized near the palindrome at the origin of replication. This site could be further separated into three bands corresponding to DNase cleavage sites at m.p. 5212, 1, and 30 of the SV40 genome. This sensitive region is followed by a stretch of 60 bp, which was refractory to nuclease action. The resistant region contains the 21-bp repeats and is followed by a highly sensitive region in which up to five hypersensitive sites could be mapped. Four of the five sites are located in the two SV40 72-bp repeats, in approximately identical positions in each of them. The positions of the hypersensitive sites were mapped to bp 110 ± 10, 165 ± 10, 185 ± 10, and 240 ± 10.

More recently, Cereghini and Yaniv (1984) have analyzed the change in SV40 DNase I-hypersensitive sites that are associated with DNA replication following transfection of CV-1 and COS-1 monkey kidney cells. Several points of interest were obtained from these studies. First, the pattern of DNase I hypersensitivity in the origin–promoter–enhancer region typical of the lytic cycle is observed in replicated DNA in the absence of late viral proteins. Second, the pattern of DNase I-hypersensitive sites differs between nonreplicating and postreplicating chromatin. In the absence of DNA replication, only three hypersensitive sites are mapped in the origin–promoter–enhancer region. Two of these were located in the 72-bp repeats. Sequences around the *Bgl*I site were in a relatively unexposed conformation. It was further determined that the presence of SV40 large T-antigen is not sufficient for the shift in the structure of nonreplicated versus replicated chromatin. These studies suggest that the physical process of DNA replication can modulate protein–DNA interaction and thus chromatin structure.

Innis and Scott (1984) have also analyzed the correlation of SV40 chromatin structure and DNA replication, but in a somewhat different manner. The SV40 origin of replication lies at one end of the segment of DNA that is preferentially sensitive to cleavage in SV40 chromatin. It has been established, using duplicated mutants of SV40, that deletion of sequences within the origin of replication had little effect on the overall nuclease sensitivity of the region. Since deletions and point mutations within *ori* abolish DNA replication, it was concluded that replication from that origin was not necessary for the establishment of the nucleosome-free region. Innis and Scott demonstrated that insertion of DNA segments into the nuclease-sensitive region of SV40 alters both replication efficiency and chromatin structure. Mutants containing insertions > 42 bp between the SV40 origin of replication and 21-bp repeats replicated poorly. This effect was *cis*-acting and independent of nucleotide sequence of the insert. The overall nuclease-sensitive chromatin structure was retained in these mutants, but the pattern of DNase I cleavage sites was displaced in the late direction from the origin of replication. Cleavage sites appeared in the inserted sequences, suggesting that the regulatory elements specifying the early border of the nuclease-sensitive region are located on the late side of the inserts, within either the SV40 21-bp repeats or 72-bp repeat enhancer element. In addition, these results support the conclusion that efficient function of the viral origin of replication is correlated with its proximity to the nucleosome-free chromatin structure.

Several laboratories have used specially constructed plasmids, containing deletions within the SV40 control region, to determine the sequences that control maintenance or induction of the nucleosome-free region in SV40 chromatin. As indicated above, the sequential displacement of the origin hypersensitive sites by insertion of sequences between the SV40 *ori* and 21-bp repeats, led Innis and Scott (1984) to conclude that sequences within the 21-bp repeats or 72-bp repeats controlled the early border of the open chromatin structure. In a detailed study of the SV40 DNase I-sensitive region, Jongstra *et al.* (1984) concluded that two SV40 control elements, the 21-bp repeats and the 72-bp repeats, induce chromatin structures of increased nuclease sensitivity when transposed elsewhere on the viral genome. In this study, the investigators defined two main regions of DNase hypersensitivity. Region I was centered over the SV40 origin of replication, while region II spanned the 72-bp repeats and approximately 75 bp to the late side of the enhancer repeats. Generation of a full-length region II nuclease sensitivity was dependent upon sequences within the 72-bp repeat and a separate set of sequences located between the *Pvu*II site (m.p. 272) and *Hpa*II site (m.p. 346) to the late side of the repeats. Region I nuclease sensitivity was induced only when the 21-bp repeats were present together with sequences between the SV40 *Hin*dIII site (m.p. 5171) and *Bgl*I (m.p. 5243/0). In addition, the 21-bp repeat had a distinct qualitative effective on the generation of increased DNase I sensitivity over the 72-bp repeat, leading to a well-defined early border.

Fromm and Berg (1983) have analyzed the appearance of DNase I-hypersensitive sites when the 72-bp repeat was translocated to other sites in the SV40 genome. Similar to the results presented above, the investigators concluded that the 72-bp repeat induced nuclease hypersensitivity. Their results differed somewhat from the conclusions of Jongstra *et al.* (1984) concerning the orientation independence of the 72-bp repeat in inducing the hypersensitive region. Jongstra *et al.* (1984) concluded that the 72-bp repeat induces an increase in DNase I sensitivity independent of orientation, while Fromm and Berg (1983) concluded that the 72-bp enhancer element induced an increase in DNase I sensitivity in only one orientation.

X. VIRUS ASSEMBLY

The first stable SV40 chromatin complex that can be isolated after DNA replication is a compact, 30-nm particle, which has a sedimentation value of 70–75 S and is characterized by the presence of cellular histones H1, H2A, H2B, H3, and H4, apparently in stoichiometric amounts (Baumgartner *et al.*, 1979; Bellard *et al.*, 1976; Coca-Prados and Hsu, 1979; Coca-Prados *et al.*, 1980; Fernandez-Munoz *et al.*, 1979; Garber *et al.*, 1978; LaBella and Vesco, 1980; Nedospasov *et al.*, 1978; Seidman *et al.*, 1979). The products of micrococcal nuclease digestion, as indicated above, are similar to those of higher-order host cell chromatin, giving a nucleosomal DNA ladder pattern with a repeat unit of approximately 187 bp of DNA (Bellard *et al.*, 1976; Shelton *et al.*, 1980; Sundin and Varshavsky, 1979). The distribution of nucleosomes within the 75 S complex is not completely random since a 350- to 400-bp length of DNA, extending from the origin of replication and extending into the late region, appears to be free of nucleosomes (Nedospasov and Georgiev, 1980; Saragosti *et al.*, 1980; Scott and Wigmore, 1978; Varshavsky *et al.*, 1979) (see Section IX).

Using procedures that permit the extraction of SV40 chromatin complexes without destroying their distinguishing structural composition, pulse–chase experiments demonstrate that DNA and protein components of the 75 S complex mature with time into a 200 S previrion, a 240 S nuclear (immature) virion, and finally into an extracellular mature virion (Baumgartner *et al.*, 1979; Coca-Prados and Hus, 1979; Coca-Prados *et al.*, 1980; Fernandez-Munoz *et al.*, 1979; Garber *et al.*, 1978; LaBella and

Vesco, 1980; Seidman *et al.*, 1979). As these transitions occur, three important changes in the viral chromatin composition have been reported. First, the acetylation level of histones H3 and H4 in the previrion, nuclear virion, and extracellular virion are significantly higher than in either 75 S chromatin or cellular chromatin. The increased level of acetylation has been reported to be the result of a decreased rate of histone deacetylation (LaBella *et al.*, 1979, 1980). Second, during the final transition of nuclear to extracellular virions, histone H1 is either specifically degraded or displaced from the virion, since nuclear, but not extracellular, virion chromatin contains H1. Third, the specific association of nonhistone viral proteins with the mature, extracellular virion chromatin complex has been demonstrated (Brady *et al.*, 1980).

Recently, the isolation of polyoma virus assembly intermediates from infected mouse embryo cells has been described (Yuen and Consigli, 1983, 1985; Garcea and Benjamin, 1983). Sucrose gradient profiles revealed the presence of 90 S, 200 S, and 240 S intermediates. These intermediates were shown to be sensitive to a number of factors, including ionic conditions of the isolation buffer, presence of chelating agents and nonionic detergents during isolation, and sonication of nuclei during extraction. The sensitivity of the polyoma assembly intermediates explains the problems in detecting these species using extraction conditions optimized for SV40 chromatin complexes (Green *et al.*, 1971; McMillen and Consigli, 1974; Seebeck and Weil, 1974). Pulse–chase experiments demonstrate that the order of formation of the polyoma intermediates proceeds from 90 S → 240 S, with the 200 S as a likely intermediate. Histone H1 was only found associated with the 90 S species. Yuen and Consigli (1985) have demonstrated by two-dimensional gel electrophoresis that the phosphorylated species of VP1 contained in the three intermediates differed. The less basic VP1 species D, E, and F (see Section VI) were present in higher percentages in the 90 S intermediate. The 240 S and 200 S intermediates contained predominantly the more basic VP1 species A, B, and C. Since the only detectable difference between the 240 S and 200 S intermediates was their isoelectric forms of VP1, these results suggest that either specific VP1 species were added to the intermediates during maturation or modification of VP1 takes place directly on the assembly intermediates.

Early complementation analysis revealed that mutations which occur in the gene coding for the major capsid protein, VP1, can be subdivided into three groups: tsB, tsC, and tsBC (Chou and Martin, 1975; Lai and Nathans, 1976). Each of these groups shows defects in virion assembly at 40°C. In tsB-infected cells, semiassembled virions are produced; in tsC-infected cells, the initiation of assembly is blocked; and in tsBC mutants, capsid–SV40 chromatin complexes accumulate (Bina *et al.*, 1983a,b; Blasquez *et al.*, 1983; Ng and Bina, 1984; Ng *et al.*, 1985a). Ng *et al.* (1985a) have determined the DNA sequence of eight tsBC mutants and compared them to wild-type SV40, polyoma virus, and BK virus. Although there are additional mutations in each virus, it is of interest that tsBC mutants

BC208, BC214, BC216, BC217, BC248, and BC274 share the same point mutation at SV40 nucleotide 2534, changing a proline to serine at VP1 amino acid residue 286. This residue is also a proline in VP1 of polyoma virus.

The data obtained from the identification of polyoma and SV40 assembly intermediates and their structural protein composition suggest that the mechansim of papovavirus assembly involves the gradual addition of structural proteins VP1, VP2, and VP3 into DNA–protein complexes. Results from *in vitro* reassembly of polyoma capsomeres and chromatin complexes into infectious polyoma virions support the same assembly mechanism (Brady *et al.*, 1979; Yuen and Consigli, 1982).

XI. CONCLUDING REMARKS

Due to the vast literature on the papovaviruses, this chapter is not a comprehensive summary. Rather, we have tried to touch on the subjects that have seen rapid growth in the past 3 to 5 years. We refer the reader to excellent reviews which may cover topics not discussed in detail here. These include:

Aloni, Y., 1981, Splicing of viral mRNAs, *Prog. Nucleic Acid Res. Mol. Biol.* **25**:1.

Basilico, C., 1984, The mechanism of cell transformation by SV40 and polyoma virus, *Pharmacol. Ther.* **26**:235.

Benjamin, T., 1982, The hr-T gene of polyoma virus, *Biochim. Biophys. Acta* **695**:69.

Consigli, R., and Center, M., 1978, Recent advances in polyoma virus research, *CRC Crit. Rev. Microbiol.* 263.

Crawford, L., 1983, The 53,000-dalton cellular protein and its role in transformation, *Int. Rev. Exp. Pathol.* **25**:1.

Das, G. C., and Niyogi, S. K., 1981, Structure, replication, and transcription of the SV40 genome, *Prog. Nucleic Acid Res. Mol. Biol.* **25**:187.

DePamphilis, M. L., and Wassarman, P. M., 1982, Organization and replication of papovavirus DNA, in: *Organization and Replication of Viral DNA* (A. S. Kaplan, ed.), pp. 37–114, CRC Press, Boca Raton, Fla.

Eckhart, W., 1981, Polyoma T antigens, *Adv. Cancer Res.* **35**:1.

Fried, M., and Griffin, B., 1977, Organization of the genome of polyomavirus, *Adv. Cancer Res.* **24**:67.

Georgiev, G. P., Bakayev, V. V., Nedospasov, S. A., Razin, S. V., and Mantieva, V. L., 1981, Studies on structure and function of chromatin, *Mol. Cell. Biochem.* **9**:29.

Graessmann, A., Graessmann, M., and Mueller, C., 1981, Regulation of SV40 gene expression, *Adv. Cancer Res.* **35**:111.

Griffin, B. E., and Dilworth, S. M., 1983, Polyomavirus: An overview of its unique properties, *Adv. Cancer Res.* **39**:183.

Hand, R., 1981, Functions of T antigens of SV40 and polyomavirus, *Biochim. Biophys. Acta* **651**:1.

Ito, Y., and Segawa, K., 1983, Mechanism of oncogenic transformation by polyoma virus, *Tampakushitsu Kakusan Koso* **28**:883.

Kingston, R. E., Baldwin, A. S., and Sharp, P. A., 1985, Transcription control by oncogenes, *Cell* **41**:3.

Levine, A. J., 1982, The nature of the host range restriction of SV40 and polyoma viruses in embryonal carcinoma cells, *Curr. Top. Microbiol. Immunol.* **101**:1.

Levine, A. J., 1982, Transformation-associated tumor antigens, *Adv. Cancer Res.* **37**:75.

Levine, A. J., Reich, N., and Thomas, R., 1983, The regulation of a cellular protein, p53, in normal and transformed cells, *Prog. Clin. Biol. Res.* **119**:159.

Liberski, P. P., 1983, Slow viruses of the central nervous system. IV. Progressive multifocal leukoencephalopathy, *Postepy Hig. Med. Dosw.* **37**:389.

Magnusson, G., 1985, Recent progress in studies of polyomavirus tumour antigens, *Exp. Cell Res.* **157**:1.

Martin, R. G., 1981, The transformation of cell growth and transmogrification of DNA syntehsis by simian virus 40, *Adv. Cancer Res.* **34**:1.

Mathis, D., Oudet, P., and Chambon, P., 1980, Structure of transcribing chromatin, *Prog. Nucleic Acid Res. Mol. Biol.* **24**:1.

Mora, P. T., 1982, The immunopathology of SV40-induced transformation, *Springer Semin. Immunopathol.* **5**:7.

Mora, P. T., and Chandrasekaran, K., 1983, Simian virus 40-coded antigens and the detection of a 55K-dalton cellular protein in early embryo cells, *Biomembranes* **11**:259.

Norkin, L. C., 1982, Papovaviral persistent infections, *Microbiol. Rev.* **46**:384.

Sack, G. H., Jr., 1981, Human cell transformation by simian virus 40—A review, *In Vitro* **17**:1.

Schaffhausen, B., 1983, Transforming genes and gene products of polyoma and SV40, *CRC Crit. Rev. Biochem.* **13**:215.

Shevliagin, V. I., 1981, New viruses of the family of Papovaviridae isolated from man, *Usp. Sovrem. Biol.* **92**:338.

Smith, A. E., 1984, Oncogenes: Growth regulation and the papovaviruses polyoma and SV40, *J. Cell. Biochem.* **2**:89.

Tjian, R., 1981, Regulation of viral transcription and DNA replication by the SV40 large T antigen, *Curr. Top. Microbiol. Immunol.* **93**:5.

Türler, H., 1980, The tumor antigens and the early functions of polyomavirus, *Mol. Cell. Biochem.* **32**:63.

Weil, R., 1978, Viral "tumor antigens." A novel type of mammalian regulatory protein, *Biochim. Biophys. Acta* **516**:301.

ZuRhein, G. M., 1983, Studies of JC virus-induced nervous system tumors in the Syrian hamster: A review, *Prog. Clin. Biol. Res.* **105**:205.

ACKNOWLEDGMENT. We wish to express our sincere gratitude to Marie Priest for editing and preparation of the manuscript.

REFERENCES

Alwine, J. C., 1982, Evidence for simian virus 40 late transcriptional control: Mixed infections of wild-type simian virus 40 and a late leader deletion mutant exhibit *trans* effects on late viral RNA synthesis, *J. Virol.* **42**:798.

Anders, D. G., and Consigli, R. A., 1983a, Chemical cleavage of polyomavirus major structural protein VP1: Identification of cleavage products and evidence that the receptor moiety resides in the carboxy-terminal region, *J. Virol.* **48**:197.

Anders, D. G., and Consigli, R. A., 1983b, Comparison of nonphosphorylated and phosphorylated species of polyomavirus major capsid protein VP1 and identification of the major phosphorylation region, *J. Virol.* **48**:206.

Baumgartner, I., Kuhn, C., and Fanning, E., 1979, Identification and characterization of fast-sedimenting SV40 nucleoprotein complexes, *Virology* **96**:54.

Bellard, M., Oudet, P., Germond, J. E., and Chambon, P., 1976, Subunit structure of simian virus 40 minichromosome, *Eur. J. Biochem.* **70**:543.

Benjamin, T. L., 1970, Host range mutants of polyoma virus, *Proc. Natl. Acad. Sci. USA* **67**:394.

Benoist, C., and Chambon, P., 1981, *In vivo* sequence requirements of the SV40 early promoter region, *Nature* **290:**304.

Bergsma, D. J., Olive, D. M., Hartzell, S. W., and Subramanian, K. N., 1982, Territorial limits and functional anatomy of the SV40 replication origin, *Proc. Natl. Acad. Sci. USA* **79:**381.

Bina, M., Ng S.-C., and Blasquez, V., 1983a, Simian virus 40 chromatin interaction with the capsid proteins, *J. Biomol. Struc. Dyn.* **1:**689.

Bina, M., Blasquez, V., Ng, S.-C., and Beccher, S., 1983b, SV40 morphogenesis, *Cold Spring Harbor Symp. Quant. Biol.* **47:**565.

Blasquez, V., Beccher, S., and Bina, M., 1983, Simian virus 40 morphogenetic pathway, *J. Biol. Chem.* **258:**8477.

Bolen, J. B., and Consigli, R. A., 1979, Differential adsorption of polyoma virions and capsids to mouse kidney cells and guinea pig erythrocytes, *J. Virol.* **32:**679.

Bolen, J. B., and Consigli, R. A., 1980, Separation of neutralizing and hemagglutination-inhibiting antibody activities and specificity of antisera to sodium dodecyl sulfate-derived polypeptides of polyoma virions, *J. Virol.* **34:**119.

Bolen, J. B., Anders, D. G., Trempy, J., and Consigli, R. A., 1981, Differences in the subpopulations of the structural proteins of polyoma virions and capsids: Biological functions of the VP1 species, *J. Virol.* **37:**80.

Bourgaux, P., 1964, The fate of polyoma virus in hamster, mouse and human cells, *Virology* **23:**46.

Brady, J. N., Winston, V. D., and Consigli, R. A., 1977, Dissociation of polyomavirus by the chelation of calcium ions found associated with purified virions, *J. Virol.* **23:**717.

Brady, J. N., Winston, V. D., and Consigli, R. A., 1978, Characterization of a DNA–protein complex and capsomere subunits derived from polyoma virus by treatment with ethyleneglycol-bis-N,N′-tetraacetic acid and dithiothreitol, *J. Virol.* **27:**193.

Brady, J. N., Kendall, J. D., and Consigli, R. A., 1979, In vitro reassembly of infectious polyoma virions, *J. Virol.* **32:**640.

Brady, J. N., Lavialle, C., and Salzman, N. P., 1980, Efficient transcription of a compact nucleoprotein complex isolated from purified simian virus 40 virions, *J. Virol.* **35:**371.

Brady, J., Lavialle, C., Radonovich, M., and Salzman, N. P., 1981, Simian virus 40 maturation: Chromatin modifications increase the accessibility of viral DNA to nuclease and RNA polymerase, *J. Virol.* **39:**603.

Brinster, R., Chen, H., Messing, A., VanDyke, T., Levine, A., and Palmiter, R., 1984, Transgenic mice harboring SV40 T-antigen genes develop characteristic tumors, *Cell* **37:**367.

Buchman, A. R., Burnett, L., and Berg, P., 1980, The SV40 nucleotide sequence, in: *DNA Tumor Viruses* (J. Tooze, ed.), pp. 799–829, Cold Spring Harbor Laboratory, Cold Spring Harbor, N.Y.

Cereghini, S., Herbomel, P., Jouanneau, J., Saragosti, S., Katinka, M., Bourachot, B., DeCrombrugghe, B., and Yaniv, M., 1983, Structure and function of the promoter–enchancer region of polyoma and SV40, *Cold Spring Harbor Symp. Quant. Biol.* **47:**935.

Cereghini, S., and Yaniv, M., 1984, Assembly of transfected DNA into chromatin: Structural changes in the origin–promoter–enhancer region upon replication, *EMBO J.* **3:**1243.

Chou, J. Y., and Martin, R. G., 1975, Products of complementation between temperature-sensitive mutants of simian virus 40, *J. Virol.* **15:**127.

Christiansen, G., Landers, T., Griffity, J., and Berg, P., 1977, Characterization of components released by alkali disruption of simian virus 40, *J. Virol.* **21:**1079.

Coca-Prados, M., and Hsu, M.-T., 1979, Intracellular forms of simian virus 40 nucleoprotein complexes. II. Biochemical and electron microscopic analysis of simian virus 40 virion assembly, *J. Virol.* **31:**199.

Coca-Prados, M., Vidali, G., and Hsu, M.-T., 1980, Intracellular forms of simian virus 40 nucleoprotein complexes. III. Study of histone modifications, *J. Virol.* **36:**353.

Consigli, R. A., Griffith, G. R., Marriott, S. J., and Ludlow, J. W., 1986, Biochemical characterization of the polyoma–receptor interaction, in: *Microbiology—1986,* in press.

Crawford, L. V., 1964, A study of Shope papilloma virus, *J. Mol. Biol.* **8**:489.
Crawford, L. V., 1965, A study of human papilloma virus DNA, *J. Mol. Biol.* **13**:362.
Crawford, L. V., and Black, P. H., 1964, The nucleic acid of simian virus 40, *Virology* **24**:388.
Cremisi, C., Pignatti, P. F., and Yaniv, M., 1976, Random location and absence of movement of the nucleosomes on SV40 nucleoprotein complexes isolated from infected cells, *Biochem. Biophys. Res. Commun.* **73**:548.
Das, G. C., Allison, D. P., and Niyogi, S. K., 1979, Sites including those of origin and termination of replication are not freely available to single-out restriction enzymes in the supercompact form of simian virus 40 minichromosome, *Biochem. Biophys. Res. Commun.* **89**:17.
Dawe, C. J., and Law, L. W., 1959, Morphologic changes in salivary gland tissue of the newborn mouse exposed to parotid-tumor agent *in vitro*, *J. Natl. Cancer Inst.* **23**:1157.
DePamphilis, M. L., and Wassarman, P. M., 1983, Organization and replication of papovavirus DNA, in: *Organization and Replication of Viral DNA* (A. S. Kaplan, ed.), pp. 37–114, CRC Press, Boca Raton, Fla.
de Villiers, J., and Schaffner, W., 1981, A small segment of polyoma virus DNA enhances the expression of a cloned β-globin gene over a distance of 1400 base pairs, *Nucleic Acids Res.* **9**:6251.
de Villiers, J., Olson, L., Tyndall, C., and Schaffner, W., 1982, Transcriptional "enhancers" from SV40 and polyoma virus show a cell type preference, *Nucleic Acids Res.* **10**:7965.
de Villiers, J., Schaffner, W., Tyndall, C., Lupton, S., and Kamen, R., 1984, Polyoma virus DNA replication requires an enhancer, *Nature* **312**:242.
di Mayorca, G. A., Eddy, B. E., Stewart, S. E., Hunter, W. S., Friend, C., and Bendich, A., 1959, Isolation of infectious deoxyribonucleic acid from SE polyoma infected tissue cultures, *Proc. Natl. Acad. Sci. USA* **55**:1805.
Dulbecco, R., and Vogt, M., 1960, Significance of continued virus production in tissue culture cells rendered neoplastic by polyoma virus, *Proc. Natl. Acad. Sci. USA* **46**:1617.
Dulbecco, R., and Vogt, M., 1963, Evidence for a ring structure of polyoma virus DNA, *Proc. Natl. Acad. Sci. USA* **50**:236.
Eddy, B., 1960, The polyoma virus, *Adv. Virus Res.* **7**:91.
Fernandez-Munoz, R., Coca-Prados, M., and Hsu, M.-T., 1979, Intracellular forms of simian virus 40 nucleoprotein complexes. I. Methods of isolation and characterization in CV-1 cells, *J. Virol.* **29**:612.
Fey, G., and Hirt, B., 1975, Fingerprints of polyoma virus proteins and mouse histones, *Cold Spring Harbor Symp. Quant. Biol.* **39**:235.
Finch, J. T., 1974, The surface structure of polyoma virus, *J. Gen. Virol.* **24**:359.
Finch, J. T., and Crawford, L. V., 1975, Structure of small DNA containing animal viruses, in *Comprehensive Virology* Vol. 5, p. 119, Plenum Press, New York.
Frearson, P. M., and Crawford, L. V., 1972, Polyoma virus basic proteins, *J. Gen. Virol.* **14**:141.
Friedmann, T., 1971, *In vitro* reassembly of shell-like particles from disrupted polyoma virus, *Proc. Natl. Acad. Sci. USA* **68**:2574.
Fromm, M., and Berg, P., 1982, Deletion mapping of DNA regions required for SV40 early region promoter function *in vivo*, *J. Mol. Appl. Genet.* **1**:457.
Fromm, M., and Berg, P., 1983, Simian virus 40 early- and late-region promoter functions are enhanced by the 72-base-pair repeat inserted at distant locations and inverted orientations, *Mol. Cell. Biol.* **3**:991.
Fujimura, F. K., and Linney, E., 1982, Polyoma mutants that productively infect F9 embryonal carcinoma cells do not rescue wild-type polyoma in F9 cells, *Proc. Natl. Acad. Sci. USA* **79**:1479.
Fujimura, F. K., Deininger, P. L., Friedmann, T., and Linney, E., 1981, Mutation near the polyoma DNA replication origin permits productive infection of F9 embryonal carcinoma cells, *Cell* **23**:809.
Garber, E. A., Seidman, M. M., and Levine, A. J., 1978, The detection and characterization of multiple forms of SV40 nucleoprotein complexes, *Virology* **90**:305.

Garcea, R. L., and Benjamin, T. L., 1983, Host range transforming gene of polyoma virus plays a role in virus assembly, *Proc. Natl. Acad. Sci. USA* **80:**3613.

Gerard, R. D., Woodworth-Gutai, M., and Scott, W. A., 1982, Deletion mutants which affect the nuclease-sensitive site in simian virus 40 chromatin, *Mol. Cell. Biol.* **2:**782.

Gluzman, Y., and Shenk, T. (eds.), 1983, *Enhancers and Eukaryotic Gene Expression—Current Communications in Molecular Biology*, Cold Spring Harbor Laboratory, Cold Spring Harbor, N.Y.

Graessmann, A., Graessmann, M., and Mueller, C., 1981, Regulation of SV40 gene expression, *Adv. Cancer Res.* **35:**111.

Green, M. H., Miller, H. J., and Hendler, S., 1971, Isolation of a polyomanucleoprotein complex from infected mouse cell cultures, *Proc. Natl. Acad. Sci. USA* **68:**1032.

Griffin, B. E., and Dilworth, S. M., 1983, Polyomavirus: An overview of its unique properties, *Adv. Cancer Res.* **39:**183.

Griffith, G. R., and Consigli, R. A., 1984, Isolation and characterization of monopinocytotic vesicles containing polyomavirus from the cytoplasm of infected mouse kidney cells, *J. Virol.* **50:**77.

Gross, L., 1953, A filterable agent, recovered from Ak leukemic extracts, causing salivary gland carcinomas in C3H mice, *Proc. Soc. Exp. Biol. Med.* **83:**414.

Gruss, P., Dhar, R., and Khoury, G., 1981, Simian virus 40 tandem repeated sequences as an element of the early promoter, *Proc. Natl. Acad. Sci. USA* **78:**943.

Habel, K., and Atanasiu, P., 1959, Transplantation of polyoma virus induced tumor in the hamster, *Proc. Soc. Exp. Biol. Med.* **102:**99.

Hanahan, D., 1985, Heritable formation of pancreatic B-cell tumors in transgenic mice expressing recombinant insulin/simian virus 40 oncogenes, *Nature* **315:**115.

Hare, D. J., and Chan, J. C., 1968, Role of hydrogen and disulfide bonds in polyoma capsid structure, *Virology* **34:**481.

Hay, N., Skolnick-David, H., and Aloni, Y., 1982, Attenuation in the control of SV40 gene expression, *Cell* **29:**183.

Herbomel, P., Saragosti, S., Blangy, D., and Yaniv, M., 1981, Fine structure of the origin-proximal DNase I-hypersensitive region in wild-type and EC mutant polyoma, *Cell* **25:**651.

Huang, E. S., Estes, M. K., and Pagano, J. S., 1972, Structure and function of the polypeptides in simian virus 40. I. Existence of subviral nucleoprotein complexes, *J. Virol.* **9:**923.

Innis, J. W., and Scott, W. A., 1984, DNA replication and chromatin structure of simian virus 40 insertion mutants, *Mol. Cell. Biol.* **4:**1499.

Jakobovitz, E. B., Bratusin, S., and Aloni, Y., 1982, Formation of a nucleosome-free region in SV40 minichromosomes is dependent upon a restricted segment of DNA, *Virology* **120:**340.

Jay, G., Nomura, S., Anderson, C. W., and Khoury, G., 1981, Identification of the SV40 agnoprotein product: A DNA binding protein, *Nature* **291:**971.

Jongstra, J., Reudelhuber, T. L., Oudet, P., Benoist, C., Chae, C.-B., Jeltsch, J.-M., Mathis, D. J., and Chambon, P., 1984, Induction of altered chromatin structures by simian virus 40 enhancer and promoter elements, *Nature* **307:**708.

Katinka, M., Yaniv, M., Vasseur, M., and Blangy, D., 1980, Expression of polyoma early functions in mouse embryonal carcinoma cells depends on sequence rearrangements in the beginning of the late region, *Cell* **20:**393.

Kelley, T. J., Jr., and Nathans, D., 1977, The genome of simian virus 40, *Adv. Virus Res.* **21:**85.

Kenney, S., Natarajan, V., Strike, D., Khoury, G., and Salzman, N. P., 1984, JC virus enhancer–promoter active in human brain cells, *Science* **226:**1337.

Khoury, G., and Gruss, P., 1983, Enhancer elements, *Cell* **33:**313.

Klug, A. J., 1965, Structure of virus of the papilloma–polyoma types. II. Comments on other work, *J. Mol. Biol.* **11:**424.

LaBella, F., and Vesco, C., 1980, Late modifications of simian virus 40 chromatin during the lytic cycle occur in an immature form of virion, *J. Virol.* **33:**1138.

LaBella, F., Vidali, G., and Vesco, C., 1979, Histone acetylation in CV-1 cells infected with simian virus 40, *Virology* **96**:564.

Lai, C.-J., and Nathans, D., 1976, The B/C gene of simian virus 40, *Virology* **75**:335.

Laimins, L. A., Khoury, G., Gorman, C., Howard, B., and Gruss, P., 1982, Host-specific activation of transcription by tandem repeats from simian virus 40 and Moloney murine sarcoma virus, *Proc. Natl. Acad. Sci. USA* **79**:6453.

Levinson, B., Khoury, G., Vande Woude, G., and Gruss, P., 1982, Activation of SV40 genome by 72-base pair tandem repeats of Moloney sarcoma virus, *Nature* **295**:568.

Li, J. J., and Kelly, T. J., 1985, Simian virus 40 DNA replication *in vitro:* Specificity of initiation and evidence for bidirectional replication, *Mol. Cell. Biol.* **5**:1238.

MacKay, R. L., and Consigli, R. A., 1976, Early events in polyomavirus infection: Attachment, penetration, and nuclear entry, *J. Virol.* **19**:620.

McMillen, J., and Consigli, R. A., 1974, Characterization of polyoma DNA–protein complexes. I. Electrophoretic identification of the proteins in a nucleoprotein complex isolated from polyoma-infected cells, *J. Virol.* **14**:1326.

Major, E. O., Miller, A. E., Mourrain, P., Traub, R. G., De Widt, E., and Sever, J., 1985, Establishment of a line of human fetal glial cells that supports JC virus multiplication, *Proc. Natl. Acad. Sci. USA* **82**:1257.

Marriott, S. J., and Consigli, R. A., 1985, Production and characterization of monoclonal antibodies to polyomavirus major capsid protein, VP1, *J. Virol.* **56**:365.

Mattern, C. F., Takemoto, K. K., and Daniel, W. A., 1966, Replication of polyoma virus in mouse embryo cells: Electron microscopic observations, *Virology* **30**:242.

Melnick, J. L., Allison, A. C., Butel, J. S., Eckhart, W., Eddy, B. E., Kit, S., Levine, A. J., Miles, J. A. R., Pagano, J. S., Sacks, L., and Nonka, V., 1974, *Papovaviridae, Intervirology* **3**:106.

Nedospasov, S. A., and Georgiev, G. P., 1980, Nonrandom cleavage of SV40 DNA in the compact minichromosome and free in solution by micrococcal nuclease, *Biochem. Biophys. Res. Commun.* **92**:532.

Nedospasov, S. A., Bakayev, V. V., and Georgiev, G. P., 1978, Chromosome of the mature virion of simian virus 40 contains H1 histone, *Nucleic Acids Res.* **5**:2847.

Negroni, G., Dourmashkin, R., and Chesterman, F. C., 1959, A "polyoma" derived from a mouse leukemia, *Br. Med. J.* **2**:1359.

Ng, S.-C., and Bina, M., 1981, Disulfide bonds protect the encapsidated chromosomes of simian virus 40, *FEBS Lett.* **130**:47.

Ng, S.-C., and Bina, M., 1984, Temperature-sensitive BC mutants of simian virus 40: Block in virion assembly and accumulation of capsid–chromatin complexes, *J. Virol.* **50**:471.

Ng, S.-C., Behm, M., and Bina, M., 1985a, DNA sensitive alterations responsible for the synthesis of thermosensitive VP1 in temperature-sensitive BC mutants of simian virus 40, *J. Virol.* **54**:646.

Ng, S.-C., Mertz, J. E., Sanden-Will, S., and Bina, M., 1985b, Simian virus 40 maturation in cells harboring mutants deleted in the agnogene, *J. Biol. Chem.* **260**:1127.

Noll, M., 1974, Subunit structure of chromatin, *Nature* **251**:249.

Nomura, S., Jay, G., and Khoury, G., 1986, Spontaneous deletion mutants resulting from a frameshift insertion in the simian virus 40 agnogene, *J. Virol.* **58**:165.

O'Farrell, P. Z., and Goodman, H. M., 1976, Resolution of simian virus 40 proteins in whole cell extracts by two-dimensional electrophoresis: Heterogeneity of the major capsid protein, *Cell* **9**:289.

Persico-DiLauro, M., Martin, R. G., and Livingston, D. M., 1977, Interaction of simian virus 40 chromatin with simian virus 40 T-antigen, *J. Virol.* **24**:451.

Polisky, B., and McCarthy, B., 1975, Location of histones on simian virus 40 DNA, *Proc. Natl. Acad. Sci. USA* **72**:2895.

Ponder, B. A. J., and Crawford, L. V., 1977, The arrangement of nucleosomes in nucleoprotein complexes from polyoma virus and SV40, *Cell* **11**:35.

Ponder, B. A. J., Crew, F., and Crawford, L. V., 1978, Comparison of nuclease digestion of polyoma virus nucleoprotein complex and mouse chromatin, *J. Virol.* **25**:175.

Rayment, I., Baker, T. S., Caspar, D. L. D., and Murakami, W. T., 1982, Polyoma virus capsid structure at 22.5 Å resolution, *Nature* **295:**110.

Roblin, R., Harle, E., and Dulbecco, R., 1971, Polyoma virus proteins. I. Multiple virion components, *Virology* **45:**555.

Rush, M. G., and Warner, R. C., 1970, Alkali denaturation of covalently closed circular duplex deoxyribonucleic acid, *J. Biol. Chem.* **245:**2704.

Sachs, L., and Winocour, E., 1959, Formation of different cell–virus relationships in tumour cells induced by polyoma, *Nature* **184:**1702.

Salzman, N. P., Sebring, E. D., and Radonovich, M., 1973, Unwinding of parental strands during simian virus 40 DNA replication, *J. Virol.* **12:**669.

Saragosti, S., Moyne, G., and Yaniv, M., 1980, Absence of nucleosomes in a fraction of SV40 chromatin between the origin of replication and the region coding for the late leader RNA, *Cell* **20:**65.

Scott, W. A., and Wigmore, D. J., 1978, Sites in simian virus 40 chromatin which are preferentially cleaved by endonucleases, *Cell* **15:**1511.

Seebeck, T., and Weil, R., 1974, Polyoma viral DNA replicated as a nucleoprotein complex in close association with the host cell chromatin, *J. Virol.* **13:**567.

Seidman, M., Garber, E., and Levine, A. J., 1979, Parameters affecting the stability of SV40 virions during the extraction of nucleoprotein complexes, *Virology* **95:**256.

Sekikawa, K., and Levine, A., 1981, Isolation and characterization of polyoma host range mutants that replicate in nullipotential embryonal carcinoma cells, *Proc. Natl. Acad. Sci. USA* **78:**1100.

Shelton, E. R., Wassarman, P. M., and DePamphilis, M. L., 1978, Structure of SV40 chromosomes in nuclei from infected monkey cells, *J. Mol. Biol.* **125:**491.

Shelton, E. R., Wassarman, P. M., and DePamphilis, M., 1980, Structure, spacing and phasing of nucleosomes of isolated forms of mature simian virus 40 chromosomes, *J. Biol. Chem.* **255:**771.

Soeda, E., Arrand, J. R., Smolar, N., Walsh, J. E., and Griffin, B. E., 1980, Coding potential and regulatory signals of the polyoma virus genome, *Nature* **283:**445.

Staneloni, R. J., Fluck, M., and Benjamin, T. L., 1977, Host range selection of transformation-defective hr-t mutants of polyoma virus, *Virology* **77:**598.

Stewart, S. E., Eddy, B. E., Gochenour, A. M., Borgese, N. G., and Grubbs, G. E., 1957, The induction of neoplasms with a substance released from mouse tumors by tissue culture, *Virology* **3:**380.

Sundin, O., and Varshavsky, A., 1979, Staphylococcal nuclease makes a single non-random cut in the simian virus 40 minichromosomes, *J. Mol. Biol.* **132:**535.

Sweet, B. H., and Hilleman, M. R., 1960, The vacuolating virus, SV40, *Proc. Soc. Exp. Biol. and Med.* **105:**420.

Tooze, J. (ed.), 1981, *DNA Tumor Viruses: Molecular Biology of Tumor Viruses*, 2nd ed., Cold Spring Harbor Laboratory, Cold Spring Harbor, N.Y.

Varshavsky, A. J., Sundin, O., and Bohn, M., 1979, A stretch of "late" SV40 viral DNA about 400 bp long which includes the origin of replication is specifically exposed in SV40 minichromosomes, *Cell* **16:**453.

Veldman, G. M., Lupton, S., and Kamen, R., 1985, Polyomavirus enhancer contains multiple redundant sequence elements that activate both DNA replication and gene expression, *Mol. Cell. Biol.* **5:**649.

Vinograd, J., and Lebowitz, J., 1966, Physical and topological properties of circular DNA, *J. Gen. Physiol.* **49:**103.

Vogt, M., and Dulbecco, R., 1960, Virus–cell interaction with a tumor-producing virus, *Proc. Natl. Acad. Sci. USA* **46:**365.

Vogt, M., and Dulbecco, R., 1962, Studies on cells rendered neoplastic by polyoma virus: The problem of the presence of virus-related materials, *Virology* **16:**41.

Waldeck, W., Fohring, B., Chowdhury, K., Gruss, P., and Sauer, G., 1978, Origin of DNA replication in papovavirus chromatin is recognized by endogenous endonuclease, *Proc. Natl. Acad. Sci. USA* **75:**5964.

Weil, R., 1981, A quantitative assay for a subviral infective agent related to polyoma virus, *Virology* **14**:46.
Weil, R., and Vinograd, J., 1963, The cyclic helix and cyclic coil forms of polyoma viral DNA, *Proc. Natl. Acad. Sci. USA* **50**:730.
Westphal, H., 1970, SV40 DNA strand selection by *Escherichia coli* RNA polymerase, *J. Mol. Biol.* **50**:407.
Winston, V. D., Bolen, J. B., and Consigli, R. A., 1980, Isolation and characterization of polyoma uncoating intermediates from nuclei of infected mouse cells, *J. Virol.* **33**:1173.
Yuen, L. K. C., and Consigli, R. A., 1982, Improved infectivity of reassembled polyoma virus, *J. Virol.* **43**:337.
Yuen, L. K. C., and Consigli, R. A., 1983, Generation of capsids from unstable polyoma virions, *J. Virol*, **47**:620.
Yuen, L. K. C., and Consigli, R. A., 1985, Identification and protein analysis of polyomavirus assembly intermediates from infected primary mouse embryo cells, *Virology* **144**:127.

CHAPTER 2

Transcription of SV40 and Polyoma Virus and Its Regulation

NORMAN P. SALZMAN, VENKATACHALA NATARAJAN, AND GERALD B. SELZER

I. RNA POLYMERASE II—ITS STRUCTURE AND GENERAL PROPERTIES

The nuclei of eukaryotic cells are known to contain three related, but distinct types of RNA polymerase (reviewed by Chambon, 1975; Roeder, 1976; Sentenac, 1985). Of these, only RNA polymerase II (also called RNA polymerase B) is thought to play a direct role in the synthesis of mRNA. Polymerases I and III (A and C) carry out synthesis of RNAs with a regulatory or enzymatic role in the cell. Some animal viruses encode their own RNA polymerase (Nevins and Joklik, 1977) or are subject to transcription by RNA polymerase III as well as II (Wu, 1978). As far as is known, however, SV40 and polyoma are transcribed only by polymerase II, a conclusion based on the sensitivity of viral mRNA synthesis to low levels of the inhibitor α-amanitin (Jackson and Sugden, 1972). Though not of direct relevance here, some data suggest that infection or transformation by SV40 stimulates transcription of host DNA by polymerases I and III (Soprano *et al.*, 1981; Singh *et al.*, 1985). In this introductory section, we examine the properties of RNA polymerase II and discuss briefly the study of factors that apparently mediate its action in the cell.

NORMAN P. SALZMAN, VENKATACHALA NATARAJAN, AND GERALD B. SELZER • Laboratory of Biology of Viruses, National Institute of Allergy and Infectious Diseases, National Institutes of Health, Bethesda, Maryland 20892.

RNA polymerase II has been purified from a variety of sources, including human (KB, HeLa, and placental), bovine, invertebrate, yeast, and plant cells (reviewed by Lewis and Burgess, 1982). The enzymes from these various sources are alike in that each consists of a large number (10 or more) of subunits of remarkably diverse size (approximately 10K to 200K daltons each) (reviewed by Paule, 1981). The ability of antibody raised against one polymerase to react with enzyme from other sources has indicated the conservation of structure and, presumably, function in a number of the subunits during evolution (Ingles, 1973; Kramer and Bautz, 1981; Huet *et al.*, 1982; Weeks *et al.*, 1982). In several cases, moreover, it appears that subunits are shared between polymerases I, II, and III from the same source, the most extensive documentation being for the yeast and plant enzymes (Buhler *et al.*, 1980; reviewed by Paule, 1981; Sentenac, 1985). This raises the possibility that the common subunits function in polymerization, while the others function in reactions unique to each class, such as promoter recognition.

Some polymerase II subunits are known to be modified by phosphorylation *in vivo* and *in vitro* (Buhler *et al.*, 1976; Dahmus, 1983) but it is not clear if this affects enzyme function. In preparations of the calf thymus enzyme, the largest subunit is found in three forms, two of which are apparently generated from the third by alternative phosphorylation or protease cleavage (Dahmus, 1983). At least one other subunit may be subject to protease action (Hodo and Blatti, 1977). In the case of the largest subunit, it is not clear that either type of modification affects enzyme function (Weil *et al.*, 1979) or that cleavage is not an artifact of extraction. The latter possibility is suggested by detection of only the largest form of the subunit in extracts of HeLa cells (Robbins *et al.*, 1984). The effect of papovavirus infection on the subunit structure of the enzyme has not been examined.

Little is known about the function of individual subunits during transcription. Recent genetic studies have identified the largest subunit as the site of alterations that confer resistance to α-amanitin (Greenleaf, 1983). Since this drug interferes with elongation (Cochet-Meilhac and Chambon, 1974), the large subunit very likely plays an active role in polymerization. Other studies have reported the isolation of hybridoma lines that produce antibodies to polymerase II (Carroll and Stollar, 1982; Christmann and Dahmus, 1981; Kramer *et al.*, 1980). Because such antibodies ought to be subunit specific, their study should provide valuable clues to the identity or function of the subunits. Thus, Carroll and Stollar (1982) have described a monoclonal antibody that interferes with DNA binding during initiation by the purified enzyme. Antibodies specific for actin bind the 43K subunit of the polymerase and can inactivate the purified enzyme, suggesting that actin, a frequent contaminant of proteins purified from eukaryotic cells, is actually part of the enzyme (Egly *et al.*, 1984). However, possible functions for actin in polymerization remain obscure.

The formation of active mRNA proceeds via three steps in which RNA polymerase is directly involved: initiation, elongation, and termination. Investigation of the regulation of gene expression in both prokaryotic and eukaryotic cells has centered on the mechanisms that determine the frequency and location of transcription initiation. Subsequent steps of transcription, as well as posttranscriptional modifications specific to eukaryotic cells (capping, splicing, and polyadenylation) must also influence the yield of gene product, but their regulatory significance is more poorly understood. Analysis of regulation has been facilitated by the development of techniques for transfection, i.e., the introduction of purified viral DNA into cells without the need for a viable virus, and of other techniques that permit the study of transcription *in vitro* (reviewed by Manley, 1983a). Obviously, both approaches have benefited from the development of recombinant DNA techniques.

Experiments that examine the expression of transfected DNA are of two types. Transient expression is usually measured 2 or 3 days after uptake of the DNA and may or may not employ vector DNAs that replicate independently of the chromosome. Alternatively, expression can be measured from genes that have inserted (integrated) into the host DNA and are thus stably maintained in the cell. The advantage of the former is that substantial transcription usually occurs because of the large amount of DNA that can be introduced into the cell, the disadvantage being that it is difficult to evaluate the precise amount of DNA that enters and the effect of this on expression. With stable integration, the situation is reversed: few gene copies and lower levels of transcript. This technique introduces the possibility of effects due to the exact site of integration, but minimizes competition for host factors necessary for expression. Experiments of either type frequently make use of gene fusions in which the promoter of interest controls the synthesis of a gene product [e.g., chloramphenicol acetyltransferase (CAT)] whose expression is easily measured (Gorman *et al.*, 1982).

Extracts of eukaryotic cells that could accurately initiate transcription were first described by Wu (1978), although the transcription was polymerase III dependent. Newer procedures yield extracts showing correct initiation by polymerase II and are thus suited for study of SV40 and polyoma (Weil *et al.*, 1979; Manley *et al.*, 1980; Dignam *et al.*, 1983). Although these extracts are usually prepared from HeLa cells, which are not permissive for SV40, correct initiation of transcription at both early and late promoters can be detected. A possible disadvantage of such *in vitro* systems is that the added template DNA does not assemble into the chromatinlike structure typical of viral DNA in infected cells (Hough *et al.*, 1982; Sinha *et al.*, 1982; Sergeant *et al.*, 1984).

One attraction of the use of cell extracts to study initiation and its regulation is that it permits the identification and purification of polymerase accessory factors necessary for transcription. Fractionation of extracts by standard protein purification procedures (Matsui *et al.*, 1980;

Tsai *et al.*, 1981; Samuels *et al.*, 1982; Dynan and Tjian, 1983a) has permitted partial or complete purification of several factors, some of which bind tightly to DNA in the promoter region of SV40. These include a factor that binds to the TATA box (Davison *et al.*, 1983; Parker and Topol, 1984) and one that binds to the 21-bp repeats (Dynan and Tjian, 1983b; Gidoni *et al.*, 1984). As yet, there is little direct information about the role of these two factors in transcription. However, considerable indirect information has come from alteration of the promoter elements to which they bind. We return to this point in greater detail below. Some factors required for the *in vitro* reaction may not play a necessary role *in vivo*. One such factor, which prevents initiation at nicks, has been shown to be poly(ADP-ribose) polymerase (Slattery *et al.*, 1983).

II. SV40 EARLY TRANSCRIPTION

The mRNAs coding for the SV40 small t and large T antigens are generated by differential splicing of RNA transcribed from the early strand of SV40 DNA. The mRNA coding for the small t antigen results from splicing the primary transcript between nucleotide position (np) 4571 and 4638, whereas the mRNA encoding the large T antigen results from a splice between np 4918 and 4571 (reviewed in Tooze, 1981). In addition to these splice sites, a minor splice donor site at nucleotide 4225 and multiple acceptor sites around nucleotide 4100 have been suggested for the production of large T antigens of reduced sizes (Sompayrac and Danna, 1985). All these different early mRNAs have a common 3′ polyadenylation site at np 2694. (See Fig. 2 of Chapter 1.) (The numbering system for SV40 that is used in this chapter is that of Buchman *et al.*, 1980).

A. 5′ Ends of SV40 Early mRNAs

1. Procedures for the Detection of 5′ Start Sites of Transcription

Three principal procedures have been used to locate the 5′ start site(s) for SV40 transcription and in each, there is a likely possibility that artifactual 5′ ends will be detected. In the first, the sequences have been determined for capped oligonucleotides generated by enzymatic digestion of radiolabeled purified transcripts. In the case of SV40, any other RNA that is present as a contaminant in the pool of mRNA that is examined will contribute false 5′ ends. In the second procedure, S1 analysis, an end-labeled SV40 DNA fragment is hybridized to RNA and unreacted RNA and DNA are digested with S1 nuclease. The length of labeled DNA that is protected from digestion in the presence of complementary RNA positions the 5′ start site of the RNA. With S1 analysis, AT-rich regions within an RNA–DNA hybrid can be cleaved (Shenk *et al.*, 1975; Handa *et al.*, 1981). Cleavage within AT-rich regions may become a major factor when the authentic 5′ end of the RNA lies upstream but close to an AT-rich region. This is the case when S1 analysis has been used to locate the SV40 early transcripts that are present at late times in the infectious cycle (Hansen and Sharp, 1983; Fromm and Berg, 1982). In the third procedure, primer extension, a short radiolabeled SV40 DNA fragment is annealed to RNA, and a cDNA copy of the RNA is synthesized using reverse transcriptase. The size of the cDNA is used to locate the 5′ end of the RNA. With primer extension, premature termination or pausing by reverse transcriptase at specific sequences will indicate false 5′ ends. Recently, the 5′ ends of SV40 late RNA synthesized *in vitro* were determined by S1 and primer extension analysis (Rio and Tjian, 1984). With S1 analysis, two major start sites were detected and the upstream site was the dominant one. In sharp contrast, primer extension yielded a much greater complexity of 5′ ends, and the major upstream start site appeared to be used to only a negligible extent. These results suggest that when primer extension is used with SV40, 5′ ends located close to the primer can be detected efficiently while sequences farther upstream are scored with low frequency. Numerous other sites that are scored as 5′ ends may arise by premature chain termination. Since additional 5′ ends are the likely consequence of artifacts in each of the procedures, undue significance should not be attached to minor start sites unless other evidence indicates that these sites function *in vivo*. In our subsequent discussion, we have focused on those start sites that have been commonly identified by several procedures.

2. Position of the 5′ Ends of SV40 Early mRNAs

Analysis of the cap structures of the mRNAs has been used to locate the 5′ ends of mRNAs (Haegeman and Fiers, 1980b; Kahana *et al.*, 1981). These studies indicated that there were multiple start sites for early mRNAs and also tentatively identified the initiation sites based on complementarity of these RNAs to the DNA of the early region. The use of several start sites for early mRNA was further supported by data obtained by primer extension (Ghosh and Lebowitz, 1981; Thompson *et al.*, 1979) and S1 nuclease mapping methods (Benoist and Chambon, 1981; Fromm and Berg, 1982).

Based on these studies, the 5′ ends of SV40 early mRNAs have been positioned around np 5231 to 5237 (Fig. 1). It is usually assumed that the 5′ end of the RNA is the start site for transcription. This has been directly confirmed for two start sites by analyzing the RNA synthesized in SV40-infected permeabilized cells using [β-^{32}P]-ATP and [β-^{32}P]-CTP (Gidoni *et al.*, 1981). In addition to RNAs beginning at np 5231 to 5237, RNA sites with 5′ ends around np 28 to 34 have been identified by reverse transcription of the RNA. Even though S1 mapping cannot localize the exact RNA start sites in this region because of its high AT content, it confirms

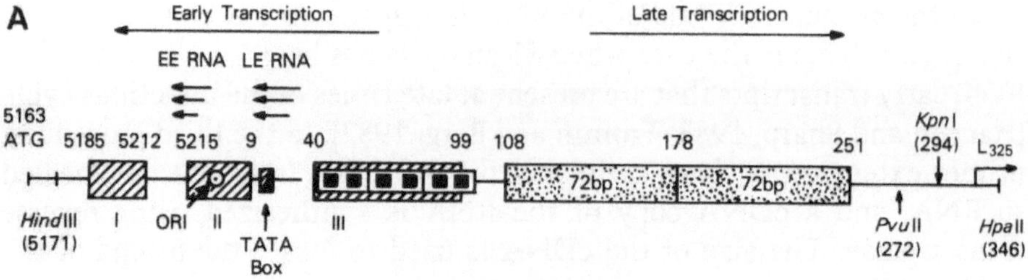

B

```
      5153       5163       5173       5183       5193
CTGTTTAAAA CTTTATCCAT CTTTGCAAAG CTTTTTGCAA AAGCCTAGGC
GACAAATTTT GAAATAGGTA GAAACGTTTC GAAAAACGTT TTCGGATCCG

                                   EE RNA
      5203       5213       5223       5233       5243
CTCCAAAAAA GCCTCCTCAC TACTTCTGGA ATAGCTCAGA GGCCGAGGCG
GAGGTTTTTT CGGAGGAGTG ATGAAGACCT TATCGAGTCT CCGGCTCCGC

                           LE RNA          "21"
        10         20         30         40         50
GCCTCGGCCT CTGCATAAAT AAAAAAAATT AGTCAGCCAT GGGGCGGAGA
CGGAGCCGGA GACGTATTTA TTTTTTTTAA TCAGTCGGTA CCCCGCCTCT

                      21                    21
        60         70         80         90        100
ATGGGCGGAA CTGGGCGGAG TTAGGGGCGG GATGGGCGGA GTTAGGGGCG
TACCCGCCTT GACCCGCCTC AATCCCCGCC CTACCCGCCT CAATCCCCGC

       110        120        130        140        150
GGACTATGGT TGCTGACTAA TTGAGATGCA TGCTTTGCAT ACTTCTGCCT
CCTGATACCA ACGACTGATT AACTCTACGT ACGAAACGTA TGAAGACGGA
                                         72 bp

       160        170        180        190        200
GCTGGGGAGC CTGGGGACTT TCCACACCTG GTTGCTGACT AATTGAGATG
CGACCCCTCG GACCCCTGAA AGGTGTGGAC CAACGACTGA TTAACTCTAC

       210        220        230        240        250
CATGCTTTGC ATACTTCTGC CTGCTGGGGA GCCTGGGGAC TTTCCACACC
GTACGAAACG TATGAAGACG GACGACCCCT CGGACCCCTG AAAGGTGTGG
              72 bp

       260        270        280        290        300
CTAACTGACA CACATTCCAC AGCTGGTTCT TTCCGCCTCA GAAGGTACCT
GATTGACTGT GTGTAAGGTG TCGACCAAGA AAGGCGGAGT CTTCCATGGA

                             L RNA
       310        320        330        340        350
AACCAAGTTC CTCTTTCAGA GGTTATTTCA GGCCATGGTG CTGCGCCGGC
TTGGTTCAAG GAGAAAGTCT CCAATAAAGT CCGGTACCAC GACGCGGCCG
```

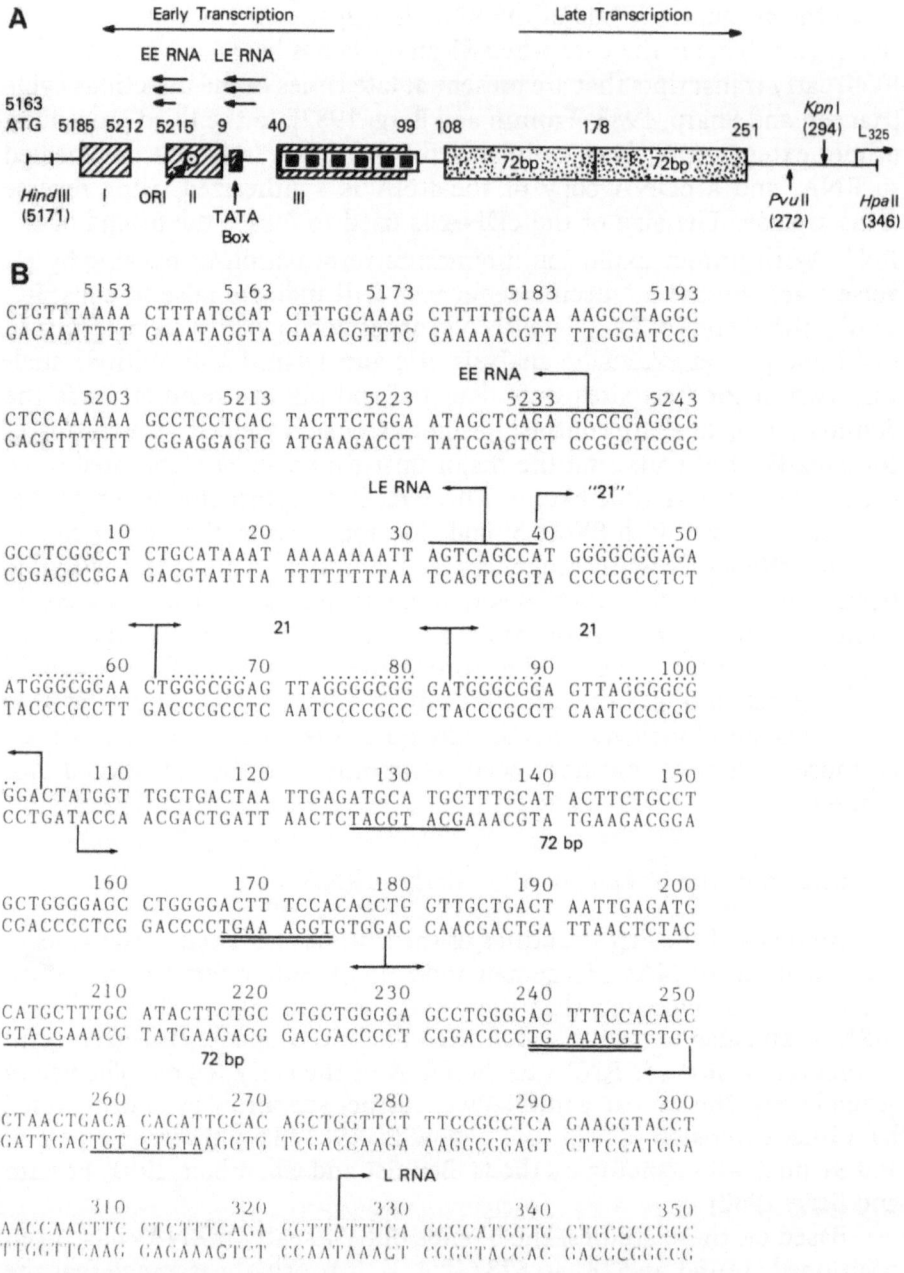

FIGURE 1. The regulatory region of the SV40 genome. (A) The *Hin*dIII–*Hpa*II fragment of SV40. Origin of replication (ORI) is shown as a circle within T antigen binding site II. The reader is referred to Chapter 3 for a detailed discussion of core and axillary sequences involved in DNA replication. T antigen binding sites I, II, and III (▨) (Tenen *et al.*, 1983; DeLucia *et al.*, 1983). GC-rich motifs (■) contained within three 21-bp repeats (▭), two 72-bp enhancer elements, and the major 5′ ends of early and late RNA are identified.

the existence of start sites upstream of the TATA box. These RNAs have been found during the late phase of the infectious cycle (Ghosh and Lebowitz, 1981; Buchman *et al.*, 1984). Their significance is discussed below.

B. Regulation of Early Transcription

Extensive studies of both *in vivo* and *in vitro* transcription have shown that the early promoter is autoregulated by SV40 large T antigen. Immediately after uncoating of the DNA, early transcription leads to the synthesis of large T antigen. Autoregulation of T antigen synthesis was first suggested by the overproduction of early viral RNA at the nonpermissive temperature following infection with SV40 with a temperature-sensitive mutation in the gene for T antigen (Tegtmayer *et al.*, 1975; Reed *et al.*, 1976; Khoury and May, 1977). More direct evidence of autoregulation of transcription was obtained by extensive *in vitro* studies. Due to the lack of purified SV40 T antigen, initial studies were carried out using a mutant T antigen that was isolated from SV40-transformed cells (SV80 T antigen), or using a T antigen hybrid (D2 protein), which is encoded by an adenovirus–SV40 hybrid virus. These studies identified three T antigen binding sites on SV40 DNA near the origin of replication. This finding has been confirmed using authentic SV40 T antigen that has been overproduced using an adenovirus promoter (Tjian, 1981). It has been shown that the purified T antigen binds specifically to three closely spaced sites that lie between np 106 and 5170 (Fig. 1). When binding was studied using purified T antigen produced during a lytic cycle of infection, there was no evidence that binding occurred in a cooperative and sequential way at these three sites (Tenen *et al.*, 1983).

Genetic and biochemical studies have shown that the interaction of T antigen with SV40 DNA is required for regulation of early transcription as well as for SV40 DNA replication (Tenen *et al.*, 1982; Wilson *et al.*, 1982; DiMaio and Nathans, 1982; Rio and Tjian, 1983). But it was not clear whether the T antigen suppressed early transcription by binding to SV40 DNA and acting as a repressor or by modifying the transcription machinery. Recent studies in support of the former proposal have come from the development of an *in vitro* transcription system responsive to T antigen addition (Manley *et al.*, 1980; Handa *et al.*, 1981). When SV40 DNA is used as the DNA template in the system, the synthesis of RNA

(B) The nucleotide sequence of the regulatory region. The 72-bp enhancer elements, and the major early and late transcription start sites are identified. The six GC clusters are indicated by dotted lines. Nucleotides with a double underline are the core enhancer elements, and those with a single underline are alternating purine–pyrimidine stretches. For additional details, see text.

starting from both early and late *in vivo* initiation sites is observed. Either D2 protein or the wild-type SV40 T antigen inhibited transcription from the early promoter, but did not affect that from SV40 or adenovirus late promoters (Rio *et al.*, 1980; Myers *et al.*, 1981; Hansen *et al.*, 1981). Transcription from SV40 mutant DNAs having an intact promoter but lacking T antigen binding sites I and/or II is not inhibited by T antigen (Rio and Tjian, 1983). These studies demonstrate that the regulation of early transcription is by the interaction of T antigen with DNA. This conclusion is further supported by evidence from *in vivo* studies of mutants in which the coding region for the T antigen was replaced by the coding region for dihydrofolate reductase. Transcription of this recombinant DNA was analyzed by transfection of COS 7 cells, which constitutively produce T antigen (Gluzman, 1981). In this host, RNA synthesis from the early promoter was stimulated four- to five-fold by deletion of T antigen binding sites I and II. This shows that the autoregulation seen *in vivo* involves the T antigen and binding sites I and II (Rio and Tjian, 1983).

Since sites I and II lie just downstream of the early promoter region, T antigen could autoregulate its synthesis by either of two possible mechanisms, an attenuation mechanism in which the bound T antigen blocks or slows the movement of RNA polymerase along the template, or by a repressor mechanism in which T antigen binding prevents binding of polymerase to DNA. The elegant experiments of Myers *et al.* (1981) support the latter mechanism. They constructed a plasmid in which the three T antigen binding sites are present just downstream of the adenovirus major late promoter. Transcription from this plasmid *in vitro* was unaffected by the addition of T antigen, while under identical conditions, transcription of the SV40 early promoter was inhibited by T antigen addition. With the adeno plasmid, the absence of short, prematurely terminated RNAs that might have accumulated if transcription was arrested at the T antigen binding sites, ruled out an attenuation mechanism. These results show that T antigen does not block the movement of RNA polymerase along the template. Autoregulation of early transcription is carried out by the binding of T antigen to its sites, which blocks the binding of RNA polymerase to DNA.

C. Late-Early RNAs

During the late phase of the SV40 lytic cycle, additional early region transcripts having 5′ ends at nucleotides 28 to 34 have been observed (Ghosh and Lebowitz, 1981). It has been estimated that the ratio of these late-early (LE) transcripts to early-early (EE) transcripts changes from 0.4 to 4 as infection progresses (Fromm and Berg, 1982). The significance of this RNA and the switch from EE to LE is not understood. Cap analysis has not been done with SV40 LE RNAs, but RNAs synthesized from similar upstream start sites in polyoma virus are not capped (Cowie *et al.*, 1982). This raises the possibility that the upstream RNA arises by processing very large RNAs whose synthesis initiates at EE start sites and extends around the entire genome. The LE RNA is spliced normally, but seems to be translated very poorly into small t and large T antigens (Buchman *et al.*, 1984). The 5′-proximal region of the LE RNA encodes an additional protein of 23 amino acids in length, but as yet no experimental evidence indicates such a protein is synthesized. If it is so, the inefficient translation of LE RNA into small t and large T antigens could be explained by the common observation that protein synthesis initiates at the furthest upstream AUG on eukaryotic mRNAs (Kozak, 1983).

D. Mechanism of the Early-Early to Late-Early Switch in RNA Synthesis

A considerable effort has been made by many groups to understand the mechanism of the EE (downstream) to LE (upstream) switch using both *in vitro* and *in vivo* transcription systems. Initial *in vivo* studies showed that the shift to upstream RNA initiation sites occurred late in infection, i.e., after the onset of viral DNA replication (Ghosh and Lebowitz, 1981). It has also been shown that a functional T antigen is needed for the switch although these experiments did not rule out a requirement for DNA replication (Ghosh and Lebowitz, 1981). However, even in the absence of DNA replication, around 10% of all the early RNA initiates from the upstream sites, and inactivation of the downstream initiation sites also leads to the increased use of upstream sites (Baty *et al.*, 1984; Wasylyk *et al.*, 1983b). Analysis of mutants with deletions in the T antigen binding sites has defined the role of T antigen and DNA replication in the shift to upstream initiation sites. Ghosh and Lebowitz (1981) have reported that the deletion of binding site III had no effect on the shift, and Hansen *et al.* (1981) have reported a similar lack of effect of mutation in binding site I. This led to the proposal that binding site II, which forms in the core of the replication origin, is involved in the shift.

The experiments of Buchman *et al.* (1984) convincingly showed that the replication *per se* is involved in the shift. They used plasmids containing SV40 DNA which had a deletion in site II or sites I and II and in which the T antigen coding region was replaced by the β-globin coding region. Transfection of COS cells with these plasmids, which are unable to replicate because of the absence of an intact replication origin, led to synthesis predominantly of EE RNA, demonstrating that functional T antigen and binding site I are not sufficient to induce the upstream RNA. Insertion of a functional replication origin at a distance away from the early promoter region restores replication in COS cells. Under these conditions, LE RNA synthesized from these deletion mutants were comparable in amount to wild type, showing that replication *per se* induces upstream RNA even in the absence of T antigen binding sites at the early promoter region (Buchman *et al.*, 1984; Buchman and Berg, 1984).

Studies with cell lines transformed by SV40 also support the proposal that replication is involved in the induction of upstream RNA. Many investigators have found that early RNA in transformed cells is mainly from the downstream start sites (Gluzman *et al.*, 1980; Benoist and Chambon, 1981; Ghosh and Lebowitz, 1981; Ghosh *et al.*, 1981a). T antigen is synthesized in transformed cells and it should function to autoregulate. However, this binding of T antigen to the integrated DNA and the inhibition of RNA synthesis from downstream sites is not sufficient for the stimulation of RNA from upstream sites.

It is interesting that there is a differential effect of the transcriptional enhancer on EE and LE transcription. The enhancer is needed for the synthesis of downstream RNA but not for the upstream RNA from replicating templates (Buchman *et al.*, 1984). With templates that are not replicating, the 72-bp repeat is essential for transcription from both upstream and downstream sites (Wasylyk *et al.*, 1983b).

As yet, it has been difficult to reproduce the shift from downstream to upstream initiation sites *in vitro*. *In vitro*, equal transcription from upstream and downstream sites has been observed in many laboratories (Hansen *et al.*, 1981; Lebowitz and Ghosh, 1982; Vigneron *et al.*, 1984), although Rio and Tjian (1983) could not detect any transcripts from the upstream sites *in vitro*. In the latter case, inhibition of downstream initiation by addition of T antigen did not induce the transcription from upstream sites (Rio and Tjian, 1983). Hansen *et al.* (1982) have reported that the upstream initiation normally present in their experiments could be inhibited by relatively high concentrations of T antigen.

Taking *in vitro* and *in vivo* results together, one can postulate that the downstream initiation sites are used during the early phase of infection, leading to the synthesis of T antigen. The binding of T antigen to the origin region depresses the transcription from downstream initiation sites as well as inducing the replication of DNA. The onset of DNA replication by some unknown mechanism activates transcription from upstream sites.

E. DNA Sequences That Control Transcription from SV40 Early Promoters

1. The Role of the TATA Box

DNA sequences that control early RNA synthesis have recently been analyzed in detail. The main focus has been on the involvement of the RNA initiation site, the TATA sequence that is located 30 bp upstream of the cap site (located at np $+1$), the GC repeats present at -40 to $\Sigma 103$, and 72-bp repeats located at -107 to -250, in controlling the levels of

RNA synthesis (Fig. 1). DNA sequences downstream of the RNA initiation site have not been implicated in the control of SV40 early transcription, even though it has been reported that downstream sequences have a role in other promoters such as those of the adenovirus E1A and globin genes (Osborne *et al.*, 1984; Charnay *et al.*, 1984; Wright *et al.*, 1984). Sequences that control accurate and efficient RNA synthesis from the early promoter have been analyzed both *in vivo* and *in vitro.* The efficiency of transcription *in vivo* has been estimated either as the transformation efficiency or as the amount of viral RNA synthesized. Since T antigen regulates RNA synthesis from early promoters, in some studies the coding region for T antigen was replaced by other genes (Fromm and Berg, 1983a; Rio and Tjian, 1983; Baty *et al.*, 1984; Das and Salzman, 1985). The deletion of RNA initiation sites and immediately adjacent sequences have no overall effect either on the amount of RNA synthesized or on the efficiency of transformation (Benoist and Chambon, 1981; Fromm and Berg, 1982, 1983a). The new RNA initiation sites are generally found to be about 27 to 34 nucleotides downstream from the first T of the TATA box sequence both *in vivo* and *in vitro* (Gluzman *et al.*, 1980; Ghosh *et al.*, 1981a; Benoist and Chambon, 1981). Hence, in the case of the SV40 EE promoter, the transcription start site and surrounding sequences are not essential for efficient transcription.

The majority of the eukaryotic genes transcribed by RNA polymerase II have a TATA box sequence (canonical sequence being TATA$^{A}_{T}$A$^{A}_{T}$) 25 to 30 nucleotides upstream of the RNA initiation site (Breathnach and Chambon, 1981). In SV40, a TATA box-like sequence, TATTTAT, is present about 30 bp upstream of the major RNA initiation site. Benoist and Chambon (1981) and Fromm and Berg (1983a) deleted this sequence from a plasmid containing the SV40 early region and used it to transform cells in culture. Deletion of the TATA sequence did not affect the efficiency of transformation, suggesting that it is dispensable for early gene expression. However, RNA isolated from the cells transformed by the mutant lacking the TATA sequence had more heterogeneous 5′ ends than that of cells transformed by wild-type DNA (Benoist and Chambon, 1981). Based on this and also on the fact that the deletion of normal initiation sites always led to new initiation sites about 30 bp downstream of the TATA sequence, it has been proposed that the function of the TATA box is to direct the initiation of RNA synthesis about 30 bp downstream.

Recently, point mutations were created in the TATA box of the early SV40 promoter to change the sequence from TATTTAT to TATCGAT, and the mutated sequence was inserted in front of either T antigen or β-globin coding sequences. The plasmids were transfected into cultured cells and RNA synthesized in a transient assay was estimated (Wasylyk *et al.*, 1983b). Because of this mutation, there was a 20- to 30-fold reduction in the amount of RNA synthesized from downstream start sites. The mutation did not affect the level of RNA synthesized from upstream start sites of the construct having a T antigen coding region. With the β-globin construct, the level of RNA synthesized from upstream start sites was stimulated by 2- to 3-fold, leading to only 40–50% reduction in the overall synthesis of RNA. Hence, it appears that the TATA box only has a crucial role in the efficiency of EE RNA synthesis *in vivo.*

Mathis and Chambon (1981) have reported that deletion of the TATA box decreases the level of initiation of RNA synthesis *in vitro* at the start sites about 30 bp downstream, while at the same time stimulating RNA synthesis from other start sites further downstream. Since the total amount of RNA synthesized was not reduced, it appears that the TATA box deletion did not have a significant effect on the efficiency of *in vitro* transcription. Similar observations have been made by others, in agreement with the proposal that the role of the TATA box *in vitro* is the same as *in vivo,* i.e., it is one element in fixing the RNA initiation sites to a narrow region (Myers *et al.*, 1981; Hansen and Sharp, 1983; Lebowitz and Ghosh, 1982). Interestingly, point mutations in the TATA box not only reduce the amount of RNA synthesized *in vitro* from the start sites 30 bp downstream of it, but do not stimulate the RNA synthesis from new start sites. The level of RNA synthesis they support was reduced 13-fold (Wasylyk *et al.*, 1983b). The reason for this discrepancy between the effects of deletion and point mutation of the TATA box is unclear.

Recently, a transcription factor has been isolated from *Drosophila* nuclear extract and shown to bind to the TATA box region of a number of promoters (Parker and Topol, 1984). The presence of a similar factor that interacts with the TATA box region of the adenovirus major late promoter has been detected in HeLa cell extracts (Davison *et al.*, 1983). It is quite possible that the TATA box sequence of the SV40 promoter also interacts with this factor.

2. The Role of GC Nucleotide Clusters

In vivo, the TATA box and downstream sequences are not sufficient to function as a promoter (Byrne *et al.*, 1983; Fromm and Berg, 1983a; Baty *et al.*, 1984). Two other sequence domains also function to regulate early transcription. One consists of two 21-bp repeats and a homologous 22-bp sequence (sometimes referred to as the three 21-bp repeats) that collectively contain six copies of a GC-rich sequence (CCGCCC). These are present 40 to 103 nucleotides upstream of the RNA initiation sites. The second lies upstream of these GC-rich repeats between np 107 and 250, and contains two identical 72-bp sequences (Fig. 1). These 72-bp repeated sequences are necessary for both early and late gene expression; their role in transcription is discussed in the section on enhancers.

The role of the GC-rich repeats in early gene transcription has been studied mainly in the presence of the 72-bp repeats. The presence of one 21-bp repeat is sufficient for virus growth. After deletion of all three 21-bp repeats, the virus, though viable, grew very poorly (Hartzell *et al.*,

1983). Mutants in which the 21-bp sequences had been deleted transformed cells less efficiently and the transformed cells had reduced levels of viral T antigen, indicating that the 21-bp repeats are required for viral early gene expression (Benoist and Chambon, 1981; Fromm and Berg, 1982, 1983a; Everett *et al.*, 1983). These studies also showed that two or three GC clusters are sufficient for the efficient transformation of cells in culture, or for synthesis of normal levels of RNA (Fromm and Berg, 1983a). The inversion of GC repeats at their normal locations did not significantly reduce the amount of T antigen synthesis, demonstrating that these repeats can activate transcription in a bidirectional manner (Everett *et al.*, 1983). Recently, the effect of 21-bp repeats on transcription has been analyzed more carefully by the use of point mutations. The T antigen coding region was replaced by the β-globin gene to avoid the regulatory effects of T antigen on transcription (Baty *et al.*, 1984). These authors conclude that all six GC clusters (i.e., three 21-bp repeats) are essential for normal gene expression. The two GC clusters proximal to the TATA box were found essential for transcription from the downstream start site (EE), whereas mutations in the four GC-rich sequences distal to the TATA box affected transcription from both upstream and downstream start sites (Baty *et al.*, 1984).

The role of GC-rich sequences in supporting efficient transcription from the early promoter has been faithfully reproduced in an *in vitro* transcription system. Myers *et al.* (1981) and Lebowitz and Ghosh (1982) have reported that sequences that control *in vitro* transcription lie about 70 to 155 nucleotides upstream of the RNA initiation sites, a region that includes some of the GC-rich sequences. Hansen and Sharp (1983) had shown that the amount of *in vitro* transcription was roughly proportional to the number of GC repeats. But the analysis of the effect of point mutations in GC blocks on *in vitro* transcription showed that the two GC-rich sequences proximal to the TATA box are not required for *in vitro* transcription, results that differ from the *in vivo* findings (Vigneron *et al.*, 1984). Mutations in the four distal GC-rich sequences affect the transcription to varying extent. In agreement with earlier *in vivo* studies, all six GC-rich repeats can be inverted from the normal position without reducing transcription. Mishoe *et al.* (1984) and Miyamoto *et al.* (1984) have reported that the GC clusters can stimulate *in vitro* transcription when joined to a heterologous promoter (adenovirus major late promoter), suggesting that these sequences can act as an independent transcriptional control element.

Dynan Tjian (1983a) have partially purified and reconstituted a transcriptional system that faithfully initiates at the SV40 early promoter. Using this system, they identified a factor, Spl, which is specifically needed for transcription of the SV40 early promoter but is not needed for transcription of other promoters including the polyoma virus early, adenovirus major late, and β-globin promoters. They also found that a deletion mutant having five GC-rich repeats was transcribed efficiently, while a mutant that has only two GC clusters was not. The Sp1 factor has been shown to bind to the GC-rich sequences by DNase I footprinting and methylation protection methods (Dynan and Tjian, 1983b; Jones and Tjian, 1984; Gidoni *et al.*, 1984). A good correlation between the DNA binding and the transcription stimulating activities during various stages of purification of Sp1 indicates that the factor activates transcription from SV40 early promoter by binding to the DNA.

In the case of SV40, the spacing between the TATA box and the GC blocks is also an element in determining the level of transcription. Deletion of four bases between the TATA box and the GC blocks stimulated the *in vitro* transcription by fourfold and the insertion of 2 to 10 bases reduced the overall *in vitro* transcription efficiency by 30 to 60% (Das and Salzman, 1985; Baty *et al.*, 1984). Based on the above-mentioned observations, one can propose that transcription from the SV40 early promoter is controlled by two different DNA elements, the GC-rich sequences and the TATA box, through their interactions with at least two different factors, the Sp1 and TATA box binding factors. In addition, the enhancer and factors that interact with it also have a major role in determining the efficiency of the early promoter (see Section VI).

III. POLYOMA VIRUS EARLY GENE TRANSCRIPTION

A. Properties of the Transcripts

A common precursor transcript from the early DNA strand is spliced to generate three mRNAs that encode the three known early proteins of polyoma. These mRNAs share common 5′ ends at the origin region and contain a stretch of poly(A) following at np 2930, but differ in location and extent of the internal regions removed by RNA splicing (reviewed in Tooze, 1981). By sequencing the cDNA copies of the three mRNAs, the splice donor and acceptor sites have been identified (Treisman *et al.*, 1981a). The large, middle, and small T antigen mRNAs are spliced between nucleotides 411 and 797, 748 and 797, and 748 and 811, respectively (see Table I). In addition to these three major species of mRNAs, a minor species of RNA polyadenylated at np 1525 has been observed in lytically infected and in some transformed cells (Triesman *et al.*, 1981a; Fenton and Basilico, 1981; Heiser and Eckart, 1982). This RNA, whose function is not known, could be translated into either small or middle T antigen. A second minor species of RNA of 1800 nucleotides has been mapped between 93 and 26 map units. At present, it is not clear whether this RNA is generated by alternative splicing of early RNA or else is transcribed from an unknown early promoter (Fenton and Basilico, 1982a).

The start sites of early region transcripts have been mapped by S1 nuclease and primer extension methods and confirmed by determining

TABLE I. SV40 and Polyoma Virus Splice Junctions[a]

	Donor exon →	← Intron	Intron →	← Acceptor exon
SV40				
Late (294–558)[b]	GAAG	GTACCTAA	TGTCTTTTATTTCAG	GTC
Late (294–435)[b]			TTGTGTTTGTTTTAG	AGC
Late, 19 S (526–558)	ACTG	GTAAGTTT		
Late, 16 S (526–1463)			TGCCTTTACTTCTAG	GCC
Late 19 S (373–558)	TAAG	GTTCGTAG		
Late (526–435)[b]				
Early (4918–4571)	TGAG	GTATTTGC	GTTTGTGTATTTTAG	ATT
Early (4638–4571)	TAAG	GTAAATAT		
Polyoma				
Late, VP1 (5022–4124)	TCAA	GTAAGTGA	TTCCTTTAATTCTAG	GGC
Early, large T (411–797)	CCAG	GTAAGAAG	CCTATATTCTTACAG	GGC
Early, middle T (748–797)	CCAA	GTAAGTAT		
Early, small t (748–811)			GGGCTCTCCCCCTAG	AAC
Consensus sequences	$^{C}_{A}AG$	$GT^{A}_{G}AGT$	$\binom{T}{C}_n N^{C}_{T}AG$	G

[a] The DNA nucleotide sequence shown in this table is the same as the RNA which is the substrate used for the splicing reaction. For each splice site, numbers designate the last nucleotide in the donor exon and the first nucleotide in the acceptor exon.
[b] Identifies the SV40 splice sites that are used with very low frequency during infection with wild-type virus.

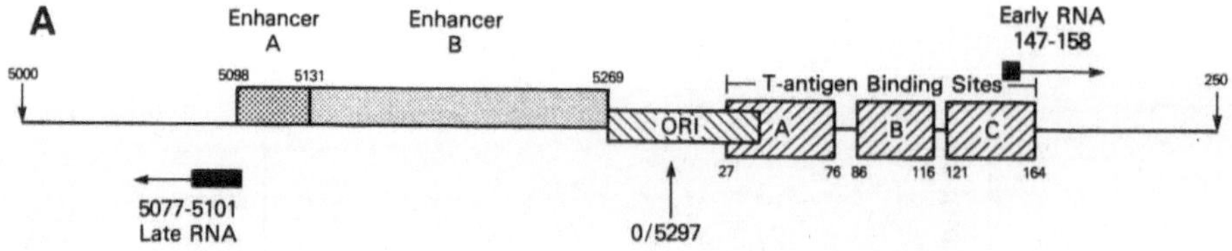

B

```
     5057       5067       5077       5087       5097
GGCCTCCAAC ACAGTAATTT TCCTCCCGAC TCTTAAAATA GAAAATGTCA
CCGGAGGTTG TGTCATTAAA AGGAGGGCTG AGAATTTTAT CTTTTACAGT
                            Late RNA <-----
       Enhancer A
     5107       5117       5127       5137       5147
AGTCAGTTAA GCAGGAAGTG ACTAACTGAC CGCAGCTGGC CGTGCGACAT
TCAGTCAATT CGTCCTTCAC TGATTGACTG GCGTCGACCG GCACGCTGTA

     5157       5167       5177       5187       5197
CCTCTTTTAA TTAGTTGCTA GGCAACTGCC CTCCAGAGGG CAGTGTGGTT
GGAGAAAATT AATCAACGAT CCGTTGACGG GAGGTCTCCC GTCACACCAA
                     Enhancer B
     5207       5217       5227       5237       5247
TTGCAAGAGG AAGCAAAAAG CCTCTCCACC CAGGCCTAGA ATGTTTCCAC
AACGTTCTCC TTCGTTTTTC GGAGAGGTGG GTCCGGATCT TACAAAGGTG

     5257       5267       5277       5287       5297
CCAATCATTA CTATGACAAC AGCTGTTTTT TTTAGTATTA AGCAGAGGCC
GGTTAGTAAT GATACTGTTG TCGACAAAAA AAATCATAAT TCGTCTCCGG

       10         20         30         40         50
GGGGGCCCCT GGCCTCCGCT TACTCTGGAG AAAAAGAAGA GAGGCATTGT
CCCCCGGGGA CCGGAGGCGA ATGAGACCTC TTTTTCTTCT CTCCGTAACA
                                       A
       60         70         80         90        100
AGAGGCTTCC AGAGGCAACT TGTCAAAACA GGACTGGCGC CTTGGAGGCG
TCTCCGAAGG TCTCCGTTGA ACAGTTTTGT CCTGACCGCG GAACCTCCGC
                                                B
      110        120        130        140        150
CTGTGGGGCC ACCCAAATTG ATATAATTAA GCCCCAACCG CCTCTTCCCG
GACACCCCGG TGGGTTTAAC TATATTAATT CGGGGTTGGC GGAGAAGGGC
             Early RNA                  C
      160        170        180
CCTCATTTCA GCCTCACCAC CATCATGGAT
GGAGTAAAGT CGGAGTGGTG GTAGTACCTA
```

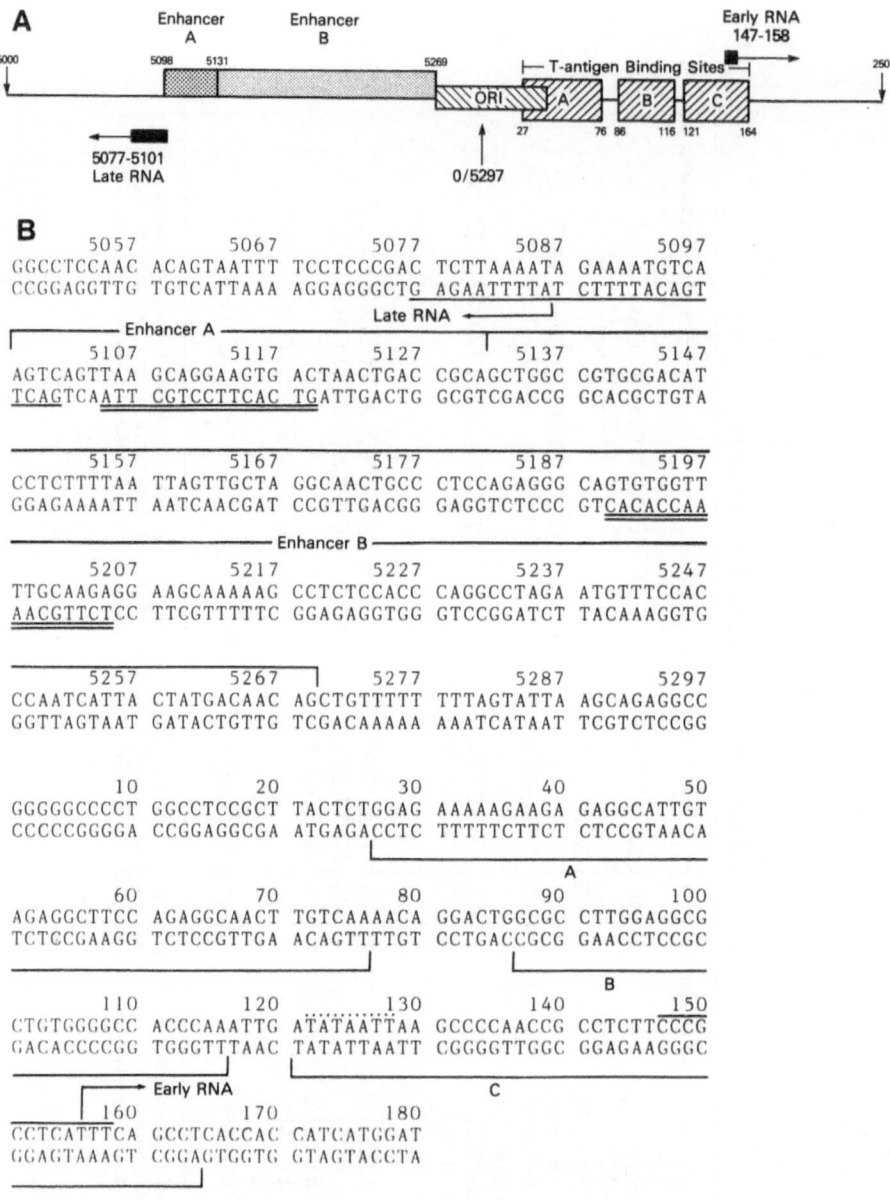

FIGURE 2. The regulatory region of polyoma virus DNA. (A) Schematic outline of various components of the polyoma virus control region. The major transcription start sites for early (Kamen *et al.*, 1982) and late (Cowie *et al.*, 1981) RNAs are shown by arrows. The three T antigen binding sites A, B, and C (Cowie and Kamen, 1984; Dilworth *et al.*, 1984) and the boundaries of two different enhancers (A and B) (Herbomel *et al.*, 1984) are identified. The outer limits of enhancer B are not well defined, although the majority of enhancer B activity is present in DNA sequences at np 5177 to 5231. The reader is referred to Chapter 3 for the boundaries of the origin of replication (ORI). For nucleotide numbering, see the Appendix. (B) The nucleotide sequence of the regulatory region. The lines with arrows

the sequences of the capped early RNAs (Heiser and Eckhart, 1982; Kamen *et al.*, 1982; Cowie *et al.*, 1982). The major termini are located around np 147 to 158 (Fig. 2). A number of minor 5′-termini have also been mapped to sites both upstream and downstream of the major 5′-termini. The most abundant minor 5′-termini were located at np 14, 22, and 302. In addition, early strand transcripts with 5′ ends corresponding to sites farther upstream of the replication origin (np 5297) were found during the late phase of polyoma virus infection (Cowie *et al.*, 1982; Kamen *et al.*, 1982; Fenton and Basilico, 1982a). Detailed analysis by Kamen and his colleagues suggests that these upstream RNAs are generated by the processing of long RNAs initiated at major start sites rather than by initiation from upstream start sites. The finding that these RNAs are not capped at their 5′ ends is consistent with this hypothesis. When cells infected with a polyoma variant that contains a ts mutation in large T antigen were shifted to a nonpermissive temperature late during infection, transcription from the early promoter was stimulated. Under these conditions, at least ten times more viral RNA was found in the nucleus than in the cytoplasm, and virtually all of this was the upstream RNA.

To explain these observations, Kamen *et al.* (1982) have proposed that late during infection, termination of transcription from early promoters is inefficient, and that this permits transcription to extend around the circular genome, generating giant transcripts. Subsequent cleavage of such giant transcripts around the origin region produced RNA with upstream 5′ ends. The hypothesis is consistent with their observation that most polyoma-transformed cell lines containing one integrated copy of the early region lack the upstream transcripts, while one particular line with two copies of the early region integrated in tandem does have the upstream transcripts. Kamen *et al.* (1982) suggest that the RNAs starting at one early promoter continue across the second promoter and are cleaved near the origin of the second early region to generate the RNAs with upstream ends.

Studies in SV40 have demonstrated that synthesis of similar upstream RNAs are linked to replication of the genome (Buchman *et al.*, 1984). Although the mechanism of generation of these polyoma and SV40 upstream transcripts is obscure, they may have a role in virus growth. As with SV40, these RNAs are spliced and polyadenylated normally and can code for protein of 53 amino acids (Fenton and Basilico, 1982a). This putative protein, however, cannot be essential for virus growth since there are viable viruses with deletions in this region (Kamen *et al.*, 1982).

---

indicate the major early and late RNA initiation sites. Enhancer A and B are bracketed and the core sequence in each of them is double underlined. The three T antigen binding sites (A, B, and C) are bracketed. The position of the TATA box sequence of the early promoter is indicated by a dotted overline.

B. Regulation of Polyoma Early Transcription by T Antigen

A ts mutation in the region encoding large T antigen derepresses polyoma early transcription both in lytic infection and in transformed cells, demonstrating that the large T antigen has a role in the regulation of early transcription (Cogen, 1978; Fenton and Basilico, 1982b). To study the mechanism of this regulation, binding of the large T antigen to polyoma DNA has been analyzed. Use of crude preparations of large T antigen identified two binding sites, a relatively strong binding site between np 39 and 78 and a weak binding site between np 106 and 190 (Gaudray *et al.*, 1981; Pomerantz *et al.*, 1983; Pomerantz and Hassell, 1984; Cowie and Kamen, 1984). Subsequently, DNase I footprinting of partially purified large T antigen has resolved three binding sites (called A, B, and C) with equal affinity that lie between np 25 and 165 (Cowie and Kamen, 1984; Dilworth *et al.*, 1984) (Fig. 2).

Deletion of binding sites B and C does not affect autoregulation of early gene expression. This suggests that binding site A, which is located upstream of the major RNA initiation site, is involved in the regulation. In addition, transcription from a heterologous promoter inserted in this deletion mutant was also regulated like wild type (Cowie and Kamen, 1984; Farmerie and Folk, 1984). In fact, Dailey and Basilico (1985) have shown that binding sites A and B are needed for the repression of early gene transcription by the large T antigen. This arrangement differs from that of SV40 where T antigen acts at binding sites downstream of the RNA initiation sites to regulate early gene transcription. This may reflect differences between these two viruses in the arrangement of their early promoter sequences and their replication origin. In SV40, the replication origin overlaps with the TATA box and early transcription initiation sites, whereas in polyoma virus the major transcription start site and TATA box are clearly separated from the origin region (Figs. 1 and 2). The availability of purified large T antigen should facilitate *in vitro* transcription studies similar to those carried out with SV40, and should provide further understanding of the mechanism of regulation of polyoma early gene transcription.

C. Control Sequences of Polyoma Early Gene Transcription

Compared to SV40, the sequences that control polyoma early gene transcription are not well defined. Analysis of viable deletion mutants has demonstrated that the RNA initiation sites and about 70 bp of upstream sequence are not essential for early gene function. Deletion of both the TATA box and the major start sites led to the synthesis of RNA from other sites that were normally minor initiation sites (Kamen *et al.*, 1982; Jat *et al.*, 1982a; Mueller *et al.*, 1984). However, when only the

TATA box and its adjacent sequences were deleted but the predominant RNA initiation site was retained (np 147–158), the majority of transcripts were initiated at the normal cap sites (Dailey and Basilico, 1985), in contrast to findings with the SV40 early promoter. On the other hand, insertion of about 130 bp of foreign DNA upstream of the TATA box led to RNA synthesis from multiple start sites (Clark *et al.*, 1984). At present, it is not clear what role, if any, the TATA box and the RNA initiation sites have in the efficiency of RNA synthesis from the polyoma early promoter.

The only upstream sequences needed for efficient transformation by the polyoma early gene are located about 200 to 400 nucleotides upstream of the RNA initiation site (Jat *et al.*, 1982a; Mueller *et al.*, 1984). This region contains the polyoma enhancer elements, the function and significance of which are discussed below.

For efficient *in vitro* transcription from the polyoma early gene promoter, only about 40 to 60 bp of DNA upstream of the RNA initiation site is needed (Jat *et al.*, 1982a,b; Mueller *et al.*, 1984). Unlike SV40, no sequences lying farther upstream are necessary, but in these studies with polyoma, a heterologous transcription system (HeLa cell extract) has been used. It is interesting to point out that HeLa cells are semipermissive for SV40 and nonpermissive for polyoma. The Sp1 transcription factor which is necessary for transcription from the SV40 early promoter is not needed for polyoma early gene transcription in the same HeLa cell-derived system (Dynan and Tjian, 1983b).

IV. SV40 LATE GENE TRANSCRIPTION

A. Location of the 5′ Start Sites of Late Transcription

Studies with nascent RNA strands should provide the most meaningful characterization of the frequency with which particular sites are used for initiation of transcription. In contrast with the measurement of steady-state levels of mRNA in the cytoplasm, characterization of nascent molecules avoids problems related to selective transport of different RNA species to the cytoplasm or to differences in the half lives of mRNA species. When the 5′ ends of nascent late RNAs were determined by S1 analysis, Lycan and Danna (1983) found three species that initiated at np 325, 260, and 195. Transcripts beginning at np 325 were twice as abundant as those initiating at np 260, while the 195 start site was only used with low frequency.

Kessler and Aloni (1984) examined nascent SV40 transcripts contained in transcription intermediates obtained from SV40-infected cells by Sarkosyl extraction. Nascent strands were extended *in vitro* and terminated at an endonuclease cleavage site. Three initiation sites at np

325, 264, and 346 were observed. The presence of a major initiation site at np 325 and a minor site at np 264 had previously been shown by primer extension and S1 mapping (see below). The frequency with which the start sites are used is: np 264, 24%; 325, 66%; and 346, 10%. Significant usage of the start site at np 346 was not observed by Lycan and Danna in their examination of nascent strands nor has it been seen in other studies except as a trivial component when cytoplasmic mRNA has been examined. Thus, it may be a short-lived nuclear transcript that is not transported to the cytoplasm. In both studies cited above, the use of a start site for transcription near np 260 occurs with a higher frequency than the steady-state level of cytoplasmic mRNA would suggest. Either this species of nuclear RNA is not transported to the cytoplasm with the same frequency as RNA initiated at np 325 or once transported, its half-life is significantly shorter (Chiu *et al.*, 1978). In any case, it is not detected at significant levels in the cytoplasm.

One recent study indicates that failure to transport RNA species from the nucleus to the cytoplasm may be determined by which 5′ start site is used for transcription. When COS cells were transfected with a recombinant containing the adult β-globin gene and its promoter, three classes of mRNA were synthesized, each of which had the same 3′ terminus and splice site. One class had 5′ start sites 200 to 1400 bp upstream from the normal +1 cap site and was restricted to the nucleus while RNAs with 5′ start sites at np +30 and +1 were located primarily in the cytoplasm (Alfonso-Pizarro *et al.*, 1984). This suggests that nucleotide sequence and/or secondary modifications in the region of the 5′ end of the transcript is one of the factors that determines if an RNA species will be transported to the cytoplasm.

Using cap analysis of cytoplasmic RNA, Canaani *et al.* (1979) observed that late 16 S and 19 S late transcripts initiated almost exclusively with adenine although only 66% of the transcripts initiated at a single site. A number of other start sites were used with low frequency. Their results using BSC1 cells differ quantitatively from those of Haegeman and Fiers (1978b), who analyzed cap structures in infected CV1 cells and found that the same major start site at np 325 was used for more than 90% of all the starts. These latter results agree with most findings in which one predominant start site for late transcription *in vivo* is observed that specifies more than 90% of the late transcripts (Piatak *et al.*, 1983; Brady *et al.*, 1982).

In vivo studies in which total cellular and cytoplasmic viral RNA have been characterized by S1 analysis (Contreras *et al.*, 1982; Ernoult-Lange and May, 1983) or primer extension (Piatak *et al.*, 1981, 1983; Ghosh *et al.*, 1982) also identify a single major start site for 16 S late RNA at np 325 and minor start sites at np 260–272. There was no significant contribution of other detectable but weak start sites near np 170, 239, 290, 302, 318, and between 428 and 481 for wild-type virus. However, some of these sites were used with high frequency when deletions were present in the leader sequence and new start sites are selected. For 19 S RNA the major start site is also np 325, but alternative start sites at np 264 and 174–178 are used to a lesser, but significant degree (Ghosh *et al.*, 1982).

In vitro transcription systems have been described in which authentic *in vivo* start sites are recognized (Weil *et al.*, 1979; Manley *et al.*, 1980). With SV40 DNA, two major 5′ start sites are used for late strand transcription. One is at np 325, while the second, which is often used with a higher frequency, is located between np 167 and 185 (Brady *et al.*, 1982; Hansen and Sharp, 1983; Rio and Tjian, 1984). We refer to this as the np 185 start site, but note that it is probably the same as the start site position reported by various authors as starting at np 167, 170, 176, or 185. The relative use of the start sites at np 325 and 185 is dependent on the assay conditions. When the whole cell extract (Manley *et al.*, 1980) is supplemented with 15 mM ammonium sulfate, there is preferential stimulation of the major *in vivo* start site at np 325 (Natarajan *et al.*, 1983).

B. Factors Involved in Positioning the 5′ Ends of Transcripts

As discussed above, an AT-rich region, the "TATA" or Goldberg–Hogness box, is located 25 to 30 bp upstream from the start site in many cloned viral and cellular genes (Corden *et al.*, 1980). The TATA box is an essential element of the promoter *in vitro* and also functions in vivo. When SV40 deletion mutants that lack a TATA box are studied either *in vivo* or *in vitro*, the level of early gene expression is not decreased. Rather, in its absence, there is greater heterogeneity of start sites (Benoist and Chambon, 1981; Ghosh *et al.*, 1981b; Mathis and Chambon, 1981). However, the unaltered levels of high expression when the TATA box is deleted are unique to the SV40 early gene. This may be related to the presence of GC clusters that lie upstream of the TATA box and that clearly have a role in promoting early transcription. In other genes that have been studied, the invariable response to deletion of the TATA box is a lowered expression of the gene that lies downstream. Point mutations within the TATA box also cause a suppression in the level of transcription, as observed with SV40. Either a T → G or a T → A replacement in the second T in the TATA box upstream of the conalbumin gene causes a drastic decrease in *in vitro* transcription (Wasylyk *et al.*, 1980; Wasylyk and Chambon, 1981). Similar findings have also been obtained with the adenovirus major late promoter (Concino *et al.*, 1983). When the conalbumin control region with point mutations in the TATA box is fused to the sea urchin H2A histone gene, the effect of the mutation is to decrease the level of specific transcripts fivefold after microinjection into frog oocytes (Grosschedl *et al.*, 1981).

If the TATA box has an obligatory function in positioning the start site of transcription, how are genes that naturally lack a TATA sequence transcribed? In the case of adenovirus type 2, the sequences preceding the mRNA cap sites have been compared for nine viral genes. Three of these (IVa_2, EII, and EIIa late) lack TATA boxes upstream from their start sites. However, their *in vivo* and *in vitro* start sites are as precisely defined as are the six genes that contain TATA boxes (Fire *et al.*, 1981).

A number of cellular genes that lack a TATA box have also been identified. Again, the transcripts do not show heterogeneity in the 5′ ends (Menne *et al.*, 1982; Brunke *et al.*, 1984; Reynolds *et al.*, 1984; Melton *et al.*, 1984). Thus, the TATA box is not an essential element in positioning the transcription start site, although when it is present, it is one element involved in performing this function.

In the case of SV40 late transcription, where a TATA box is absent, Brady *et al.* (1982) searched for signals in DNA that could function in defining the start sites, assuming that such a surrogate TATA box would lie about 25–30 bp upstream of the major late start site (np 325). Point mutations were introduced in an 11-bp region between np 294 and 304. The sequence upstream of the start site starting at np 294 is 5′-GGTACCTAACC-3′. The replacement of cytosines at np 298 and 299 converted this sequence to one that is similar to the TATA box present in the SV40 early promoter. These base changes specifically increased *in vitro* transcription from the np 325 start site by eightfold while transcription from other start sites was unchanged. These findings indicate that sequences other than an easily identifiable TATA box can serve as a promoter element. The presence of similar surrogate TATA boxes in the control regions of other genes has been reported for the hepatitis B surface antigen promoter (Cattaneo *et al.*, 1983) and for the adenovirus-2 EIIa early gene (Elkaim *et al.*, 1983).

Brady *et al.* (1982) also have observed that a deletion at np 295–298 reduces transcription from the np 325 start site (Fig. 1). Piatak *et al.* (1983), using the same deletion mutant, did not see this decrease in transcription. One possible basis for these different findings is that the first group used S1 analysis while the second used primer extension. If a deletion alters the secondary structure near np 325 in RNAs that start farther upstream, the reverse transcriptase might indicate that their 5′ ends were at np 325.

While, as has been noted above, there is a major start site for late transcription, other start sites are used with low frequency. This cannot be attributed simply to the absence of a classic TATA box upstream of the major late start site since there is an analogous diversity of capped structures in both early and late SV40 RNAs (Haegeman and Fiers, 1980a,b; Kahana *et al.*, 1981; Canaani *et al.*, 1979). Similarly, the 5′ ends of polyadenylated spliced 16 S and 19 S late mRNAs in transformed cells are heterogeneous although they represent a subset of 5′ ends used in the lytic cycle (Ernoult-Lange and May, 1983).

C. Changes in the DNA within the Leader Cause a Change in the Start Sites for Transcription

Haegeman *et al.* (1979) examined transcription with an SV40 variant deleted from np 305 to 344. In this mutant, loss of the major start site at np 325 resulted in frequent use of a start site at np 290 which is also used by wild-type virus although with a much lower frequency. Similar findings have been reported by Villareal *et al.* (1979).

Ghosh *et al.* (1982) have studied a series of viable mutant viruses containing deletions within the DNA that specifies the leader (i.e., the 5′ untranslated region) of the 16 S and 19 S RNAs. There was no obvious relationship between the size or position of the deleted sequences in these mutants and the ability of the virus to replicate (Barkan and Mertz, 1981). In several mutants, the deleted regions contain the major transcription start site at np 325 but the major donor splice site at np 526 is intact. The consequence of deletions within the leader region is the utilization of start sites upstream from the preferred start site (Ghosh *et al.*, 1982). This includes enhanced use of normally minor start sites at np 276, 264, 239, and 192. Even two mutants with deletions that differ by only 5 bp at their 3′ ends produce different arrays of 16 S and 19 S mRNAs. Insertion or deletion some 20 bp downstream of the np 325 start site also produces a shift in the start sites (Piatak *et al.*, 1983). Such results suggest that sequences downstream from the start sites help locate the start sites for transcription by polymerase II as has been demonstrated for transcription by polymerase III. However, the lack of a relationship between the size and location of the deletions and insertions and the location of the new 5′ ends, argues strongly against such a role.

If the GC-rich nucleotide clusters (np 40–103) (Fig. 1) function for both early and late transcription and serve as the site where a transcription complex is formed, then sequences lying downstream at np 190–276 would normally be scanned and rejected as start sites before the preferred start site at np 325 is selected. In this scheme, what would be the consequence of deletion of the preferred start site? In a single scan the alternative upstream start sites should be used at the same frequency as they are normally used, and in the absence of a preferred site, the enzyme should then dissociate and recycle. In this way, minor start sites would be used with higher relative frequencies, but the absolute rate of initiation of RNA synthesis would be lower. If a sharp decrease in the rate of transcription is not observed, that would suggest that the deletion had caused structural changes in the DNA being scanned so that minor start sites were now being seen as strong start sites.

In the cases where the deletion or insertion is downstream of the np 325 start site, an altered configuration of the transcription origin may lead to the selection of new 5′ start sites. Alternatively, selective degradation of primary transcripts may depend on the secondary structure of the RNA and transcripts starting at np 325 but containing an altered leader sequence may be perferentially degraded.

D. The Structure of the Leader Determines the Metabolic Fate of the Transcript

We noted earlier that β-globin transcripts with far upstream start sites failed to migrate from the nucleus to the cytoplasm. Additional studies indicate that the sequence of the leader can play a role in determining the metabolic fate of the transcript. Barkan and Mertz (1984) characterized the 5′ ends of SV40 mRNA molecules and found that 16 S species with shorter leaders were preferentially incorporated into polysomes. In other studies, the 5′ noncoding region of the herpes thymidine kinase gene has been modified by the introduction of varying numbers of oligonucleotide linkers, which generate increased secondary structure in the mRNA. The consequence of these modifications is to reduce the efficiency with which these RNAs function as mRNAs both *in vitro* and *in vivo,* indicating that the secondary structure contributes significantly to the intrinsic translational efficiency of different mRNAs (Pelletier and Sonenberg, 1985), findings that may be related to those of Barkan and Mertz (1984). As yet, however, no particular sequences in the untranslated 5′ ends of mRNA molecules have been shown to serve as recognition signals.

E. The Relationship between Viral DNA Replication and Late Transcription

The proposal that viral DNA replication is necessary for late gene expression is attractive since late transcription is only evident after the start of viral DNA replication. Contreras *et al.* (1982) reported a sharp curtailment of late transcription when a plasmid that is defective in DNA replication was injected into *Xenopus* oocytes. However, several studies clearly show that late transcription can proceed in the absence of SV40 DNA replication. About 70% of the nascent RNA in transcription complexes obtained from cells infected with tsA58 under conditions where this mutant is unable to synthesize DNA, is late mRNA, indicating that late transcription can proceed in the absence of DNA replication. Despite this, 80 to 90% of the total nuclear viral RNA labeled *in vivo* during a 15-min pulse hybridized to early gene sequences. This suggests that there is selective degration of late gene transcripts at early times *in vivo* (Birkenmeier *et al.,* 1979). Infection of mouse cells, which do not support SV40 DNA replication, also gives rise to spliced 16 S and 19 S late mRNAs. The same results were obtained when the mouse cells were infected with the tsA58 mutant at the restrictive temperature to provide further assurance that DNA replication was blocked (Lange *et al.*, 1981). Similarly, with recombinants that lack the origin of SV40 replication, the late promoter can function to initiate transcription from authentic late 5′ start sites (Ernoult-Lange *et al.*, 1984). While there is no obligatory requirement for DNA synthesis for use of the late promoter, substantial late synthesis is only seen after DNA replication has begun. More definitive answers on the relationship between transcription and replication await further studies.

F. The Role of the GC-Rich Domain in Regulating Late Transcription

Two GC-rich perfect 21-bp repeats at np 62–82 and 83–103 and a third GC-rich 22-bp region at np 40–61 that shares strong homology with the other two regions are positioned between the major transcription start sites for both early and late SV40 promoters (Fig. 1). While technically imprecise, they are often called the "three 21-bp repeats." Within each of these repeats, the GC cluster 5′-GGGCGG-3′ occurs twice. Studies with cells transfected with viral DNA that contains deletions within this region have demonstrated that the region is essential for *in vivo* transcription of the early genes (Fromm and Berg, 1982; Hartzell *et al.*, 1983). It is not clear what their role is in supporting late gene expression. Fromm and Berg (1982) had reported that *in vivo* transcription of the late genes is partially suppressed when the 21-bp repeats are deleted. In more recent study, Omilli *et al.* (1986) found that after transfection of HeLa cells with plasmids containing the two 72-bp repeats and the late promoter sequences, late transcription occurs at least as efficiently whether or not the 21-bp repeats are present. These same results were obtained in the absence of and during DNA replication. The structure of the hypersensitive region is affected by both the enhancer and 21-bp sequences (see below), and levels of early and late transcription with deletion mutants may be altered as a consequence of their presence or absence. This may be one factor in explaining the different results noted above on the role of the 21-bp repeats on late transcription *in vivo*. *In vitro*, the 21-bp repeats are essential for both early and late transcription (see below).

The role of the GC clusters has been evaluated in a number of *in vitro* studies (Hansen and Sharp, 1983; Brady *et al.*, 1984b; Rio and Tjian, 1984; Vigneron *et al.*, 1984). One GC cluster will support low levels of transcription from the np 325 start site whereas three are sufficient to maintain a wild-type level of expression. *In viro*, there is strong expression of a start site at np 185 (P185) whose use is also dependent on the presence of the GC clusters. When only one GC cluster is present, there is an almost total block in the use of P185, suggesting that start sites closer to the GC clusters are more strongly dependent on them. Consistent with this is the finding that removal of sequences between the GC base pairs and the np 325 start site enhances use of the start site. Removal of one of the two 72-bp repeats, which moves the GC region 72 bp closer, is not sufficient for the effect (Brady *et al.*, 1984b), but when 159 bp of intervening sequence is deleted, there is a three- to sixfold increase in P325 usage (Hansen and Sharp, 1983). Similarly, a deletion that fuses np 72 to 273 and brings three GC clusters to within 52 bp of the P325 start site, results in a three- to sevenfold increase in transcription from the P325 start site (Rio and Tjian, 1984).

The enhanced level of late transcription from P325 that is observed when GC clusters are moved closer to P325 has been shown to be a consequence of the change in position and not because deletion of the intervening sequences has removed inhibitory sequences. Replacement of SV40 sequences between np 104 and 254 by pBR322 DNA did not give the stimulation of late transcription if inhibitory sequences had been lost by this sequence substitution (Vigneron *et al.*, 1984).

While binding of factor Spl to the GC repeats is essential for transcription from P185, as well as for SV40 early genes, it is not essential for P325 transcription in a reconstituted system (Dynan and Tjian, 1983b). In those cases where Spl is required *in vitro*, the most proximal GC cluster is located 50–80 bp from the 5′ start site for transcription. It has been shown that altering the distance between the most proximal GC cluster and the TATA box in the SV40 early promoter has a strong effect on the level of transcription both *in vivo* and *in vitro* (Das and Salzman, 1985). Similar results have been obtained by K. Khalili (personal communication), who found that a 4-bp insert at np 37 reduced chloramphenicol acetyltransferase (CAT) expression that was under the control of the early promoter. The effect of changes in the spacing between two promoter elements, the GC clusters and the TATA box, suggests a requirement for *simultaneous* contact of the transcription machinery with GC clusters and the TATA box in a complex that may also include the cap site. When the transcription machinery functions in this way, Spl factor is probably required. It seems likely that there is a second way in which the transcription machinery can also define a start site that is based on two elements, a TATA or a "surrogate" TATA box and a cap site. In this mode, the cap site is a critical element which is not the case in the alternate mode. The transcription machinery would function in this way when the distance of the GC clusters from the start site precludes simultaneous binding. The np 325 start site is a likely candidate that falls into this latter category since it does not require Spl for its expression. In a construct XSIO, where intervening sequences between the GC clusters and the np 325 start site have been deleted (Fromm and Berg, 1982), the most proximal GC is now 52 bp from P325. It would be interesting to determine if because of the altered spacing between the GC clusters and the 5′ start site, Spl is now required for transcription of P325, as it is for the early promoter and the late P185 start site in wild-type virus.

The critical role of GC clusters in early SV40 transcription *in vivo* has also been studied by Vigneron *et al.* (1984) and Baty *et al.* (1984) These investigators have shown that when a fragment containing the 21-bp repeats is excised and reintroduced in the opposite orientation, the repeats continue to function as a promoter element. In addition, they have examined both deletions which change the number of GC clusters upstream from early and late start sites and point mutations generated within the GC clusters. They conclude that the six GC clusters are not equivalent and that only GC clusters 1 and 2, which are proximal to early transcription start sites, have a specific role in early transcription. It seems likely that as a consequence of base changes of C's to T's in clusters 1 and 2, the repeats are no longer functional and that inhibition of transcription is due to the increased distance between the TATA box and the nearest functional GC cluster. This altered spacing would preclude simultaneous contact between these two domains, results similar to those reported by Das and Salzman (1985).

The GC-rich domain can function as an independent promoter control element in regulating the level of transcription, and this may reflect its ability to bind essential proteins. Mishoe *et al.* (1984) have shown that under *in vitro* conditions, both early and late transcription from SV40 DNA were inhibited by adding a competitor DNA fragment that contained GC clusters. This competition was only seen when three GC clusters were present in the competitor fragment. A single GC cluster was not sufficient to block transcription, in agreement with the finding that a single GC cluster is not sufficient to support transcription in deletion mutants.

In addition to this undefined mechanism by which GC clusters function to support the levels of late transcription, the GC base pairs are capable of setting transcription start sites (Mishoe *et al.*, 1984). When pBR sequences are inserted between three GC clusters and the P325 start site, new initiation transcription start sites that are located 57 and 73 bp downstream from the most proximal GC cluster are used and use of the np 325 start site is suppressed by 90%. This is not observed when only one GC cluster is present. It is likely that selection of these start sites may also involve simultaneous use of three promoter elements: the GC clusters, a surrogate TATA box (Brady *et al.*, 1982; Nandi *et al.*, 1985), and a cap site that exists within the pBR sequences.

If a mechanism involving simultaneous recognition of the GC clusters, the TATA box, and the cap site operates in selecting the SV40 early promoter, then the findings that new initiation sites are used when the TATA box is deleted might be interpreted as indicating that in the absence of the TATA box, the transcription machinery dissociates from the GC region and scans the DNA for alternative start sites specified by two elements, a surrogate TATA box and cap site.

The critical role of the GC clusters and the TATA box has also been discussed by Gabarro-Arpa *et al.* (1984).

G. T Antigen Affects SV40 Late Transcription

Because T antigen is necessary for viral DNA replication, it will enhance the level of late transcription *in vivo* simply by increasing the number of SV40 molecules that are available to serve as templates. In addition to affecting late transcription in this way, several recent studies have demonstrated that late transcription is directly enhanced by T antigen. When SV40 mutants that are unable to replicate viral DNA are used, a stimulatory effect of T antigen on late transcription is still observed (Brady *et al.*, 1984a; Keller and Alwine, 1984). Using transformed cells that produce altered T antigens with decreased abilities to bind to DNA, late promoter activation is seen to be dependent on a functional T antigen.

T antigen-dependent positive regulation of late transcription requires DNA sequences at T antigen binding site II and within the enhancer. If cells are transfected with SV40 and a plasmid DNA that contains the T antigen binding sites (np 5171–5243) and the 72-bp tandem repeat (np 128–272), the level of late transcription is reduced. However, cotransfection with two separate plasmids, one containing T antigen binding sites I and II and the other, the 72-bp repeats, does not show this effect. This suggests that binding of *trans*-acting factors requires two sites in *cis*, and that a specified distance between time is required (Brady *et al.*, 1985).

In vivo competition between SV40 and a plasmid containing most of the late region sequence (np 180 to 2533) increased transcription of late genes in COS 1 cells. This has been interpreted to suggest that there are factors that suppress late transcription and that they can be dissociated by binding to the competitor (Brady and Khoury, 1985). Below, we discuss in greater detail the possibility that VP2 and VP3 have such an effect on late transcription.

V. POLYOMA VIRUS LATE TRANSCRIPTION

Most of the late polyoma primary transcripts contain RNAs that are tandem repeats of the entire viral genome. Molecules examined by electron microscopy can be 3.5 times the genome length (Acheson, 1978). When RNase T1 digests of polyoma mRNAs are examined, the region located between 65 and 18 nucleotides preceding the initiation codon for protein VP2 is present in three- to fivefold molar excess. This is also true for the fractionated RNAs that specify VP1, VP2, and VP3. These sequences which are reiterated in the mRNA molecules are present in only a single copy in the viral genome. They are presumed to arise by post-transcriptional processing of long primary transcripts that extend many times around the genome. Multiple copies of the leader region and only a single copy of the coding region are retained in the mRNA molecule (Legion *et al.*, 1979).

Three distinct classes of mRNA molecules that accumulate in the cytoplasm code for VP1, VP2, and VP3. The 16 S RNA species codes for VP1, while 18 S and 19 S RNAs code for VP3 and VP2 (see Fig. 3 of Chapter 1). There appear to be at least 15 different start sites within the 6-bp region from np 5075 to 5170. However, 90% of the start sites are in a 25-bp region (np 5077–5101). Start sites in this 25-base sequence are present at nearly every purine, but pyrimidines are not used (Treisman *et al.*, 1981a; Cowie *et al.*, 1981). The start site at np 5129, which is located 32 bp downstream from a "TATA" box, is the only start site used *in vitro* in a whole cell extract (Jat *et al.*, 1982b).

The exclusive use of purine start sites for late polyoma transcription is also true for late SV40 transcription (Canaani *et al.*, 1979). This differs from early SV40 transcription where cytosine is used with high frequency (Gidoni *et al.*, 1981), indicating that there is a lower stringency in selecting a start site for early transcription. Baker and Ziff (1981) have suggested that RNA polymerase II may measure a fixed distance from the TATA box and then scan a 2- to 7-bp region for one or more suitable start sites. Das and Salzman (1985) have proposed that during initiation of early transcription, there is simultaneous contact of the transcription machinery at both the GC clusters and the TATA box. When the transcription complex has effected a two-point attachment, a specific binding at the cap site may not be required. In the case of the late start sites which are defined by a surrogate TATA box and a cap site in the absence of a proximal GC binding site, the higher affinity of the transcription machinery at the DNA sequences that specify the cap site may be required in order to properly bind the enzyme.

While late transcripts are not detected in cells transformed by polyoma virus, chimeric plasmids containing the polyoma late promoter region linked to the thymidine kinase or neomycin-resistant genes are able to transform cells to a TK^+ or NEO^{RES} phenotype (Kern and Basilico, 1985). Transcription of these integrated genomes is from a 5′ start site also used during the lytic cycle. The site lies upstream of the region present as tandem repeats in viral mRNAs. The absence of these repeats in the chimeric RNA indicates that the late mRNA need not contain these structures to be active. Furthermore, the synthesis of functional late transcripts by integrated chimeric plasmids under control of the polyoma late promoter suggests that the absence of late transcripts in polyoma-transformed cells may reflect posttranscriptional degradation rather than failure to use the integrated late promoter.

VI. SV40 ENHANCER

The two 72-bp repeats function as an enhancer element within the transcriptional control region (Fig. 1). Deletion of both of these repeats drastically reduces the level of transcription from the SV40 early promoter

(Gruss *et al.*, 1981; Benoist and Chambon, 1981; Fromm and Berg, 1982). These repeats activate transcription *cis* from homologous and heterologous promoters, relatively independent of position or orientation (Moreau *et al.*, 1981; Banerji *et al.*, 1981), and also enhance transformation of cells in culture by other genes (Luciw *et al.*, 1983; Berg and Anderson, 1984).

Enhancer sequences are needed for viability of the virus (Levinson *et al.*, 1982; Weber *et al.*, 1984; Fromm and Berg, 1983b; Swimmer and Shenk, 1984; Herr and Gluzman, 1985). They can provide this function in either orientation at several different locations except in one particular orientation at a distal location, in maintaining viability (Fromm and Berg, 1983b). Transcriptional control elements having similar properties have been identified in a number of viral and cellular DNAs (Walker *et al.*, 1983; Emorine *et al.*, 1983; Ott *et al.*, 1984; Picard and Schaffner, 1984; Weiher and Botchan, 1984; Kenney *et al.*, 1984; Shaul *et al.*, 1985; Mosthaf *et al.*, 1985; Rosen *et al.*, 1985; Grosschedl and Baltimore, 1985; reviewed by Khoury and Gruss, 1983; Gluzman and Shenk, 1983; Gruss, 1984).

A. The Effect on Activation of the Distance Separating the Promoter and Enhancer

In addition to the SV40 early promoter, a number of viral and cellular promoters are also activated in *cis* by the SV40 enhancer in an orientation-independent manner (de Villiers *et al.*, 1982; Banerji *et al.*, 1981; Humphries *et al.*, 1982; Treisman *et al.*, 1983; Natarajan and Salzman, 1985). The distance between the enhancer and promoter has a variable effect on the strength of the promoter, depending on the nature of the DNA sequences separating them and also on the particular promoter used (Picard and Schaffner, 1983; Berg *et al.*, 1984). Relatively equal amounts of β-globin gene transcription were observed when the enhancer was present either at np −425 or +2500 relative to the RNA initiation site (de Villiers *et al.*, 1983). But when the SV40 enhancer was separated by 3.7 kb of DNA from both 5′ and 3′ ends of the β-globin promoter, no enhancing activity was observed (Banerji *et al.*, 1981). Further investigations showed that presence of an additional promoter between the enhancer and the β-globin promoter interfered in the enhancer effect on the globin promoter (de Villiers *et al.*, 1983).

Kadesch and Berg (1983) have studied the effect of the SV40 enhancer on multiple transcription units (SV40 early promoters) present in the same plasmid. Their results confirm that when several promoters are present, the enhancer stimulates transcription from the more proximal promoter. More recent studies show that apart from the effect of intervening sequences which contain a promoter, the enhancer function can also be decreased by short intervening segments without any obvious promoter function. For example, the activation of SV40 early and conalbumin promoters is decreased tenfold by insertion of about 150 bp of pBR322 plasmid DNA (Wasylyk *et al.*, 1984). Hence, for maximum activation of these promoters, the enhancer has to be present within a critical distance (Moreau *et al.*, 1981; Wasylyk *et al.*, 1984). Similar observations have also been made with Rous sarcoma virus enhancer (Cullen *et al.*, 1985).

B. Activation of the SV40 Late Promoter by Enhancer

The 72-bp repeated sequences are located about 75 bp upstream of the major late RNA initiation site. For efficient expression of SV40 late genes, the presence of T antigen and DNA replication are necessary. Recently, the role of T antigen on late gene expression has been evaluated using transient expression systems. Plasmids with the late promoter (including the 72-bp repeats) linked either to viral late genes or to a heterologous gene coding for an assayable enzyme were transfected into cells and the amount of gene product synthesized was measured (Hartzell *et al.*, 1984a; Brady *et al.*, 1984a; Keller and Alwine, 1984). Under these conditions, the late promoter was stimulated by about 5- to 15-fold by T antigen in the absence of any DNA replication. With DNA replication, the late promoter was activated up to 1000-fold (Brady *et al.*, 1984a; Hartzell *et al.*, 1984b). The sequences necessary for the T antigen-mediated stimulation of late promoter have been localized within the 72-bp repeats (Brady and Khoury, 1985; Keller and Alwine, 1985).

Apart from the T antigen, other factors that are associated with SV40 minichromosomes may also be required for *trans* activation of the late promoter (Cereghini and Yaniv, 1984; Tack and Beard, 1985). Because the late promoter is activated by enhancer only when other *trans*-acting factors are present, the enhancer functions conditionally. This is reminiscent of what has been observed with mouse mammary tumor virus enhancer, where the glucocorticoid hormone is needed for the enhancer function (Scheidereit *et al.*, 1983; Ponta *et al.*, 1985).

C. Cell-Type Specificity of the SV40 Enhancer

The SV40 72-bp repeat region can be replaced by 72-bp repeats from Moloney murine sarcoma virus long terminal repeats without affecting viability of the virus in monkey cells (Levinson *et al.*, 1982). However, expression of the T antigen by this hybrid virus was about fivefold less than that of wild-type virus in monkey cells. In mouse cells, roughly equal amounts of T antigen were synthesized by wild-type and hybrid viruses (Laimins *et al.*, 1982). These results suggested a relationship between enhancer activity and the host cell. Further studies with different promoters confirmed that the SV40 enhancer has higher activity in primate cells than in murine, rat, bovine, or human cells (de Villiers *et al.*, 1982; Byrne *et al.*, 1983; Berg *et al.*, 1983; Spandidos and Wilkie, 1983;

Augereau and Wasylyk, 1984). But unlike many other cellular or viral enhancers, the SV40 enhancer is active in a wide variety of cell types (Spandidos and Wilkie, 1983; Walker *et al.*, 1983; Kenney *et al.*, 1984). Since most of the viral enhancers tested showed maximum activity in their normal host cells, it has been suggested that the enhancers could be a factor in determining the host range of a virus. This hypothesis was tested by replacing the polyoma enhancer region with the SV40 enhancer region. Despite this replacement, the hybrid virus showed DNA replication and gene expression in mouse cells but not in primate cells (Campbell and Villarreal, 1985). Similarly, a hybrid SV40 virus having polyoma enhancer sequences grew in monkey cells but not in mouse cells (Weber *et al.*, 1984). Thus, even though the viral enhancer has highest activation of promoters in its usual host cells, it alone is not able to alter the host range of a second virus. Previous work showed that the type and amount of T antigen produced plays a role in determining the host range of SV40 (Graessmann *et al.*, 1981; Pipas, 1985). In addition, apart from enhancers, promoters also show cell-type specificity (Berg *et al.*, 1984; Foster *et al.*, 1985; Grosschedl and Baltimore, 1985). It is likely that the host range of a virus is determined by multiple genetic elements.

D. Critical Sequences Needed for Enhancer Activity

Although there are two 72-bp repeats, one is sufficient for enhancer activity and removal of both abolishes the enhancer function (Moreau *et al.*, 1981; Banerji *et al.*, 1981; Weiher *et al.*, 1983; Herr and Gluzman, 1985). The sequences required for enhancer activity were initially identified by deleting portions of a single 72-bp repeat and measuring the activity of a promoter linked to it. Deletion of the 20 bp from the 5′ end of the single 72-bp repeat (i.e., with respect to the early promoter) drastically reduced its activity (Moreau *et al.*, 1981; Banerji *et al.*, 1981). Deletion of the 3′ end or internal segments also reduced enhancer activity, demonstrating that the whole 72-bp sequence is important for efficient enhancer function (Moreau *et al.*, 1981; Sassone-Corsi *et al.*, 1985). By creating point mutants in the 72-bp repeat sequence and comparing their effect on enhancer activity and also based on their earlier deletion mutant data, Weiher *et al.* (1983) have identified the sequence TGGAAAGT as important for enhancer activity (Fig. 1b). Similar core elements with a consensus sequence TGG$^{AAA}_{TTT}$G have been found in a number of viral and cellular enhancers (Weiher *et al.*, 1983; Laimins *et al.*, 1983).

In addition to the core sequence, three 8-bp segments of alternating purines and pyrimidines have also been found to be important for enhancer function. One is present in each of the 72-bp repeats and another lies near the late promoter at np 261 (Fig. 1b). These segments can assume Z-DNA conformation which has been implicated in the enhancer function (Nordheim and Rich, 1983). A survey of a number of viral enhancers showed the presence of similar 8 bp of alternating purines and pyrimidines about 50 to 80 bp apart (Nordheim and Rich, 1983). Alternating purine–pyrimidine sequences have also been found in the eukaryotic genome and these sequences can also enhance promoter activity in transient assays (Hamada *et al.*, 1984a,b).

Several observations suggest that the alternating purine–pyrimidine sequences have a role in SV40 enhancer function. Analysis of a number of SV40 deletion mutants indicates that at least one of the three alternating purine–pyrimidine blocks is always present in viable mutants (Fromm and Berg, 1982; Gheysen *et al.*, 1983). Interestingly, deletion mutants in which one 72-bp sequence is sufficient for enhancer function have retained two alternating purine–pyrimidine blocks (Moreau *et al.*, 1981; Banerji *et al.*, 1981). Some deletion mutants containing an intact core and two alternating purine–pyrimidine blocks, but with the distance between them reduced by 20 bp, had much less enhancer activity (Moreau *et al.*, 1981).

To study the importance of these sequences in enhancer function, they were altered by point mutation of an SV40 variant that had only one 72-bp repeat (and therefore only two alternating purine–pyrimidine blocks, one within the 72-bp repeat and the other at np 261) and their enhancer activity estimated (Herr and Gluzman, 1985). Two transversions in each of these blocks abolished the activity of the enhancer and the viability of the virus, demonstrating that even when an intact core sequence is present, it is not sufficient for enhancer function. Potential Z-DNA forming sequences are also needed for enhancer activity as well as viability. However, when revertants were selected from these mutants, all of the revertants contained a tandem duplication of 45 to 135 bp of the enhancer region, all of which included the enhancer core sequence. The revetants with two core sequences separated by about 45 to 135 bp have varying degrees of enhancer activity (Herr and Gluzman, 1985), suggesting that when the Z-DNA forming sequences are absent, duplication of the core sequence can restore the enhancer function.

Other studies have shown that duplication of sequences around the origin region of SV40 that do not include either the core sequence or the alternating purine–pyrimidine blocks can also restore the enhancer function as well as viability to mutants lacking the entire enhancer region (Weber *et al.*, 1984; Swimmer and Shenk, 1984). In these studies, enhancerless SV40 genomes were created by deleting about 200 bp of DNA between np 100 and 300 (including both 72-bp repeats) and transfected into cells in culture. Revertants of this nonviable mutant that were characterized all had duplicated a segment of SV40 DNA that either spanned the deleted region or extended from the deletion, and always included 70 nucleotides from the late side of the deletion (np 301–370). This duplication restores viability of the virus and has enhancer function in a transient assay (Weber *et al.*, 1984; Swimmer and Shenk, 1984). The duplicated region includes the start site of late mRNA and encodes the agnoprotein, which may be involved in maturation of the virus (Ng *et al.*, 1985). However, neither the start site nor the agnoprotein coding sequences are essential for virus growth since mutants lacking this region are viable (Barkan and Mertz, 1981; Subramanian, 1979; Piatak *et al.*, 1981; Gheysen *et al.*, 1983).

The above observations demonstrate that the SV40 enhancer contains multiple elements and deletion of one element can be compensated by duplication of remaining elements.

E. Factors That Interact with the SV40 Enhancer

The cell-specific nature of the enhancers suggested that specific cellular factors may interact with enhancers. By using transient assays, Scholer and Gruss (1984) were able to demonstrate that SV40 enhancer sequences interact with cellular factors. They used plasmids with the SV40 enhancer and promoter regions cloned in front of a gene coding for a bacterial enzyme whose activity can be measured (Gorman *et al.*, 1982). The enhancer activity was estimated by the amount of enzyme synthesized after transfecting plasmids into cells. By use of plasmids with identical SV40 transcriptional control regions but each encoding a different enzyme, a competition for the limiting amount of cellular factors was shown (Scholer and Gruss, 1984). Mutations that reduced enhancer activity also prevented competition in this assay, suggesting that factors limiting transcription interacted with the enhancer region. In addition, a host cell preference of different viral enhancers was suggested by this assay. The cell-type-specific interaction between the enhancer and factor(s) has been shown more clearly with an immunoglobulin gene enhancer. By using an *in vivo* footprinting method, Ephrussi *et al.* (1985) have demonstrated that this enhancer interacts with factor(s) in B cells but not in other cells.

Azorin and Rich (1985) have isolated Z-DNA binding proteins from SV40 minichromosomes and demonstrated that these proteins interact with the 72-bp repeat region of SV40. At present, it is not known whether these proteins are involved in activation of transcription.

Proteins encoded by the adenovirus E1A region have been shown to stimulate transcription in *trans* from many promoters (Yaniv, 1984). However, results obtained from two different laboratories suggest that the E1A products can inhibit enhancer-dependent activation of transcription, raising the possibility that these proteins can both activate and depress transcription (Borrelli *et al.*, 1984; Velcich and Ziff, 1985). The data suggest that repression involves interaction between the enhancer and a *trans*-acting factor(s), probably the E1A products. But it is not clear how these factors interact with the enhancers from SV40, polyoma, and adenovirus, which do not have any primary sequence homology. This is further complicated by the observations made in many laboratories that presence of the E1A gene in *trans* and certain other enhancers in *cis* can have an additive effect on RNA synthesis (Kingston *et al.*, 1984; Allan *et al.*, 1984; Imperiale *et al.*, 1985; Natarajan and Salzman, 1985). Since different promoters having varying lengths of upstream control sequences have been used in these studies, it is difficult to compare the results and further studies are needed.

Recently, existing *in vitro* transcription systems have been modified so that enhancer sequences can stimulate transcription *in vitro* (Sassone-Corsi *et al.*, 1984; Wildeman *et al.*, 1984; Sergeant *et al.*, 1984). *In vitro*, as *in vivo*, the SV40 enhancer activates transcription from both homologous and heterologous promoters. This activation is absent when the DNA template is assembled into chromatin (Sergeant *et al.*, 1984). The mutations that decreased *in vivo* activity of the enhancer also affected the *in vitro* activity. *In vitro* stimulation of transcription from a promoter by an enhancer can be abolished by addition of a competitor DNA containing enhancer sequences, suggesting that the enhancer interacts with factors present in the transcription system. The competitor DNA must contain both 5′ and 3′ parts of the enhancer to be effective. Based on this competition assay, other enhancers like polyoma and adenovirus and immunoglobulin enhancer seems to interact with the same set of factors (Wildeman *et al.*, 1984; Sassone-Corsi *et al.*, 1985). The availability of an *in vitro* assay should permit the purification of this factor.

F. Mechanism of Enhancer Action

Several hypotheses, which are not mutually exclusive, have been proposed to explain the stimulation of transcription by enhancers. To explain how the enhancers activate transcription independent of orientation and relatively independent of distance, and also enhance transcription from the proximal promoter in preference to the distal, Chambon and his colleagues have advanced the hypothesis that enhancers are the entry sites for RNA polymerae and/or transcription factors (Moreau *et al.*, 1981; Wasylyk *et al.*, 1983a, 1984). This hypothesis is also supported by the fact that the SV40 enhancer region has nuclease-hypersensitive sites and is free of nucleosomes (see Section X). According to the entry site hypothesis, the transcription machinery enters at the nucleosome-free 72-bp repeat and reaches the promoter region either by scanning in both directions for a promoter or by a direct intramolecular transfer (Moreau *et al.*, 1981; Wasylyk *et al.*, 1983a, 1984). Other models proposed to explain the enhancer action are that enhancer sequences may be involved in changing the topology of the template, or in binding the active template to the nuclear matrix, where active transcriptional complexes may be located (Banerji *et al.*, 1981; Cereghini *et al.*, 1983; de Villiers *et al.*, 1983; Jackson and Cook, 1985). That the presence of SV40 enhancer sequences can increase the number of transcribing RNA polymerase II molecules on the promoter linked to it and that SV40 enhancer-like sequences facilitate the formation of stable transcriptional complexes are compatible with any of the above-mentioned mechanisms (Treisman and Maniatis, 1985; Weber and Schaffner, 1985; Mattaj *et al.*, 1985).

VII. POLYOMA VIRUS ENHANCER

The replication origin–promoter region of polyoma virus is depicted in Fig. 2. Sequences needed for efficient expression of early genes are located about 200 to 400 nucleotides upstream of the major early RNA initiation site (Tyndall *et al.*, 1981; Jat *et al.*, 1982a; Mueller *et al.*, 1984). These sequences are also capable of stimulating transcription from heterologous promoters in *cis.* Since they function relatively independently of position and/or orientation, they constitute the enhancer of polyoma virus (de Villiers and Schaffner, 1981; Cereghini *et al.*, 1983; Ruley and Fried, 1983). Unlike the SV40 enhancer, which has two 72-bp repeats, the polyoma enhancer lacks repeated sequences. The polyoma regulatory region has been resolved into two distinct enhancers, enhancers A and B (Fig. 2 and Herbomel *et al.*, 1984). Enhancer A has a core sequence similar to that of the adenovirus enhancer and enhancer B has a core sequence homologous to that of the SV40 enhancer (Herbomel *et al.*, 1984).

Polyoma virus grows in differentiated mouse cells (NIH 3T6 cells) but its growth is restricted in embryonic and embryonal carcinoma cells. Polyoma virus variants that can grow in embryonal carcinoma cell lines have been isolated (reviewed by Levine, 1983; also Tanaka *et al.*, 1982; Dandolo *et al.*, 1983; Melin *et al.*, 1985a,b) and all have either point mutations, deletions, or duplications in the enhancer region. The variants selected with a particular line of embryonal carcinoma cells (e.g., F9) are still restricted for growth in another and thus provide a system to study the cell-type specificity of enhancers. With heterologous promoters, the wild-type enhancer A is three times more active in 3T6 cells than in F9 cells, whereas wild-type enhancer B is equally active in both types of cells, although at much lower levels than enhancer A (Herbomel *et al.*, 1984). One variant of polyoma virus (PyF441) selected to grow in F9 cells has only a single point mutation in the enhancer B region (A to G at position 5235) (Fujimura *et al.*, 1981). This point mutation increases enhancer B activity in 3T6 cells as well as in F9 cells (Linney and Donerly, 1983; Herbomel *et al.*, 1984). Comparison of many polyoma virus variants that can grow in both embryonal carcinoma cells and differentiated cells shows that the sequences that constitute enhancer A are retained or duplicated in all of them, whereas enhancer B has deletions and/or duplications, suggesting that a modified enhancer B is active in embryonal carcinoma cells (Melin *et al.*, 1985b).

In the case of SV40, the enhancer sequences are separated from the sequences needed for replication and deletion of the enhancer does not affect the efficiency of replication (Fromm and Berg, 1982; Hartzell *et al.*, 1984a). But in polyoma virus, the enhancer sequences overlap with the sequences necessary for replication (Tyndall *et al.*, 1981; Muller *et al.*, 1983). Recent studies have established that the enhancer is needed in *cis* for the replication of polyoma DNA. Heterologous enhancers can also substitute for the polyoma enhancer in a position- and orientation-independent manner and confer the ability to replicate (de Villiers *et al.*, 1984; Veldman *et al.*, 1985). At present, it is not known whether the two different functions of polyoma enhancer are interconnected.

VIII. SPLICING OF VIRAL TRANSCRIPTS

Studies with adenoviruses led to the surprising observation that mRNA molecules did not represent faithful copies of viral DNA segments but instead were specified by discontinuous DNA sequences. It was first observed that the 5′-terminal oligonucleotide derived from an adenovirus population that contained at least 12 distinct species of mRNA contained a single predominant oligonucleotide. Direct evidence that the same 5′ oligonucleotide was present in different mRNA molecules was obtained when a subset of late mRNA molecules were preselected by hybridization to different restriction fragments and each was shown to yield this oligonucleotide (Gelinas and Roberts, 1977). Similar conclusions were reached when hybrids of adenovirus RNA and DNA molecules were examined by electron microscopy. Chow *et al.* (1977) found that the 5′-termini of several late mRNAs, including fiber and 100K, are complementary to sequences near adenovirus map positions 17, 20, and 27; findings similar to these were described by Berget *et al.* (1977). Examination of two populations of purified late mRNAs coding for the fiber and 100K proteins provided direct chemical evidence that different adenovirus mRNA molecules share an identical nucleotide sequence at their 5′ ends and that a single mRNA molecule contains sequences specified by discontinuous parts of the viral genome (Klessig, 1977). These adenovirus mRNAs have a common 5′-terminal leader sequence specified by a region at least 10 kb upstream from their structural genes.

When primary transcripts are processed to yield mRNA molecules, the uninterrupted regions that are retained in the mRNAs are termed exons and the intervening sequences between exons that are excised are termed introns.

When the late mRNA in SV40 cells was examined, discrete RNA populations were also found to consist of RNA sequences that came from spatially separated parts of the viral genome. Aloni *et al.* (1977) examined the pattern that was obtained when preselected populations of 16 S and 19 S RNAs were hybridized to SV40 DNA fragments that had been generated with restriction enzymes, gel fractionated, and transferred to nitrocellulose. 16 S mRNA, which specifies VP1, was known to be transcribed from DNA sequences located at 0.95 to 0.17 map units, and yet and yet it was also able to hybridize near map position 0.67, almost 1400 nucleotides away from the main coding sequences. Subsequent studies (Lavi and Groner, 1977; Hsu and Ford, 1977; Ghosh *et al.*, 1978a; Haegeman and Fiers, 1978a) provided additional evidence for splicing of both 16 S and 19 S mRNAs, and located the sites at which this occurred. Similar findings were made for the late mRNA species in polyoma-infected cells (Srivatsan *et al.*, 1981), and for both viruses the early mRNA species were also found to consist of RNA transcripts that had been spliced. However, in early transcripts the splices join together coding regions (Treisman *et al.*, 1981a) in distinction to late transcripts where splicing joins RNAs in leader sequences that are not translated. An approximate location of the RNA splice site can be determined by S1 analysis. Nucleotide sequence analysis of transcripts by reverse transcriptase primer extension locates the specific nucleotides involved in the formation of the spliced junction. The splice junctions and the sequences at the exon–intron donor sites and at the intron–exon acceptor sites are shown in Table I for SV40 and for polyoma.

As was predicted by the scientists who first observed splicing, it is a general cell mechanism and almost all species of mRNA from higher eukaryotes contain one and in most cases numerous splice junctions. It was originally noted (Seif *et al.*, 1979; Breathnach *et al.*, 1980) that the sequence at the 5′ end of introns was GT and at the 3′ end, AG. Mount (1982) has examined the sequences at 139 exon–intron boundaries at the donor domain of nuclear and viral genes encoding proteins and of 130 intron–exon boundaries at acceptor sequences. By examination of this large number of splice sites, a longer consensus sequence has been identified. The consensus sequences at the 5′ donor and 3′ acceptor regions are also shown in Table I.

Where it has been studied, the same recognition signals are used for defining splice sites in cellular and viral genes, and so it is not surprising that studies on the detailed mechanism of splicing with a cellular and a viral gene have yielded totally consistent results. Following the discovery of splicing, there were major efforts to obtain *in vitro* systems that would carry out the splicing reaction. The failure of these initial attempts was due not to the cell extract that was used, but rather to the fact that specific precursor mRNAs (pre-mRNAs) are required. Suitable pre-mRNAs can be formed *in vitro* or *in vivo* and then added to cell extracts. Adenovirus or globin pre-mRNAs containing the first two exons and one intron are efficiently spliced when added to HeLa extracts. Based on isolation of intermediates in the splicing reaction, the initial step is cleavage of the pre-mRNA at the 5′ splice site to produce free exon I. The 5′ end of the intron is not free but forms a covalent 2′:5′ phosphodiester bond with an adenine nucleotide within the intron near its 3′ end. The intron with this looped, or lariat, structure is still linked to exon II, but is then cleaved at the 3′ splice site to give a lariat structure containing only intron sequences and exon II covalently linked to exon I. It has been proposed that the alignment of the 5′ end of the intron with the region of the intron where the covalent branch is formed is based on a complementary base structure. Ruskin *et al.* (1984) have noted a consensus sequence (PyXPyTPuAPy) near the 3′ end within introns, and Keller (1984) has also reported a similar consensus sequence (CTGAC) within the 3′ ends of mammalian introns. The model derived from studies with adenovirus and globin genes proposes base pairing of the 5′ exon sequences with sequences within the intron, prior to or simultaneous with lariat formation. Small nuclear ribonucleoprotein particles (SNRPs) have also been shown to bind to the 5′ splice site (Mount *et al.*, 1983), and antiserum that precipitates these particles blocks splicing. It is not clear how these SNRPs function, but if they do form a complementary structure, they must then dissociate prior to or simultaneous with lariat formation. The evidence for this model comes from studies by M. Edmonds, M. Green, P. Sharp, T. Maniatis, W. Keller and their co-workers. We present the model in detail since it is likely that similar pathways are followed to generate SV40 and polyoma mRNAs.

The mechanism for splicing that is discussed above implies an obligatory role for the intron in lariat formation. When deletions were introduced into the rabbit β-globin gene large intron, the minimum sequence requirement for splicing was 5 nucleotides within the intron at the 5′ donor site and 12 nucleotides at the 3′ site. In addition, there was a requirement for a minimum intron size of 80 nucleotides. In this study, the authors concluded that the nucleotides that made up the spacer between the required 5′ and 3′ bases did not have to contain specific sequences (Wieringa *et al.*, 1984). There is a discrepancy between those data that indicate a minimum intron size of 80 and the use in SV40 late 19 S RNA of a splice with an intron that contains only 28 bp.

In other studies, extensive deletion within introns has not blocked the formation of spliced mRNA molecules (Khoury *et al.*, 1979; Gruss and Khoury, 1980; Wieringa *et al.*, 1983). These results appear inconsistent with a mechanism requiring significant homology between the 5′ end of this intron and sequences within the body of the intron. This apparent discrepancy may reflect a low level of stringency in the sequences within introns where the 5′ end joins to form a lariat or the utilization of alternative sequences when the natural splice site has been altered by the introduction of either point or deletion mutations. While the model described above implies homology between the 5′ end of the intron and a cluster of bases within the intron, it does not indicate how the proper site for lariat formation is selected when several candidate sites are contained within the same intron. Since the site for lariat formation is positioned close to the 3′ acceptor splice site, the latter may function in positioning the site at which the 2′:5′ phosphodiester bond is formed.

A. The Secondary Structure of Pre-mRNA Has a Role in Splice Site Selection

During the late period of SV40 infection, the most abundant mRNA species is the 16 S RNA, which contains a single splice between np 526 and 1463 (Celma *et al.*, 1977; Ghosh *et al.*, 1978b; Haegeman and Fiers, 1978a; Van de Voorde *et al.*, 1977; Bina-Stein *et al.*, 1979). More than 90% of the spliced 16 S mRNA is initiated at np 325. There are much lower levels of the two 19 S RNA species, which are generated using donor splice sites at np 373 and 528 and an acceptor site at np 558. These 19 S RNAs also initiate at the np 325 start site as well as at start sites near np 260 and 195 (Ghosh *et al.*, 1978b). This suggests that the start site for transcription is one determinant of whether a primary SV40 transcript will be spliced to form either 16 S or 19 S mRNA. There are also other factors that function in determing the choice of splice sites. Examination of DNA sequences reveals that potential splice sites with consensus donor and acceptor sites are present but not normally used unless the regular splice site is inactivated. Mutations outside of the consensus sequences are also known to abolish correct splicing. The most consistent interpretation of the present data is that an RNA primary transcript has alternative secondary configurations. A particular RNA conformation could determine which of several alternative consensus donor and acceptor sites are used. In RNA folding is an important determinant of the selection of splice sites, nonfunctional consensus splice sites are probably excluded because they are not oriented properly. These ideas are in agreement with the observation that proper splicing is only obtained *in vitro* if a particular pre-mRNA is used as substrate while some other transcripts containing the same splice site sequences are not spliced.

To determine if splicing involves a scanning mechanism, hybrid RNAs containing duplicate donor and acceptor splice sites have been examined. In a scanning mechanism, the first of two donor and acceptor sites nearest to the 5′ end of the transcript would be selected for splicing. In two such studies (Lang and Spritz, 1983; Kuhne *et al.*, 1983), opposite conclusions were reached. However, these results could be reconciled if there are secondary structural differences in the different RNAs that are being processed. The adenovirus E1A gene gives rise to two mRNAs (12 S and 13 S) that have common 5′ and 3′ ends but differ in their splicing pattern. This is similar to the way in which the mRNAs encoding large T and small t antigen are generated. Both argue against a scanning mechanism but rather suggest alternative configurations of a primary transcript in which different acceptor sequences are used.

Much of the current work is providing additional information about intermediates in the splicing scheme. In this scheme, there is not a definite role for the SNRPs although they have been demonstrated to be necessary for splicing (Lerner *et al.*, 1980; Yang *et al.*, 1981; Kramer *et al.*, 1984). This indicates that additional reactions that yield spliced mRNA molecules are yet to be described.

IX. TERMINATION OF TRANSCRIPTION

The primary SV40 transcripts extend beyond the site at which poly(A) is added to mRNAs (Ford and Hsu, 1978), and similar observations have been made with adenovirus (Nevins *et al.*, 1980). Thus, in order to specify the 3′ end of the mRNA molecule, the primary transcript must undergo endonucleolytic cleavage and then polyadenylation. Both of these modifications occur prior to splicing and both events are common to almost all mRNA precursors. When sV40 late mRNAs are examined, the site at which polyadenylation occurs is at np 2674. While it seems likely that endonucleolytic cleavage and polyadenylation occur at the same site, there is no evidence for this.

When the sequences of mRNA molecules are examined, the sequence AAUAAA is almost invariably present 10–30 nucleotides before the polyadenylation site (Proudfoot and Brownlee, 1976; McLauchlan *et al.*, 1985). If deletions are introduced in SV40 DNA at a location between the AAUAAA sequence and the authentic polyadenylation site, the poly(A) site shifts so that it always remains 11–19 bp downstream from the AAUAAA signal. When the AAUAAA sequence was deleted, the site used for polyadenylation then shifted to downstream of a second AAUAAA signal present in the recombinant being studied (Fitzgerald and Shenk, 1981). While this sequence is required for polyadenylation, its presence within the coding sequences of genes indicates that it is not itself sufficient to signal cleavage and/or polyadenylation. Recently, Sadofsky and Alwine (1984) have further studied the deletion mutants used by Fitzgerald and Shenk. They found that a second effect of deletions downstream of the AAUAAA sequence was to decrease the efficiency with which the normal polyadenylation site functioned. Under the conditions used, there was an accumulation of longer, nonpolyadenylated RNAs whose 3′ ends (np 2794 and 2848) map to areas where the secondary structure of the transcript might facilitate pausing or termination of transcription (Sadofsky and Alwine, 1984).

Polyadenylation of the cloned hepatitis B surface antigen gene occurs 12 to 19 bp 3′ from a signal UAUAAA, which is similar to the AAUAAA signal usually found. By deletion studies, it was shown that the signal hexanucleotide itself cannot specify cleavage and polyadenylation. For proper processing of the RNA, an additional 30 nucleotides downstream is required (Simonsen and Levinson, 1983). Similar results have also been reported by McLauchlan *et al.* (1985) in their studies of herpesvirus transcripts. They have identified a consensus sequence PyGUGUUPyPy which has preferred location 24–30 nucleotides downstream from the AAUAAA signal. This consensus sequence is present in 67% of the genes examined, and it is present in both SV40 and polyoma early and late genes (McLauchlan *et al.*, 1985).

The primary function of the hexanucleotide signal appears to be positioning the site-specific endonucleolytic cleavage of the RNA rather than to function in poly(A) addition (Montell *et al.*, 1983; Wickens and Stephenson, 1984). When point mutations were generated in the AAUAAA chain termination signal, most of the transcipts were not cleaved at the normal polyadenylation site. However, the small fraction of primary transcripts that were cleaved contained poly(A) tails, and so the primary effect of the point mutants was to inhibit endonucleolytic cleavage.

Endonucleolytic cleavage and polyadenylation at an authentic *in vivo* site can also be affected *in vitro.* Processing of the RNA at the 3′ end is not coupled to active transcription and is strongly inhibited by antibodies to U1 RNP, an SNRP that has also been shown to have a role in RNA splicing (Moore and Sharp, 1984). A second *in vitro* reaction has been described in which poly(A) is added to the 3′ ends of RNAs that have been generated in vitro using a runoff transcription system. Under these conditions, there is no requirement for endonucleolytic cleavage (Manley, 1983a). For this reaction to occur, the signal AAUAAA must be present; however, it can lie 400 bp upstream of the 3′ end of the RNA molecule and still function (Manley *et al.*, 1985).

X. THE VIRAL CHROMOSOME

It is now about 15 years since polyoma and SV40 DNAs were found to exist in the cell complexed with protein (Green *et al.*, 1971; White and Eason, 1971). In this section we examine what is known about the formation and structure of the nucleoprotein complex (NPC) and its relevance to the control of virus transcription, relying mainly, unless otherwise noted, on studies carried out with SV40. Other chapters will deal with the significance of the complex in DNA replication and packaging.

A. Isolation of Nucleoprotein Complexes

The major protein components of the complexes consist of the usual complement of histones that are found on chromosomal DNA (Varshavsky *et al.*, 1976). When the NPC present in infected cells is digested with micrococcal nuclease, the DNA products contain 150–200 bp (or multiples thereof) (Shelton *et al.*, 1978). These results indicate that the DNA and histone are organized into nucleosomes, structures typical of the

DNA in chromatin, and therefore the complex has been referred to as a minichromosome (Griffith, 1975). When examined in the electron microscope, the minichromosome appears to contain up to 26 nucleosomes, nearly the maximum number that could be accommodated by circular DNA the size of SV40. However, most NPCs contain fewer nucleosomes so that the average number is about 24 (Griffith, 1975; Bellard *et al.*, 1976; Muller *et al.*, 1978; Saragosti *et al.*, 1980). Below, we discuss the arrangement of nucleosomes on the minichromosome in greater detail.

The original method used to prepare NPCs involved use of buffers containing detergent, EDTA, and relatively high salt (Green *et al.*, 1971; White and Eason, 1971). Both these and subsequent investigations have shown that the chemical and physical conditions of extraction influence the yield, the composition of the protein portion, and some physical properties of the complexes. In particular, two classes of complexes can be obtained from the cell at late times of infection (Garber *et al.*, 1978; Baumgartner *et al.*, 1979; Fernandez-Munoz *et al.*, 1979). One class contains NPCs that have sedimentation coefficients of about 75 S and that consist primarily of histones and DNA. This class of complexes includes (but is not limited to) molecules actively involved in transcription or replication, the latter having sedimentation coefficients of between 75 and 200 S (Garber *et al.*, 1978). The other class has a much higher sedimentation coefficient (250 S) and includes both mature virions and an intermediate (provirion) in the assembly of the mature virus particle (Garber *et al.*, 1978; Baumgartner *et al.*, 1979; LaBella and Vesco, 1980). If these 250 S complexes are incubated in the presence of EGTA or other agents that chelate calcium ions, two complexes are generated: one a 75 S NPC that comes from the provirions, the other a 110 S NPC released by the mature virion (Brady *et al.*, 1978). The stability of the virion and provirion are also affected by increased salt, temperature, or pH so that the relative yields of the 250 S and 75 S NPCs vary (Jakobovits and Aloni, 1980). The effect of extraction conditions may be mediated, in part, by enzymatic activities present in the extract (Garber *et al.*, 1978; Seidman *et al.*, 1979). Surprisingly, the virion complex (110 S) lacks histone H1 but has an otherwise normal complement of histones (Lake *et al.*, 1973; Fey and Hirt, 1974; Sen *et al.*, 1974; Fernandez-Munoz *et al.*, 1979). The loss of H1 occurs subsequent to the packaging of the 75 S NPC into the provirion and is accompanied by an increase in the acetylation level of the H2A and H3 histones (LaBella *et al.*, 1979; LaBella and Vesco, 1980). The 110 S complex also contains VP1, VP2, and VP3 (Griffith *et al.*, 1975; Mann and Hunter, 1979; Brady *et al.*, 1981a; Llopis and Stark, 1981).

Most recent work has made use of hypotonic buffers (0.1 M or less) to prepare NPCs from isolated nuclei (Su and DePamphilis, 1976; Fernandez-Munoz *et al.*, 1979; Jakobovits and Aloni, 1980). Systematic investigation of the effect of ionic strength on 75 S NPCs has shown that increasing the NaCl concentration beyond 0.3 M leads to progressive loss of histone H1 from the complex, with an accompanying decrease in the sedimentation coefficient to 40 S (Fey and Hirt, 1974; Meinke *et al.*, 1975; Muller *et al.*, 1978). At NaCl concentrations greater than 0.8 M, the NPC begins to lose other histones. Electron microscopy of these complexes shows that the 75 S form has a compact appearance possibly analogous to chromatin in the 100-Å nucleofilament form (Finch and Klug, 1976) and that the more extended beads-on-a-string form (Olins and Olins, 1974; Oudet *et al.*, 1975) is present in 0.5–0.7 M NaCl (Muller *et al.*, 1978). Further increases in the ionic strength lead to the loss of nucleosomes from the structure.

A significant fraction of the complexes also contain T antigen (Mann and Hunter, 1979; Reiser *et al.*, 1980; Segawa *et al.*, 1980; Tack and Beard, 1985). Although some T antigen is lost from the complex in 1.0 M salt, a fraction of T antigen molecules display salt-resistant binding (Rudolph and Mann, 1983). Similar results have been reported for the T antigen on polyoma NPCs (Clertant *et al.*, 1984).

When the NaCl is lowered to 10 mM or less, the 75 S NPC again changes to a 40 S form which has the appearance of bead-on-a-string in the electron microscope. Unlike the change caused by high salt, this alteration is readily reversed by a return to 0.1 M NaCl (Muller *et al.*, 1978). A possible explanation for the effect of very low salt is that the condensed structure of the 75 S NPC results from the interaction of H1 molecules of adjacent nucleosomes. Muller *et al.* (1978) suggest that this reflects a cooperative interaction of free H1 molecules which is sensitive to low salt concentration (Renz *et al.*, 1977). These various effects of salt on condensation of the NPC parallel closely the effects of salt on nuclear chromatin where the exact role of H1 is not yet well understood (reviewed by McGhee and Felsenfeld, 1980).

About 60–70% of the intracellular viral DNA at late times is present in provirions. Depending on conditions of extraction, the 75 S NPC preparation may consist largely of complexes derived from provirions (Garber *et al.*, 1978; Jakobovits and Alone, 1980; Fernandez-Munoz *et al.*, 1979). Moreover, even in the most carefully made preparations, relatively few of the NPCs are actually complexes that were engaged in transcription (see below). Thus, the aggregate properties of the 75 S complexes need not reflect those of the complexes of greatest interest here.

B. Structure of the Complexes

As mentioned above, extensive digestion of the minichromosome with nonspecific DNase gives rise to a collection of DNA fragments whose size is similar to that obtained when bulk chromatin is digested. However, if the extent of digestion is severely limited, preferential cleavage of the region around the origin of replication occurs (Varshavsky *et al.*, 1978; Scott and Wigmore, 1978; Waldeck *et al.*, 1978). In general, no such preference is seen with naked DNA or with the NPC present in (pro-)virions (Hartman and Scott, 1981, 1983). This was one of the first observations of what now appears to be a common feature of active genes, namely the presence of a region of the DNA which is more readily attacked by nuclease than is neighboring DNA (Weintraub and Groudine, 1976). These regions of nuclease hypersensitivity are usually found at the beginning of the structural gene where the upstream regulatory sequences are located (reviewed by Elgin, 1981).

When examined in the electron microscope, approximately 25% of the NPCs appear to have a nucleosome-free region, or gap, whose position coincides with that of the region of hypersensitivity (Jakobovits *et al.*, 1980; Saragosti *et al.*, 1980). Although histones are found in the NPC from virions, the NPC is apparently not organized in the usual nucleosome structure (Brady *et al.*, 1981b; Moyne *et al.*, 1982), as was suggested by earlier results (Polisky and McCarthy, 1975; Ponder and Crawford, 1977).

In general, NPCs used in studies of hypersensitivity have been prepared from SV40-infected CV-1 or BSC cells, both of which are permissive for the virus. Studies of defective viruses or of plasmids containing the SV40 origin have used the COS cell line which makes SV40 T antigen (see above). Nuclei are prepared and lysed at 40 hr postinfection, well after the onset of replication and late gene expression. The NPCs are isolated from the lysate by sucrose gradient centrifugation and then treated with nuclease. Alternatively, the NPCs are digested *in situ* by addition of nuclease to the intact nuclei. This latter procedure should minimize the effect of proviral complexes.

Some of the earliest descriptions of hypersensitivity depended on digestion of the NPC by an endogenous nuclease present in nuclear extracts (Scott and Wigmore, 1978; Waldesk *et al.*, 1978). A variety of other DNases have been used for digestion, including micrococcal nuclease, Staphylococcal nuclease, DNase II, various restriction nucleases, and, most commonly, DNase I (Persico-DiLauro *et al.*, 1977; Shelton *et al.*, 1978; Varshavsky *et al.*, 1978; Sundin and Varshavsky, 1979; Shakov *et al.*, 1982). Many investigations of the gap have relied on an "indirect" labeling technique in which DNA is purified from the nuclease-treated NPC (or nuclei), further digested with a convenient restriction endonuclease, and then analyzed by gel electrophoresis and Southern blotting (Nedospasov and Georgiev, 1980; Wu, 1980). The radioactive probe used contains a segment of the viral genome lying between the site of restriction endonuclease cleavage and the site(s) cleaved during nuclease treatment. By using a restriction enzyme that cleaves close by, the exact position of nuclease action can be estimated to within a few base pairs.

C. Location and Extent of the Hypersensitive Region

Studies of 75 S NPC prepared 40 hr (or later) after infection of cells with wild-type virus indicate that the HS region is about 400 bp in size and that it includes the major start sites for both early and late transcription (Varshavsky *et al.*, 1979; Scott *et al.*, 1984). The corresponding region of polyoma NPC also displays hypersensitivity, though in this case the HS region is somewhat smaller (Herbomel *et al.*, 1981). As shown in Fig. 3, the SV40 HS region includes both 72-bp repeats, the three 21-bp repeats, the three T antigen binding sites, and the origin of replication (ORI). Thus, the region contains control elements necessary for normal initiation of replication and of early and late transcription, although it is not itself transcribed.

Just how large is the gap? Data obtained by a variety of techniques bear on this question. Varshavsky *et al.* (1979) treated 75 S NPCs with a series of restriction enzymes to map the endpoints of the region. Prior to digestion, the structure of the complex was fixed by treatment with formaldehyde to prevent changes following initial cleavage of the DNA. The data indicate that the region is about 400 bp in length and that it extends over the segment corresponding to *Alu*I fragments F and P (np 5226 to 406). Scott *et al.* (1984) came to a similar conclusion by first cleaving in the gap with a restriction nuclease, and then further digesting with the exonuclease *Bal*31. If the initial cleavage was at the *Bgl*I site at the origin of replication (see Fig. 3), then the exonuclease readily removed about 400 bp in the direction of the late genes, but only about 50 bp in the early direction. Conversely, if the initial cleavage was at np 346, which is near the start site for late transcripts, the exonuclease attacked mainly DNA on the early gene side of the breakpoint. Careful examination of the products present at intermediate times of digestion suggested that the exonuclease was impeded at the borders of the 72-bp repeat.

In fact, not all bases in the HS region are equally sensitive to nuclease.

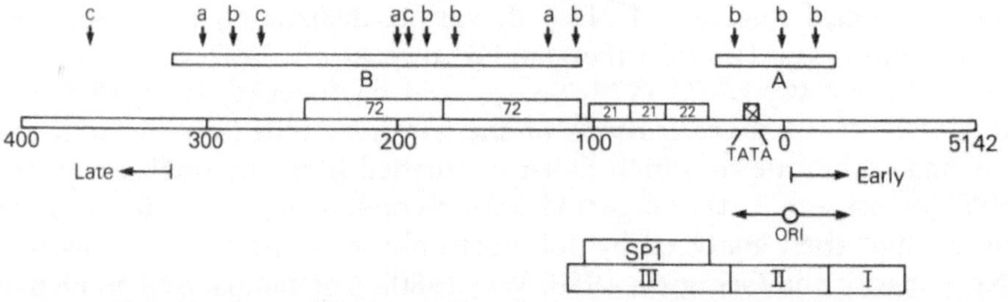

FIGURE 3. Sites of DNase I hypersensitivity in the SV40 chromosome. This presents a schema of the region of the SV40 chromosome that includes the origin of replication (ORI) and the promoters for early and late transcription. The arrows at the top indicate sites of DNase I hypersensitivity according to (a) Jongstra *et al.* (1984), (b) Cereghini and Yaniv (1984), and (c) Cremisi (1981). The bars marked A and B indicate the two regions of hypersensitivity discussed by Jongstra *et al.* (1984) (see text). Below this, the location and extent of the 72- and 21-bp repeats and the TATA box are shown. The barred line shows nucleotide positions as indicated on the SV40 DNA according to Buchman *et al.* (1980). The locations of the major start sites for transcription are marked (Early, np 5235; Late, np 325); the location and extent of regions protected from DNase I by the transcription factor Spl (SP1) and by T antigen (I, II, III) are indicated by boxes (see text for exact positions and references).

Most of the favored sites occur in two segments that lie to either side of the 21-bp repeats (Saragosti *et al.*, 1982; Innis and Scott, 1984; Jongstra *et al.*, 1984). There appears to be reasonable agreement among several investigators about the locations of sites most sensitive to DNase I, with some minor differences possibly due to imprecision in estimating the sizes of larger DNA fragments. Of the three predominant sites to the late side of the 21-bp repeats (region B of Fig. 3), two lie in analogous positions within each of the 72-bp repeats, while the third lies just upstream of the major start site for late transcription. Another favored site at position 370 has been reported (Cremisi, 1981), although the site in the origin-proximal 72-bp repeat was not seen in these experiments. To the early side of the 21-bp repeats, the region of sensitivity coincides with the TATA box and the start site for early transcription (region A of Fig. 3). Again, it appears that at least one site lies near the beginning but within the early transcribed region (Cremisi, 1981; Cereghini and Yaniv, 1984) (not in Fig. 3).

A similar pattern is seen when NPCs are digested with the nuclease endogenous to nuclear extracts. In this case, however, cleavage occurs in the 21-bp repeat region (positions 40 to 103) as well as within the 72-bp repeats and to the early side of the origin (Innis and Scott, 1984). Thus, the failure of DNase I to digest in the 21-bp repeat region may be due to a known preference of the enzyme for particular sequences or structures in DNA (Dingwall *et al.*, 1981).

Studies of polyoma have shown a roughly similar organization of the HS region, which is about 300 bp in its overall extent (Herbomel *et al.*, 1981). The sites most sensitive to DNase I lie within either of two regions (centered on np 5207 and 5097) at either side of a palindrome halfway between the start sites for early and late transcription. Although not a direct repeat, this sequence element shares the GC-rich property and approximate total size of the 21-bp repeats. The HS region to the late side of the palindrome coincides wih the region of (multiple) start sites for late transcription. As discussed above, this region, like the 72-bp repeat, serves as an enhancer for transcription (de Villiers *et al.*, 1982; Cereghini *et al.*, 1983; Ruley and Fried, 1983).

D. Genetic Determination of the Hypersensitive Region

The results described above suggest that the 21-bp repeats play an important role in determining the location of the HS region. This conclusion is consistent with results of experiments with derivatives of a mutant virus containing two copies of the origin of replication. The mutant, int(Or)1411, contains a second copy of the segment between np 5180 and 215 inserted at an *Hpa*I site (np 2666) (Shenk, 1978). As was already known, the insertion creates a second HS region although only one complete copy of the 72-bp repeat is contained within the inserted segment (Wigmore *et al.*, 1980; Jakobovits *et al.*, 1982). Gerard *et al.* (1982) generated several variants by deletion of portions of the inserted segment and examined them for sensitivity to the endogenous nuclease present in nuclear extracts. A deletion which removed all the 21-bp repeats but left the (single) 72-bp repeat caused the loss of the second HS region, while another which removed the distal portion of the 72-bp repeat but retained the 21-bp repeats resulted in reduced but significant sensitivity of the second HS region. It may be significant that the effect of the latter deletion is to create a direct repeat consisting of the first (approximately) 40 bp of two copies of the 72-bp repeat. Thus, the second HS region of this variant contains the two sites of cleavage present in the normal enhancer region (Fig. 3). Furthermore, Swimmer and Shenk (1984) have recently shown that a segment containing a direct repeat of the region adjacent to the 72-bp repeat has enhancerlike activity.

Using this same double origin mutant, Innis and Scott (1984) have also examined the effect of insertions of varying extents between positions 37 and 38, i.e., between the origin and the 21-bp repeats. These insertions reduced the ability of the region to act as a replication origin (when present on a plasmid), but did not decrease nuclease sensitivity. However, the largest insertions (90 to 260 bp) noticeably displaced HS region B (see Fig. 3) away from the origin and toward the late side, just as it displaced the 21- and 72-bp repeats.

The importance of the 72-bp repeat in creating the HS sites has been shown by studies that use virus deleted for the 72-bp repeats and part of the 21-bp repeat region (np 73 to 272) (Fromm and Berg, 1983b). Insertion of a segment containing the 72-bp repeats (positions 95 to 272) at either of two sites (np 2666 and 4740) in the deleted genome conferred local sensitivity to DNase I and the ability to replicate in CV-1 cells. Both sites were distant from the usual origin of replication. These effects occurred irrespective of the orientation of the inserted region, except for insertion at np 2666 where one orientation conferred greater sensitivity than the other. These results suggest that the 21-bp repeats are not required for formation of an HS region, although a contribution by the residual portion of one repeat present in the insert (np 95–103) cannot be eliminated.

A third detailed study of the effect of deletions on the HS region has provided a slightly different viewpoint on required sequences (Jongstra *et al.*, 1984). Again a double origin variant was used in which a segment containing both 72-bp repeats (np 5171 to 346) was inserted at np 2666 of an otherwise wild-type virus. The results showed that at least one of the 21-bp repeats was absolutely required for generation of HS sites in region A (i.e., to the early side of the 21-bp repeats; see Fig. 3) but not in region B (i.e., to the late side). Deletion of the 21-bp repeats decreased somewhat the extent of region B, but did not abolish it. Conversely, loss of the 72-bp repeats affected region B but not A. Increasing the number of copies of the 72-bp repeat to four increased the size of region B proportionately. In this study, insertion variants with a single copy of the

72-bp sequence (only) in either orientation displayed similar nuclease sensitivity, a result that differs from that reported in the two similar studies cited above. The explanation for the discrepancy is unclear. Examination of these variants in the electron microscope indicated that, as had been shown previously (Jakobovits *et al.*, 1982), a significant number of NPCs containing double origin molecules have two gapped regions. Although the various deletions affected the size of the HS region, they appeared not to affect the frequency of molecules containing either one or two gaps.

In summary, these various results indicate that hypersensitivity can be conferred by the 72-bp or 21-bp repeats. The location and extent of the HS region appears to be determined by which of the two is present, and to be increased in a slightly nonadditive fashion when both are present. The location of the HS region correlates well with the location of a nucleosome-free region, or gap, in the minichromosome. Thus, hypersensitivity appears to result from the absence of the nucleosomes which protect other regions of the DNA in assays for nuclease sensitivity. The 21-bp and 72-bp repeats act not only to specify the location of the nucleosome-free region, but may also specify particular sites of nuclease sensitivity within it. As we discuss below, it seems possible that both effects are mediated by the interaction of nonhistone proteins with the DNA in the region of the gap, but this remains to be established.

E. Proteins Binding to the Hypersensitive Region

From analysis of the chromatin containing the chicken β^A-globin and the *Xenopus* 5 S RNA genes, it seems likely that HS regions occur because specific nonhistone proteins bind to the DNA and thereby locally block the formation of nucleosomes. In the case of the chicken β^A-globin gene, it has been possible to reconstitute the HS region *in vitro* if particular site-specific binding proteins are present during reconstitution of nucleosomes on DNA containing the 5′-upstream region of the gene (Emerson and Felsenfeld, 1984). In both cases, the factor which excludes histones has been shown to be site-specific binding proteins (Emerson *et al.*, 1985; Bogenhagen *et al.*, 1982), and, in the case of the *Xenopus* 5 S RNA gene, to be required for initiation of transcription *in vitro* (Engelke *et al.*, 1980).

Although this suggests a mechanism that positions the HS region, it does not provide an explanation for why particular bases within the region are more susceptible to DNase than others. These may simply be positions that are exposed to nuclease because they lie between one binding protein and an adjacent binding protein or nucleosome. In some cases, however, bases internal to the binding domain of a protein may remain unprotected, and sometimes the sensitivity of bases lying either within or just outside the domain may actually be enhanced (Galas and Schmitz, 1978). These effects can, of course, be dependent on the nuclease used and the particular conditions employed.

Besides nonhistone proteins, the structure of the DNA may itself favor particular locations (i.e., phasing) of the nucleosomes. In particular, Trifanov and Sussman (1980) have proposed that the regular occurrence in the SV40 genome of sequence elements that bend the axis of the helix, such as neighboring purines, would lead to phasing of the nucleosomes. This could be an important structural factor, especially in the determination of the borders of the HS region. As yet little is known with respect to the role of DNA structure in formation of the SV40 HS region. Some data bearing on the structure of DNA in viral transcription complexes are discussed below.

Proteins known to bind specifically to the region include T antigen and the transcription factor Sp1. From both direct (Dynan and Tjian, 1983b; Azorin and Rich, 1985) and indirect (Tack and Beard, 1985; Sergeant *et al.*, 1984) evidence, it appears likely that there are several factors that can bind to the 72-bp repeat. As discussed earlier, there is a transcription factor which has been shown to bind to the TATA box of some *Drosophila* genes (Parker and Topol, 1984). Other proteins which may be present include topoisomerase II (Yang *et al.*, 1985), poly(ADP-ribose) polymerase (Prieto-Soto *et al.*, 1983), and the RNA and DNA polymerases. It is clear that viral capsid proteins are not required for gap formation (Saragosti *et al.*, 1982; Innis and Scott, 1983).

In the case of SV40, attempts to reconstitute the gapped region *in vitro* have been unsuccessful (Sergeant *et al.*, 1984). Thus, we can only speculate about which proteins contribute both to formation of the gap and to the presence of particular sites of hypersensitivity.

Because T antigen appears to be present on many NPCs, it seems a likely candidate for a protein that is bound to DNA in the region of the gap. This has been confirmed for a small fraction of the gapped NPCs by electron microscopy of NPCs treated with antibody specific for T antigen (Harper *et al.*, 1984). The domain of T antigen binding to purified DNA, as revealed by DNase I protection experiments (Tegtmeyer *et al.*, 1983; DeLucia *et al.*, 1983), is shown in Fig. 3. Three binding sites are present, of which sites I and II appear to be the strongest (reviewed by Tjian, 1981). There is at least one site whose DNase I sensitivity is enhanced when T antigen is bound, this located around np 40 just within region III. The enhancement is seen at low ionic strengths (5 mM NaCl), but not in 150 mM NaCl. A number of bases in all three sites retain some sensitivity to DNase I, particularly at the borders between the sites and at one position within site II just to the late side of the *Bgl*I site (np 5243/0). For the most part, these sites would lie within region A in Fig. 3. Despite this coincidence, however, it seems likely that the HS region can form in the absence of any T antigen, as is discussed below (Cremisi, 1981; Saragosti *et al.*, 1982).

The binding domain of Sp1 covers the GC-rich region in the 21-bp repeats (Dynan and Tjian, 1983b; Gidoni *et al.*, 1984). DNase I digestion of SV40 DNA in the presence of Sp1 shows several sites of enhanced cleavage, most of which lie at the borders of the protected region. Of particular interest is the strong enhancement of cleavage at np 112–114. This corresponds to the approximate location of the HS site seen by several investigators at the beginning of the origin-proximal 72-bp repeat (Fig. 3). Several other sites of enhanced cleavage were found to the early side of the binding site, close to or within HS region A. Whether or not these sites of enhanced cleavage actually correspond to HS sites in the NPC, the apparent importance of the 21-bp repeat region in conferring nuclease sensitivity (Gerard *et al.*, 1982; Jongstra *et al.*, 1984) suggests that Sp1 plays an important role in formation of the HS region.

F. Transcription Complexes

Soon after the initial descriptions of NPCs, it was reported that a small fraction of the complexes could carry out synthesis of viral RNA with a high degree of specificity (Green and Brooks, 1976). Analysis of the density of nascent *in vitro* transcripts still attached to these viral transcription complexes (VTCs) showed that the protein content of the complex is identical to that of the bulk of NPCs (Green and Brooks, 1977; Gariglio *et al.*, 1979). Examination of complexes by electron microscopy confirmed that their chromatin structure (beads-on-a-string) is the usual one seen for NPCs (Gariglio *et al.*, 1979).

Because the transcription complexes constitute only about 1% of all the 75 S NPCs (Gariglio *et al.*, 1979; Llopis *et al.*, 1981), their analysis has been a demanding task. One approach to dealing with the small quantity of VTCs has been to use the detergent Sarkosyl to increase the efficiency of extraction. The original extraction procedure (Triton X-100 and 0.4 M NaCl) left about 80% of the viral transcription activity in a pelleted fraction that also contains cellular chromatin (Green and Brooks, 1976). This is avoided when the transcription complexes are obtained from isolated nuclei by lysis with Sarkosyl (Gariglio and Mousset, 1975, 1977). As had been shown in total chromatin, exposure to Sarkosyl removes histones and most other protein from the DNA, but leaves RNA polymerase II molecules that have initiated transcription (Green *et al.*, 1975; Gariglio and Mousset, 1975). The treatment with Sarkosyl also increases the transcription activity of complexes (Brooks and Green, 1977; Gariglio *et al.*, 1979). The usual interpretation of this effect is that removal of the histones permits more rapid polymerse movement on the template (Gariglio *et al.*, 1979). In fact, hypertonic salt (>100 mM ammonium sulfate), which opens the chromatin structure as discussed above, has an effect on activity similar to that of Sarkosyl (Brooks and Green, 1977). The treatment appears not to favor particular transcription units since about 80% of the RNA synthesized by Sarkosyl-treated complexes isolated late after infection hybridizes to the late strand of SV40 DNA (Ferdinand *et al.*, 1977).

An important problem has been to identify the form of viral DNA present in the VTC. Analysis of the template DNA was facilitated by the discovery that nascent transcripts labeled *in vivo* or *in vitro* could be purified as RNA–DNA hybrids (Girard *et al.*, 1974; Gariglio and Mousset, 1977; Birkenmeier *et al.*, 1977). Such hybrids have also been observed among *in vitro* transcription products obtained with *E. coli* RNA polymerase, but only when the template is negatively supercoiled DNA (Richardson, 1975). In the case of SV40, only about 60–150 nucleotides of the most recently synthesized portion of the nascent transcript is found hybridized to the DNA, the remainder of the transcript is found hybridized to the DNA, the remainder of the transcript being susceptible to single-strand-specific RNase (Birkenmeier *et al.*, 1977). Although the exact mechanism of its formation is somewhat obscure, the hybrid serves to mark template molecules actively involved in transcription at the time of extraction.

Based on an analysis of hybrids containing nascent RNA labeled *in vivo*, Girard *et al.* (1974) had proposed that the template for transcription was an intermediate in replication, although concurrent replication was known not to be necessary for late transcription (Cowan *et al.*, 1973). Birkenmeier *et al.* (1977) performed a more extensive analysis of template molecules found as hybrids following *in vitro* synthesis of RNA by Sarkosyl-treated VTCs. The results indicated that the structure of the template is the same as that of form I DNA and therefore unlike that of most replication intermediates.

The results of a more recent study suggest that this question may need reexamination. NPCs prepared by standard techniques were treated with topoisomerase and their DNA then analyzed. A minor fraction (2–5%) of the DNA was found to differ in its superhelicity as a result of the topoisomerase treatment, thus indicating the existence of complexes whose DNA is torsionally strained prior to removal of histones (Luchnik *et al.*, 1982). In the bulk of the NPCs, the nucleosomes stabilize the negative supercoiling of the DNA and protect it from the action of the topoisomerase (Germond *et al.*, 1975). When analyzed on sucrose gradients, the topoisomerase-sensitive complexes sedimented more quickly than the majority of NPCs, but identically to transcription complexes. This suggests that they contain RNA polymerase which had initiated transcription. The mechanism of formation of these apparently torsionally strained complexes remains unclear. The possibility that some VTCs do not contain form I DNA can be reconciled with the data cited above if the use of Sarkosyl activates blocked transcription complexes whose template DNA differs from that of complexes active *in vivo*.

In fact, the existence of more than one class of VTC has been suggested by experiments which examine the salt and Sarkosyl sensitivities of VTCs that leach from intact nuclei. Llopis and Stark (1982) reported that complexes released by the nuclei within a few minutes after isolation were not activated by increased ammonium sulfate (100–300 mM), whereas those released by 60 min were. If the cells were briefly labeled with radioactive uridine just prior to preparation of the nuclei, most VTCs containing nascent (i.e., labeled) RNA were released from the nuclei within a few minutes. The two types of complexes could also be distinguished by their behavior on sucrose gradients. An analogous result was obtained when VTCs prepared with Sarkosyl were sedimented in sucrose gradients. These moved more rapidly than the bulk of viral DNA, but were themselves found in two populations. One contained the nascent RNA, while the other contained most of the transcription activity detected during subsequent *in vitro* assay. Finally, transcription complexes isolated from nuclei of cells infected with the SV40 mutants dl1261 or dl1262 were not activated by increased ammonium sulfate, but were by Sarkosyl (Llopis and Stark, 1981). Both these mutants have small in-frame deletions within the structural gene for VP2/VP3.

Llopis and Stark proposed that an important fraction of intracellular VTCs have been inactivated by the binding of VP2 (and/or VP3), possibly as an initial step in formation of the provirion. These complexes are released slowly from the nuclei and are activated when salt or Sarkosyl provokes the release of the bound VP2 (VP3). By this model, most of the transcription activity present in VTCs prepared with Sarkosyl or high salt is that of polymerase molecules whose progress along the template had been blocked by VP2 (VP3). They suggest that the mutant VP2 (VP3) proteins bind more tightly to the DNA and are harder to release with salt.

The model suggests that a large fraction of the VTCs can come from blocked complexes not active *in vitro*. Besides presenting a caveat for the interpretation of studies that use Sarkosyl-treated VTCs, it also provides an explanation for data indicating the formation of blocked late transcription complexes when DNA synthesis is prevented. For example, VTCs prepared with Sarkosyl from cells infected with a tsA mutant of SV40 synthesize predominantly late RNA, even though little synthesis of late RNA *in vivo* can be detected (Birkenmeier *et al.*, 1979). Previous experiments of Ferdinand *et al.* (1977) had also pointed to the existence of blocked complexes which are apparently activated by replication.

The model of Llopis and Stark may also be relevant to the analysis of data which suggest the occurrence of attenuation during late transcription of SV40. For this reason we briefly describe the data here, even though they are not directly related to the structure of the VTC.

Aloni and his collaborators have analyzed a short transcript (94 nucleotides) which extends from at or near the major start site of late transcription through a region whose sequence has a potential for extensive secondary structure (Hay *et al.*, 1982; Hay and Aloni, 1984, 1985). The secondary structures bear a striking resemblance to those responsible for premature termination (attenuation) of the transcription of several operons ultimately responsible for amino acid biosynthesis in *E. coli* (reviewed by Yanofsky and Kolter, 1982). These workers have detected the transcript among RNAs made by nuclei or VTCs isolated from infected cells at late times, but only if transcription is carried out in low salt (e.g., 30 mM ammonium sulfate) (Hay and Aloni, 1984). Preincubation in 100 mM ammonium sulfate destroys the ability of the nuclei to make the short RNA, possibly by altering the balance between one of two possible secondary structures in the transcript. Surprisingly, synthesis of the RNA can be restored by a factor present in nuclear extracts from uninfected cells (Hay and Aloni, 1984). This factor was assumed to aid transcription termination since it had been previously shown that a nucleoside analogue which provokes pausing and (premature) termination in some systems could increase the yield of short late transcripts (Laub *et al.*, 1980). More recent experiments of Hay and Aloni (1985) have suggested that the SV40 agnoprotein somehow stabilizes the secondary structure required for attenuation of the transcript. Thus, they explain the increase in ability to make the small RNA as infection progresses by an increase in nuclear content of the agnoprotein, and suggest that this mechanism is important in coordinating the synthesis of late gene products with production of viral DNA.

Although this scheme for production of the short RNA and regulation of late transcription has a number of attractive features, there are alternative interpretations of much of the data that have been presented. One possibility is that the short RNA arises by breakdown of longer late transcripts, as has been observed for a different part of the late message by Alwine (1982). Data presented by Hay and Aloni (1984) argue against this, but do not exclude it. A second possibility is presented by the model of Llopis and Stark which suggests that the small RNA is synthesized by the VP2/VP3 blocked complexes unless activated *in vitro* by salt or Sarkosyl.

G. Role of Chromatin Structure in the Early to Late Switch

In a previous section, we discussed in some detail the mechanisms that cause a change in relative transcription of early and late genes as infection progresses. Virtually all the data discussed to this point have been obtained with NPCs present in the cell at late times (i.e., after DNA replication has begun). Here we deal briefly with experiments that attempt to relate the structure of the NPC to the switch to late expression.

Cremisi (1981) has located sites of DNase I hypersensitivity in NPCs isolated at early times, and has also examined the sensitivity of integrated viral DNA present in a cell line transformed with SV40. When nuclei isolated 7 hr postinfection were treated with DNase I, most cleavages occurred around position 200, which is slightly to the late side of a site (near positions 180–190; see Fig. 3) cleaved frequently in late NPCs. Major sites of cleavage lying between the distal 72-bp repeat and the origin were not detected even in late NPCs possibly for technical reasons. Some sites farther to the late side of position 200, which were cleaved in late NPCs, were rarely cleaved in the early complex. Conversely, some cleavage did occur at sites to the early side of the replication origin, a region that appeared refractory to DNase I in late NPCs (see Fig. 3). Digestion of chromatin in nuclei isolated from SV40-transformed rat cells showed a pattern similar to that observed with the early NPC. Remarkably, a flat revertant in which the T antigen gene is partially deleted appeared largely unchanged in its hypersensitivity. These results indicate that the structure of the NPC at early times is roughly similar to that at late, although definite differences are seen. In particular, transcribed regions are somewhat more accessible to nuclease than are untranscribed ones, a result consistent with observations in a number of nonviral systems. The data obtained with the flat revertant indicate that the T antigen does not play an obligatory role in formation of a nuclease-sensitive region. It may, nonetheless, play a role in the change from early to late configuration of the chromatin, either directly or, as suggested by Cremisi, through its role in replication.

The reader is also referred to other reviews that detail earlier studies with SV40 and polyoma virus (Das *et al.*, 1985; DePamphilis and Wassarman, 1983; Griffin and Dilworth, 1983; Hansen and Sharp, 1985; Khoury and Gruss, 1983; Manley, 1983b; Nevins, 1983; Darnell, 1982; Das and Niyogi, 1981; Kelley and Nathans, 1977).

ACKNOWLEDGMENT. The authors are deeply indebted to Ms. Jeanne Carolan for the invaluable editorial assistance she provided throughout the preparation of the chapter.

REFERENCES

Acheson, N. H., 1978, Polyoma giant RNAs contain tandem repeats of the nucleotide sequence of the entire viral genome, *Proc. Natl. Acad. Sci. USA* **75**:4574.

Alfonso-Pizarro, A., Carlson, D. P., and Ross, J., 1984, Subcellular localization of RNAs in transfected cells: Role of sequences at the 5′ terminus, *Nucleic Acids Res.* **12**:8363.

Allan, M., Zhu, J., Montague, P., and Paul, J., 1984, Differential response of multiple ε-globin cap sites to *cis-* and *trans*-acting controls, *Cell* **38**:399.

Aloni, Y., Dhar, R., Laub, O., Horowitz, M., and Khoury, G., 1977, Novel mechanism for RNA maturation: The leader sequences of simian virus 40 mRNA are not transcribed adjacent to the coding sequences, *Proc. Natl. Acad. Sci. USA* **74**:3686.

Alwine, J. C., 1982, Hybrid selection of small RNAs by using simian virus 40 DNA: Evidence that the simian virus 40-associated small RNA is synthesized by specific cleavage from large viral transcripts, *J. Virol.* **43**:987.

Augereau, P., and Wasylyk, B., 1984, The MLV and SV40 enhancers have a similar pattern of transcriptional activation, *Nucleic Acids Res.* **12**:8801.

Azorin, F., and Rich, A., 1985, Isolation of Z-DNA binding proteins from SV40 minichromosomes: Evidence for binding to the viral control region, *Cell* **41**:365.

Baker, C. C., and Ziff, E. B., 1981, Promoters and heterogeneous 5′-termini of the messenger RNAs of adenovirus serotype 2, *J. Mol. Biol.* **149**:189.

Banerji, J., Rusconi, S., and Schaffner, W., 1981, Expression of a β-globin gene is enhanced by remote SV40 DNA sequences, *Cell* **27**:299.

Barkan, A., and Mertz, J. E., 1981, DNA sequence analysis of SV40 mutants with deletions mapping in the leader region of the late viral mRNAs: Mutants with deletions similar in size and position exhibit varied phenotypes, *J. Virol.* **37**:730.

Barkan, A., and Mertz, J. E., 1984, The number of ribosomes on simian virus 40 late 16S mRNA is determined in part by the nucleotide sequence of its leader, *Mol. Cell. Biol.* **4**:813.

Baty, D., Barrera-Saldana, H. A., Everett, R. D. Vigneron, M., and Chambon, P., 1984, Mutational dissection of the 21 bp repeat region of the SV40 early promoter reveals that it contains overlapping elements of the early–early and late–early promoters, *Nucleic Acids Res.* **12**:915.

Baumgartner, I., Kuhn, C., and Fanning, E., 1979, Identification and characterization of fast-sedimenting SV40 nucleoprotein complexes, *Virology* **96**:54.

Bellard, M., Oudet, P., Germond, J. E., and Chambon, P., 1976, Subunit structure of simian virus 40 minichromosome, *Eur. J. Biochem.* **70**:543.

Benoist, C., and Chambon, P., 1981, *In vivo* sequence requirements of the SV40 early promoter region, *Nature* **290**:304.

Berg, P. E. and Anderson, W. F., 1984, Correlation of gene expression and transformation frequency with the presence of an enhancing sequence in the transforming DNA, *Mol. Cell. Biol.* **4**:368.

Berg, P. E., Yu, J. K., Popovic, Z., Schumperli, D., Johansen, H., Rosenberg, M., and Anderson, W. F., 1983, Differential activation of the mouse β-globin promoter by enhancers, *Mol. Cell. Biol.* **3**:1246.

Berg, P. E., Popovic, Z., and Anderson, W. F., 1984, Promoter dependence of enhancer activity, *Mol. Cell. Biol.* **4**:1664.

Berget, S. M., Moore, C., and Sharp, P. A., 1977, Spliced segments at the 5′ terminus of adenovirus 2 late mRNA, *Proc. Natl. Acad. Sci. USA* **74**:3171.

Bina-Stein, M., Thoren, M., Salzman, N. P., and Thompson, J. A., 1979, Rapid sequence determination of late SV40 16S mRNA leader by using inhibitors of reverse transcriptase, *Proc. Natl. Acad. Sci. USA* **76**:731.

Birkenmeier, E. H., Radonovich, M., Shani, M., and Salzman, N. P., 1977, The SV40 DNA template for transcription of late mRNA in viral nucleoprotein complexes, *Cell* **11**:495.

Birkenmeier, E. H., Chiu, N., Radonovich, M. F., May, E., and Salzman, N. P., 1979, Regulation of simian virus 40 early and late gene transcription without viral DNA replication, *J. Virol.* **29**:983.

Bogenhagen, D. F., Wormington, W. M., and Brown, D. D., 1982, Stable transcription complexes of *Xenopus* 5S RNA genes: A means to maintain the differentiated state, *Cell* **28**:413.

Borrelli, E., Hen, R., and Chambon, P., 1984, Adenovirus-2 E1A products repress enhancer-induced stimulation of transcription, *Nature* **312**:608.

Brady, J. N., and Khoury, G., 1985 Trans activation of the simian virus 40 late transcription unit by T-antigen, *Mol. Cell. Biol.* **5**:1391.

Brady, J. N., Loeken, M. R., and Khoury, G., 1985, Interaction between two transcriptional control sequences required for T-antigen mediated SV40 late gene expression, *Proc. Natl. Acad. Sci. USA* **82**:7299.

Brady, J. N., Winston, V. D., and Consigli, R. A., 1978, Characterization of a DNA–protein complex and capsomere subunits derived from polyoma virus by treatment with EGTA and dithiothreitol, *J. Virol.* **27**:193.

Brady, J. N., Lavialle, C. A., Radonovich, M. F., and Salzman, N. P., 1981a, Stable association of viral protein VP_1 with simian virus 40 DNA, *J. Virol.* **39**:432.

Brady, J. N., Lavialle, C. A., Radonovich, M. F., and Salzman, N. P., 1981b, Simian virus 40 maturation: Chromatin modifications increase the accessibility of viral DNA to nuclease and RNA polymerase, *J. Virol.* **39**:603.

Brady, J., Radonovich, M., Vodkin, M., Natarajan, V., Thoren, M., Das, G., Janik, J., and

Salzman, N. P., 1982, Site specific base substitution and deletion mutations that enhance or suppress transcription of the SV40 major late promoter, *Cell* **31**:625.

Brady, J. N., Bolen, J. B., Radonovich, M., Salzman, N., and Khoury, G., 1984a, Stimulation of simian virus 40 late gene expression by simian virus 40 tumor antigen, *Proc. Natl. Acad. Sci. USA* **81**:2040.

Brady, J., Radonovich, M., Thoren, M., Das, G., and Salzman, N. P., 1984b, Simian virus 40 major late promoter: An upstream DNA sequence required for efficient *in vitro* transcription, *Mol. Cell. Biol.* **4**:133.

Breathnach, R., Chambon, P., 1981, Organization and expression of eucaryotic split genes coding for proteins, *Annu. Rev. Biochem.* **50**:349.

Breathnach, R., Benoist, C., O'Hare, K., Gannon, F., and Chambon, P., 1980, Ovalbumin gene: Evidence for a leader sequence in mRNA and DNA sequences at the exon–intron boundaries, *Proc. Natl. Acad. Sci. USA* **77**:4853.

Brooks, T. L., and Green, M., 1977, The SV40 transcription complex. I. Effect of viral chromatin proteins on endogenous DNA polymerase activity, *Nucleic Acids Res.* **4**:4261.

Brunke, K. J., Anthony, J. G., Sternberg, E. J., and Weeks, D. P., 1984, Repeated consensus sequence and pseudopromoters in the four coordinately regulated tubulin genes of *Chlamydomonas reinhardi*, *Mol. Cell. Biol.* **4**:1115.

Buchman, A. R., and Berg, P., 1984, Unusual regulation of SV40 early-region transcription in genomes containing two origins of DNA replication, *Mol. Cell. Biol.* **4**:1915.

Buchman, A. R., Burnett, L., and Berg, P., 1980, The SV40 nucleotide sequence, in: *DNA Tumor Viruses* (J. Tooze, ed.), pp. 799–829, Cold Spring Harbor Laboratory, Cold Spring Harbor, N.Y.

Buchman, A. R., Fromm, M., and Berg, P., 1984, Complex regulation of SV40 early-region transcription from different overlapping promoters, *Mol. Cell. Biol.* **4**:1900.

Buhler, J.-M., Iborra, F., Sentenac, A., and Fromageot, P., 1976, The presence of phosphorylated subunits in yeast RNA polymerase A and B, *FEBS Lett.* **72**:37.

Buhler, J.-M., Huet, J., Davies, K. E., Sentenac, A., and Fromageot, P., 1980, Immunological studies of yeast nuclear RNA polymerases at the subunit level, *J. Biol. Chem.* **255**:9949.

Byrne, B. J., Davis, M. S., Yamaguchi, J., Bergsma, D. J., and Subramanian, K. N., 1983, Definition of the SV40 early promoter region and demonstration of a host range bias in the enhancement effect of the SV40 72-base-pair repeat, *Proc. Natl. Acad. Sci. USA* **80**:721.

Campbell, B. A., and Villarreal, L. P., 1985, Host species specificity of polyoma virus DNA replication is not altered by the SV40 72 base-pair repeats, *Mol. Cell. Biol.* **5**:1534.

Canaani, D., Kahana, C., Mukamel, A., and Groner, Y., 1979, Sequence heterogeneity at the 5′ termini of late simian virus 40 19S and 16S mRNAs, *Proc. Natl. Acad. Sci. USA* **76**:3078.

Carroll, S. B., and Stollar, B. D., 1982, Inhibitory monoclonal antibody to calf thymus RNA polymerase II blocks formation of enzyme–DNA complexes, *Proc. Natl. Acad. Sci. USA* **79**:7233.

Cattaneo, R., Will, H., Hernandez, N., and Schaller, H., 1983, Signals regulating hepatitis B surface antigen transcription, *Nature* **305**:336.

Celma, M. L., Dhar, R., Pan, J., and Weissman, S. M., 1977, Comparison of the nucleotide sequence of the messenger RNA for the major structural protein of SV40 with the DNA sequence encoding the amino acids of the protein, *Nucleic Acids Res.* **4**:2549.

Cereghini, S., and Yaniv, M., 1984, Assembly of transfected DNA into chromatin: Structural changes in the origin–promoter–enhancer region upon replication, *EMBO J.* **3**:1243.

Cereghini, S., Herbomel, P., Jouanneau, J., Saragosti, S., Katinka, M., Bourachot, B., DeCrombrugghe, B., and Yaniv, M., 1983, Structure and function of the promoter–enhancer region of polyoma and SV40, *Cold Spring Harbor Symp. Quant. Biol.* **47**:935.

Chambon, P., 1975, Eukaryotic nuclear RNA polymerases, *Annu. Rev. Biochem.* **44**:613.

Charnay, P., Treisman, R., Mellow, P., Chao, M., Axel, R., and Maniatis, T., 1984, Differences in human α- and β-globin gene expression in mouse erythroleukemia cells: The role of intragenic sequences, *Cell* **38**:251.

Chiu, N., Radonovich, M., Thoren, M., and Salzman, N. P., 1978, Selective degradation of newly synthesized non-messenger SV40 transcripts, *J. Virol.* **28**:590.

Chow, L. T., Gelinas, R. E., Broker, T. R., and Roberts, R. J., 1977, An amazing sequence arrangement at the 5′ ends of adenovirus 2 messenger RNA, *Cell* **12**:1.

Christmann, J. L., and Dahmus, M. E., 1981, Monoclonal antibody specific for calf thymus RNA polymerases II_O and II_A, *J. Biol. Chem.* **256**:11798.

Clark, K. L., Bendig, M. M., and Folk, W. R., 1984, Isolation of a polyomavirus with an insertion of foreign DNA in the early gene promoter region, *J. Virol.* **52**:1032.

Clertant, P., Gaudray, P., and Cuzin, F., 1984, Salt-stable binding of the large T protein to DNA in polyoma virus chromatin, *EMBO J.* **3**:303.

Cochet-Meilhac, M., and Chambon, P., 1974, Animal DNA-dependent RNA polymerases. II. Mechanism of the inhibition of RNA polymerases B by amatoxins, *Biochim. Biophys. Acta* **353**:160.

Cogen, B., 1978, Virus-specific early RNA in 3T6 cells infected by a tsA mutant of polyoma virus, *Virology* **85**:222.

Concino, M., Goldman, R. A., Caruthers, M. H., and Weinmann, R., 1983, Point mutations of the adenovirus major late promoter with different transcriptional efficiencies *in vitro*, *J. Biol. Chem.* **258**:8493.

Contreras, R., Gheysen, D., Knowland, J., Van de Voorde, A., and Fiers, W., 1982, Evidence for the direct involvement of DNA replication origin in synthesis of late SV40 RNA, *Nature* **300**:500.

Corden, J., Wasylyk, B., Buchwalder, A., Sassone-Corsi, P., Kedinger, C., and Chambon, P., 1980, Promoter sequences of eukaryotic protein-coding genes, *Science* **209**:1406.

Cowan, K., Tegtmeyer, P., and Anthony, D. D., 1973, Relationship of replication and transcription of simian virus 40 DNA, *Proc. Natl. Acad. Sci. USA* **70**:1927.

Cowie, A., and Kamen, R., 1984, Multiple binding sites for polyomavirus large T antigen within regulatory sequences of polyomavirus DNA, *J. Virol.* **52**:750.

Cowie, A., Tyndall, C., and Kamen, R., 1981, Sequences at the capped 5′-ends of polyoma virus late region mRNAs: An example of extreme heterogeneity, *Nucleic Acids Res.* **9**:6305.

Cowie, A., Jat, P., and Kamen, R., 1982, Determination of sequences at the capped 5′-ends of polyomavirus early region transcripts synthesized *in vivo* and *in vitro* demonstrates an unusual microheterogeneity, *J. Mol. Biol.* **159**:225.

Cremisi, C., 1981, The appearance of DNase I hypersensitive sites at the 5′-end of the late SV40 genes is correlated with the transcriptional switch, *Nucleic Acids Res.* **9**:5949.

Cullen, B. R., Raymond, K., and Ju, G., 1985, Functional analysis of the transcription control region located within the avian retroviral long terminal repeat, *Mol. Cell. Biol.* **5**:438.

Dahmus, M. E., 1983, Structural relationship between the large subunits of calf thymus RNA polymerase II, *J. Biol. Chem.* **258**:3956.

Dailey, L., and Basilico, C., 1985, Sequences in the polyomavirus DNA regulatory region involved in viral DNA replication and early gene expression. *J. Virol.* **54**:739.

Dandolo, L., Blangy, D., and Kamen, R., 1983, Regulation of polyoma virus transcription in murine embryonal carcinoma cells, *J. Virol.* **47**:55.

Darnell, J. E., Jr., 1982, Variety in the level of gene control in eukaryotic cells, *Nature* **297**:365.

Das, G. C., and Niyogi, S. K., 1981, Structure, replication and transcription of the SV40 genome, *Prog. Nucleic Acid Res. Mol. Biol.* **25**:187.

Das, G. C., and Salzman, N. P., 1985, Simian virus early promoter mutations that affect promoter function and autoregulation by large T antigen, *J. Mol. Biol.* **182**:229.

Das, G. C., Niyogi, S. K., and Salzman, N. P., 1985, SV40 promoters and their regulation, *Prog. Nucleic Acid Res. Mol. Biol.* **32**:217.

Davison, B. L., Egly, J.-M., Mulvihill, E. R., and Chambon, P., 1983, Formation of stable pre-initiation complexes between eukaryotic class B transcription factors and promoter sequences, *Nature* **301**:680.

DeLucia, A. L., Lewton, B. A., Tjian, R., and Tegtmeyer, P., 1983, Topography of simian virus 40 A protein–DNA complexes: Arrangement of pentanucleotide interaction sites at the origin of replication, *J. Virol.* **46**:143.

DePamphilis, M. L., and Wassarman, P. M., 1983, Organization and replication of papovavirus DNA, in: *Organization and Replication of Viral DNA* (A. S. Kaplan, ed.), pp. 37–114, CRC Press, Boca Raton, Fla.

de Villiers, J., and Schaffner, W., 1981, A small segment of polyoma virus DNA enhances the expression of a cloned β-globin gene over a distance of 1400 base pairs, *Nucleic Acids Res.* **9**:6251.

de Villiers, J., Olson, L., Tyndall, C., and Schaffner, W., 1982, Transcriptional "enhancers" from SV40 and polyoma virus show a cell type preference, *Nucleic Acids Res.* **10**:7965.

de Villiers, J., Olson, L., Banerji, J., and Schaffner, W., 1983, Analysis of the transcriptional enhancer effect, *Cold Spring Harbor Symp. Quant. Biol.* **47**:911.

de Villiers, J., Schaffner, W., Tyndall, C., Lupton, S., and Kamen, R., 1984, Polyoma virus DNA replication requires an enhancer, *Nature* **312**:242.

Dignam, J. D., Lebovitz, R. M., and Roeder, R. G., 1983, Accurate transcription initiation by RNA polymerase II in a soluble extract from isolated mammalian nuclei, *Nucleic Acids Res.* **11**:1475.

Dilworth, S. M., Cowie, A., Kamen, R. I., and Griffin, B. E., 1984, DNA binding activity of polyoma virus large tumor antigen, *Proc. Natl. Acad. Sci. USA* **81**:1941.

DiMaio, D., and Nathans, D., 1982, Regulatory mutants of SV40: Effect of mutations at a T-antigen binding site on DNA replication and expression of viral genes, *J. Mol. Biol.* **156**:531.

Dingwall, C., Lomonossoff, G. P., and Laskey, R. A., 1981, High sequence specificity of micrococcal nuclease, *Nucleic Acids Res.* **9**:2659.

Dynan, W. S., and Tjian, R., 1983a, Isolation of transcription factors that discriminate between different promoters recognized by RNA polymerase II, *Cell* **32**:669.

Dynan, W. S., and Tjian, R., 1983b, The promoter-specific transcription factors Spl binds to upstream sequences in the SV40 early promoter, *Cell* **35**:79.

Egly, J., Miyamoto, N. G., Moncollin, V., and Chambon, P., 1984, Is actin a transcription initiation factor for RNA polymerase B?, *EMBO J.* **3**:2363.

Elgin, S. C. R., 1981, DNase-I hypersensitive sites of chromatin, *Cell* **27**:413.

Elkaim, R., Goding, C., and Kedinger, C., 1983, The adenovirus-2 EIIa early gene promoter: Sequences required for efficient *in vitro* and *in vivo* transcription, *Nucleic Acids Res.* **11**:7105.

Emerson, B. M., and Felsenfeld, G., 1984, Specific factor conferring nuclease hypersensitivity at the 5′-end of the chicken adult β-globin gene, *Proc. Natl. Acad. Sci. USA* **81**:95.

Emerson, B. M., Lewis, C. D., and Felsenfeld, G., 1985, Interaction of specific nuclear factors with the nuclease-hypersensitive region of the chicken adult β-globin gene: Nature of the binding domain, *Cell* **41**:21.

Emorine, L., Kuehl, M., Weir, L., Leder, P., and Max, E. E., 1983, A conserved sequence in the immunoglobulin Jk–Ck intron: Possible enhancer element, *Nature* **304**:447.

Engelke, D. R., Ng, S.-Y., Shastry, B. S., and Roeder, R. G., 1980, Specific interaction of a purified transcription factor with an internal control region of 5S RNA genes, *Cell* **19**:717.

Ephrussi, A., Church, G. M., Tonegawa, S., and Gilbert, W., 1985, B lineage-specific interaction of an immunoglobulin-enhancer with cellular factors *in vivo*, *Science* **227**:134.

Ernoult-Lange, M., and May, E., 1983, Evidence of transcription from the late region of the integrated simian virus 40 genome in transformed cells: Location of the 5′ ends of late transcripts in cells abortively infected and in cells transformed by simian virus 40, *J. Virol.* **46**:756.

Ernoult-Lange, M., May, P., Moreau, P., and May, E., 1984, Simian virus 40 late promoter region able to initiate simian virus 40 early gene transcription in the absence of the simian virus 40 origin sequence, *J. Virol.* **50**:163.

Everett, R. D., Baty, D., and Chambon, P., 1983, The repeated GC-rich motifs upstream from the TATA box are important elements of the SV40 early promoter, *Nucleic Acids Res.* **11**:2447.
Farmerie, W. G., and Folk, W. R., 1984, Regulation of polyoma virus transcription by large tumor antigen, *Proc. Natl. Acad. Sci. USA* **81**:6919.
Fenton, R. G., and Basilico, C., 1981, Viral gene expression in polyoma virus-transformed rat cells and their cured revertants, *J. Virol.* **40**:150.
Fenton, R. G., and Basilico, C., 1982a, Changes in the topography of early region transcription during polyoma virus lytic infection, *Proc. Natl. Acad. Sci. USA* **79**:7142.
Fenton, R. G., and Basilico, C., 1982b, Regulation of polyoma virus early transcription in transformed cells by large T-antigen, *Virology* **121**:384.
Ferdinand, F. J., Brown, M., and Khoury, G., 1977, Synthesis and characterization of late lytic simian virus 4 from transcriptional complexes, *Virology* **78**:150.
Fernandez-Munoz, K., Coca-Prados, M., and Hsu, M.-T., 1979, Intracellular forms of simian virus 40 nucleoprotein complexes. I. Methods of isolation and characterization in CV-1 cells, *J. Virol.* **29**:612.
Fey, G., and Hirt, B., 1974, Fingerprints of polyoma virus proteins and mouse histones, *Cold Spring Harbor Symp. Quant. Biol.* **39**:235.
Finch, J. T., and Klug, A., 1976, Solenoidal model for superstructure in chromatin, *Proc. Natl. Acad. Sci. USA* **73**:1897.
Fire, A., Baker, C. C., Manley, J. L., Ziff, E. B., and Sharp, P. A., 1981, *In vitro* transcription of adenovirus, *J. Virol.* **40**:703.
Fitzgerald, M., and Shenk, T., 1981, the sequence 5′-AAUAAA-3′ forms part of the recognition site for polyadenylation of late SV40 mRNAs, *Cell* **24**:251.
Ford, J. P., and Hsu, M.-T., 1978, Transcription pattern of *in vivo* labeled late simian virus 40 RNA: Equimolar transcription beyond the mRNA 3′ terminus, *J. Virol.* **28**:95.
Foster, J., Stafford, J., and Queen, C., 1985, An immunoglobulin promoter displays cell-type specificity independent of the enhancer, *Nature* **315**:423.
Fromm, M., and Berg, P., 1982, Deletion mapping of DNA regions required for SV40 early region promoter function *in vivo, J. Mol. Appl. Genet.* **1**:457.
Fromm, M., and Berg, P., 1983a, Transcription *in vivo* from SV40 early promoter deletion mutants without repression by large T antigen, *J. Mol. Appl. Genet.* **2**:127.
Fromm, M., and Berg, P., 1983b, Simian virus 40 early- and late-region promoter functions are enhanced by the 72-base-pair repeat inserted at distant locations and inverted orientations, *Mol. Cell. Biol.* **3**:991.
Fujimura, F. K., Deininger, P. L., Friedmann, T., and Linney, E., 1981, Mutation near the polyoma DNA replication origin permits productive infection of F9 embryonal carcinoma cells, *Cell* **23**:809.
Gabarro-Arpa, J., Ehrlich, R., and Reiss, C., 1984, Local stability involved in characterizing and controlling promoters in eukaryotes, *Folia Biol. (Prague)* **30** Spec. No.:93.
Galas, D. J., and Schmitz, A., 1978, DNase footprinting: A simple method for the detection of protein–DNA binding specificity, *Nucleic Acids Res.* **5**:3157.
Garber, E., Seidman, M., and Levine, A., 1978, The detection and characterization of multiple forms of SV40 nucleoprotein complexes, *Virology* **90**:305.
Gariglio, P., and Mousset, S., 1975, Isolation and partial characterization of a nuclear RNA polymerase SV40 DNA complex, *FEBS Lett.* **56**:149.
Gariglio, P., and Mousset, S., 1977, Characterization of a soluble simian virus 40 transcription complex, *Eur. J. Biochem.* **76**:583.
Gariglio, P., Llopis, R., Oudet, P., and Chambon, P., 1979, The template of the isolated native simian virus 40 transcription complexes is a minichromosome, *J. Mol. Biol.* **131**:75.
Gaudray, P., Tyndall, C., Kamen, R., and Cuzin, F., 1981, The high affinity binding site on polyoma virus DNA for the viral large-T protein, *Nucleic Acids Res.* **9**:5697.
Gelinas, R. E., and Roberts, R. J., 1977, One predominant 5′-undecanucleotide in adenovirus 2 late messenger RNAs, *Cell* **11**:533.

Gerard, R., Woodworth-Gutai, M., and Scott, W., 1982, Deletion mutants which affect the nuclease-sensitive site in simian virus 40 chromatin, *Mol. Cell. Biol.* **2**:782.

Germond, J. E., Hirt, B., Oudet, P., Gross-Bellard, M., and Chambon, P., 1975, Folding of the DNA double helix in chromatin-like structures from simian virus 40, *Proc. Natl. Acad. Sci. USA* **72**:1843.

Gheysen, D., Van de Voorde, A., Contreras, R., Vanderhegden, J., Duerinck, F., and Fiers, W., 1983, Simian virus 40 mutants carrying extensive deletions in the 72 base pair repeat region, *J. Virol.* **47**:1.

Ghosh, P. K., and Lebowitz, P., 1981, Simian virus 40 early mRNA's contain multiple 5'-termini upstream and downstream from a Hogness–Goldberg sequence: A shift in 5'-termini during the lytic cycle is mediated by large "T" antigen, *J. Virol.* **40**:224.

Ghosh, P. K., Reddy, V. B., Swinscoe, J., Choudary, P. V., Lebowitz, P., and Weissman, S. M., 1978a, The 5' terminal leader sequence of late 16S mRNA from cells infected with simian virus 40, *J. Biol. Chem.* **253**:3643.

Ghosh, P. K., Reddy, V. B., Swinscoe, J. S., Lebowitz, P., and Weissman, S. M., 1978b, Heterogeneity and 5'-terminal structures of the late RNAs of simian virus 40, *J. Mol. Biol.* **126**:813.

Ghosh, P. K., Lebowitz, P., Frisque, R. J., and Gluzman, Y., 1981a, Identification of a promoter component involved in postitioning the 5' termini of simian virus 40 early mRNAs, *Proc. Natl. Acad. Sci. USA* **78**:100.

Ghosh, P. K., Roy, P., Barkan, A., Mertz, J. E., Weissman, S. M., and Lebowitz, P., 1981b, Unspliced functional late 19S mRNAs containing intervening sequences are produced by a late leader mutant of simian virus 40, *Proc. Natl. Acad. Sci. USA* **78**:1386.

Ghosh, P. K., Piatak, M., Mertz, J. E., Weissman, S. M., and Lebowitz, P., 1982, Altered utilization of splice sites and 5' termini in late RNAs produced by leader region mutants of simian virus 40, *J. Virol.* **44**:610.

Gidoni, D., Kahana, C., Canaani, D., and Groner, Y., 1981, Specific *in vitro* initiation of transcription of SV40 early and late genes occurs at the various cap nucleotides including cytidine, *Proc. Natl. Acad. Sci. USA* **78**:2174.

Gidoni, D., Dynan, W. S., and Tjian, R., 1984, Multiple specific contacts between a mammalian transcription factor and its cognate promoters, *Nature* **312**:409.

Girard, M., Marty, L., and Manteuil, S., 1974, Viral DNA–RNA hybrids in cells infected with simian virus: The simian virus 40 transcriptional intermediates, *Proc. Natl. Acad. Sci. USA* **71**:1267.

Gluzman, Y., 1981, SV40-transformed simian cells support the replication of early SV40 mutants, *Cell* **23**:175.

Gluzman, Y., and Shenk, T. (eds.), 1983, *Enhancers and Eukaryotic Gene Expression*, Cold Spring Harbor Laboratory, Cold Spring Harbor, N.Y.

Gluzman, Y., Sambrook, J. F., and Frisque, R. J., 1980, Expression of early genes of origin-defective mutants of SV40, *Proc. Natl. Acad. Sci. USA* **77**:3898.

Gorman, C. M., Moffat, L. F., and Howard, B. H., 1982, Recombinant genomes which express chloramphenicol acetyl transferase in mammalian cells, *Mol. Cell. Biol.* **2**:1044.

Graessmann, A., Graessmann, M., and Mueller, C., 1981, Regulation of SV40 gene expression, *Adv. Cancer Res.* **35**:111.

Green, M. H., and Brooks, T. L., 1976, Isolation of two forms of SV40 nucleoprotein containing RNA polymerase from infected monkey cells, *Virology* **72**:110.

Green, M. H., and Brooks, T. L., 1977, The SV40 transcription complex. II. Non-dissociation of protein from SV40 chromatin during transcription, *Nucleic Acids Res.* **4**:4279.

Green, M. H., Miller, H., and Hendler, S., 1971, Isolation of a polyoma–nucleoprotein complex from infected mouse-cell cultures, *Proc. Natl. Acad. Sci. USA* **68**:1032.

Green, M. H., Buss, J., and Gariglio, P., 1975, Activation of nuclear RNA polymerase by Sarkosyl, *Eur. J. Biochem.* **53**:217.

Greenleaf, A. L., 1983, Amanitin-resistant RNA polymerase II mutations are in the enzyme's largest subunit, *J. Biol. Chem.* **258**:1343.

Griffin, B. E., and Dilworth, S. M., 1983, Polyoma virus: An overview of its unique properties, *Adv. Cancer Res.* **39**:183.

Griffith, J. D., 1975, Chromatin structure: Deduced from minichromosome, *Science* **187**:1202.

Griffith, J., Dieckmann, M., and Berg, P., 1975, Electron microscope localization of a protein bound near the origin of simian virus 40 DNA replication, *J. Virol.* **15**:167.

Grosschedl, R., and Baltimore, D., 1985, Cell-type specificity of immunoglobulin gene expression is regulated by at least three DNA sequence elements, *Cell* **41**:885.

Grosschedl, R., Wasylyk, B., Chambon, P., and Birnstiel, M. L., 1981, Point mutation in the TATA box curtails expression of sea urchin H2A histone gene *in vivo*, *Nature* **294**:178.

Gruss, P., 1984, Magic enhancers, *DNA* **3**:1.

Gruss, P., and Khoury, G., 1980, Rescue of a splicing defective mutant by insertion of a heterologous intron, *Nature* **286**:634.

Gruss, P., Dhar, R., and Khoury, G., 1981, Simian virus 40 tandem repeated sequences as an element of the early promoter, *Proc. Natl. Acad. Sci. USA* **78**:943.

Haegeman, G., and Fiers, W., 1978a, Evidence for "splicing" of SV40 16S mRNA, *Nature* **273**:70.

Haegeman, G., and Fiers, W., 1978b, Location of the 5′ terminus of late SV40 mRNA, *J. Virol.* **25**:824.

Haegeman, G., and Fiers, W., 1980a, Characterization of the 5′-terminal capped structures of late simian virus 40-specific mRNA, *J. Virol.* **35**:824.

Haegeman, G., and Fiers, W., 1980b, Characterization of the 5′-terminal cap structures of early simian virus 40 RNA, *J. Virol.* **35**:955.

Haegeman, G., Van Heuverswyn, H., Gheysen, D., and Fiers, W., 1979, Heterogeneity of the 5′ terminus of late mRNA induced by a viable simian virus 40 deletion mutant, *J. Virol.* **31**:484.

Hamada, H., Petrino, M. G., Kakunaga, T., Seidman, M., and Stollar, B. D., 1984a, Characterization of genomic poly (dT-dG). Poly(dc-dA) sequences: Structure, organization and conformation, *Mol. Cell. Biol.* **4**:2610.

Hamada, H., Seidman, M., Howard, B. H., and Gorman, C., 1984b, Enhanced gene expression by the poly (dT-dG). poly(dc-dA) sequence, *Mol. Cell. Biol.* **4**:2622.

Handa, H., Kaufman, R. J., Manley, J. M., Gefter, M., and Sharp, P. A., 1981, Transcription of simian virus 40 DNA in a HeLa whole cell extract, *J. Biol. Chem.* **256**:478.

Hansen, U., and Sharp, P. A., 1983, Sequences controlling *in vitro* transcription of SV40 promoters, *EMBO J.* **2**:2293.

Hansen, U., and Sharp, P. A., 1985, Transcription by RNA polymerase II, *Compr. Virol.* **19**:65.

Hansen, U., Tenen, D. G., Livingston, D. M., and Sharp, P. A., 1981, T antigen repression of SV40 early transcription from two promoters, *Cell* **27**:603.

Harper, F., Florentin, Y., and Puvion, E., 1984, Localization of T-antigen on simian virus 40 minichromosomes by immunoelectron microscopy, *EMBO J.* **3**:1235.

Hartman, J. P., and Scott, W. A., 1981, Distribution of DNase I-sensitive sites in simian virus 40 nucleoprotein complexes from disrupted virus particles, *J. Virol.* **37**:908.

Hartman, J. P., and Scott, W., 1983, Nuclease-sensitive sites in the two major intracellular simian virus 40 nucleoproteins, *J. Virol.* **46**:1034.

Hartzell, S. W., Yamaguchi, J., and Subramanian, K. N., 1983, SV40 deletion mutants lacking the 21-bp repeated sequences are viable, but have non-complementable deficiencies, *Nucleic Acids Res.* **11**:1601.

Hartzell, S. W., Byrne, B. J., and Subramanian, K. N., 1984a, Mapping of the late promoter of simian virus 40, *Proc. Natl. Acad. Sci. USA* **81**:23.

Hartzell, S. W., Byrne, B. J., and Subramanian, K. N., 1984b, The SV40 minimal origin and the 72 base pair repeat are required simultaneously for efficient induction of late gene expression with large tumor antigen, *Proc. Natl. Acad. Sci. USA* **81**:6335.

Hay, N., and Aloni, Y., 1984, Attenuation in SV40 as a mechanism of transcription-termination by RNA polymerase B, *Nucleic Acids Res.* **12**:1401.

Hay, N., and Aloni, Y., 1985, Attenuation of late simian virus 40 mRNA synthesis is enhanced by the agnoprotein and is temporally regulated in isolated nuclear systems, *Mol. Cell. Biol.* **5:**1327.

Hay, N., Skolnik-David, H., and Aloni, Y., 1982, Attenuation in the control of SV40 gene expression, *Cell* **29:**183.

Heiser, W. C., and Eckhart, W., 1982, Polyoma virus early and late mRNA's in productively infected mouse 3T6 cells, *J. Virol.* **44:**175.

Herbomel, P., Saragosti, S., Blangy, D., and Yaniv, M., 1981, Fine structure of the origin-proximal DNase I-hypersensitive region in wild-type and EC mutant polyoma, *Cell* **25:**651.

Herbomel, P., Bourachot, B., and Yaniv, M., 1984, Two distinct enhancers with different cell specificities coexist in the regulatory region of polyoma, *Cell* **39:**653.

Herr, W., and Gluzman, Y., 1985, Duplications of a mutated simian virus 40 enhancer restore its activity, *Nature* **313:**711.

Hodo, H., and Blatti, S. P., 1977, Purification using polyethylenimine precipitation and low molecular weight subunit analyses of calf thymus and wheat germ DNA-dependent RNA polymerase II, *Biochemistry* **16:**2334.

Hough, P. U. C., Mastrangelo, I. A., Wall, J. S., Hainfeld, J. F., Simon, M. N., and Manley, J. L., 1982, DNA–protein complexes spread on N2-discharged carbon film and characterized by molecular weight and its projected distribution, *J. Mol. Biol.* **160:**375.

Hsu, M.-T., and Ford, J., 1977, Sequence arrangement of the 5′ ends of simian virus 40 to 16S and 19S mRNAs, *Proc. Natl. Acad. Sci. USA* **74:**827.

Huet, J., Sentenac, A., and Fromageot, P., 1982, Spot-immunodetection of conserved determinants in eukaryotic RNA polymerases: Study with antibodies to yeast RNA polymerase subunits, *J. Biol. Chem.* **257:**2613.

Humphries, R. K., Ley, T., Turner, P., Moulton, A. D., and Nienhuis, A. W., 1982, Differences in human α-, β- and δ-globin gene expression in monkey kidney cells, *Cell* **30:**173.

Imperiale, M., Hart, R. P., and Nevins, J. R., 1985, An enhancer-like element in the adenovirus E2 promoter contains sequences essential for uninduced and E1A-induced transcription, *Proc. Natl. Acad. Sci. USA* **82:**381.

Ingles, C. J., 1973, Antigenic homology of eukaryotic RNA polymerases, *Biochem. Biophys. Res. Commun.* **55:**364.

Innis, J., and Scott, W., 1983, Chromatin structure of simian virus 40–pBR322 recombinant plasmids in Cos-1 cells, *Mol. Cell. Biol.* **3:**2203.

Innis, J., and Scott, W., 1984, DNA replication and chromatin structure of simian virus 40 insertion mutants, *Mol. Cell. Biol.* **4:**1499.

Jackson, A. H., and Sugden, B., 1972, Inhibition by α-amanitin of simian virus 40 specific ribonucleic acid synthesis in nuclei of infected monkey cells, *J. Virol.* **10:**1086.

Jackson, D. A., and Cook, P. R., 1985, Transcription occurs at a nucleoskeleton, *EMBO J.* **4:**919.

Jakobovits, E. B., and Aloni, Y., 1980, Isolation and characterization of various forms of simian virus 40 DNA protein complexes, *Virology* **102:**107.

Jakobovits, E. B., Bratosin, S., and Aloni, Y., 1980, A nucleosome-free region in SV40 minichromosomes, *Nature* **285:**263.

Jakobovits, E. B., Bratosin, S., and Aloni, Y., 1982, Formation of a nucleosome-free region in SV40 minichromosomes is dependent upon a restricted segment of DNA, *Virology* **120:**340.

Jat, P., Novak, U., Cowie, A., Tyndall, C., and Kamen, R., 1982a, DNA sequences required for specific and efficient initiation of transcription at the polyoma virus early promoter, *Mol. Cell. Biol.* **2:**737.

Jat, P., Roberts, J. W., Cowie, A., and Kamen, R., 1982b, Comparison of the polyomavirus early and late promoters by transcription *in vitro*, *Nucleic Acids Res.* **10:**871.

Jones, K. A., and Tjian, R., 1984, Essential contact residues within SV40 large T antigen binding sites I and II identified by alkylation-interference, *Cell* **36:**155.

Jongstra, J., Reudelhuber, T. L., Oudet, P., Benoist, C., Chae, C.-B., Jeltsch, J.-M., Mathis,

D. J., and Chambon, P., 1984, Induction of altered chromatin structures by simian virus 40 enhancer and promoter elements, *Nature* **307**:708.

Kadesch, T. R., and Berg, P., 1983, Effects of the position of the 72 bp enhancer segment on transcription from the SV40 early region promoter, in: *Enhancers and Eukaryotic Gene Expression* (Y. Gluzman and T. Shenk, eds.), pp. 21–27, Cold Spring Harbor Laboratory, Cold Spring Harbor, N.Y.

Kahana, C., Gidoni, D., Canaani, D., and Groner, Y., 1981, Simian virus 40 early mRNA's in lytically infected and transformed cells contain six 5′-terminal caps, *J. Virol.* **37**:7.

Kamen, R., Jat, P., Treisman, R., and Favaloro, J., 1982, 5′ termini of polyoma virus early region transcripts synthesized *in vivo* by wild-type virus and viable deletion mutants, *J. Mol. Biol.* **159**:189.

Keller, J. M., and Alwine, J. C., 1984, Activation of the SV40 late promoter: Direct effects of T antigen in the absence of viral DNA replication, *Cell* **36**:381.

Keller, J. M., and Alwine, J. C., 1985, Analysis of an activatable promoter: Sequences in the simian virus 40 late promoter required for T-antigen-mediated *trans* activation, *Mol. Cell. Biol.* **5**:1859

Keller, W., 1984, The RNA lariat: A new ring to the splicing to mRNA precursors, *Cell* **39**:423.

Kelley, T. J., Jr., and Nathans, D., 1977, The genome of simian virus 40, *Adv. Virus Res.* **21**:85.

Kenney, S., Natarajan, V., Strike, D., Khoury, G., and Salzman, N. P., 1984, JC virus enhancer–promoter active in human brain cells, *Science* **26**:1337.

Kern, F. G., and Basilico, C., 1985, Transcription from the polyoma late promoter in cells stably transformed by chimeric plasmids, *Mol. Cell. Biol.* **5**:797.

Kessler, M., and Aloni, Y., 1984, Mapping *in vivo* initiation sites of RNA transcription and determining their relative use, *J. Virol.* **52**:277.

Khoury, G., and Gruss, P., 1983, Enhancer elements, *Cell* **33**:313.

Khoury, G., and May, E., 1977, Regulation of early and late SV40 transcription: Overproduction of early viral RNA in the absence of functional 'T' antigen, *J. Virol.* **23**:167.

Khoury, G., Gruss, P., Dhar, R., and Lai, C. J., 1979, Processing and expression of early SV40 mRNA: A role of RNA conformation in splicing, *Cell* **18**:85.

Kingston, R. E., Kaufman, R. J., and Sharp, P. A., 1984, Regulation of adenovirus E11 promoter by E1a gene products: Absence of sequence specificity, *Mol. Cell. Biol.* **4**:1970.

Klessig, D. F., 1977, Two adenovirus mRNAs have a common 5′ terminal leader sequence encoded at least 10 kb upstream from their main coding regions, *Cell* **12**:9.

Kozak, M., 1983, Translation of insulin-related polypeptides from messenger RNAs with tandemly reiterated copies of the ribosome binding site, *Cell* **34**:971.

Kramer, A., and Bautz, E. K. F., 1981, Immunological relatedness of subunits of RNA polymerase II from insects and mammals, *Eur. J. Biochem.* **117**:449.

Kramer, A., Haars, R., Kabisch, R., Will, H., Bautz, F. A., and Bautz, E. K. F., 1980, Monoclonal antibody directed against RNA polymerase II of *Drosophila melanogaster, Mol. Gen. Genet.* **180**:193.

Kramer, A., Keller, W., Appel, B., and Luhrmann, R., 1984, The 5′ terminus of the RNA moiety of U1 small nuclear ribonucleoprotein particles is required for the splicing of messenger RNA precursors, *Cell* **38**:299.

Kuhne, T., Wieringa, B., Reiser, J., and Weissmann, C., 1983, Evidence against a scanning model of RNA splicing, *EMBO J.* **2**:727.

LaBella, F., and Vesco, C., 1980, Late modifications of simian virus 40 chromatin during the lytic cycle occur in an immature form of virion, *J. Virol.* **33**:1138.

LaBella, F., Vidali, G., and Vesco, C., 1979, Histone acetylation in CV-1 cells infected with simian virus 40, *Virology* **96**:564.

Laimins, L. A., Khoury, G., Gorman, C., Howard, B., and Gruss, P., 1982, Host-specific activation of transcription by tandem repeats from SV40 and Moloney murine sarcoma virus, *Proc. Natl. Acad. Sci. USA* **79**:6453.

Laimins, L. A., Kessel, M., Rosenthal, N., and Khoury, G., 1983, Viral and cellular enhancer elements, in: *Enhancers and Eukaryotic Gene Expression* (Y. Gluzman and T. Shenk, eds.), pp. 28–37, Cold Spring Harbor Laboratory, Cold Spring Harbor, N.Y.

Lake, R. S., Barban, S., and Salzman, N. P., 1973, Resolution and identification of the core deoxynucleoproteins of the simian virus 40, *Biochem. Biophys. Res. Commun.* **54**:640.

Lang, K. M., and Spritz, R. A., 1983, RNA splice site selection: Evidence for a 5′ leads to 3′ scanning model, *Science* **220**:1351.

Lange, M., May, E., and May, P., 1981, Ability of nonpermissive mouse cells to express a simian virus 40 late function(s), *J. Virol.* **38**:940.

Laub, O., Jakobovits, E. B., and Aloni, Y., 1980, 5,6-Dichloro-1-β-ribofuranosylbenzimidazole enhances premature termination of late transcription of simian virus 40 DNA, *Proc. Natl. Acad. Sci. USA* **77**:3297.

Lavi, S., and Groner, Y., 1977, 5′ terminal sequences and coding region of late simian virus 40 mRNAs are derived from noncontinguous segments of the viral genome, *Proc. Natl. Acad. Sci. USA* **74**:5323.

Lebowitz, P., and Ghosh, P. K., 1982, Initiation and regulation of SV40 early transcription *in vitro*, *J. Virol.* **40**:449.

Legion, S., Flavell, A. J., Cowie, A., and Kamen, R., 1979, Amplification in the leader sequence of late polyoma virus mRNAs, *Cell* **16**:373.

Lerner, M. R., Boyle, J. A., Mount, S. M., Wolin, S. L., and Steitz, J. A., 1980, Are snRNPs involved in splicing?, *Nature* **283**:220.

Levine, A. J., 1983, The nature of the host range restriction of SV40 and polyoma viruses in embryonal carcinoma cells, *Curr. Top. Microbiol. Immunol.* **101**:1.

Levinson, B., Khoury, G., Vande Woude, G., and Gruss, P., 1982, Activation of SV40 genome by 72-base pair tandem repeats of Moloney sarcoma virus, *Nature* **295**:568.

Lewis, M. K., and Burgess, R. R., 1982, Eukaryotic RNA polymerases, in: *The Enzymes* (P. D. Boyer, ed.), Volume 15, pp. 109–153, Academic Press, New York.

Linney, E., and Donerly, S., 1983, DNA fragments from F9 PyEc mutants increase expression of heterologous genes in transfected F9 cells, *Cell* **35**:693.

Llopis, R., and Stark, G. R., 1981, Two deletions within genes for simian virus 40 structural proteins VP2 and VP3 lead to formation of abnormal transcriptional complexes, *J. Virol.* **38**:91.

Llopis, R., and Stark, G. R., 1982, Separation and properties of two kinds of simian virus 40 late transcription complexes, *J. Virol.* **44**:864.

Llopis, R., Perrin, F., Bellard, F., and Gariglio, P., 1981, Quantitation of transcribing native simian virus 40 minichromosomes extracted from CV1 cells late in infection, *J. Virol.* **38**:82.

Luchnik, A. N., Bakayev, V. V., Zbarsky, I. B., and Georgiev, G. P., 1982, Elastic torsional strain in DNA within a fraction of SV40 minichromosomes: Relation to transcriptionally active chromatin, *EMBO J.* **1**:1353.

Luciw, P. A., Bishop, J. M., Varmus, H. E., and Capecchi, M. R., 1983, Location and function of retroviral and SV40 sequences that enhance biochemical transformation after microinjection of DNA, *Cell* **33**:705.

Lycan, D. E., and Danna, K. J., 1983, Characterization of the 5′ termini of purified nascent simian virus 40 late transcripts, *J. Virol.* **45**:264.

McGhee, J. D., and Felsenfeld, G., 1980, Nucleosome structure, *Annu. Rev. Biochem.* **49**:1115.

McLauchlan, J., Gaffney, D., Whitton, J. L., and Clements, J. B., 1985, The consensus sequence YGTGTTYY located downstream from the AATAAA signal is required for efficient formation of mRNA 3′ termini, *Nucleic Acids Res.* **13**:1347.

Manley, J. L., 1983a, Accurate and specific polyadenylation of mRNA precursors in a soluble whole cell lysate, *Cell* **33**:595.

Manley, J. L., 1983b, Analysis of the expression of genes encoding animal mRNA by *in vitro* techniques, *Prog. Nucleic Acid Res. Mol. Biol.* **30**:195.

Manley, J. L., Fire, A., Cano, A., Sharp, P. A., and Gefter, M. L., 1980, DNA-dependent transcription of adenovirus genes in a soluble whole-cell extract, *Proc. Natl. Acad. Sci. USA* **77**:3855.

Manley, J. L., Yu, H., and Ryner, L., 1985, RNA sequence containing hexanucleotide AAUAAA directs efficient mRNA polyadenylation *in vitro*, *Mol. Cell. Biol.* **5**:373.

Mann, K., and Hunter, T., 1979, Association of simian virus 40 T antigen with simian virus 40 nucleoprotein complexes, *J. Virol.* **29**:232.

Mathis, D. J., and Chambon, P., 1981, The SV40 early region TATA box is required for accurate *in vitro* initiation of transcription, *Nature* **290**:310.

Matsui, T., Segall, J., Weil, P. A., and Roeder, R. G., 1980, Multiple factors required for accurate initiation of transcription by purified RNA polymerase II, *J. Biol. Chem.* **255**:11992.

Mattaj, I. W., Lienhard, S., Jiricny, J., and DeRoberts, E. M., 1985, An enhancer-like sequence within the Xenopus U2 gene promoter facilitates the formation of stable transcription complexes, *Nature* **316**:163.

Meinke, W., Hall, M. R., and Golstein, D. A., 1975, Proteins in intracellular simian virus 40 nucleoprotein complexes: Comparison with simian virus 40 core proteins, *J. Virol.* **15**:439.

Melin, F., Pinon, H., Kress, C., and Blangy, D., 1985a, Isolation of polyomavirus mutants multiadapted to murine embryonal carcinoma cells, *J. Virol.* **53**:862.

Melin, F., Pinon, H., Reiss, C., Kress, C., Montreau, N., and Blangy, D., 1985b, Common features of polyomavirus mutants selected on PCC4 embryonal carcinoma cells, *EMBO J.* **4**:1799.

Melton, D. W., Konecki, D. S., Brennand, J., and Caskey, C. T., 1984, Structure, expression, and mutation of the hypoxanthine phosphoribosyl-transferase gene, *Proc. Natl. Acad. Sci. USA* **81**:2147.

Menne, C., Suske, G., Arnemann, J., Wenz, M., Cato, A. C. B., and Beato, M., 1982, Isolation and structure of the gene for the progesterone-inducible protein uteroglobin, *Proc. Natl. Acad. Sci. USA* **79**:4853.

Mishoe, H., Brady, J. N., Radonovich, M., and Salzman, N. P., 1984, Simian virus 40 guanine–cytosine-rich sequences function as independent transcriptional control elements *in vitro*, *Mol. Cell. Biol.* **4**:2911.

Miyamoto, N. G., Moncollin, V., Wintzerith, M., Hen, R., Egly, J. M., and Chambon, P., 1984, Stimulation of *in vitro* transcription by the upstream element of the adenovirus major late promoter involves a specific factor, *Nucleic Acids Res.* **12**:8779.

Montell, C., Fisher, E. F., Caruthers, M. H., and Berk, A. J., 1983, Inhibition of RNA cleavage but not polyadenylation by a point mutation in mRNA 3′ consensus sequence AAUAAA, *Nature* **305**:600.

Moore, C. L., and Sharp, P. A., 1984, Site specific polyadenylation in a cell free reaction, *Cell* **36**:581.

Moreau, P., Hen, R., Wasylyk, B., Everett, R., Gaub, M. P., and Chambon, P. 1981, The SV40 72 base pair repeat has a striking effect on gene expression both in SV40 and other chimeric recombinants, *Nucleic Acids Res.* **9**:6047.

Mosthaf, L., Pawlita, M., and Gruss, P., 1985, A viral enhancer element specifically active in human haematopoietic cells, *Nature* **315**:597.

Mount, S. M., 1982, A catalogue of splice junction sequences, *Nucleic Acids Res.* **10**: 459.

Mount, S. M., Pettersson, I., Hinterberger, M., Karmas, A., and Steitz, J. A., 1983, The U1 small nuclear RNA–protein complex selectively binds a 5′ splice site *in vitro*, *Cell* **33**:509.

Moyne, G., Harper, F., Saragosti, S., and Yaniv, M., 1982, Absence of nucleosomes in a histone-containing nucleoprotein complex obtained by dissociation of purified SV40 virions, *Cell* **30**:123.

Mueller, C. R., Mes-Masson, A., Bouvier, M., and Hassell, J. A., 1984, Location of sequences in polyoma virus DNA that are required for early gene expression *in vivo* and *in vitro*, *Mol. Cell. Biol.* **4**:2594.

Muller, U., Zentgraf, H., Eicken, I., and Keller, W., 1978, Higher order structure of simian virus 40 chromatin, *Science* **201**:406.

Muller, W. J., Mueller, C. R., Mes, A. M., and Hassell, J. A., 1983, Polyomavirus origin for DNA replication comprises multiple genetic elements, *J. Virol.* **47**:586.

Myers, R. M., Rio, D. C., Robbins, A. K., and Tjian, R., 1981, SV40 gene expression is modulated by the cooperative binding of T antigen to DNA, *Cell* **25**:373.

Nandi, A., Das, G., and Salzman, N. P., 1985, Characterization of a surrogate TATA box promoter that regulates *in vitro* transcription of the simian virus 40 major late gene, *Mol. Cell. Biol.* **5**:591.

Natarajan, V., and Salzman, N. P., 1985, Cis and trans activation of adenovirus IVa_2 gene transcription, *Nucleic Acids Res.* **13**:4067.

Natarajan, V., Madden, M. J., and Salzman, N. P., 1983, Preferential stimulation of transcription from simian virus late and adeno IVa_2 promoters in a HeLa cell extract, *J. Biol. Chem.* **258**:14652.

Nedospasov, S. A., and Georgiev, G. P., 1980, Nonrandom cleavage of SV40 DNA in the compact minichromosmome and free in solution by micrococcal nuclease, *Biochem. Biophys. Res. Commun.* **92**:532.

Nevins, J. R., 1983, The pathway of eukaryotic mRNA formation, *Annu. Rev. Biochem.* **52**:441.

Nevins, J. R., and Joklik, W. K., 1977, Isolation and properties of the vaccinia virus DNA-dependent RNA polymerase, *J. Biol. Chem.* **252**:6930.

Nevins, J. R., Blanchard, J. M., and Darnell, J. E., Jr., 1980, Transcription units of adenovirus type 2: Termination of transcription beyond the poly(A) addition site in early regions 2 and 4, *J. Mol. Biol.* **144**:377.

Ng, S.-C., Mertz, J. E., Sanden-Will, S., and Bina, M., 1985, Simian virus 40 maturation in cells harboring mutants deleted in the agnogene, *J. Biol. Chem.* **260**:1127.

Nordheim, A., and Rich, A., 1983, Negatively supercoiled SV40 DNA contains Z-DNA segments within transcriptional enhancer sequences, *Nature* **303**:674.

Olins, A. L., and Olins, D. E., 1974, Spheroid chromatin units (δ bodies), *Science* **183**:330.

Omilli, F., Lange, M., Borde, J., and May, E., 1986, Sequences involved in initiation of simian virus 40 late transcription in the absence of T-antigen, *Mol. Cell. Biol.* **6**:1875.

Osborne, T. F., Arvidson, D. N., Tyav, E. S., Dunsworth, M., and Berk, A. J., 1984, Transcription control region within the protein-coding portion of adenovirus E1A genes, *Mol. Cell. Biol.* **4**:1293.

Ott, M., Sperling, L., Herbomel, P., Yaniv, M., and Weiss, M. C., 1984, Tissue-specific expression is conferred by a sequence from the 5′-end of the rat albumin gene, *EMBO J.* **3**:2505.

Oudet, P., Gross-Bellard, M., and Chambon, P., 1975, Electron microscopic and biochemical evidence that chromatin is a repeating unit, *Cell* **4**:281.

Parker, C. S., and Topol, J., 1984, A Drosophila RNA polymerase II transcription factor contains a promoter-region-specific DNA-binding activity, *Cell* **36**:357.

Paule, M. R., 1981, Comparative subunit composition of the eukaryotic nuclear RNA polymerases, *Trends Biochem. Sci.* **6**:128.

Pelletier, J., and Sonenberg, N., 1985, Insertion mutagenesis to increase secondary structure within the 5′ noncoding region of a eukaryotic mRNA reduces translational efficiency, *Cell* **40**:515.

Persico-DiLauro, M., Martin, R. G., and Livingston, D. M., 1977, Interaction of SV40 chromatin with SV40 T-antigen, *J. Virol.* **24**:451.

Piatak, M., Subramanian, K. N., Roy, P., and Weissman, S. M., 1981, Late messenger RNA production by viable simian virus 40 mutants with deletions in the leader region, *J. Mol. Biol.* **153**:589.

Piatak, M., Ghosh, P. K. Norkin, L. C., and Weissman, S. M., 1983, Sequences locating the 5′ ends of the major simian virus 40 late mRNA forms, *J. Virol.* **48**:503.

Picard, D., and Schaffner, W., 1983, Correct transcription of a cloned mouse immunoglobulin gene *in vivo*, *Proc. Natl. Acad. Sci. USA* **80**:417.

Picard, D., and Schaffner, W., 1984, A lymphocyte-specific enhancer in the mouse immunoglobulin K gene, *Nature* **307**:80.

Pipas, J. M., 1985, Mutations near the carboxyl terminus of the SV40 large tumor antigen alter viral host range, *J. Virol.* **54**:569.

Polisky, B., and McCarthy, B. J., 1975, Location of histones on SV40 DNA, *Proc. Natl. Acad. Sci. USA* **72**:2895.

Pomerantz, B. J., and Hassell, J. A., 1984, Polyomavirus and simian virus 40 large T antigens bind to common DNA sequences, *J. Virol.* **49**:925.

Pomerantz, B. J., Mueller, C. R., and Hassell, J. A., 1983, Polyomavirus large T antigen binds independently to multiple, unique regions on the viral genome, *J. Virol.* **47**:600.

Ponder, B. A. J., and Crawford, L. V., 1977, The arrangement of nucleosomes in nucleoprotein complexes from polyoma virus and SV40, *Cell* **11**:35.

Ponta, H., Kennedy, N., Skroch, P., Hynes, N. E., and Groner, B., 1985, Hormonal response region in the mouse mammary tumor virus long terminal repeat can be dissociated from the proviral promoter and has enhancer properties, *Proc. Natl. Acad. Sci. USA* **82**:1020.

Prieto-Soto, A., Gourlie, B., Miwa, M., Pigiet, V., Sugimura, T., Malik, N., and Smulson, M., 1983, Polyoma virus minichromosomes: Poly ADP-ribosylation of associated chromatin proteins, *J. Virol.* **45**:600.

Proudfoot, N. J., and Brownlee, G. G., 1976, 3′ non-coding region sequences in eukaryotic messenger RNA, *Nature* **263**:211.

Reed, S., Stark, G. R., and Alwine, J. C., 1976, Autoregulation of SV40 gene A by T antigen, *Proc. Natl. Acad. Sci. USA* **73**:3083.

Reiser, J., Renart, J., Crawford, L., and Stark, G. R., 1980, Specific association of simian virus 40 tumor antigen with simian virus 40 chromatin, *J. Virol.* **33**:78.

Renz, M., Nehls, P., and Hozier, J., 1977, Involvement of histone H1 in the organization of the chromatin fiber, *Proc. Natl. Acad. Sci. USA* **74**:1897.

Reynolds, G. A., Basu, S. K., Osborne, T. F., Chin, D. J., Gil, G., Brown, M. S., Goldstein, J. L., and Luskey, K. L., 1984, HMB CoA reductase: A negatively regulated gene with unusual promoter and 5′ untranslated regions, *Cell* **38**:275.

Richardson, J. P., 1975, Attachment of nascent RNA molecules to superhelical DNA, *J. Mol. Biol.* **98**:565.

Rio, D. C., and Tjian, R., 1983, SV40 T antigen binding site mutations that affect autoregulation, *Cell* **32**:1227.

Rio, D. C., and Tjian, R., 1984, Multiple control elements involved in the initiation of SV40 late transcription, *J. Mol. Appl. Genet.* **2**:423.

Rio, D., Robbins, A., Myers, R., and Tjian, R., 1980, Regulation of SV40 early transcription *in vitro* by a purified tumor antigen, *Proc. Natl. Acad. Sci. USA* **77**:5706.

Robbins, A., Dynan, W. S., Greenleaf, A., and Tjian, R., 1984, Affinity-purified antibody as a probe of RNA polymerase II subunit structure, *J. Mol. Appl. Genet.* **2**:343.

Roeder, R. G., 1976, Eukaryotic nuclear RNA polymerases, in: *RNA Polymerase* (R. Losick and M. Chamberlin, eds.), pp. 285–329, Cold Spring Harbor Laboratory, Cold Spring Harbor, N.Y.

Rosen, C. A., Sodroski, J. G., and Haseltine, W. A., 1985, The location of cis-acting regulatory sequences in the human T cell lymphotropic virus III long terminal repeat, *Cell* **41**:813.

Rudolph, K., and Mann, K., 1983, Salt-resistant association of simian virus 40 T antigen with simian virus 40 DNA in nucleoprotein complexes, *J. Virol.* **47**:276.

Ruley, H. E., and Fried, M., 1983, Sequence repeats in a polyoma virus DNA region important for gene expression, *J. Virol.* **47**:233.

Ruskin, B., Krainer, A. R., Maniatis, T., and Green, M. R., 1984, Excision of an intact intron as a novel lariat structure during pre-mRNA splicing *in vitro*, *Cell* **38**:317.

Sadofsky, M., and Alwine, J. C., 1984, Sequences on the 3′-side of the hexanucleotide AAUAAA affect the efficiency of cleavage at the polyadenylation site, *Mol. Cell. Biol.* **4**:1460.

Samuels, M., Fire, A., and Sharp, P. A., 1982, Separation and characterization of factors mediating accurate transcription by RNA polymerase II, *J. Biol. Chem.* **257**:14419.

Saragosti, S., Moyne, G., and Yaniv, M., 1980, Absence of nucleosomes in a fraction of SV40 chromatin between the origin of replication and the region coding for the late leader RNA, *Cell* **20**:65.

Saragosti, S., Cereghini, S., and Yaniv, M., 1982, Fine structure of the regulatory region of simian virus 40 minichromosomes revealed by DNAse I digestion, *J. Mol. Biol.* **160**:133.

Sassone-Corsi, P., Dougherty, J. P., Wasylyk, B., and Chambon, P., 1984, Stimulation of *in vitro* transcription from heterologous promoters by the SV40 enhancer, *Proc. Natl. Acad. Sci. USA* **81**:308.

Sassone-Corsi, P., Wildeman, A., and Chambon, P., 1985, A trans-acting factor is responsible for the SV40 enhancer activity *in vitro, Nature* **313**:458.

Scheidereit, C., Geisse, S., Westphal, H. M., and Beato, M., 1983, The glucocorticoid receptor binds to defined nucleotide sequences near the promoter of mouse mammary tumor virus, *Nature* **304**:749.

Scholer, H. R., and Gruss, P., 1984, Specific interaction between enhancer-containing molecules and cellular components, *Cell* **36**:403.

Scott, W. A., and Wigmore, D. J., 1978, Sites in simian virus 40 chromatin which are preferentially cleaved by endonucleases, *Cell* **15**:1511.

Scott, W. A., Walter, C., and Cryer, B., 1984, Barriers to nuclease *Bal*31 digestion across specific sites in simian virus 40 chromatin, *Mol. Cell. Biol.* **4**:604.

Segawa, M., Sugano, S., and Yamaguchi, N., 1980, Association of simian virus 40 T antigen with replicating nucleoprotein complexes of simian virus 40, *J. Virol.* **35**:320.

Seidman, M., Garber, E., and Levine, A. J., 1979, Parameters affecting the stability of SV40 virions during the extraction of nucleoprotein complexes, *Virology* **95**:256.

Seif, I., Khoury, G., and Dhar, R., 1979, BKV splice sequences based on analysis of preferred donor and acceptor sites, *Nucleic Acids Res.* **6**:3387.

Sen, A., Hancock, R., and Levine, A. J., 1974, The properties and origin of the proteins in the SV40 nucleoprotein complex, *Virology* **61**:11.

Sentenac, A., 1985, Eucaryotic RNA polymerases, *CRC Crit. Rev. Biochem.* **18**:31.

Sergeant, A., Bohman, D., Zentgraf, H., Weiher, H., and Keller, W., 1984, A transcription enhancer acts *in vitro* over distances of hundreds of base-pairs of both circular and linear templates but not on chromatin-reconstituted DNA, *J. Mol. Biol.* **180**:577.

Shakov, A. N., Nedospasov, S. A., and Georgiev, G. P., 1982, Deoxyribonuclease II as a probe to sequence-specific chromatin organization: Preferential cleavage in the 72-bp modulator sequence of SV40 minichromosome, *Nucleic Acids Res.* **10**:3951.

Shaul, Y., Rutter, W. J., and Laub, O., 1985, A human hepatitis B viral enhancer element, *EMBO J.* **4**:427.

Shelton, E. R., Wassarman, P. M., and DePamphilis, M. L., 1978, The structure of simian virus 40 chromosomes in nuclei from infected monkey cells, *J. Mol. Biol.* **125**:491.

Shenk, T., 1978, Construction of a viable SV40 variant containing two functional origins of DNA replication, *Cell* **13**:791.

Shenk, T. E., Rhodes, C., Rigby, P. W., and Berg, P., 1975, Biochemical method for mapping mutational alterations in DNA with S1 nuclease: The location of deletions and temperature-sensitive mutations in simian virus 40, *Proc. Natl. Acad. Sci. USA* **72**:989.

Simonsen, C. C., and Levinson, A. D., 1983, Analysis of processing and polyadenylation signals of the hepatitis B virus surface antigen gene by using simian virus 40–hepatitis B virus chimeric plasmids, *Mol. Cell. Biol.* **3**:2250.

Singh, K., Carey, M., Saragosti, S., and Botchan, M., 1985, Expression of enhanced levels of small RNA polymerase III transcripts encoded by the B2 repeats in simian virus 40-transformed mouse cells, *Nature* **314**:553.

Sinha, S. N., Hellwig, R. J., Allison, D. P., and Niyogi, S. K., 1982, Conversion of simian virus 40 DNA to ordered nucleoprotein structures by extracts that direct accurate initiation by eukaryotic RNA polymerase II, *Nucleic Acids Res.* **10**:5533.

Slattery, E., Dignam, J. D., Matsui, T., and Roeder, R. G., 1983, Purification and analysis of a factor which suppresses nick-induced transcription by RNA polymerase II and its identity with poly(ADP-ribose) polymerase, *J. Biol. Chem.* **258**:5955.

Sompayrac, L., and Danna, K. J., 1985, The SV40 T-antigen gene can have two introns, *Virology* **142**:432.

Soprano, K. J., Jonak, G. J., Galanti, N., Floros, J., and Baserga, R., 1981, Identification of an SV40 DNA sequence related to the reactivation of silent rRNA genes in human greater than mouse hybrid cells, *Virology* **109**:127.

Spandidos, D. A., and Wilkie, N. M., 1983, Host-specificities of papilloma virus, Moloney murine sarcoma virus and simian virus 40 enhancer sequences, *EMBO J.* **2**:1193.

Srivatsan, E. S., Deininger, P. L., and Freidmann, T., 1981, Nucleotide sequence at polyoma VP1 mRNA splice sites, *J. Virol.* **37**:244.

Su, R. T., and DePamphilis, M. L., 1976, *In vitro* replication of simian virus 40 DNA in a nucleoprotein complex, *Proc. Natl. Acad. Sci. USA* **73**:3466.

Subramanian, K. N., 1979, Segments of SV40 DNA spanning most of the leader sequences of the major late mRNA are dispensable, *Proc. Natl. Acad. Sci. USA* **76**:2556.

Sundin, O., and Varshavsky, A., 1979, Staphylococcal nuclease makes a single non-random cut in the simian virus 40 minichromosome, *J. Mol. Biol.* **132**:535.

Swimmer, C., and Shenk, T., 1984, A viable simian virus 40 variant that carries a newly generated sequence reiteration in place of the normal duplicated enhancer element, *Proc. Natl. Acad. Sci. USA* **81**:6652.

Tack, L., and Beard, P., 1985, Both trans-acting factors and chromatin structure are involved in the regulation of transcription from the early and late promoters in simian virus 40 chromosomes, *J. Virol.* **54**:207.

Tanaka, K., Chowdhury, K., Chang, K. S. S., Israel, M., and Ito, Y., 1982, Isolation and characterization of polyoma virusmutants which grow in murine embryonal carcinoma and trophoblast cells, *EMBO J.* **1**:1521.

Tegtmeyer, P., Schwartz, M., Collins, J. K., and Rundell, K., 1975, Regulation of tumor antigen synthesis by SV40 gene A, *J. Virol.* **16**:168.

Tegtmeyer, P., Lewton, B. A., DeLucia, A. L., Wilson, V. G., and Ryder, K., 1983, Topography of SV40 A protein–DNA complexes: Arrangement of protein bound to the origin of replication, *J. Virol.* **46**:151.

Tenen, D. G., Haines, L. L., and Livingston, D. M., 1982, Binding of an analog of SV40 T antigen to wild-type and mutant viral replication origins, *J. Mol. Biol.* **157**:473.

Tenen, D. G., Taylor, T. S., Haines, L. L., Bradley, M. K., Martin, R. G., and Livingston, D. M., 1983, Binding of simian virus 40 large T antigen from virus-infected monkey cells to wild-type and mutant viral replication origins, *J. Mol. Biol.* **168**:791.

Thompson, J. A., Radonovich, M. F., and Salzman, N. P., 1979, Characterization of the 5′ terminal structure of SV40 early mRNAs, *J. Virol.* **31**:437.

Tjian, R., 1981, Regulation of viral transcription and DNA replication by the SV40 large T antigen, *Curr. Top. Microbiol. Immunol.* **93**:5.

Tooze, J. (ed.), 1981, *DNA Tumor Viruses*, Cold Spring Harbor Laboratory, Cold Spring Harbor, N.Y.

Treisman, R., and Maniatis, T., 1985, Simian virus 40 enhancer increases number of RNA polymerase II molecules on linked DNA, *Nature* **315**:72.

Treisman, R., Cowie, A., Favaloro, J., Jat, P., and Kamen, R., 1981a, The structure of the spliced mRNA's encoding polyoma virus early region proteins, *J. Mol. Appl. Genet.* **1**:83.

Treisman, R., Novak, U., Favaloro, J., and Kamen, R., 1981b, Transformation of rat cells by an altered polyoma virus genome expressing only the middle-T protein, *Nature* **292**:595.

Treisman, R., Green, M. R., and Maniatis, T., 1983, Cis and trans activation of globin gene transcription in transient assays, *Proc. Natl. Acad. Sci. USA* **80**:7428.

Trifanov, E. N., and Sussman, J. L., 1980, The pitch of chromatin DNA is reflected in its nucleotide sequence, *Proc. Natl. Acad. Sci. USA* **77**:3816.

Tsai, S. Y., Tsai, M.-J., Kops, L. E., Minghetti, P. P., and O'Malley, B. W., 1981, Transcription factors from oviduct and HeLa cells are similar, *J. Biol. Chem.* **256**:13055.

Tyndall, C., La Mantia, G., Thacker, C. M., Favaloro, J., and Kamen, R., 1981, A region of the polyoma virus genome between the replication origin and late protein coding sequences is required in *cis* for both early gene expression and viral DNA replication, *Nucleic Acids Res.* **9**:6231.

Van de Voorde, A., Contreras, R., Haegeman, G., Rogiers, R., Van Heuverswyn, H. Van Herreweghe, J., Volckaert, G., Ysebaert, M., and Fiers, W., 1977, Structural organization of the SV40 genome, in: *Early Proteins on Oncogenic DNA Viruses* (P. May, R. Monier, and R. Weil, eds.), pp. 17–30, INSERM, Paris.

Varshavsky, A. J., Bakayev, V. V., Chumarkov, P. M., and Georgiev, G. P., 1976, Minichromosomes of simian virus 40: Presence of histone H1, *Nucleic Acids Res.* **2**:2101.

Varshavsky, A. J., Sundin, O., and Bohn, M. J., 1978, SV40 viral minichromosome: Preferential exposure of the origin of replication as probed by restriction endonucleases, *Nucleic Acids Res.* **5**:3469.

Varshavsky, A. J., Sundin, O., and Bohn, M. J., 1979, A stretch of "late" SV40 viral DNA about 400 bp long which includes the origin of replication is specifically exposed in SV40 minichromosomes, *Cell* **16**:453.

Velcich, A., and Ziff, E., 1985, Adenovirus E1a proteins repress transcription from the SV40 early promoter, *Cell* **40**:705.

Veldman, G. M., Lupton, S., and Kamen, R., 1985, Polyomavirus enhancer contains multiple redundant sequence elements that activate both DNA replication and gene expression, *Mol. Cell. Biol.* **5**:649.

Vigneron, M., Barrerra-Saldana, H. A., Baty, D., Everett, R. E., and Chambon, P., 1984, Effect of the 21 bp repeat upstream element on *in vitro* transcription from the early and late SV40 promoters, *EMBO J.* **3**:2373.

Villareal, L., White, R., and Berg, P., 1979, Mutational alterations within the simian virus 40 leader segment generate altered 16S and 19S mRNAs, *J. Virol.* **29**:209.

Waldeck, W., Fohring, B., Chowdhury, K., Gruss, P., and Sauer, G., 1978, Origin of DNA replication in papovavirus chromatin is recognized by endogenous endonuclease, *Proc. Natl. Acad. Sci. USA* **75**:5964.

Walker, M. D., Edlund, T., Boulet, A. M., and Rutter, W. J., 1983, Cell-specific expression controlled by the 5′-flanking region of insulin and chymotrypsin genes, *Nature* **305**:557.

Wasylyk, B., and Chambon, P., 1981, A T to A base substitution and small deletions in the conalbumin TATA box drastically decrease specific *in vitro* transcription, *Nucleic Acids Res.* **9**:1813.

Wasylyk, B., Darbyshire, R., Guy, A., Molko, D., Roget, A., Teoule, R., and Chambon, P., 1980, Specific *in vitro* transcription of conalbumin gene is drastically decreased by single point mutation in T-A-T-A box homology sequence, *Proc. Natl. Acad. Sci. USA* **77**:7024.

Wasylyk, B., Wasylyk, C., Augereau, P., and Chambon, P., 1983a, The SV40 72 bp repeat preferentially potentiates transcription starting from proximal natural or substitute promoter elements, *Cell* **32**:503.

Wasylyk, B., Wasylyk, C., Matthes, H., Wintzerith, M., and Chambon, P., 1983b, Transcription from the SV40 early–early and late–early overlapping promoters in the absence of DNA replication, *EMBO J.* **2**:1605.

Wasylyk, B., Wasylyk, C., and Chambon, P., 1984, Short and long range activation by the SV40 enhancer, *Nucleic Acids Res.* **12**:5589.

Weber, F., and Schaffner, W., 1985, Simian virus 40 enhancer increases RNA polymerase density within the linked gene, *Nature* **315**:75.

Weber, F., de Villiers, J., and Schaffner, W., 1984, An SV40 "enhancer trap" incorporates exogenous enhancers or generates enhancers from its own sequences, *Cell* **36**:983.

Weeks, J. R., Coulter, D. E., and Greenleaf, A., 1982, Immunological studies of RNA polymerase II using antibodies to subunits of *Drosophila* and wheat germ enzyme, *J. Biol. Chem.* **257**:5884.

Weiher, H., and Botchan, M. R., 1984, An enhancer sequence from bovine papilloma virus DNA consists of two essential regions, *Nucleic Acids Res.* **12**:2901.

Weiher, H., König, M., and Gruss, P., 1983, Multiple point mutations affecting the SV40 enhancer, *Science* **219**:626.
Weil, P., Luse, D., Segall, J., and Roeder, R., 1979, Selective and accurate initiation of transcription at the Ad2 major late promoter in a soluble system dependent on purified RNA polymerase II and DNA, *Cell*, **18**:469.
Weintraub, H., and Groudine, M., 1976, Chromosomal subunits in active genes have an altered conformation, *Science* **193**:848.
White, M., and Eason, R., 1971, Nucleoprotein complexes in simian virus 40-infected cells, *J. Virol.* **8**:363.
Wickens, M., and Stephenson, P., 1984, Role of the conserved AAUAAA sequence: Four AAUAAA point mutants prevent messenger RNA 3′ end formation, *Science* **226**:1045.
Wieringa, B., Meyer, F., Reiser, J., and Weissmann, C., 1983, Unusual splice sites revealed by mutagenic inactivation of an authentic splice site of the rabbit beta-globin gene, *Nature* **301**:38.
Wieringa, B., Hofer, E., and Weissmann, C., 1984, A minimal intron length but no specific internal sequence is required for splicing the large rabbit α-globin intron, *Cell* **37**:915.
Wigmore, D., Eaton, R. W., and Scott, W. A., 1980, Endonuclease-sensitive regions in SV40 chromatin from cells infected with duplicated mutants, *Virology* **104**:462.
Wildeman, A. G., Sassone-Corsi, P., Grundstrom, T., Zenke, M., and Chambon, P., 1984, Stimulation of *in vitro* transcription from the SV40 early promoter by the enhancer involves a specific trans-acting factor, *EMBO J.* **3**:3129.
Wilson, V. G., Tevethia, M. J., Lewton, B. A., and Tegtmeyer, P., 1982, DNA binding properties of SV40 temperature-sensitive A proteins, *J. Virol.* **44**:458.
Wright, S., Rosenthal, A., Flavell, R., and Grosveld, F., 1984, DNA sequences required for regulated expression of β-globin genes in murine erythroleukemia cells, *Cell* **38**:265.
Wu, C., 1980, The 5′-ends of *Drosophila* heat shock genes in chromatin are hypersensitive to DNase I, *Nature* **286**:854.
Wu, G. J., 1978, Adenovirus DNA-directed transcription of 5.5S RNA *in vitro*, *Proc. Natl. Acad. Sci. USA* **75**:2175.
Yang, L., Rowe, T. C., Nelson, E. M., and Liu, L. F., 1985, *In vivo* mapping of DNA topoisomerase II-specific cleavage sites on SV40 chromatin, *Cell* **41**:127.
Yang, V. W., Lerner, M. R., Steitz, J. A., and Flint, S. J., 1981, A small nuclear ribonucleoprotein is required for splicing of adenoviral early RNA sequences, *Proc. Natl. Acad. Sci. USA* **78**:1371.
Yaniv, M., 1984, Regulation of eukaryotic gene expression by trans-activating proteins and cis-activating DNA elements, *Biol. Cell.* **50**:203.
Yanofsky, C., and Kolter, R., 1982, Attenuation in amino acid biosynthetic operons, *Annu. Rev. Genet.* **16**:113.

CHAPTER 3

Replication of SV40 and Polyoma Virus Chromosomes

MELVIN L. DEPAMPHILIS AND
MARGARET K. BRADLEY

I. INTRODUCTION

The replication and structure of simian virus 40 (SV40) and polyoma virus (PyV) chromosomes have been reviewed during the past 5 years by Acheson (1980), DePamphilis and Wassarman (1980, 1982), Das and Niyogi (1981), Levine (1982), Seidman and Salzman (1983), and Richter and Otto (1983). This review attempts to collate and interpret the large amount of data available on the replication of papovavirus chromosomes with particular emphasis on (1) the structure of large tumor-antigen (T-Ag) and its interaction with the origin of DNA replication, (2) the major DNA replication intermediates and their relationship to the mechanism for separating sibling molecules and terminating DNA replication, and (3) the sequence of events at replication forks and their relationship to nucleosome organization and initiation of DNA replication at *ori*.

A. Significance of Papovavirus DNA Replication

There are four features of papovavirus DNA replication that make it an attractive system for study. The first is that the high-molecular-weight viral T-Ag is required for initiation of viral DNA replication, induction

MELVIN L. DEPAMPHILIS • Department of Biological Chemistry, Harvard Medical School, Boston, Massachusetts 02115. MARGARET K. BRADLEY • Department of Pathology, Harvard Medical School, and the Dana Farber Cancer Institute, Boston, Massachusetts 02115.

of cellular DNA replication, and transformation of normal cells into cancer cells. All three of these functions reside in SV40 T-Ag, although transformation and viral DNA replication can be separated by genetic mutations. In PyV, the primary transforming function resides in the middle-sized T-Ag, but large T-Ag is required to immortalize the transformed phenotype and for DNA replication. Thus, viral T-Ag provides a link between regulation of DNA replication and regulation of cell proliferation. The second feature is that papovaviruses have a highly restricted host-range, suggesting that a specific interaction with cellular factors is involved in control of viral DNA replication. Each papovavirus replicates only in the cells (or specific cell types) of one mammalian species, but each can express their early genes in the cells of many other species even though viral DNA replication is blocked. Host range specificity can be altered by mutations in the viral gene enhancer or promoter regions, and such mutations appear to affect independently both the transcription and the replication of viral genes. Thus, papovaviruses can be used to identify the specific cellular factors involved in these processes. The third feature is that some papovaviruses are human pathogens that replicate so slowly in specific cell types that their life cycle may be analogous to that of a bacterial plasmid. For example, papilloma viruses cause epidermal tumors (warts) that are quite dangerous when located in the laryngeal or vaginal areas, and PML viruses are associated with progressive multifocal leukoencephalopathy, a rare but lethal demyelinating disease. Thus, control of viral DNA replication may be related to the establishment of certain human diseases. Finally, papovavirus chromosomes provide a model for the replication of mammalian chromosomes. The viral genomes replicate in the nucleus of their host cell as "minichromosomes," relying (with the exception of viral T-Ag) exclusively on the cell to provide the materials needed for DNA replication and chromatin assembly.

B. SV40 and PyV Chromosomes as Models for the Replication of Mammalian Chromosomes

Viral DNA replication is highly restricted with respect to the host cell (Acheson, 1980). SV40 replicates efficiently in African green monkey kidney cells (*permissive* host), poorly in human cell lines (*semipermissive* host), and not at all in rodent and other cells tested (*nonpermissive* host). The cells that are permissive for SV40 are nonpermissive for PyV. PyV replicates efficiently in most mouse cells (Dubensky and Villarreal, 1984), and poorly in other rodent cells. The sequence of events in a lytic infection begins with adsorption of the virus to the cell envelope, followed by penetration of the virus into the nucleus, uncoating of the viral genome, synthesis of early viral mRNA, production of viral large and small T-antigens, induction of cellular enzymes involved in nucleic acid metabolism and cellular DNA replication, viral DNA replication, synthesis of large quantities of late viral mRNA, production of viral capsid proteins and agnoprotein, assembly of virions, and cell lysis. The entire process takes 60–70 hr and is not well synchronized among the cells in an infected population. The maximum rate of viral DNA replication generally occurs between 30 and 40 hr postinfection. Induction of cellular events does not require viral DNA replication, but viral DNA replication precedes extensive late gene transcription. Both viral DNA replication and viral T-Ag play a role in late gene expression.

Nonpermissive cells allow only the expression of early viral genes and these induce cell proliferation; viral DNA replication is not detected and expression of late genes remains at its basal level. Although all of the cells in a nonpermissive population may undergo this abortive infection, only a small fraction are permanently transformed to take on the characteristics of a cancer cell. Transformation requires integration of viral DNA into the cellular genome and expression of viral T-antigens. Most of the infected nonpermissive cells undergo an abortive transformation and rid themselves of viral DNA within three or four cell divisions.

Papovaviruses are excellent models for investigating the replication of mammalian chromosomes (DePamphilis and Wassarman, 1980, 1982). SV40 and PyV genomes consist of small circular, covalently closed, dsDNA molecules that have been sequenced (Fig. 1C). Although the actual number of nucleotides per genome as well as their sequence can vary among viral strains [Hay *et al.*, 1984 (SV40 wt800); van Heuverswyn and Fiers, 1979 (SV40 Rh911); Rothwell and Folk, 1983; Ruley and Fried, 1983 (PyV A1, A2, A3)], the prototype viruses, SV40 776 and PyV A2, contain 5243 and 5295 bp, respectively (references in Fig. 10). These viruses replicate in the nucleus of their host as circular chromosomes whose nucleosome structure and histone composition are indistinguishable from those of their host (Fig. 1A,B). With the exception of large T-Ag, which is required for initiation of viral DNA replication, all of the events in the replication of viral DNA and its organization into a chromosome are carried out by the host. In fact, the structure, enzymology, and mechanistic properties of replication forks in SV40 and PyV replicating DNA intermediates are remarkably similar to those in eukaryotic cell chromosomes (Sections IV and V, and DePamphilis and Wassarman, 1980). Even the final stages in replication, the separation of sibling molecules and completion of DNA synthesis, may also be the same for both cells and virus, since the topological problem involved in unwinding DNA in front of two converging replication forks is the same in a small covalently closed, circular DNA molecule as it is in a large linear DNA molecule containing multiple replication bubbles. Neither situation allows free rotation of one arm of a replication fork about the other (Section III.I). The most significant differences between viral and cellular DNA replication will likely occur in the mechanism for initiation of new rounds of DNA replication. Cellular chromosomes must replicate all of their DNA once and only once

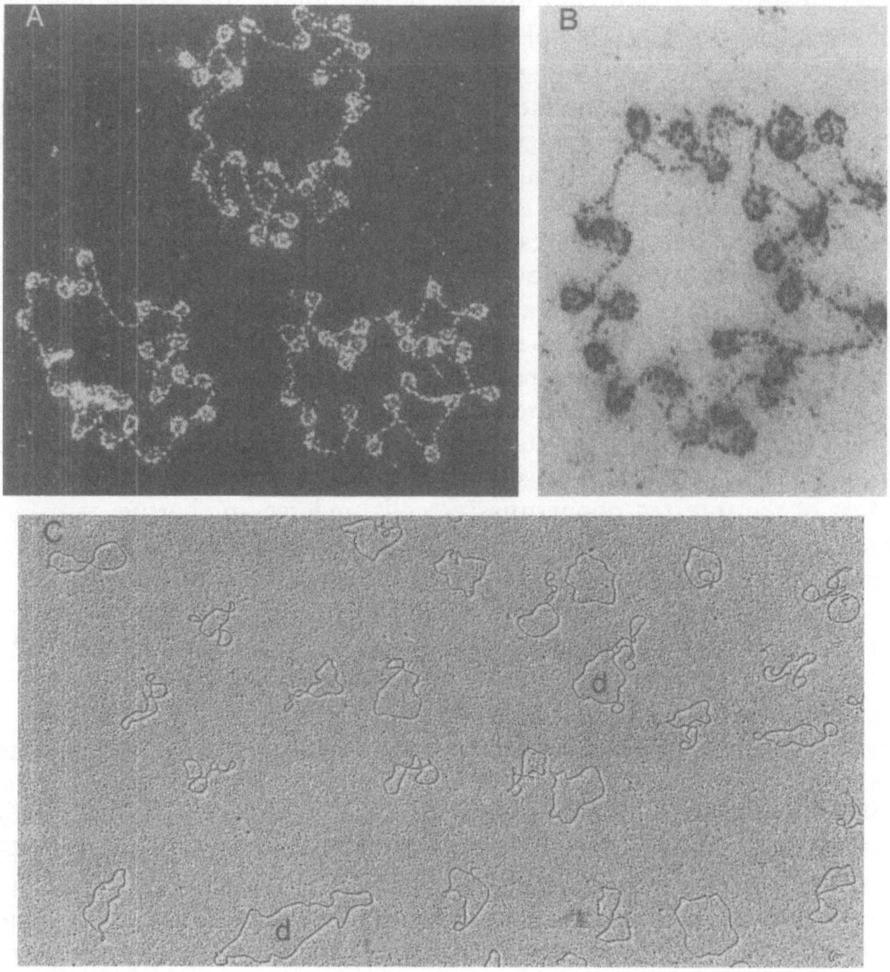

FIGURE 1. SV40 chromosomes and DNA. (A) Darkfield electron micrograph of SV40 chromosomes isolated under isotonic salt conditions and stained with uranyl acetate (Moyne *et al.*, 1981). (B) Transmission electron micrograph of the same sample used in (A). (C) Transmission electron micrograph of SV40 DNA shadowed with platinum/palladium (Tapper and DePamphilis, 1980; Tapper *et al.*, 1982). A catenated and a circular dimer are indicated by "d." The micrographs of SV40 chromosomes were kindly provided by Dr. Moshe Yaniv, and those of SV40 DNA by Dr. Douglas Tapper.

during each S-phase cycle of proliferating cells, whereas papovavirus DNA replicates many times during a single S-phase. Thus, T-Ag may be the viral equivalent of specific cellular proteins that are expressed under certain conditions (e.g., terminally differentiated cells, or cells under selective pressure) to amplify specific cellular genes (Varshavsky, 1981). The fact that each papovavirus replicates its DNA only in certain permissive cell types indicates that the normal regulation of cellular DNA replication also involves highly specific interactions between proteins and DNA sequences, and that the components from one species may not be compatible with those from another.

II. LARGE T-ANTIGEN

A. Summary

T-Ag is a multifunctional regulatory protein synthesized by papovavirus early after infection of the host cell. T-Ag is found predominantly in the nucleus where it is required for initiation of viral DNA replication and for stimulation of cellular DNA synthesis. Purified T-Ag binds viral DNA specifically, has associated ATPase and protein kinase activities, can bind a host protein p53, and is found associated with replicating viral chromosomes. Temperature-sensitive mutations in the gene encoding T-Ag (tsA) affecting the initiation of viral DNA replication, produce defects in all of these *in vitro* activities. Well known as a transformation protein, T-Ag also induces the synthesis of host mRNA and rRNA in addition to regulating viral transcription.

T-Ag is highly modified after translation and has structural characteristics that probably affect the active state as well as the migration properties of the purified protein. It can have 8–10 phosphorylated sites at serine and threonine residues, which are phosphorylated specifically in either the cytoplasm or the nucleus of the host cell. In addition, a few percent of T-Ag molecules are either adenylated, acylated, ADP-ribosylated, glycosylated, or methylaminated. Monomeric, dimeric, and tetrameric species of T-Ag have been identified. The smaller forms of T-Ag have a higher specific activity for binding to viral DNA *in vitro* and consist of newly synthesized, partially phosphorylated T-Ag. Oligomeric T-Ag can form a complex with cellular p53 in the nucleus 1–2 hr after T-Ag is synthesized. This relationship may be important for stimulating the cell to enter S phase. A nuclear localization signal that resides near the N-terminus of SV40 T-Ag (around Lys^{128}) is required for the antigen to function as a replication protein in the nucleus.

Some of these modifications correlate with active species of T-Ag. (1) Newly synthesized T-Ag that is underphosphorylated has a high affinity for viral DNA and may be the T-Ag species responsible for initiation of viral DNA replication. (2) Acylation of SV40 T-Ag, which contains a cluster of hydrophobic residues, appears responsible for attachment of T-Ag to the plasma membrane lamina. (3) Oligomerization of T-Ag may regulate its binding to DNA as well as its ATPase activity. Neither property was necessary for transformation activity. (4) Replication mutants mapping near the putative adenylation site on T-Ag suggest that adenylation may be necessary for initiation of viral DNA synthesis.

T-Ag has both nonspecific and specific binding affinities for DNA. About 10% of T-Ag is able to bind viral DNA specifically. Both SV40 T-Ag and PyV T-Ag bind to the consensus pentanucleotide 5′-GAGGC-3′ located within specific sites at and near their respective origins of replication (*ori*). SV40 T-Ag binds to DNA site 1 with the highest affinity. However, site 1 is on the early gene side of *ori* while site 2 includes the minimal origin of replication. SV40 DNA site 2 contains both a high-affinity (2A, to the early side) and a low-affinity (2B, to the late side) T-Ag binding site. PyV DNA high-affinity binding sites A, B, and C all lie to the early gene side of *ori;* a tenfold higher concentration of PyV T-Ag was required to detect viral DNA binding sites within the origin region (PyV DNA sites 1 and 2). T-Ag is preferentially associated with early replicating viral chromatin, and a salt-stable interaction between T-Ag and viral chromosomes may be involved in T-Ag's replication activity.

DNA-independent ATPase activity is probably intrinsic to the protein, and has been correlated with the replication function by genetic analyses. T-Ag-associated kinase may be responsible for the regulation of T-Ag and p53 functions. There is also some evidence for an adenyl transferase activity associated with the adenylation of T-Ag that specifically dissociates AMP from the antigen, suggesting that T-Ag donates the first ribonucleotide in initiation of viral DNA replication.

Certain functional domains of T-Ag can be hypothesized based on analyses of mutations in T-Ag that affect viral DNA replication, activities associated with T-Ag *in vitro,* and predicted secondary structure. Of the 708 amino acids of SV40 T-Ag, the region that appears directly involved in binding to viral DNA extends from amino acids 133 to 228 (nonspecific DNA binding activity may include residues 83 to 272). SV40 tsA mutations, resulting in the loss of the DNA binding activity, occur in a more C-terminal region (229 to 447), suggesting that interaction between domains is required for initiation of DNA replication. SV40 T-Ag and PyV T-Ag retain a sequence of highly conserved amino acids in the C-terminal domains, especially near the specific sites for ATPase and adenylation (SV40 412 to 528). Finally, the helper function or host range control region in the last 35 amino acids of SV40 T-Ag is not required for DNA synthesis and it can function independently. Therefore, only a small region of SV40 T-Ag appears to be dispensible in the replication function.

B. Background

Papovaviruses each have a set of early genes that are expressed in the infected host cell, but these proteins are not found in the virion. Their DNA coding sequences are overlapping, forming separate mRNAs. The genes in SV40 (the monkey papovavirus) are known as A and F with reference to particular fragments of restriction enzyme-digested viral DNA. Large T-Ag is a product of the A gene, hence the common name A protein

(708 amino acids in SV40 T-Ag and 785 amino acids in PyV T-Ag). The F gene, which includes the coding information for the N-terminal 82 amino acids of the SV40 A gene plus 97 unique amino acids from within the T-Ag intron, codes for the small t-antigen (t-Ag; Tooze, 1981). There is also a late gene in SV40 which codes for a nonvirion protein called the agnoprotein; this is the agnogene (0.7–0.73 *Eco*RI map units; Jay *et al.*, 1981b). In the polyoma virus (PyV) of the mouse, a third mRNA is derived from the same early DNA sequence that codes for T-Ag and t-Ag messages, and it translates into the middle T-Ag protein (mT-Ag, 56K) using the t-Ag code at the N-terminal region and an alternate reading frame from that used for T-Ag for the rest (Feunteun *et al.*, 1976; Simmons *et al.*, 1979; Schaffhausen, 1983).

The term antigen refers to the discovery of these proteins by immunological means [see Tooze (1981) for a more complete review]. "T" was derived from "tumor," and the T-Ag was visualized in the nuclei of the virally infected or transformed cells. Much of the early biochemical work on T-Ag depended upon anti-"T" sera generated in hamsters bearing SV40-induced tumors, and proof that the antigens were viral products has been documented in detail (Tooze, 1981). Experiments demonstrating that T-Ag was required for viral DNA replication were performed by Tegtmeyer (1972) and Francke and Eckhart (1973), using temperature-sensitive, mutant viruses. The A gene protein was thus defined as an antigen with functional pleiotropy, since transformation and many events in the lytic cycle of the virus in addition to viral DNA replication, were affected by mutations in that region of the genome.

The process of defining T-Ag as a protein with multiple functional domains, multiple sites of action in the cell, and both active and inactive states has been somewhat dependent upon advances in research technology. Combined genetic, biochemical, and immunological investigation distinguished the SV40 F from the A gene product (Shenk *et al.*, 1976; Prives *et al.*, 1977; Paucha *et al.*, 1978), and defined the unique PyV mT-Ag (Ito *et al.*, 1977a,b; Schaffhausen *et al.*, 1978; Hunter *et al.*, 1978). The advent of recombinant DNA technology and the development of monospecific antibodies (including monoclonal antibodies), provided the reagents necessary for studying the separate, viral gene products as proteins in specific antigenic subclasses. There exist both nondefective ($Ad2^{+}ND_{1\text{–}5}$; Lewis *et al.*, 1973; Tooze, 1981) and defective viruses (e.g., $Ad2^{+}D2$; Hassell *et al.*, 1978) which overproduce versions of SV40 T-Ag under adenoviral controls, ranging from adeno–SV40 fusion proteins that include 23% to 85% of the C-terminal amino acids of SV40 T-Ag. Such viruses coding for wild-type (wt) T-Ag have recently been constructed *in vitro* to overproduce the T-Ag peptide and make possible its biochemical analyses (Thummel *et al.*, 1983; Gluzman *et al.*, 1982). The truncated T-Ag's have often been used to map modified residues, antibody epitopes (specific antigenic sites such as those recognized by monoclonal antibodies), and specific functions of SV40 T-Ag (Tooze, 1981).

C. Initiation of Viral and Cellular DNA Replication

Temperature-sensitive A gene (tsA) mutations in SV40 and tsA mutations in PyV were shown to prevent new initiations of viral DNA replication at elevated temperatures, although molecules already in the process of replicating completed their synthesis at the high temperature (Tegtmeyer, 1972; Francke and Eckhart, 1973; Chou *et al.*, 1974; Dinter-Gottlieb and Kaufmann, 1982). Some of these mutations mapped to a homologous region in each of the two T-Ag's (Fig. 2; SV40, amino acids 350 to 450; PyV, 500 to 600). Induction of cellular DNA synthesis was not necessarily affected by these mutations (Chou and Martin, 1975; Hiscott and Defendi, 1979a,b; Setlow *et al.*, 1980), but in many cases the replication defect was also seen as a transformation defect. Since the

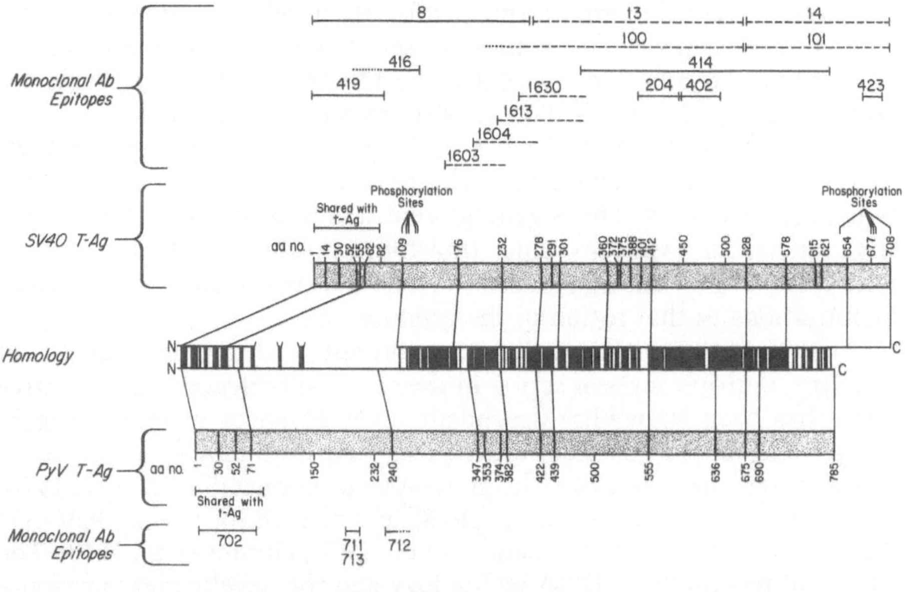

FIGURE 2. Amino acid homologies between SV40 and PyV large T-antigens (adapted from Clertant and Cuzin, 1982). Black areas in the central bar indicate regions of amino acid homology. A more detailed map can be found in Griffin and Dilworth (1983). Each speckled bar represents the appropriate length of either SV40 or PyV T-Ag with a divison at every ethionine residue (i.e., cyanogen bromide cleavage map). The proteins are aligned with respect to the most homologous region near their C-termini. Shared amino acid sequences for t-Ag are shown for each T-Ag, and the maximum number of phosphorylation sites for SV40 T-Ag are indicated (Scheidtmann *et al.*, 1982). Monoclonal antibody epitopes have been mapped to various extents. Solid lines represent peptide fragments immunoprecipitated by the indicated antibody [Ab700 series, Dilworth and Griffin (1982); Ab204, D. Lane (personal communication); Ab400 series, Schwyzer *et al.* (1983)]. Dotted lines show possible extensions of the observed peptide, and dashed lines represent estimates of the epitopes derived by immunoprecipitation of N-terminally truncated T-Ag's encoded by adeno–SV40 hybrid viruses, $Ad2^{+}ND_{1-5}$ [Ab1 series, Mercer *et al.* (1983); Ab100 series, Deppert *et al.* (1981); Ab400 series, Harlow *et al.* (1981a); Ab1600 series, Ball *et al.* (1984)].

coding regions for mT-Ag and T-Ag overlap, PyV mutations were often expressed in both antigens (see reviews: Schaffhausen, 1983; Griffin and Dilworth, 1983; Rigby and Lane, 1983; Eckhart, 1983). Uncoupling of the replication and transformation functions was accomplished more recently by isolation of mutant viral DNA from transformed cells or by specific construction of mutants in both SV40 and PyV T-Ag's (Gluzman *et al.*, 1977; Cosman and Tevethia, 1981; Pintel *et al.*, 1981; Stringer, 1982; Gluzman and Ahrens, 1982; Hayday *et al.*, 1983; Nilsson and Magnusson, 1984; Carmichael *et al.*, 1984; Templeton and Eckhart, 1984).

Microinjection of either T-Ag or early viral genes directly into mammalian cells showed that a complete T-Ag was required for the initiation of viral DNA synthesis (Graessmann and Graessmann, 1976; Kriegler *et al.*, 1978; Tjian *et al.*, 1978; Graessmann *et al.*, 1979). Similarly, microinjection of tumor serum, containing antibodies to SV40 T-Ag, blocked SV40 DNA synthesis in virus-infected cells (Antman and Livingston, 1980). However, when a truncated SV40 A gene was microinjected into monkey cells, it was discovered that an N-terminal domain of T-Ag was sufficient for stimulation of cellular DNA synthesis (Mueller *et al.*, 1978). This was confirmed by the ability of C-terminal deletion mutants of SV40 T-Ag (to amino acid 272) to stimulate cellular DNA synthesis (Shenk *et al.*, 1976; Tjian *et al.*, 1978; Soprano *et al.*, 1983; Rawlins *et al.*, 1983). Therefore, induction of both viral and cellular DNA synthesis appeared to require some capacity for T-Ag/DNA binding, but truncated, SV40 protein with a short half-life ($t_{1/2}$ 10 min) and no detectable enzyme activity was sufficient for the stimulation of cellular DNA synthesis (Rawlins *et al.*, 1983).

Point mutations in T-Ag have been shown to destroy the ability of T-Ag both to specifically bind viral DNA and to initiate viral DNA replication (Stringer, 1982; Kalderon and Smith, 1984; Manos and Gluzman, 1985; E. Paucha, personal communication), but these same T-Ag's bound to cellular DNA and induced cellular DNA synthesis (Scheller *et al.*, 1982). In other studies, SV40 DNA and early mRNA were microinjected into hamster cells that were temperature-sensitive for cellular DNA replication. T-Ag stimulated cellular DNA synthesis at both permissive and nonpermissive temperatures (Floros *et al.*, 1981), although to a lesser extent at the nonpermissive temperature (Christiansen and Brockman, 1982). It appeared that T-Ag "dispensed" with two or more cellular functions required for progression of cell division from the G_1 to the S phase of the cell cycle. Not all cell lines show induction of DNA synthesis by T-Ag (Schaffhausen, 1983), perhaps because they are already maximally stimulated. This activity can be blocked by monoclonal antibodies against the N-terminal region of T-Ag (Mercer *et al.*, 1983). In summary, stimulation of cellular DNA synthesis by T-Ag could be considered an independent functional domain from initiation of viral DNA replication, although it may be a necessary part of the latter (see review: Martin, 1981).

Although T-Ag is the only viral protein *required* for viral DNA replication, other proteins may regulate the process. For example, transfection of mouse cells with PyV DNA coding either for T-Ag alone or for t-Ag plus mT-Ag revealed that viral DNA synthesis increased three- to four-fold when T-Ag was supplemented with t-Ag and mT-Ag (Nilsson and Magnusson, 1984). SV40 t-Ag was located in the nuclei as well as the cytoplasm of cells (Ellman *et al.*, 1984). In addition, the agnoprotein of SV40, synthesized 48 hr postinfection, is a basic peptide (7.9K; histone-like) that binds tightly to cellular DNA and may regulate late transcription, virion assembly or down-regulate the replication function of T-Ag (Jay *et al.*, 1981b). While it has been established that T-Ag is required for replication, mT-Ag in PyV (Lania *et al.*, 1980; Rassoulzadegan *et al.*, 1981, 1982) and T-Ag in SV40 (Tooze, 1981) are responsible for oncogenic transformation of mammalian cells. Since the peptidyl domains responsible for viral DNA replication and cellular transformation overlap in SV40 T-Ag, and PyV mT-Ag did not appear necessary for the replication of polyoma virus (Carmichael *et al.*, 1984; Templeton and Eckhart, 1984), a careful comparison of SV40 T-Ag with PyV T-Ag may serve to sort out some of the replication-specific data.

Other early viral genes may also be involved in the stimulation of cellular DNA replication. The studies mentioned above (Floros *et al.*, 1981; Mercer *et al.*, 1983) did not separate the functions of the A from the F gene of SV40. Therefore, the actions of both T-Ag and t-Ag might have been involved in stimulating cells to enter S phase. Also, SV40-infected cells which lacked t-Ag did not induce cellular DNA synthesis in the presence of theophylline, whereas cells containing both antigens were unaffected by this inhibitor of cAMP (Rundell and Cox, 1979). The type of cell studied was significant. Primary mouse cells were induced to synthesize their DNA through only one cycle if T-Ag alone was present (Schlegel and Benjamin, 1978; Hiscott and Defendi, 1979b), whereas a flat, normal line of rat cells required only SV40 T-Ag for multiple rounds of DNA synthesis (Okuda *et al.*, 1984). The role played by t-Ag, or mT-Ag in PyV, in the induction of DNA synthesis may be an indirect one, more similar to a hormonal stimulation. The cellular host apparently has the ability to provide some part of this function as evidenced by the ability of papovavirus defective for expression of t-Ag (and mT-Ag) to replicate in particular cell lines (Shenk *et al.*, 1976; Silver *et al.*, 1978).

Initiation of viral DNA synthesis requires a T-Ag with some unique conformation and probably more than one biochemical activity. In addition to a defect in T-Ag's ability to bind viral DNA specifically, the ability of the protein to oligomerize appeared to be absent in replication-defective mutants (e.g., tsA58; Montenarh *et al.*, 1984). Biochemical analyses of T-Ag derived from mutants of SV40 showed that T-Ag/ATPase activity correlated with the induction of viral DNA synthesis (Clark *et al.*, 1983; Manos and Gluzman, 1985), however, the ATPase activity was apparently not required for the transformation function in SV40 (mutant pC2; Manos and Gluzman, 1984). Some T-Ag activities were not necessary for initiation of DNA synthesis. Host range functions or late gene interactions perhaps related to virion assembly, were genetically mapped by mutational analyses to the N-terminal region of PyV (Schaffhausen, 1983) and to the C-terminus of SV40 (Tornow and Cole, 1983b). In addition, some part of the SV40 transformation function was not necessary for replication activity (tsA1642; Cosman and Tevethia, 1981). Each of these functions could be inactivated by mutation without significantly affecting the ability of T-Ag to induce viral DNA replication. Other than these regions, the available data indicated that some complex or dynamic form of T-Ag was required, with several domains of the protein defined as necessary but not sufficient.

It is not clear whether T-Ag is required for the actual synthesis of viral DNA, or just to activate the origin of replication. Another mutation in the A gene, masked by the requirements for the initiation function, could have an independent effect on elongation, a possibility not excluded by the available genetic evidence (Tegtmeyer, 1972; Chou *et al.*, 1974; Dinter-Gottlieb and Kaufmann, 1982). T-Ag has been found associated with the replicating chromosomes of SV40 (Section II.F.5; Segawa *et al.*, 1980; Reiser *et al.*, 1980; Rudolph and Mann, 1983; Stahl and Knippers, 1983; Tack and DePamphilis, 1983; Clertant *et al.*, 1984a), although some of these data may be related to other T-Ag functions such as control of transcription or virion assembly. Stahl *et al.* (1983) reported isolating a T-Ag associated with replication forks. Since the interaction of T-Ag with chromatin was shown to decay within a 2-hr period (Staufenbiel and Deppert, 1983), T-Ag may be inactivated as a replication protein after one cycle, and the initiation and elongation species of the antigen may be inseparable.

D. Posttranslational Modifications

T-Ag is a multifunctional protein, which may explain the large number of posttranslational modifications associated with it. Two points must be considered in analyses of T-Ag and replication. (1) Since viral DNA replication is host specific (Tooze, 1981), PyV and SV40 T-Ag's purified from cells other than mouse or monkey, respectively, might prove to be inefficient or defective in induction of viral DNA synthesis due to differences in the modification of the antigen or by irreversible association with specific host cell factors (Section II.F.4). (2) Newly synthesized T-Ag has been found associated with viral chromatin (Staufenbiel and Deppert, 1983) and was most efficient in binding to viral DNA (Gidoni *et al.*, 1982). Therefore, T-Ag purified from a lytically infected cell source soon after its synthesis would probably be the most biologically active protein.

Most of the T-Ag's in a transformed cell would have been synthesized long before the cells were harvested for extraction and possibly inactivated as a replication protein ($t_{1/2}$ 10–12 hr; Prives *et al.*, 1978; Jay *et al.*, 1981b). Methods for quick purification of SV40 T-Ag by immunoaffinity chromatography have been reported (Dixon and Nathans, 1985; Simanis and Lane, 1985), and the antigen was active in all *in vitro* assays including that for initiation of viral DNA replication *in vitro* (Li and Kelly, 1984).

Listed below are all the known modifications of SV40 and PyV T-Ag's where data were available. Each species might represent only a small fraction of the protein synthesized by the host cell. Even though all the forms of T-Ag mentioned here are not directly associated with the replication function of the protein, they may interact at some level with the replication-active T-Ag with respect to cellular or viral DNA synthesis.

1. Phosphorylation

Eight to ten phosphorylated residues on SV40 T-Ag have been mapped (Fig. 2; Schwyzer *et al.*, 1980; Scheidtmann *et al.*, 1982; van Roy *et al.*, 1983). t-Ag has no phosphorylated amino acids, and the N-terminal 82 amino aids shared by the two proteins were not phosphorylated on T-Ag. However, a prominent cluster of phosphoserines and a phosphothreonine were located at the unique N-terminal portion of T-Ag within a 17K tryptic peptide fragment. Therefore, the modified amino acids were located between residues 83 and 128. Another cluster of phosphorylated serines appeared near the C-terminus and there was a phosphothreonine within the last 8 amino acids which was easily lost through proteolytic digestion of the protein. Some sites were known to be phosphorylated in the cytosol while others were not found unless the protein was located in the nucleus (Scheidtmann *et al.*, 1984b).

Turnover of the phosphorylated residues had a shorter half-life than that of the protein itself (Edwards *et al.*, 1979; Greenspan and Carroll, 1981), and an important nuclear phosphorylation site at the N-terminus has been shown to regulate T-Ag activity (Scheidtmann *et al.*, 1984a). Specifically, the DNA binding properties of the protein were highest when the peptide was relatively new and the phosphate content was low (Oren *et al.*, 1980), particularly at N-terminal sites (Scheidtmann *et al.*, 1984a). Consistent with these data were *in vitro* studies by Shaw and Tegtmeyer (1981) and Baumann and Hand (1982) which showed no decrease in viral DNA binding capacity after T-Ag had been enzymatically dephosphorylated. No increase in binding could be detected, but that is consistent with the inability as yet to reconstitute any T-Ag activity after it has been lost. Phosphorylation of T-Ag increased over time (Scheidtmann *et al.*, 1984b; Simmons, 1984) starting at Ser^{111}, Thr^{124}, Thr^{701} and perhaps two more sites (Scheidtmann *et al.*, 1984a,b). These sites were found on cytoplasmically located T-Ag and were characteristic of T-Ag having the highest affinity for DNA. Serines 106, 123, 639, 676, 677, and 679 were only found to be phosphorylated when T-Ag was in the nucleus. Ser^{112} and Thr^{701} were found to be phosphorylated in both the cytoplasm and the nucleus. A high-salt extract (0.5 M NaCl) of nuclear T-Ag was found to have the characteristic phosphorylation pattern of newly synthesized T-Ag and it bound with high affinity to viral DNA. The investigators suggested that this was the chromatin-bound T-Ag described by Staufenbiel and Deppert (1983) having a half-life of about 2 hr. This was not to be confused with T-Ag stably associated with the nuclear matrix.

2. Fatty Acid Acylation

Klockmann *et al.* (1984) have shown a correlation of an acylated residue on SV40 T-Ag with its association with the plasma membrane lamina. It was already known that about 5% of SV40 T-Ag was located outside the nucleus (Deppert and Walter, 1976; Soule and Butel, 1979; for review see Schaffhausen, 1983). A 28K hybrid adeno–SV40, C-terminal T-Ag associated with the plasma membrane, probably as a result of a highly hydrophobic domain of SV40 T-Ag lying between amino acids 568 to 582 which is contained in that hybrid protein (there is a homologous region in PyV T-Ag). A tight association with the plasma membrane lamina correlated with acylation of a site somewhere between amino acids 340 and 511 as defined by acylated, hybrid T-Ag proteins produced by $Ad2^{+}ND_5$ and $Ad2^{+}ND_4$, and the nonacylated 28K protein from $Ad2^{+}ND_1$ (Klockmann and Deppert, 1983; Klockmann *et al.*, 1984). Acylation seems most relevant to the study of induction of transformation by T-Ag (Verderame *et al.*, 1983; Lanford and Butel, 1984), unless there is some nuclear membrane interaction required for replication. There is evidence that SV40 and PyV T-Ag could be found tightly associated with the nuclear matrix (Deppert, 1978; Buckler-White *et al.*, 1980; Covey *et al.*, 1984), but no strong data for replication activity have been associated with it.

Regions of T-Ag have been found exposed on the cell surface (Tooze, 1981; Pan and Knowles, 1983; Gooding and O'Connell, 1983; Tevethia *et al.*, 1984; Santos and Butel, 1984). In addition, a "minor subclass" of T-Ag bound covalently to lipids on the surface of cells *in vitro* (on palmitate residues; Henning and Lange-Mütschler, 1983; Lange-Mütschler and Henning, 1983, 1984). There are no clear data that define how the newly synthesized T-Ag is partitioned into the nuclear and the membranous fractions of the cell. Perhaps the C-terminal hydrophobic amino acid sequence paired with fatty acid acylation are sufficient for transport of T-Ag to the plasma membrane in the presence of the nuclear localization signal (see Section II.E.4), and that species of T-Ag induces changes in the cell's response to growth factors.

3. ADP-Ribosylation

A covalent linkage of ADP-ribose was detected in a few percent of SV40 T-Ag molecules by Goldman *et al.* (1981a). They were also successful in ADP-ribosylating T-Ag in a nuclear cell extract *in vitro* with radioactive NAD. It is unclear what purpose this modification serves, but it has been associated with nuclear enzymes involved in DNA replication and repair, and possibly with nuclear transport (Hayaishi and Ueda, 1977). Papovavirus chromatin does have an associated ADP-ribosylase (Otto and Fanning, 1978; Prieto-Soto *et al.*, 1983), which may be responsible for this modification.

4. Methylamination

Methylamine could be bound to SV40 T-Ag *in vitro* (L. Tack, personal communication). So far the adduct has been mapped to the N-terminal half of the molecule. Such modification of T-Ag appeared to have no effect on ATPase activity. However, *in vitro* adenylation (see below) was inhibited. The incorporation of methylamine suggested that an active ester bond may be formed which would serve to link T-Ag covalently to other proteins or itself. This has not been established as yet, but is provocative given that T-Ag may require a specific association with replication enzymes or other cellular factors.

5. Glycosylation

Galactose has been found bound to T-Ag (Jarvis and Butel, 1985). T-Ag could be radioactively labeled *in vivo* wih galactose or glucosamine but not mannose or fucose. The relevant location of glycosylated T-Ag or its effect on functions *in vivo* or *in vitro* has not been discerned.

6. N-Acetylation

The N-terminal methionines of T-Ag and t-Ag are acetylated (Paucha *et al.*, 1978).

7. Adenylation

The affinity label, dialdehyde-ATP, could be bound specifically to crude and purified preparations of papovavirus T-Ag (Clertant and Cuzin, 1982; Bradley and Livingston, unpublished observations). A replication protein of papilloma virus also bound to dialdehyde-ATP (Clertant and Seif, 1984). Clertant and Cuzin (1982) reported that the labeling of temperature-sensitive mutants of PyV and SV40 T-Ag with dialdehyde-ATP was defective at the restrictive temperature; therefore, this was an interesting modification with respect to the viral DNA replication function. In retrospect, the tsa (and tsA) mutants chosen did not map to the adenylation sites, which probably indicated that conformational changes affected that modification of T-Ag. Further analyses of purified SV40 T-Ag (Bradley *et al.*, 1984) revealed that the protein could be directly adenylated with unmodified ATP in the presence of magnesium. The peptide-bound moiety was AMP and it was covalently linked to the protein via a phosphodiester bond to a serine residue.

Clertant *et al.* (1984b) located the adenine nucleotide affinity label on particular cyanogen bromide fragments in PyV and SV40 T-Ag's (confirmed by M. Bradley, unpublished observations). The fragments were derived from highly homologous regions in the two T-Ag's, between amino acids 614 and 653 in PyV and between amino acids 412 and 528 in SV40 (Fig. 2). The site of adenylation has been separated from the ATP binding site for T-Ag/ATPase (Clertant *et al.*, 1984b). T-Ag from the PyV6 mutant, truncated near amino acid 550 (out of 785), retained ATPase activity but could not be labeled with the dialdehyde-ATP. Therefore, the active site of the ATPase lies closer to the N-terminus than the adenylation site. There was also an indication that ATPase and adenylation were functionally related. That is, SV40 or PyV T-Ag's defective in their ATPase activities bound dialdehyde-ADP rather than -ATP, indicating a possible processing of ATP to ADP for adenylation by T-Ag itself (P. Clertant, personal communication).

Adenylated T-Ag may be important in the replication of viral DNA. The region where AMP binds to T-Ag has been shown to be sensitive to mutations resulting in defects in viral DNA replication (e.g., 7axB in PyV T-Ag and pC2 and dl2462 in SV40; see Table I). The SV40 mutant pC2 (Manos and Gluzman, 1984) encodes a conservative change from lysine to arginine at amino acid 516; nevertheless, it produced a T-Ag that retained ATPase activity but was defective in viral DNA replication. Adenylation of pC2 T-Ag has not been measured. Other mutations in the region of the protein near the putative adenylation site (SV40 amino acids 448 to 538, Section II.I.6) affect ATPase and viral DNA replication together. A monoclonal antibody to SV40 T-Ag (pAb204) directly inhibited T-Ag/ATPase (Clark *et al.*, 1981). The antibody binding epitope has been mapped to approximately 395 to 447 (D. Lane, personal communication). It may be that the active site for ATPase lies within that fragment and that adenylation occurs farther away at the highly conserved Ser^{504} in SV40 and Ser^{766} in PyV, close to the pC2, dl2462 and 7axB mutations, respectively.

It has been reported (Bradley *et al.*, 1984) that a monoclonal antibody (pAb402, Harlow *et al.*, 1981a) that bound T-Ag at 450–500 (Schwyzer *et al.*, 1983) near the putative adenylation site, failed to block adenylation. However, this study utilized a C-terminally truncated T-Ag which was not immunoprecipitated by this antibody (M. Bradley, unpublished ob-

TABLE I. Viral DNA Replication Mutants of SV40

Domains[a]	Amino acid changed	Name	Replication	T-Ag binds ORI	Nuclear localization of T-Ag[b]	Transformation	T-Ag/ATPase activity	Reference
1	17–27	d11135	neg	neg		pos	neg	Clark *et al.* (1983)
	30, 51	pC6-1	decr	pos		pos	pos	Gluzman and Ahrens (1982)
2	106	C8A	neg	pos	pos	pos		Kalderon and Smith (1984), E. Paucha[c]
	126	C61	neg	pos	pos	pos		Kalderon and Smith (1984)
	128	cT	neg		neg			Lanford and Butel (1984)
	128	d10	neg	pos	neg	pos		Kalderon and Smith (1984), E. Paucha[c]
3a	139	W75	neg	neg		pos		Kalderon and Smith (1984)
	143–146	d12411	neg	neg			slight decr	C. Cole[c]
	147	SV80	neg?	decr			pos	M. Botchen[c], Kriegler *et al.* (1978), Gruss *et al.* (1984), Bradley *et al.* (1982)
	149	U24	neg	neg		incr		Kalderon and Smith (1984), E. Paucha[c]
	152,154	U19	neg	neg		incr		Kalderon and Smith (1984), E Paucha[c]
	153	pC6-2	neg	neg		pos	pos	Gluzman and Ahrens (1982)
	157	sp1030 SR/2	pos[d]	pos				Shortle *et al.* (1979)
	166	sp1030 SR/3	pos[d]	pos				Shortle *et al.* (1979)

3b	186	tsA3900[e]	neg		pos		Hutchinson *et al.* (1985)
	203	T22	neg	neg	pos	pos	Manos and Gluzman (1985)
	214	SVR9D	neg	neg			Stringer (1982)
	224	pC8-A	neg	pos	pos	pos	Manos and Gluzman (1985)
4	300	tsA30	neg	neg		neg	Seif *et al.* (1980a)
5	417	5030	decr		pos		K. Peden[c]
	422	tsA255	neg	neg	neg		Seif *et al.* (1980a)
	427	tsA209	neg	neg	neg		Seif *et al.* (1980a)
	438	tsA58	neg	neg	neg		Bourre and Sarasin (1983)
6	453	tsA1642	slight decr	pos	neg		Cosman and Tevethia (1981)
	(495)[f]	7axB	neg	pos			Hayday *et al.* (1983)
	(495)[f]	bc1071	neg	pos			Nilsson and Magnusson (1984)
	509	dl2462	neg	pos		decr	C. Cole[c]
	516	pC2	neg	pos	pos	pos	Manos and Gluzman (1985)
	522	pC11-A	neg	pos	pos	neg	Manos and Gluzman (1985)
7	549	pC11-B	decr	pos	pos	pos	Manos and Gluzman (1985)
	660 (stop codon)	pC8-B	decr	pos	pos	pos	Manos and Gluzman (1985)

[a] Hypothesized domains as discussed in Section II:I.
[b] All other mutants are assumed to be positive for nuclear localization. Some mutations of C-terminal phosphoserines (not listed) appeared to lose the ability to locate in the nucleus (E. Fanning, personal communication).
[c] Personal communication.
[d] A revertant selected for wild-type replication activity. The original mutant virus sp1030 was defective in replication and produced pinpoint plaques.
[e] Point mutant derived from double mutant tsA1499 (Pintel *et al.*, 1981).
[f] The mutation is actually in PyV amino acid 642 which corresponds to SV40 amino acid 495, within a region of homology between SV40 and PyV T-Ag (Fig. 2).

servations). Accordingly, when full-sized T-Ag was used in this experiment, adenylation was blocked.

So far the significance of adenylated T-Ag to viral DNA replication lies in its apparent location with respect to replication-defective mutants in T-Ag. However, a role for this complex is suggested by analogy with other replication proteins. For example, AMP may act indirectly as an allosteric regulator of T-Ag binding to the *ori* region as suggested for dnaB protein in *E. coli* DNA replication (Arai and Kornberg, 1981a). Alternatively, AMP may function directly as a primer for DNA synthesis as shown for the terminal protein–dCMP complex of adenovirus (Section IV.H; Chalberg and Kelly, 1979; Stillman *et al.*, 1981; Ikeda *et al.*, 1981; Desiderio and Kelly, 1981). Parvovirus DNA also has a terminal protein covalently attached to its 5′-end (Revie *et al.*, 1979). Since the noncovalent binding of T-Ag to viral DNA has not been sufficient to explain the replication function of the protein, perhaps adenylation and the enzyme activities associated with it will help define a replication initiator species of T-Ag.

E. Structural Characteristics

1. Size of the Peptide

Initial studies on the molecular weight of T-Ag's gave conflicting results. There are 708 amino acids in SV40 T-Ag and 785 in PyV T-Ag which suggested that the molecular weights would be approximately 81.5K and 89.0K, respectively. Both isolated proteins did appear to have the predicted molecular weight when analyzed on 6 M guanidine–Sepharose chromatography (Griffin *et al.*, 1978; Schaffhausen *et al.*, 1978). However, SDS–polyacrylamide gel electrophoresis resolved SV40 T-Ag at 88–94K and PyV T-Ag at 94–100K depending on the gel composition. Posttranslational modifications may account for some of this heterogeneity in molecular weight. *In vitro* translation of PyV early mRNA produced the faster-migrating T-Ag species (94K), but the size was shown to increase posttranslationally (Schaffhausen, 1983).

SV40 T-Ag often appeared as two bands on an SDS–polyacrylamide gel. This proteolytic digestion was once thought to be a specific modification of T-Ag (Tooze, 1981). However, it has been shown that most faster-migrating species of SV40 T-Ag were due to proteolytic digestion during the extraction of T-Ag from the cellular host (Tegtmeyer *et al.*, 1977; Smith *et al.*, 1978). A deletion of the 12–13 proline-rich amino acids at the C-terminus was sufficient to effect a migration change similar to the loss of 2–3K. In addition, a more amino-terminal modification or structure may also contribute to the slower migration of T-Ag in polyacrylamide gels, because a truncated T-Ag, coded for by the am404 mutation (at amino acid 538; Rawlins *et al.*, 1983), migrated more slowly than expected for its predicted molecular weight.

2. Oligomerization

The activities of T-Ag may be modulated by its ability to oligomerize (Türler, 1980; Fanning *et al.*, 1981, 1982; Greenspan and Carroll, 1981; Bradley *et al.*, 1982; Gidoni *et al.*, 1982). SV40 T-Ag has been examined by zonal sedimentation and three species of antigen have been identified. A light form of 5–7 S was labeled with a 15-min pulse of [^{35}S]methionine and then chased into a 15–16 S species that, in turn, became a 21–23 S complex (Tooze, 1981; Fanning *et al.*, 1981; Gidoni *et al.*, 1982; Montenarh and Henning, 1983). The heaviest species was associated with the cellular protein p53 (Lane and Crawford, 1979; see Section II.E.3). Electron microscopy and immunoprecipitation studies indicated that the 15–16 S species represented a tetrameric form of T-Ag (Myers *et al.*, 1981c; Bradley *et al.*, 1981). A 5.5 S form was correlated with monomeric T-Ag molecules, and a 7 S species was proposed to be a dimeric form (Bradley *et al.*, 1982).

Although the mechanism of oligomerization is not clear, a number of important observations have been reported. Heavier species were more highly phosphorylated than lighter species, and T-Ag from temperature-sensitive mutants in the A gene (tsA58) could not form the higher-order structures at the restrictive temperature while the previously formed aggregates remained intact (Fanning *et al.*, 1981; Greenspan and Carroll, 1981; Scheidtmann *et al.*, 1984a; Montenarh *et al.*, 1984). Oligomerization did not occur *in vitro* when the 5–7 S form was simply concentrated (Bradley *et al.*, 1982) or supplemented with divalent cations (Montenarh and Henning, 1983). However, after 15–16 S complexes were dissociated with EDTA into 5–7 S forms, reconstitution into the tetrameric species occurred upon addition of divalent cations (Montenarh and Henning, 1983).

There is some debate as to the site of oligomerization in the cell. Lanford and Butel (1984) observed that only half of the antigen found in the nucleus (synthesized in a 15-min, radioactive pulse) was isolated in the 15–16 S form. They proposed that T-Ag might remain in the cytoplasm as a result of premature oligomerization.

The smaller forms of SV40 T-Ag had a higher specific activity for binding to the origin of SV40 DNA than did the larger species (Gidoni *et al.*, 1982; Dorn *et al.*, 1982; Fanning *et al.*, 1982; Burger and Fanning, 1983; Scheidtmann *et al.*, 1984a). Furthermore, a 7 S form of T-Ag was shown to have ATPase and DNA binding activities, while the 5.5 S form of T-Ag did not (Bradley *et al.*, 1982). Other reports (Scheidtmann *et al.*, 1984a; R. Hand, personal communication) that monomeric (5.5 S) T-Ag also possessed these activities did not exclude the possibility that the monomeric form was converted into dimeric (7 S) T-Ag under their assay conditions.

3. Association with p53

SV40 T-Ag has been found to bind to the cellular protein p53 (Lane and Crawford, 1979; Linzer and Levine, 1979; Kress *et al.*, 1979; Melero *et al.*, 1979; DeLeo *et al.*, 1979; McCormick and Harlow, 1980) 1–2 hr after its own synthesis (Gurney *et al.*, 1980; Carroll and Gurney, 1982), and T-Ag could form the complex *in vitro* (McCormick *et al.*, 1981). T-Ag stimulated the synthesis of p53 (Harlow *et al.*, 1981b) and the formation of the T-Ag/p53 complex was responsible for increasing the half-life of p53 (Oren *et al.*, 1981; Reich *et al.*, 1983). The complex formed in the nucleoplasm and in the nuclear matrix, but not with the T-Ag associated with chromatin (Staufenbiel and Deppert, 1983) even though the T-Ag/53 complex could bind to viral DNA *in vitro* (Reich and Levine, 1982). The specific DNA binding activity of the T-Ag/p53 multimer was lower than that of T-Ag alone (Burger and Fanning, 1983). Complexes of p53 with T-Ag may also exist at the cell surface (Santos and Butel, 1984).

Two lines of evidence suggest that p53 is involved in the regulation of cell proliferation. First, microinjection of antibodies against p53 into cell nuclei prevented the onset of S phase (Mercer *et al.*, 1982). Second, induction of cell transformation by a variety of agents first increased the intracellular level of p53 (Rotter, 1983). Although truncated T-Ag (McCormick *et al.*, 1981; Galanti *et al.*, 1981) and tsA58 T-Ag at the restrictive temperature (Montenarh *et al.*, 1984) were not associated with p53 and failed to initiate SV40 DNA replication, these T-Ag proteins were still able to induce cellular DNA synthesis (Section II.C). Perhaps T-Ag represents a homologue of the cellular protein, as suggested by the isolation of monoclonal antibodies against p53 that cross-react with oligomers of T-Ag (Leppard and Crawford, 1984).

4. Nuclear Localization Sequence

Lanford and Butel (1980a,b) originally demonstrated that an N-terminal sequence of SV40 T-Ag was required for T-Ag localization in the nucleus. More recently, a point mutation that converted Lys^{128} to an Asn (cT-Ag) prevented nuclear accumulation of T-Ag (Lanford and Butel, 1984). Kalderon *et al.* (1984b) have shown that the SV40 amino acid sequence Pro-Lys-Lys-Lys-Arg-Lys-Val (amino acids 126–132) which surrounds Lys^{128}, is required for T-Ag to accumulate in the nucleus of the host cell after synthesis (mutant d10; Kalderon *et al.*, 1984a). This short peptide could be fused to a cytoplasmic protein, and the majority of the fused peptide was found in the nucleus. The nuclear accumulation process was far less efficient if certain of these amino acids were changed and the process was totally disrupted if Lys^{128} was altered (e.g., to Thr). An interaction with some transport protein may depend upon the integrity of this peptide sequence and a hydrophilic, local secondary structure.

PyV T-Ag amino acids near Lys^{282} have some homology with the SV40 T-Ag signal sequence that might form the required secondary structure necessary for nuclear localization. Although Nilsson and Magnusson (1984) have identified a PyV mutant T-Ag missing amino acids 191–209 (d12208) that was defective in its ability to enter the nucleus, the actual minimal sequence for PyV T-Ag nuclear localization has yet to be identified.

A defective nuclear localization signal was dominant over the wt T-Ag signal. That is, the SV40 mutant cT-Ag appeared to inhibit the transport of wt T-Ag to the nucleus when both were present in the same cell, unless the concentration of wt T-Ag was significantly greater than cT-Ag (Lanford and Butel, 1980b; Scheidtmann *et al.*, 1984b). wt T-Ag's truncated at the C-terminus (33K or 70K) were also prevented from entering the nucleus if cT-Ag was present in sufficient concentration (Kalderon and Smith, 1984). Lanford and Butel (1984) postulated that cT-Ag may saturate a cellular protein necessary for transport, or that premature oligomerization occurs as cT-Ag accumulates in the cytoplasm. Interestingly, cT-Ag was underphosphorylated compared to wt T-Ag (Jarvis *et al.*, 1984), and certain truncated T-Ag's (less than ca. 620 amino acids) with defective nuclear localization signals diffused into the nucleus without accumulating in either the cytoplasmic or the nuclear compartment (Kalderon *et al.*, 1984b). In yeast, a similar localization sequence appears on histones, but in contrast to t-Ag its presence had a dominant role over the defective signal (B. Moreland, personal communication).

Since T-Ag must be present in the nucleus for viral or cellular replication to occur, mutations in the nuclear localization sequence can prevent T-Ag-dependent DNA replication. Using a monkey cell line which constitutively produces SV40 T-Ag (COS cells; Gluzman, 1981), one can introduce the gene for cT-Ag into the cells and deplete the concentration of wt T-Ag in the nucleus, resulting in the eventual inhibition of DNA replication (J. Butel, personal communication). For PyV, cotransfection of wt PyV and d12208 DNA's resulted in a decrease in viral DNA replication (Nilsson and Magnusson, 1984).

5. Association with RNA

A variety of RNA sequences have been identified in association with purified SV40 T-Ag (Khandjian *et al.*, 1982; Darlix *et al.*, 1984). About 10% of the total cellular T-Ag may be associated with messenger ribonucleoprotein particles (mRNPs; Michel and Schwyzer, 1982) that could not be immunoprecipitated with a specific monoclonal antibody against T-Ag (pAb402), suggesting a specific binding site on T-Ag for mRNPs (Schwyzer *et al.*, 1983).

F. T-Ag Binding to Cellular and Viral DNA

1. Cellular DNA

T-Ag binds nonspecifically to many cellular DNA sequences. Calf thymus DNA linked to cellulose (DNA cellulose) has been used to analyze the binding affinity of T-Ag to cellular DNA (Oren *et al.*, 1980; Tooze, 1981). "Loose binding" was defined by Prives *et al.* (1982) as SV40 T-Ag that bound to DNA cellulose at pH 6 in low concentrations of salt. Bound T-Ag could be eluted by increases in salt and/or pH. t-Ag (which shares amino acids 1–82 with T-Ag; Fig. 2) did *not* bind calf thymus DNA (Prives *et al.*, 1980), and several, N-terminally truncated fragments of T-Ag coded for by the Ad2^{+}ND hybrid viruses, did bind. Although it was determined that amino acids on the N-terminal side of SV40 T-Ag residue 333 were required for "loose binding," *only* wt T-Ag bound with 100% efficiency. Analyses of C-terminal truncations of SV40 T-Ag by Chaudry *et al.* (1982a,b) showed that peptide fragments extending only to 272 could bind to DNA cellulose at pH 6, and that shorter, phosphorylated 15K and 22K N-terminal fragments of T-Ag could not form the loose complex. However, the 17K tryptic peptide extending to Lys^{130}, has been shown to bind to DNA cellulose at pH 6 as long as it was not phosphorylated (Morrison *et al.*, 1983). These data were consistent with those of Scheidtmann *et al.* (1984a) that described a higher affinity of SV40 T-Ag for DNA at a lower level of phosphorylation. Therefore, the region of SV40 T-Ag from amino acids 83 to 333 has the ability to interact weakly with DNA. This region which gives nonspecific or "loose binding" includes the domain required for specific T-Ag/DNA binding as determined by mutational analyses (domain 3; Section I). Smaller peptide fragments within the region might have some independent affinity for cellular DNA.

A "tight binding" fraction of SV40 T-Ag was defined as T-Ag that bound to DNA cellulose at pH 7 (Prives *et al.*, 1982). Only about 10% of full-sized antigen bound to DNA cellulose under these conditions, and the complex could not be dissociated without the addition of high concentrations of salt ($>$ 0.8 M). In addition, under the same conditions, SV40 T-Ag could bind to dsDNA as well as ssDNA cellulose (Spillman *et al.*, 1979). Furthermore, newly synthesized T-Ag ($<$ 2 hr old) bound to calf thymus DNA at pH 7 with a higher affinity than did older T-Ag (Oren *et al.*, 1980), and 5–7 S forms of SV40 T-Ag bound to calf thymus DNA more tightly than did the higher-order oligomers (Gidoni *et al.*, 1982). Therefore, a small fraction of full-sized, newly synthesized, monomeric or dimeric SV40 T-Ag binds strongly to DNA, apparently without regard for DNA sequence specificity.

Specific binding of T-Ag to the origin region of viral DNA ("site-specific binding") could be differentiated from "tight binding" of T-Ag to calf thymus DNA. Mutant T-Ag's, such as that coded for by SV40 pC6 (Table I), bound to DNA cellulose at pH 7 as tightly as wt T-Ag, but the mutant T-Ag did not bind to viral origin-containing DNA fragments in an immunoprecipitation assay (Scheller *et al.*, 1982). pC6 T-Ag did form complexes with cellular p53, transform animal cells, and stimulate cellular DNA synthesis. However, it was unable to initiate viral DNA replication (Manos and Gluzman, 1985). Therefore, the "site-specific binding" is required for initiation of viral DNA synthesis, while the "loose" and/or "tight" interactions with DNA may be sufficient for other functions associated with T-Ag.

2. SV40 DNA

Figures 3 and 4 provide a summary of the experimental evidence that defines the regions of SV40 DNA which are specifically bound by T-Ag. The DNA binding sites have been named 1, 2, and 3 with site 1 proximal to the ATG codon for early mRNA transcription (Fig. 4; Tjian, 1978). Dependent upon the assay employed and the type of T-Ag studied, the actual boundaries of each of the sites varied extensively. Figure 4 shows the broadest boundaries for wt T-Ag binding to SV40 DNA and for historical reasons the original boundaries as determined by D2 T-Ag binding. T-Ag/DNA affinity constants of 10^{-12} M for SV40 and 10^{-11} M for PyV have been reported (Jessel *et al.*, 1976; Gaudray *et al.*, 1981). There is an origin core within the binding region corresponding to SV40 site 2, which has been shown to be the minimal DNA sequence necessary for viral DNA replication *in vivo* (Section III.B). It has been established that specific binding of T-Ag to the origin region was required for initiation of viral DNA replication *in vivo.* Shortle and Nathans (1979) made mutations in the origin region of SV40 DNA (site 2) that produced virus which was partially defective in its ability to replicate viral DNA. They then mutagenized the DNA and selected second-site revertants for virus that could replicate as well as wild type (Shortle *et al.*, 1979). The mutants recovered which suppressed the *ori* defects were mapped to the gene for T-Ag to within the T-Ag domain described above for nonspecific DNA binding (sp1030 SR/2 and SR/3, Table I). These mutant T-Ag's were shown to have relaxed site-specific DNA binding to the origin sequences, therefore indicating that T-Ag interactions with viral DNA were essential to the initiation of viral DNA synthesis (Margolskee and Nathans, 1984).

Different assays have been used to map the T-Ag binding sites on SV40 DNA and to ascertain the relative binding affinities (Fig. 3). DNase I protection experiments have produced detailed maps of T-Ag protected bases, and have identified those bases which became hypersensitive to DNase I after the T-Ag/DNA complex was formed (Tjian, 1978; Myers *et al.*, 1981a; Tenen *et al.*, 1982, 1983b; DeLucia *et al.*, 1983; Tegtmeyer *et al.*, 1983; R. Dixon, personal communication). Fragment analyses, which included isolation of DNA fragments after treatment with DNase I or restriction enzymes in the presence of T-Ag (Tjian, 1978; Myers and Tjian, 1980; Tenen *et al.*, 1983b; Tegtmeyer *et al.*, 1983), yielded similar results

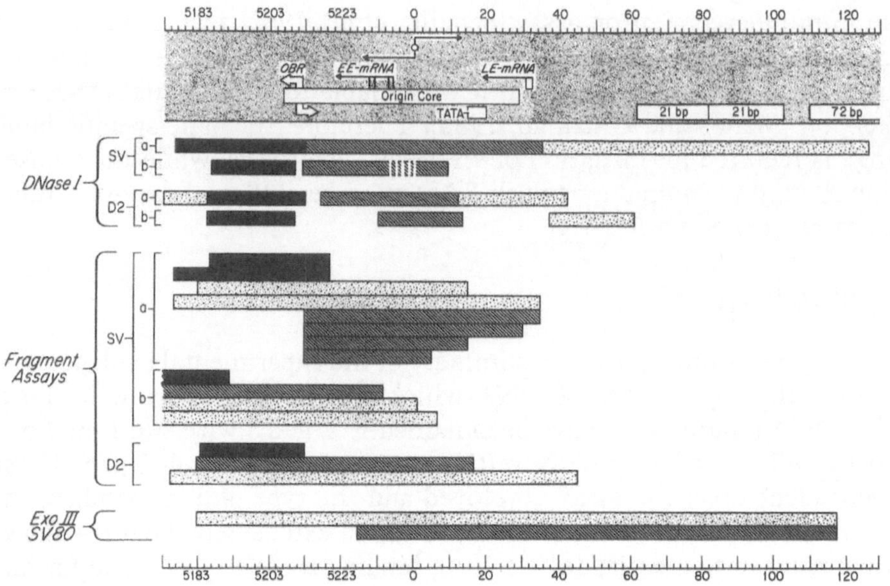

FIGURE 3. SV40 T-antigen binding domains. *DNase I:* Those SV40 nucleotides that are protected by T-Ag in the presence of DNase I are depicted by shaded bars. Solid black bars indicate the highest-affinity binding site which requires the minimum concentration of T-Ag for protection (site 1), cross-hatched bars indicate the next set of lower-affinity binding sites (site 2), and speckled bars have the lowest affinity for T-Ag (site 3). Studies have been performed with SV40 T-Ag [SV(a), Tegtmeyer *et al.* (1981), DeLucia *et al.* (1983), Tenen *et al.* (1983b); SV(b), R. Dixon (personal communication)], or with D2 T-Ag [D2(a), Tenen *et al.* (1983a); D2(b), Tjian (1978), Myers and Tjian (1980)]. *Fragment assays:* Fragment assays include incubation of specific, restriction enzyme DNA fragments with T-Ag, digestion of the complex with DNase I, and then recovery of the protected DNA fragment [SV(a), Tegtmeyer *et al.* (1983); D2, Tjian (1978), Myers and Tjian (1980)], and recovery of DNA fragments protected from restriction endonuclease digestion (*Hae*III) by the presence of T-Ag [SV(b), Tenen *et al.* (1983b)]. Solid black bars are fragments recovered at the lowest concentrations of T-Ag, and speckled bars indicate fragments recovered at the highest T-Ag concentrations tested. *ExoIII SV80:* This result refers to the SV40 origin region protected from digestion with *E. coli* exonuclease III by T-Ag from SV80 cells at low (cross-hatched bar) and high (speckled bar) concentrations (Shalloway *et al.*, 1980). Scales at the top and bottom of the figure provide the nucleotide map positions. The minimal origin of replication (*ori*-core, Figs. 7, 10), the origin of bidirectional replication (OBR, Figs. 15, 16A, 17), early mRNA start sites observed early after infection (EE-mRNA), early mRNA start sites observed late after infection (LE-mRNA, Ghosh and Lebowitz, 1981; Hansen *et al.*, 1981), TATA box (Fig. 4), 27-bp palindromic sequence (Fig. 10), and 21-bp repeated sequences are indicated as points of reference.

and additional information on relative binding affinities. Lastly, origin DNA could be protected from digestion with exonuclease III in the presence of T-Ag from SV80 cells (Shalloway *et al.*, 1980).

Not included in Fig. 3 are DNA immunoprecipitation assays (McKay and DiMaio, 1981; McKay, 1981, 1983). This assay was dependent upon

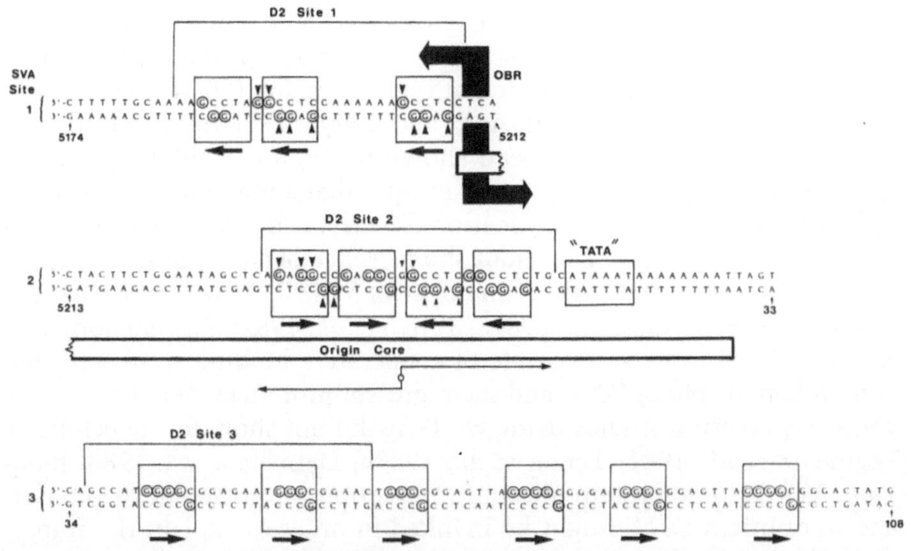

FIGURE 4. Proposed DNA contact points for SV40 T-antigen. Three DNA binding sites for wt T-Ag (SVA sites 1, 2, and 3) based on data in Fig. 3 (DNase I, SV, a) and D2 T-Ag (D2 sites 1, 2, and 3) based on data in Fig. 3 (DNase I, D2, b). Guanines protected by bound T-Ag against methylation have been circled (Tjian, 1978; DeLucia *et al.*, 1983; Cohen *et al.*, 1984). Guanines whose methylation interferes with T-Ag binding (Jones and Tjian, 1984) have been indicated by arrowheads (large arrowheads indicate strong sites and small arrowheads indicate weak sites). The pentanucleotide sequence 5′-(G > T)(A > G)GGC-3′ (boxed sequences) has been proposed as the recognition sequences for T-Ag (Tjian, 1978; Tegtmeyer *et al.*, 1981; DeLucia *et al.*, 1983). Arrows underneath boxes indicate polarity of the pentanucleotide. The *ori*-core, origin of bidirectional replication (OBR), TATA box and 27-bp palindrome are also indicated.

the antibody used and whether it was bound to T-Ag before or after addition of DNA (Scheller *et al.*, 1982; E. Paucha, personal communication; M. Bradley, unpublished observations). Also, Tenen *et al.* (1983b) observed changes in the binding activity of wt T-Ag to origin DNA fragments using this assay if T-Ag was preincubated before the addition of DNA. These variables in the experimental data have made it difficult to compare the findings among different investigators.

T-Ag binds to site 1 with the highest affinity (sequences defined by solid bars, Fig. 3; Tjian, 1978; Myers *et al.*, 1981a; Tenen *et al.*, 1982, 1983b; DeLucia *et al.*, 1983; Jones and Tjian, 1984). Site 1 was not absolutely required for replication of viral DNA *in vivo*, but it was required for T-Ag control of early transcription (Rio *et al.*, 1980; Hansen *et al.*, 1981). Sizeable deletions in site 1 resulted in mutant viruses that were cold sensitive for replication, although point mutations had no effect (DiMaio and Nathans, 1980, 1982). Therefore, this site as a whole could have an important relationship to replication. Data from SV40 DNA replication studies performed *in vitro* have shown that point mutations in site 1 had no effect on the efficiency of viral DNA replication as would be predicted from the *in vivo* data, but the DNA's of the cold-sensitive (cs) mutants did not support efficient replication (cs1097 replication was 3% of that supported by wt DNA; Li and Kelly, 1985). *In vitro* initiation of SV40 DNA synthesis has been shown to require SV40 T-Ag (Li and Kelly, 1984), and therefore these data imply that a major disruption of T-Ag binding to site 1 affects replication directly. It has been observed that T-Ag could not bind site 2 without the presence of site 1; however, these studies were performed with an analogue of SV40 T-Ag from the Ad2$^+$D2 virus (D2 T-Ag; Myers *et al.*, 1981a; Tenen *et al.*, 1982). The phenomenon was considered to be an example of cooperative binding similar to that seen for lambda phage DNA and the repressor protein (Sauer *et al.*, 1979). DNase I protection studies using wt T-Ag did not show this dependence (Tegmeyer *et al.*, 1981; Tenen *et al.*, 1983b; DeLucia *et al.*, 1983; Jones *et al.*, 1984). Some protein–protein interaction or change in the conformation of origin DNA might be invoked in order to explain the importance of site 1 binding, but its exact nature will require further analyses.

Site 2 is a weaker but independent binding site for wt T-Ag that is required for viral DNA replication (sequences defined by cross-hatched bars minus solid bars, Fig. 3; Myers and Tjian, 1980; Tegtmeyer *et al.*, 1981, 1983; DiMaio and Nathans, 1982; Tenen *et al.*, 1983b; Jones *et al.*, 1984). Point mutations in the site destroyed the replication function of T-Ag (Fig. 10). It can be divided into two domains: 2A (proximal to site 1), and 2B (proximal to site 3). T-Ag bound to site 2A with a higher affinity than to site 2B under certain conditions of DNase I analyses (Tegtmeyer *et al.*, 1983; R. Dixon, personal communication). This was confirmed by DNA fragment analyses (Tenen *et al.*, 1983b), and by immunoprecipitation of DNA/T-Ag complexes (McKay and DiMaio, 1981; Scheller *et al.*, 1982). The division of site 2 into two components was also consistent with differences in the interference with T-Ag binding caused by DNA methylation (Fig. 4). Stronger protection occurs to the early side of nucleotide 5238 (larger arrows, Fig. 4; Jones and Tjian, 1984), corresponding to the border between sites 2A and 2B. Note that start sites for early viral mRNA transcription were mapped to site 2A (EE-mRNA, Fig. 3; Hansen *et al.*, 1981). But more relevant to the requirement for site 2 in replication is that two prevalent *in vivo* initiation sites for newly synthesized SV40 DNA were mapped to site 2B (nucleotides 4–5, and 12, Figs. 10 and 16; Hay and DePamphilis, 1982).

Site 3 has the lowest affinity for T-Ag and its boundaries are strongly influenced by salt concentrations (sequences defined by speckled bars minus solid and cross-hatched bars, Fig. 3; Tjian, 1978; DeLucia *et al.*, 1983). Portions of site 3 facilitated SV40 DNA replication *in vivo* (Bergsma *et al.*, 1982), although they were not needed for the minimum replication function. *In vitro,* replication was not affected by site 3 (Li and Kelly, 1985). The significance of T-Ag binding to site 3 has not been determined and the *in vivo* case may involve host protein interactions with T-Ag or with site 3 DNA.

A pentanucleotide consensus sequence 5′-GAGGC-3′ has been described as central to T-Ag/DNA binding sites (Tjian, 1978; Tegtmeyer *et al.*, 1983). Although the number and orientation of pentanucleotides were different in each of the three sites (Fig. 4), changes in the spacing of the pentanucleotides did affect the affinity of T-Ag binding to DNA (Lewton *et al.*, 1984; Cohen *et al.*, 1984). Since cellular DNA, containing an identical arrangement of pentanucleotides, had a lower affinity for T-Ag than viral DNA (Wright *et al.*, 1984), site-specific binding of T-Ag must involve more of the DNA sequence than that encompassed by the pentanucleotides. Furthermore, Conrad and Botchan (1982) selected cellular DNA for its ability to bind T-Ag, and although the fragments isolated were rich in G and C nucleotides, they did not contain pentanucleoides. It has also been found that formation of stable cruciform structures at the palindrome in the origin disrupted T-Ag binding *in vitro* (Tenen *et al.*, 1985), and some mutations in the origin region that did not affect the binding of T-Ag apparently disrupted some chromatin structure necessary for initiation (DiMaio and Nathans, 1982; Innis and Scott, 1984). Therefore, the initiation of viral DNA replication requires more than binding of T-Ag on a particular set of pentanucleotide sequences on a preformed structure (Tegtmeyer *et al.*, 1983).

3. PyV DNA

Figure 5 summarizes the data for PyV T-Ag binding to the origin region of PyV DNA. DNA fragment-recovery assays mapped T-Ag binding sites to the early side of the minimal origin of replication (Fig. 5A; Gaudray *et al.*, 1981; Pomerantz *et al.*, 1983; Cowie and Kamen, 1984), and DNase I protection studies have defined the borders of the highest affinity binding sites at A, B, and C on the early side of the *ori* core sequences (Fig. 5B; Dilworth *et al.*, 1984; Cowie and Kamen, 1984). Pomerantz and Hassell (1984) showed a requirement for the consensus pentanucleotide sequence 5′-GAGGC-3′ within the binding sites, as they appeared in sites A, B, and C (bold arrows). Lower-affinity binding sites have been defined as 1, 2, and 3 (Cowie and Kamen, 1984).

The PyV origin core is homologous to the origin core of SV40 (Figs. 8 and 9; Section III.B), and includes a palindromic sequence and a similar arrangement of pentanucleotide consensus sequences to that found in SV40 T-Ag/DNA binding site 2 (site 1 in Fig. 5B). PyV DNA sequences required for initiation of viral DNA replication included the origin core plus one of two sequences to the late side known as alpha and beta (Pomerantz *et al.*, 1983). Binding sites A, B, and C were not required for replication, although they may have some effect on the efficiency. Site 3 is somewhat similar to SV40 site 3 in its location and low affinity for T-Ag. The significance of these binding sites on the late side of *ori* remains

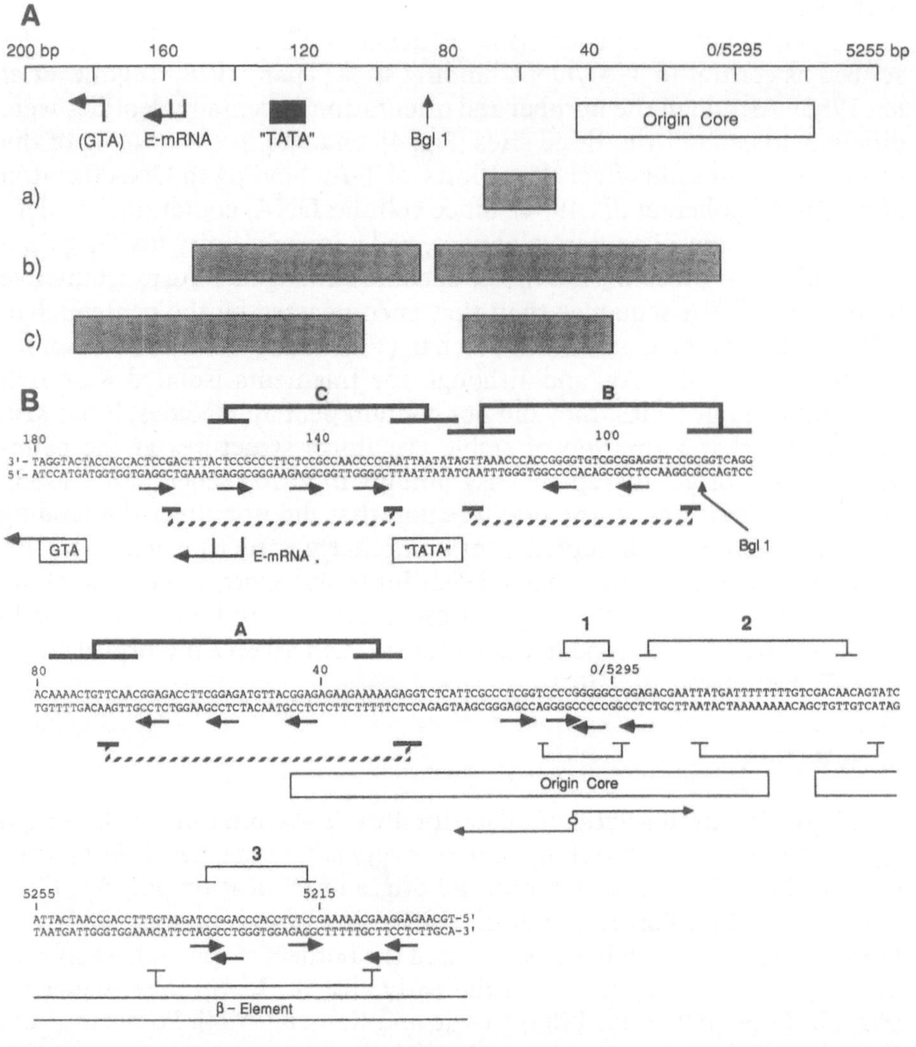

FIGURE 5. DNA binding sites for PyV T-antigen. (A) DNA fragments specifically bound to PyV T-Ag were identified by their ability to elute bound T-Ag from PyV DNA (line a, Gaudray *et al.*, 1981), and by immunoprecipitation of T-Ag-bound PyV DNA restriction fragments (line b, Pomerantz *et al.*, 1983; line c, Cowie and Kamen, 1984). The PyV genome is orientated with early genes on the left and late genes on the right so that it can be compared with SV40 data (Figs. 3, 4). The *ori*-core, early mRNA start sites, TATA box, and T-Ag start codon are indicated with distances given in PyV base pairs. (B) DNase I footprinting analyses were used to identify sequences bound to T-Ag. The strongest T-Ag binding sites are designated A, B, and C. The average of two sets of data is shown on the 3′–5′ strand (heavy brackets, Dilworth *et al.*, 1984; Cowie and Kamen, 1984); only one set of data was available for the 5′–3′ strand (dashed brackets, Cowie and Kamen, 1984). Lower-affinity sites 1, 2, and 3 identified by Cowie and Kamen (1984) are also shown (light brackets). Hexanucleotide sequences 80% consistent with 5′-GAGGC-3′ (the common element in PyV T-Ag binding, Pomerantz and Hassell, 1984) are indicated by arrows. The initiation codon for T-Ag, initiation sites for early mRNA, TATA box, *BglI* restriction site, *ori*-core, 34-bp palindrome, and β-element are also indicated.

unclear. T-Ag control of cellular replication and transcription factor binding to this region has been suggested (Chapter 2; DePamphilis and Wassarman, 1982; Pomerantz *et al.*, 1983).

There are some data to support a T-Ag specific replication activity in addition to T-Ag/DNA binding. A mutant PyV T-Ag was isolated that repressed replication induced by wt T-Ag (bc1071, Table I; Nilsson and Magnusson, 1984). Since the mutant was able to bind specifically to PyV DNA, the defect was in some other T-Ag activity. Defective T-Ag might have bound to the entire origin region thus blocking wt T-Ag interactions, and/or defective T-Ag might have formed mixed oligomers with the wt T-Ag resulting in inactive complexes.

4. Virus-Specific Interactions between T-Ag and DNA

Interactions of T-Ag with *ori* are specific to the different papovaviruses even though their *ori* regions are remarkably similar (Fig. 10), and their T-Ag amino acid sequences are quite homologous (e.g., SV40 and PyV T-Ag's; Fig. 2). However, there does exist cross-reactive T-Ag/DNA binding. SV40 T-Ag bound to BKV origin DNA (Ryder *et al.*, 1983) and PyV origin DNA (Pomerantz and Hassell, 1984), but not to the origin DNA from the human papovavirus JCV (Frisque, 1983). PyV T-Ag likewise bound to SV40 origin DNA (Pomerantz and Hassell, 1984; Scheller and Prives, 1985). There is even some evidence for a functional cross-reaction between SV40 T-Ag and human adenovirus origin DNA (Rabek *et al.*, 1981; Goldman *et al.*, 1981b).

Nevertheless, T-Ag and DNA did show virus-specificity for initiation of viral DNA replication. Although both BKV and JCV DNA could be replicated *in vitro* with the addition of purified SV40 T-Ag (Li and Kelly, 1985), the results obtained from *in vivo* studies of viral DNA replication were not completely consistent with these findings. BKV T-Ag did complement the tsA58 mutant of SV40 (Mason and Takemoto, 1976), and human cells expressing SV40 T-Ag replicated BKV DNA (Major and Matsumura, 1984). However, initiation of the replication of JCV DNA in human cells expressing SV40 T-Ag was not detected (Major and Matsumura, 1984), nor did mouse cells expressing SV40 T-Ag support the replication of PyV DNA (Pomerantz and Hassell, 1984). Perhaps these deficiencies were due to subtle differences in the T-Ag binding to similar but not virus-specific DNA sequences, or to the absence of species-specific T-Ag interactions with host factors. For example, extracts prepared from human cells replicated SV40 DNA *in vitro* with the addition of SV40 T-Ag, but extracts from mouse cells were unable to replicate any detectable SV40 DNA in this system (Li and Kelly, 1985). But interestingly, SV40 T-Ag made in a mouse cell did support SV40 DNA replication in a monkey cell extract (B. Weiner and M. Bradley, unpublished observations).

There has also been reported an interference between nonhomologous T-Ag's. Coinfection of tsA58 SV40 with wt BKV repressed the replication of BKV DNA (Lai *et al.*, 1979). Therefore, even though wt T-Ag could complement a defective T-Ag, the latter could also upset some balance necessary for the initiation of viral DNA synthesis.

5. Binding to Viral Chromatin

T-Ag has been found in association with replicating viral chromatin. Specifically, SV40 T-Ag was immunoprecipitated in a complex with SV40 replicating chromosomes and that complex was found to be missing in tsA mutants at the restrictive temperature (Mann and Hunter, 1979). T-Ag was found in association with the nucleosome-free origin region of mature SV40 chromosomes (Reiser *et al.*, 1980; Harper *et al.*, 1984). It was proposed that T-Ag was recycled from mature chromatin (Segawa *et al.*, 1980) or that T-Ag bound to the nuclear matrix was active in replication (Buckler-White *et al.*, 1980). However, most SV40 T-Ag was found to be associated with early replicating chromosomes (Segawa *et al.*, 1980; Tack and DePamphilis, 1983; Milthorp *et al.*, 1984), i.e., within 60 min of initiation of viral DNA replication, and it was dissociated after 70% of the DNA strands were replicated (Tack and DePamphilis, 1983). Newly synthesized T-Ag could be chased from chromatin after 2 hr to accumulate in the nuclear matrix or nucleoplasm (Staufenbiel and Deppert, 1983). Although some newly synthesized T-Ag did enter the nuclear matrix directly, that association appeared not to be important for the replication function (Covey *et al.*, 1984).

A complex of T-Ag with early replicating chromosomes that was stable in high salt has also been described (0.5–1 M; Reiser *et al.*, 1980; Rudolph and Mann, 1983; Stahl *et al.*, 1983; Clertant *et al.*, 1984a). In contrast, T-Ag dissociated from mature chromatin in 0.2–0.3 M salt, and parallel studies *in vitro* showed that SV40 T-Ag binding to naked viral DNA could be dissociated in 0.2–0.3 M salt (Stahl *et al.*, 1983). The salt-resistant T-Ag complex with chromatin appeared to be associated with replication forks. This T-Ag was bound to DNA sequences throughout the genome, and a monoclonal antibody against T-Ag (pAb1630; Ball *et al.*, 1984) inhibited the continuation of SV40 replication *in vitro* (Stahl and Knippers, 1983; Stahl *et al.*, 1985).

In support of this hypothesis, it is worth noting that the properties of the T-Ag that remains associated with replicating chromatin are similar to the properties of that species of T-Ag most likely to be involved in replication. Although the phosphorylation pattern of T-Ag isolated from SV40 nucleoprotein complexes was not unique (Mann and Hunter, 1980), Scheidtmann *et al.* (1984a) found that SV40 T-Ag extracted from nuclei in 0.5 M salt was equivalent to newly synthesized, cytoplasmic T-Ag in its phosphorylation pattern and its ability to bind viral DNA. Furthermore, Chen and Hsu (1983) found that SV40 chromatin contained newly synthesized T-Ag that could be additionally phosphorylated *in vitro* and labeled with the affinity reagent dialdehyde-ATP, and Clertant *et al.* (1984a) showed T-Ag associated ATPase in PyV chromatin complexes isolated in high salt. SV40 T-Ag can be extracted from cells at high salt, but the ability to assay T-Ag activity *in vitro* for DNA binding, ATPase, and adenylation decreased as the salt was increased above 0.5 M (D. Tenen and M. Bradley, unpublished observations). If T-Ag cannot be recycled once it has dissociated from the replicating chromosome, it is possible that only the prereplicative species of T-Ag can be purified and tested *in vitro*.

What then is the relationship between T-Ag binding to the *ori* region and its ability to initiate viral DNA replication? Tegtmeyer *et al.* (1983) have suggested that high-affinity binding of T-Ag to DNA may interfere with the replication function. Scheidtmann *et al.* (1984a) have endorsed that proposal and added that a primary, tight interaction could be modified over time by the phosphorylation of N-terminal residues on T-Ag, resulting in a decrease in the binding affinity of the antigen for DNA, and perhaps in an activation of the T-Ag replication function.

G. *In Vitro* Associated Enzyme Activities

Enzyme activities of T-Ag may reside either in the molecule itself or in another protein bound to T-Ag. Some host proteins are associated with T-Ag such as p53 (Section II.E.3), and have not been associated with any enzyme activity as yet (van Roy *et al.*, 1984). A 68K peptide from infected monkey cells shares antigenic sites with SV40 T-Ag, as defined by the monoclonal antibody pAb204 (Lane and Hoeffler, 1980), and another protein of approximately 90K was copurified with SV40 T-Ag by Tegtmeyer and Anderson (1981). Although some preparations of T-Ag contained topoisomerase activity (Giacherio and Hager, 1980), further purification apparently removed it (Myers *et al.*, 1981c).

1. ATPase

Phosphohydrolase activity, releasing free phosphate and adenosine diphosphate from ATP in the presence of magnesium, was found to be closely associated with T-Ag of both SV40 and PyV (Tjian and Robbins, 1979; Giacherio and Hager, 1979; Griffin *et al.*, 1979; Gaudray *et al.*, 1980). ATP and dATP were the preferred substrates (Giacherio and Hager, 1979; Clark *et al.*, 1983), and the enzyme activity did not depend upon DNA but was stimulated by poly(dT) under certain assay conditions (Giacherio and Hager, 1979). A ssDNA-dependent ATPase has been isolated from SV40 chromatin, but it could be clearly distinguished from the viral enzyme (approximately 60K; Brewer *et al.*, 1983). tsA mutants of SV40 and PyV were inactive in ATPase activity at the restrictive temperature (J. D. Griffin *et al.*, 1980; Gaudray *et al.*, 1980), and this activity has not been dissociated from the replication function as yet (Clark *et al.*, 1983; Manos and Gluzman, 1984, 1985).

ATPase activity was considered to be intrinsic to T-Ag since monoclonal antibodies to SV40 T-Ag have been isolated that blocked T-Ag/ATPase [Fig. 2, pAb204 (Clark *et al.*, 1983), pAb402 (E. Harlow, personal communication)]. Attempts to map the active site of the ATPase with an affinity label (dialdehyde-ATP) have identified an adenylation site (Clertant and Cuzin, 1982; Bradley *et al.*, 1984), but recent reports showed that site to be an unlikely candidate for the active site of ATPase (Section II.D.7; Clertant *et al.*, 1984b). Three possibilities remain: (1) the site turns over much more rapidly than that for adenylation and has remained undetected, (2) the site is located on a tightly associated host protein whose interaction with T-Ag is disrupted by monoclonal antibodies to a specific epitope, or (3) the site is constituted by more than one molecule of T-Ag (a dimer; Bradley *et al.*, 1982). ATPase has been reconstituted from a monomeric species of T-Ag isolated by polyacrylamide gel electrophoresis under nondenaturing conditions (R. Hand, personal communication), but only the dimeric (7 S) and tetrameric (15.5 S) species of T-Ag separated by zonal sedimentation were active as ATPase (Bradley *et al.*, 1982). So far the evidence strongly suggests that the ATPase activity is intrinsic to T-Ag.

2. Protein Kinase

The ability to transfer phosphate from ATP to amino acids on T-Ag itself or to an exogenous substrate has been attributed to SV40 T-Ag (Tjian and Robbins, 1979; Griffin *et al.*, 1979; Baumann and Hand, 1979). Later, some investigators found that protein kinase activity did not copurify with the antigen (Giacherio and Hager, 1979), while others found that the kinase activity was affected by tsA mutations of T-Ag and that some kinase activity remained associated with T-Ag upon purification (Griffin *et al.*, 1979; Bradley *et al.*, 1981). No monoclonal antibody to T-Ag has been found that blocks kinase activity directly, such as pAb204, which blocked the T-Ag/ATPase activity (Clark *et al.*, 1983). It is clear that the kinase activity associated with T-Ag of SV40 and PyV modifies serine and threonine residues only, and is distinct from the kinase activity associated with PyV mT-Ag (Schaffhausen, 1983).

The pattern of phosphorylation has been shown to affect the ability of T-Ag to bind DNA and has been correlated with specific species of T-Ag (Sections II.D.1 and II.E.2). The higher-order oligomers of T-Ag (> 21 S) are more heavily phosphorylated, and contain the phosphorylated protein p53 (Jay *et al.*, 1981a). The only strong piece of evidence that suggests that the kinase is an intrinsic property of T-Ag was reported by van Roy *et al.* (1984); the phosphorylation of p53 could be blocked by the addition of antibodies to T-Ag, suggesting that a T-Ag associated kinase activity was directly involved in that modification.

3. Nucleotidyl Transferase

There is some evidence for an association of T-Ag with an adenyl transferase activity that is specifically stimulated by ssDNA. Nucleotidyl transferases bind specific nucleotides and release them in the presence of cation and acceptor molecules (e.g., guanyl transferase of HeLa cells is responsible for the addition of 5′-guanyl residues to mRNAs; Venaskatsen and Moss, 1980). The SV40 T-Ag/AMP complex (Section II.D.6) could be specifically dissociated by the addition of magnesium cation, pyrophosphate, and poly(dT) or fragments of SV40 ssDNA, but not by the addition of magnesium and pyrophosphate alone (Bradley *et al.*, 1984). Therefore, the release reaction was not simply a phosphatase activity that cleaved the labeled phosphate from the protein, but possibly a transferase that used ssDNA as the acceptor molecule. The adenylation of T-Ag is not necessarily due to the same enzyme that removes the AMP from T-Ag. The importance of an exchangeable 5′-AMP on T-Ag that is associated with the origin region of viral DNA is discussed in Section IV.H under alternate mechanisms for initiation of viral DNA replication.

H. *In Vivo* Associated Activities

1. Immortalization of Mammalian Cells

Primary cultures of cells can be induced to grow in tissue culture continuously by infecting them with a nonpermissive papovavirus. More specifically, C-terminally truncated mutants of SV40 and PyV T-Ag's contained sufficient activity to "immortalize" cells in tissue culture (Colby and Shenk, 1982; Rassoulzadegan *et al.*, 1982), although the C-terminus may contribute to the efficiency of the function (Tevethia, 1984). The continued synthesis of the protein might not be required for survival of the cultured cells once they were "immortalized," but the actual induction may depend upon a DNA binding function since these truncated T-Ag's contained sufficient amino acid sequences for nonspecific binding to DNA (Section II.F.1). Substitution of T-Ag for cellular factors necessary for progression from G_1 to the S phase of the cell cycle may be relevant here (Section II.C; Floros *et al.*, 1981), and the involvement of t-Ag in this activity has not been ruled out (Schaffhausen, 1983; Ellman *et al.*, 1984).

2. Stimulation of Cellular mRNA and Protein Synthesis

T-Ag induced transcription of cellular mRNA within an hour after T-Ag was synthesized (Khandjian *et al.*, 1980). Specifically, thymidine kinase and 100 other messages were induced (Postel and Levine, 1976; Schutzbank *et al.*, 1982). Increased synthesis of mRNA for replication enzymes may be important in the efficiency of the replication function of T-Ag, or one of the requirements for stimulation of cellular DNA synthesis.

3. Activation of rRNA Synthesis

Silent genes coding for rRNA were stimulated to transcribe by the N-terminal domain of T-Ag (to SV40 amino acid 509; Galanti *et al.*, 1981; Soprano *et al.*, 1983; Adams *et al.*, 1983; Learned *et al.*, 1983). Regions of T-Ag required for the stimulation of cellular DNA synthesis appeared to be necessary but not sufficient for induction of rRNA synthesis. Mutants able to induce synthesis were inactive as ATPases, although amino acids near the putative ATPase active site were required. A protein binding site has been proposed in this region (Chaudry *et al.*, 1982b).

4. Regulation of Early and Late Viral RNA Synthesis

T-Ag decreased the synthesis of early viral mRNA (Rio *et al.*, 1980; Hansen *et al.*, 1981), and stimulated the synthesis of late viral mRNA (Keller and Alwine, 1984; Brady *et al.*, 1984). Since these functions of T-Ag were affected by tsA mutations, they might rely on one or many activities of the antigen (Tooze, 1981). Regulation of early viral mRNA requires specific binding to viral DNA (Rio *et al.*, 1980; Hansen *et al.*, 1981) and stimulation of late message may also require binding to DNA and/or to certain cellular factors, but neither function requires viral DNA replication (Keller and Alwine, 1984; Brady *et al.*, 1984).

5. Excision of Viral DNA from Host Chromosomes

Miller *et al.* (1984) have shown that T-Ag was required for the excision of SV40 DNA from cellular DNA (Section III.D).

6. Acetylation of Cellular Histones

Schaffhausen and Benjamin (1976) and others (Coca-Prados *et al.*, 1980) described a change of acetylation patterns in papovavirus-infected cells. A recent report has shown that hyperacetylation of histones in SV40-infected cells inhibits the maturation of the viral RIs (Roman, 1982). The maturation of RIs and assembly of the virion does involve T-Ag (Garcea and Benjamin, 1983a,b), but further analyses are required in order to understand the details and relationship of these activities in virion production.

7. Induction of Heat Shock Proteins

Khandjian and Türler (1983) showed that both SV40 and PyV T-Ag's could induce the synthesis of peptides in papovavirus-infected animal cells that could also be induced by incubation of the cells at 43.5°C for

30–60 min (70K, 72K, and 92K peptides). No coimmunoprecipitation of these proteins with T-Ag was seen. Interestingly, a prokaryotic, replication protein dnaK has been shown to be involved in regulation of the heat shock response in *E. coli* (Tilley *et al.*, 1983). Similar to T-Ag, dnaK possessed both ATPase and protein kinase activities and was required for the initiation of *E. coli* DNA replication at the origin (Zylicz *et al.*, 1983). An early protein of adenovirus (ElA) induced heat shock proteins as well (Nevins, 1982). There may be some advantage to animal viruses in stimulating this response or it may be induced in parallel with deregulation of other cellular machinery.

8. Initiation and Maintenance of Cellular Transformation

These activities warrant a lengthy discussion that can be found in Chapter 4. As mentioned earlier, PyV T-Ag was not required for transformation but SV40 T-Ag was (Tooze, 1981; Rassoulzadegan *et al.*, 1982). Duplication of an internal region of the T-Ag gene was often found in SV40 and PyV T-Ag's isolated from transformed cells (super SV40 T-Ag's with duplications of amino acids 246 to 403; Lovett *et al.*, 1982; Chaudry *et al.*, 1982a; May *et al.*, 1983). The resultant protein was defective in its ability to induce viral DNA replication and to bind viral DNA specifically (Clayton *et al.*, 1982). It was able to bind p53, but it appeared unable to oligomerize with itself (May *et al.*, 1983). Transient expression of a replication-competent T-Ag was required for the partial gene duplications in both SV40- and PyV-transformed cells (Muller *et al.*, 1983b; Chen *et al.*, 1983; Dailey *et al.*, 1984; Manos and Gluzman, 1985).

9. Helper Function

The "helper function" refers to the ability of a C-terminal peptide of SV40 T-Ag encoded by $Ad2^{+}ND_{1}$, to complement the late functions of human adenovirus in monkey cells, a semipermissive host (Kelly and Lewis, 1973; Tooze, 1981). C-terminal truncations of SV40 T-Ag (35 amino acids; Polvino-Bodnar and Cole, 1982) rendered SV40 unable to efficiently synthesize virion capsid protein in either the CV-1 line of monkey cells or primary monkey cells while not affecting viral replication in BSC-1 monkey cells (Tornow and Cole, 1983a,b; Tornow *et al.*, 1985; Pipas, 1985). This host range function could act independently in SV40-infected cells as a fusion peptide with SV40 VP1 to complement the truncated T-Ag mutant (Tornow *et al.*, 1985). The C-terminal domain of SV40 T-Ag has no homology with PyV T-Ag (Fig. 2), and so far, PyV host range mutations have been mapped to the N-terminal region of the early gene associated with mT-Ag and t-Ag, not T-Ag (Schaffhausen, 1983). Interestingly, mutants of adenovirus able to compensate for the helper function had alterations in the DNA binding protein (Rice and Klessig, 1984), and SV40 T-Ag truncations made only 14 amino acids upstream of the helper function domain, decreased the efficiency of SV40 DNA replication as well as that of virion protein synthesis (pC8-B; Manos and Gluzman, 1985).

I. Genetic Alterations That Define Domains Functional in Replication

The stringent requirement for full-sized, fully active T-Ag for initiation of viral DNA replication (Section II.C), is reflected in the long list of replication mutations located throughout the A gene of SV40 (Table I). Considering the activities required for T-Ag-dependent viral DNA replication, an outline of what could constitute a domain in SV40 T-Ag is as follows: (1) the nuclear localization signal, (2) those amino acid sequences that contact nonspecific DNA sequences directly, (3) those sequences that contact specific pentanucleotides within a viral DNA binding site, (4) a T-Ag oligomerization site, (5) perhaps a p53 binding site, (6) an active site for the T-Ag-associated ATPase activity, (7) an active site for adenylation, (8) sequences responsible for interactions between functional domains of the protein, (9) sequences which serve to stabilize and protect functional domains of the protein from denaturation. Seven hypothesized domains are presented here in order to discuss the replication mutants listed in Table I (for the C-terminal domain 8, see Section II.H.9). Only SV40 T-Ag mutants sustaining minimal changes in the amino acid sequence were listed here in order to avoid problems with discussion of overlapping domains.

The replication mutants could be placed initially into domains according to the location of the amino acid change and the resultant *in vivo* or *in vitro* characteristics of the mutant T-Ag. Additional information regarding the predicted secondary structure of the protein was used to decide the borders of the domains. Secondary structural features were mentioned when relevant as "turns," "α helices," and "β strands" or "sheets," although they may not be real. The computer program used to obtain this information (Fig. 6) was adapted from Chou and Fasman (1978) by T. F. Smith and W. Ralph for the Molecular Biology Computer Research Resource (MBCRR) at the Dana-Farber Cancer Institute/Harvard School of Public Health (July 1984; updated January 1985). A conservative interpretation of such data is recommended for proteins such as SV40 T-Ag that have not been crystallized and studied by X-ray diffraction or by other three-dimensional means as discussed by Richardson (1981). It must be stressed that Fig. 6 is a two-dimensional rendering of three-dimensional information, and certain structural choices were made as indicated in the figure legend.

1. Amino Acids 1 to 82

This T-Ag peptide has all the amino acids that are shared by t-Ag and it can be replaced by adenovirus sequences and retain most wild-type

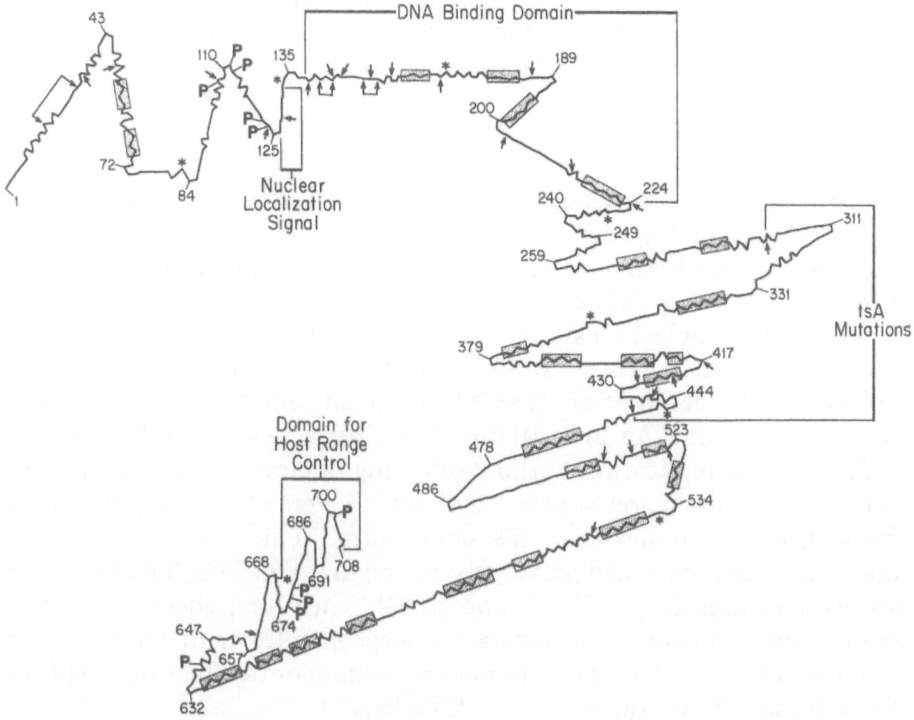

FIGURE 6. Proposed functional domains in SV40 T-antigen. Since the crystallographic structure of T-Ag is not yet available, a computer program was used to obtain secondary structural information (adapted from Chou and Fasman, 1978, by T. F. Smith and W. Ralph for the Harvard Medical School GENE Computer Resource). The line drawn represents a two-dimensional rendering of three-dimensional information for the sequence of 708 amino acids in T-Ag. It was derived from a secondary structure program ("Smartcom") and visualized by a graphics program ("Draw3D") as developed by D. V. Faulkner and T. F. Smith at the MBCRR. β strands are enclosed in rectangles, α helices are represented as squiggles, and random coils are depicted as straight lines. Random coil sequences would be less rigid than in the actual molecule, allowing formation of a more compact structure. The turns predicted by "Smartcom" occurred at specific amino acids (numbered), but were rotated arbitrarily to fit within the confines of this figure. Turns located in the region rich in β strands (259–550) were rotated to indicate the presence of a β sheet (Richardson, 1981). Asterisks mark the borders of the domains discussed in Section II.I. Arrows mark the positions of the mutations listed in Table I. A general map of T-Ag activities derived from larger deletions in SV40 T-Ag is available in Clark *et al.* (1983).

activities (D2 T-Ag; Hassell *et al.*, 1978; Schaffhausen, 1983). But deletions and point mutations in the wt T-Ag have deleterious effects on the *in vitro* activities of the protein, and the mutant virus is defective in its replication function and therefore nonviable (Clark *et al.*, 1983; Pipas *et al.*, 1983; Manos and Gluzman, 1984). In addition, an N-terminally truncated T-Ag was not able to bind "tightly" to cellular DNA (Prives *et al.*, 1982). However, at least one monoclonal antibody, binding to an epitope in this N-terminal region, does not significantly affect the *in vitro* DNA binding or ATPase activities of T-Ag (pAb419; Harlow *et al.*, 1981a; M. Bradley, unpublished observations). Therefore, this region might simply donate stability to the DNA binding activity.

2. Amino Acids 83 to 133

This next hypothesized domain is bordered at the C-terminus by a specific DNA binding region (Fig. 6). Around Lys^{128} is the seven-amino-acid sequence necessary for nuclear localization of the antigen, and the N-terminal cluster of differentially phosphorylated residues (Sections II.D.1 and II.E.4). The nuclear localization signal could be exposed on the surface of the protein via a "turn" at 125 initiated by proline residues, but as indicated by Kalderon *et al.* (1984b) the signal peptide can be moved to the N-terminus of T-Ag and still function. Changing some of these amino acids did affect replication without affecting nuclear localization (Thr^{124} and Pro^{126}; Kalderon and Smith, 1984), which were at and near the phosphorylation sites found near this same "turn." Differential phosphorylation has been mentioned previously as a regulator of T-Ag/DNA binding affinity (Sections II.D.1, II.F.1, and II.F.4). Therefore, these "exposed" amino acid sequences may interact nonspecifically with DNA and/or with the actual DNA binding domain to initiate or to affect the stability of the specific T-Ag interaction with DNA.

3. Amino Acids 134 to 228

Domain 3 was delineated by a series of mutant T-Ag's that were unable to bind specifically to viral DNA *in vitro,* which was the apparent reason for their inability to replicate *in vivo* (DNA binding domain, Fig. 6; Kalderon and Smith, 1984; E. Paucha, personal communication). The ability of T-Ag to bind nonspecifically to DNA was retained (Prives *et al.*, 1983) and the T-Ag/ATPase activity was not affected. As discussed in Section II.C, this region is necessary but not sufficient for the specific binding activity of SV40 T-Ag. In addition, the domain apparently has a series of "α helices" and "β strands," but it is not similar to those reported for prokaryotic, DNA binding proteins (Sauer *et al.*, 1982; Pabo and Sauer, 1984). A particular secondary structure is apparently formed either before T-Ag binds specifically to DNA or as a result of that binding. In support of this hypothesis, a monoclonal antibody (pAb100 or Mcl 7) binds to most if not all of the specific DNA binding species of T-Ag (apparently only 10% of the T-Ag in a cell), and the mutant pC6 T-Ag could not be immunoprecipitated by it (Scheller *et al.*, 1982). The site of the pAb100 epitope (Fig. 2) was not in 3, but closer to the C-terminus and the area of tsA mutations (Deppert *et al.*, 1981; Scheller *et al.*, 1982). Since Moore and Williams (1984) have claimed that antigenic sites may be related to transient conformations, especially on "charged surfaces of proteins likely involved in their flexibility," one could hypothesize that an epitope is recognized by the antibody (pAb100) only when a particular conformation, characteristic of the species of T-Ag active in specific DNA binding, is attained. That epitope likely includes some more C-terminal domain of T-Ag.

Within region 3 there appeared to be an N-terminal and a C-terminal cluster of replication mutations, with these two portions of the domain in SV40 having considerable amino acid homology with sequences in PyV T-Ag separated by a nonhomologous region (Fig. 2; E. Paucha, personal communication). Considering the large number of mutations that have been mapped within this domain, it has been divided here into 3a and 3b for purposes of discussion, and to suggest a possible difference between the two ends in their potential interactions with viral DNA.

a. Amino Acids 134 to 167

Many sites in this part of domain 3 were exquisitely sensitive to mutation, resulting in destruction of the specific DNA binding activity of T-Ag (Kalderon and Smith, 1984). In fact, the revertant described by Shortle *et al.* (1979; Section II.F.2) mapped to this stretch of amino acids. They were mutants of SV40 T-Ag that were capable of binding specifically to viral DNA mutated at the origin (mutant sp1030 revertants SR/2 and SR/3), indicating that the specific affinity of T-Ag for the origin region depending on a direct interaction of the protein and viral DNA (Margolskee and Nathans, 1984). Some of the other mutants were recognized *in vivo* by detection of replication-defective T-Ag's encoded by integrated viral DNA in transformed cell lines. In some cases the mutant SV40 DNA was rescued and propagated in *E. coli* plasmids (e.g., pC6-2; Gluzman and Ahrens, 1982). Therefore, domain 3a, lying close to an important series of phosphorylated residues, may be the region that comes into direct contact with SV40 origin DNA.

b. Amino Acids 168 to 228

One could propose that this region is a repeat of domain 3a and is also in contact with DNA, or that 3b interacts with the 3a region to establish the specificity and stabilize the T-Ag/DNA interaction. The only tsA mutant recognized so far within the DNA binding domain 3 was tsA3900 (Hutchinson *et al.*, 1985), which is in the region that is not homologous with PyV T-Ag. That region contains two potential "turns" and two potential "β strands." It is possible that the structure, but not the amino acid sequence may be conserved between SV40 and PyV T-Ag's. Finally, 3b may interact with some other domain of the protein that is required in conjunction with association with the origin of replication, since the pC8-A T-Ag was shown to retain specific DNA binding and ATPase activities without being able to replicate (Manos and Gluzman, 1985).

4. Amino Acids 229 to 361

Domain 4 may describe a transitional region between an N-terminal peptide of T-Ag, comprising domains 1 to 4, and the rest of the protein, stabilizing the overall conformation. The borders of the domain were based upon the gene duplication data (Section II.H.8) that describe a common repeat of amino acids 246 to 403, a classification of tsA DNA binding mutants suggested by Wilson *et al.* (1982), and by considering a series of four "turns" at the N-terminus that potentially separates domain 4 from 3b (Fig. 6).

Most of the tsA mutations resulting in replication defects have been mapped to domains 4 and 5 (Fig. 6). This domain contains the mutant tsA30 (Table I) plus the tsA mutants 40, 47, 57, 1609, 1612, 1619, 1634, and 1637 that have not been mapped at the amino acid level (Lai and Nathans, 1974; Seif *et al.*, 1980a; Tevethia *et al.*, 1981). Wilson *et al.* (1982) tested the purified T-Ag of tsA30 for its ability to bind DNA and found that it was apparently so fragile that only minimal binding to SV40 DNA site 1 was detected even at permissive temperature. This mutation also affected the T-Ag/ATPase activity as measured in purified tsA30 T-Ag at the high temperature (J. D. Griffin *et al.*, 1980). It must be noted that a C-terminally truncated SV40 T-Ag was capable of binding SV40 DNA *in vitro* with 1.5-fold the affinity of the wt T-Ag (having amino acids 1 to 399, dl1055; Clark *et al.*, 1983), and that a DNA binding peptide from the 72K adenovirus DNA binding protein was crystallized independently (Tsernoglou *et al.*, 1984). Therefore, a similar purified T-Ag peptide ending with domain 4 may also form crystals.

5. Amino Acids 362 to 447

Domain 5 could be the N-terminal region of a C-terminal domain of T-Ag that is responsible for the activation of the replication function of T-Ag by regulating the binding affinity of T-Ag for viral DNA, and/or coordinating some interaction with host replication factors. The domain itself includes sequences necessary for ATPase and the sites of a series of tsA mutations affecting the DNA binding activity. Wilson *et al.* (1982) tested the T-Ag's of tsA58 as well as tsA28, 207, and 241 (all mapped to within this peptidyl domain), and found them to be temperature-sensitive in their ability to bind DNA *in vitro.* However, the reason for the DNA binding defect was probably not destruction of primary, DNA-contact residues of T-Ag, but disruption of a domain that affected stability of the T-Ag/DNA interaction.

All mutants listed in this section map within the peptide immunoprecipitated by pAb204 (Fig. 2), which blocked the ATPase activity of the protein directly, perhaps by blocking the active site (Clark *et al.*, 1981). In addition, tsA58 T-Ag was shown unable to oligomerize (Montenarh and Henning, 1984), which may be important for the T-Ag/ATPase activity as well as for the ability of T-Ag to bind to DNA (Section II.E.2; Bradley *et al.*, 1982). It is not clear whether the actual protein binding site on T-Ag lies within domain 5 or within the preceding domain 4. The antibody pAb204 blocked the *in vitro* binding of p53 to T-Ag (D. Lane, personal communication), but the large immunoglobulin might interfere with a nearby binding site as well as block a specific site in domain 5 (e.g., in domains 4 or 6).

As mentioned above, the C-terminal half of T-Ag starting with domain 5 could be a single, structural domain separable from the DNA binding domain. Amino acid 362 marks a point on the protein where there begins a potential "β sheet" (continuing through amino acid 600 approximately), containing a minimum 5–6 β strands of peptide required for formation of a "β sheet" (as aligned in Fig. 6; Richardson, 1981). tsA58 maps to within an "α helical" structure that lies between β strands. Revertants to that mutation were found that changed an amino acid within the same structure at 441, and six other revertants of tsA58 mapped at widespread intervals, amino acids 296, 408, 472, 495, 551, 628 (Bourre and Sarasin, 1983), throughout regions 4, 5, and 6.

6. Amino Acids 448 to 538

Domain 6 was separated from domain 5 because mutants mapped to this region were still able to bind an origin DNA fragment specifically (Wilson *et al.*, 1982; Manos and Gluzman, 1985). tsA1642, a mutant completely defective in transformation, had only a threefold decrease in its ability to synthesize viral DNA, and it was defective in some activity relative to virion production (Cosman and Tevethia, 1981). Its mutation site lies close to those listed in the previous section, but the T-Ag apparently has sufficient ATPase activity to separate it functionally from the mutants in that set. The rest of the mutants in domain 6 were completely defective in viral DNA replication, even though they could bind to viral DNA.

This domain contains the putative site for the adenylation of SV40 T-Ag, Ser^{504} (Bradley *et al.*, 1984; Clertant *et al.*, 1984b; M. Bradley, unpublished observations). T-Ag mutants mapping very close to this site, pC2 and d12462, retained at least some ATPase activity (Manos and Gluzman, 1985; C. Cole, personal communication), and thus some explanation of their replication defect is necessary. These mutants may be defective in their ability to form AMP–serine complexes. In a highly homologous region of PyV T-Ag (Fig. 2), PyV T-Ag mutants 7axB and bc1071 mapped to Asp^{642} (Hayday *et al.*, 1983; Nilsson and Magnusson, 1984). The dialdehyde-ATP binding site was within this peptide, suggesting that the inability of 7axB and bc1071 to replicate might also be attributed to a defect in AMP/T-Ag complex formation. The proximity of the adenylation site to the ATPase site (domain 5) may be significant since Clertant *et al.* (1984b) found that an ADP might be the preferred substrate for the adenylation reaction.

7. Amino Acids 539 to 670

This domain was apparently antigenic and therefore exposed since antibodies have found to bind to this region (e.g., especially antibodies made against SDS gel-purified T-Ag; M. Bradley, unpublished observations). But it is not clear whether this is a domain responsible only for conformational stability of T-Ag involved in DNA replication, or whether it contains some specific activity associated with virion assembly or some activity associated with cytoplasmic T-Ag. Domain 7 does contain a highly hydrophobic region between amino acids 568 and 582, which may be responsible for the association of T-Ag with the membrane (Section II.D.2), but which may not be involved in the role played by T-Ag in initiating viral DNA replication.

Domain 7 also contains the replication mutant tsA7 (Cosman and Tevethia, 1981), which was not listed in Table I since its specific mutation has not been identified. tsA7 T-Ag appeared to have the same defects in DNA binding as those described for domain 5 (e.g., tsA58). However, it is one of the "leakiest" mutants isolated (Wilson *et al.*, 1982) that may indicate a general disruption of this highly structured region of T-Ag. More characteristic of the region are the T-Ag mutants pC11-B and pC8-B (Table I), which were able to replicate but formed pinpoint plaques of virus, indicative of an inefficient replication function (Manos and Gluzman, 1985). pC11-B T-Ag mapped as a truncation of the last 49 amino acids. These deleted residues contain the helper function or host range control domain (amino acids 682 to 708; Tornow *et al.*, 1985), and are often proteolytically digested during the extraction of T-Ag from the cell (Section II.E.1), suggesting that they were apparently on the surface of the protein (Fig. 6). It must be noted that tsA1499, as mapped at the C-terminus, did not prove to be a replication mutant, but was still defective in transformation. tsA3900, a replication mutant mapped to domain 3b, was isolated from the original double mutant tsA1499 (Hutchinson *et al.*, 1985).

III. THE DNA REPLICATION CYCLE

A. Summary

Of the 11 forms of viral DNA identified in a lytic infection (DePamphilis and Wassarman, 1982), only 5 appear as major intermediates or products of replication (Fig. 11A). Initiation of SV40 or PyV DNA replication requires the interaction of at least three components: a unique, *cis*-acting DNA sequence called the origin of replication (*ori*), the homologous viral T-Ag which binds to specific DNA sequences in and around *ori*, and permissive cell factors. The SV40 origin of bidirectional replication (the point from which the two replication forks emerge) is located at the junction of *ori* and T-Ag binding site 1. Initiation of SV40 DNA synthesis at *ori* appears identical to initiation of Okazaki fragment synthesis throughout the genome, and does not require either RNA polymerase II or III. PyV, BKV, and JCV papovaviruses each have a region homologous to the SV40 *ori* sequence: a 64- to 66-bp sequence containing a 14-bp inverted repeat, a 23- to 34-bp GC-rich palindrome, and a 15- to 20-bp A/T sequence in the same relative order (Fig. 10). However, in contrast to SV40, the larger PyV *ori* sequence includes regions that also function as enhancer elements for gene transcription, and may impart cell specificity to PyV *ori* activation. Furthermore, the major sites for PyV T-Ag binding, the TATA box, and the early mRNA initiation sites lie outside of *ori*, while some or all of these sequences are included within the SV40 *ori* (Fig. 9).

Replication begins with covalently closed, superhelical, circular viral DNA (form I) and results in replicating intermediates (RI) with covalently closed parental strands and two forks traveling at similar, but not identical rates. A topoisomerase activity presumably releases the torsional strain that accumulates as DNA is unwound at replication forks. As replication continues past 70% completion, T-Ag is no longer bound to replicating chromosomes, and RI begin to accumulate until late replicating intermediates that are about 90% completed (RI*) are about threefold more abundant than an equivalent sample of RI at earlier stages in replication. Replication forks accumulate at several specific sites in the normal region where separation of sibling molecules occurs and DNA replication terminates. Many of these sites correspond to sites that arrest DNA polymerase α on the same template *in vitro*. However, a unique DNA sequence is not required for either separation of sibling molecules or termination of replication, although the termination region can determine which of two possible pathways is used for separation. In what appears to be the major pathway, parental DNA in the termination region is completely unwound to release two circular monomers containing a gap in the nascent DNA strand within the termination region (form II*). DNA replication terminates in the production of form I molecules when this gap is filled in and the nascent strand sealed. Superhelical DNA is a consequence of nucleosome assembly that occurs on both arms of replication forks throughout the replication process. An alternate pathway for separation is observed when unwinding of parental DNA in the termination region is inhibited but replication continues (Fig. 11B). This results in accumulation of RI* and formation of two multiply intertwined, circular monomers (catenated dimers). If the nascent DNA strands in these dimers are sealed prior to decatenation, form I is the product. If decatenation occurs before DNA replication terminates, then form II* is the product. The choice of pathway is strongly influenced by the sequence at which termination occurs, and topoisomerase II may be specifically involved in relaxing torsional strain that occurs in the termination region as well as separating catenated dimers.

B. *cis*-Acting DNA Sequences Required for Replication (*ori*)

During the period of maximum DNA synthesis, DNA replication in both SV40 and PyV proceeds bidirectionally from a unique location on the genome. This was demonstrated in two ways. First, SV40 form I [SV40(I)] (Danna and Nathans, 1972; Tapper *et al.*, 1979) or PyV(I) (Crawford *et al.*, 1973, 1974, 1975) DNA, radiolabeled for varying periods of time in whole cells or isolated nuclei (Tapper *et al.*, 1979), was digested with restriction endonucleases and the relative specific-radioactivity was determined for each DNA fragment. The products of DNA replication contained the most label at the site where termination occurred and the least label at the site where replication began. The resulting gradients of radioactivity per base pair showed that SV40 DNA replication began near the *Hind*II and III A/C junction, that PyV DNA replication began 29 ± 2% from the *Eco*RI site, and that replication in both viruses proceeded bidirectionally until the two forks met 180° from the origin. Second, the genomic position of DNA replication bubbles was measured by electron microscopic analysis of replicating DNA that had been cut at a unique restriction site. These studies demonstrated that 90% (Fareed *et al.*, 1972; Robberson *et al.*, 1975) to 98% (Martin and Setlow, 1980; Tapper *et al.*, 1982) of SV40 RI DNA had initiated replication bidirectionally 33% from the *Eco*RI site (Fareed *et al.*, 1972; Robberson *et al.*, 1975), and 10 to 35 bp from the *Bgl*I site (Tapper and DePamphilis, 1980; Martin and Setlow, 1980). PyV began 29% from the *Eco*RI site (Robberson *et al.*, 1975; Bjursell *et al.*, 1979) close to the *Hpa*II(3/5) cleavage site (Griffin *et al.*, 1974). Genetic analysis of these origins of replication has identified the boundaries of a *cis*-acting sequence required for initiation of viral DNA, the *ori* sequence.

1. SV40

In the case of SV40, single base-pair changes near the single *Bgl*I restriction site (Figs. 7, 10), with one exception, either increased or decreased the rate of viral DNA replication without significantly affecting its bidirectional nature (Shortle and Nathans, 1979). Similarly, insertion of a single G residue between nucleotides 5242 and 5243 (Cohen *et al.*, 1984), and small deletions at the *Bgl*I site prevented replication except when propagated either as a bacterial plasmid (Gluzman *et al.*, 1980; Myers and Tjian, 1980) or as a virus (Shenk, 1978) containing a second *ori* region. Since these mutations could not be complemented by the presence of wt SV40 or through transfection of COS cells which provide

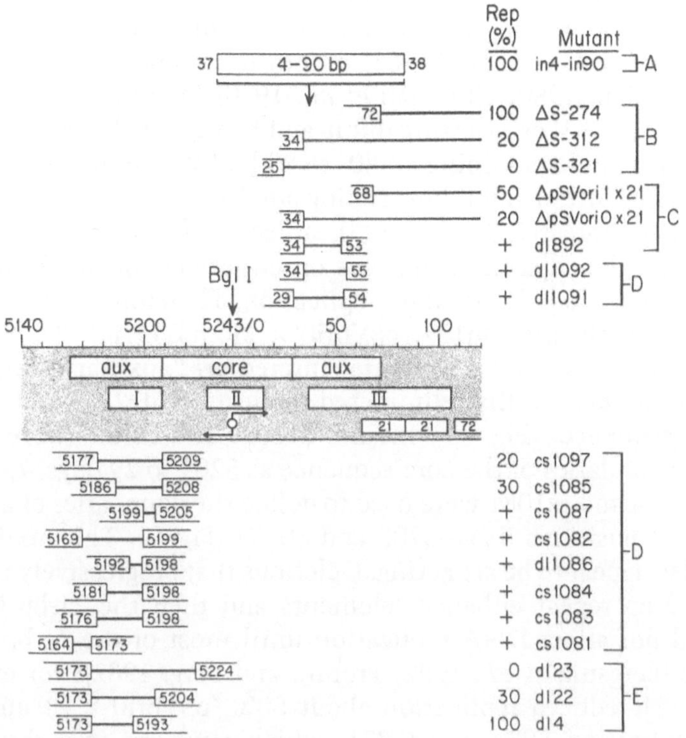

FIGURE 7. SV40 mutants that define the SV40 origin of replication. Insertion mutants were selected from the data of (A) Innis and Scott (1984), and deletion mutants were selected from the data of (B) Fromm and Berg (1982), (C) Bergsma *et al.* (1982) and Subramanian and Shenk (1978), (D) DiMaio and Nathans (1980, 1982), and (E) Myers and Tjian (1980) that defined the SV40 *cis*-acting sequence required for initiation of viral DNA replication (*ori*). The sequences deleted from plasmids carrying the SV40 *ori* region are indicated by bars with an open box (SV40 sequence) at each end; the number in the box indicates the SV40 nucleotide bordering the deletion. DNA sequences 4–90 bp long (indicated by rectangle) were also inserted between SV40 nucleotides 37 and 38. The percent replication reported for each mutant is listed on the right. Where numbers were not reported, a "+" is used to indicate good replication and a "−" to indicate no replication. The shaded box in the center of the figure represents the SV40 genome and reveals the *ori*-"core" sequence flanked by two *ori*-"auxiliary" (aux) sequences, one on each side of the core. In addition, the locations of the three T-Ag binding sites (I, II, and III), the two 21-bp repeats (21), one of the 72-bp repeats (72), a 27-bp palindrome (palindrome symbol), and the *BglI* restriction site are indicated. The numbering system, in bp, is from Buchman *et al.* (1980), and runs from the *BglI* site toward the late genes.

SV40 T-Ag from an integrated genome, they are referred to as *cis*-acting mutations, and, as such, define a *cis*-acting sequence required for DNA replication.

The minimum size of the SV40 *ori*, 64 bp (Fig. 7), was defined by mutations on either side that still allowed DNA replication, even if it was reduced from that of wild-type virus or was temperature-sensitive. These mutants were selected either as viable deletions (Cole *et al.*, 1977;

Subramanian and Shenk, 1978; Shortle and Nathans, 1979; DiMaio and Nathans, 1980, 1982) or as evolutionary variants (Gutai and Nathans, 1978; Woodworth-Gutai *et al.*, 1983), or they were constructed by deleting (Myers and Tjian, 1980; Gluzman *et al.*, 1980; Fromm and Berg, 1982; Bergsma *et al.*, 1982) or inserting (Innis and Scott, 1984) DNA segments. A bacterial plasmid containing an 80- to 85-bp SV40 segment consisting of T-Ag binding sites I and II but lacking site III, replicated in permissive cell lines that expressed SV40 T-Ag (Learned *et al.*, 1981).

Although the adjacent sequences on either side of the minimal *ori* are not required, they do facilitate replication. Therefore, those deletions that reduced replication (d122, pSVori0 × 21, pSVori1 × 21, S-312) or that were cold-sensitive (cs mutants) entered *ori* "auxiliary" sequences, while those mutations that eliminated replication (d123, S-321) entered *ori* "core" sequences (Fig. 7). Mutants d11091 and cs1097 were used to define the boundaries of the core sequence as 5209 to 29 (Figs. 7, 10), and mutants S-274 and cs1081 were used to define the boundaries of auxiliary sequences at positions 5164–5208 and 30–72 (Fig. 7). The auxiliary sequences also appear to be *cis*-acting. Deletions that progressively removed first the 72-bp repeat enhancer elements and then the 21-bp GC-rich repeats did not affect DNA replication until most of the 21-bp repeats were gone (Bergsma *et al.*, 1982; Fromm and Berg, 1982). For example, pSVori1 × 21 reduced replication about 50%, pSVori0 × 21 and S-312 reduced replication 80%, and S-321, which extended into the AT sequence, inhibited replication completely (Fig. 7). However, virus missing sequences 35–177 was still viable although it replicated poorly and could not be complemented by coinfecting the cells with wt SV40 (Hartzell *et al.*, 1983). The presence of an auxiliary sequence was also demonstrated by inserting from 4 to 390 bp of DNA between nucleotides 37 and 38 (Innis and Scott, 1984). Separating the *ori*-core sequence from its late gene side auxiliary region by 4 bp had no effect, but insertions of 43 to 90 bp reduced synthesis of form I by about threefold and insertion of DNA segments of 260–390 bp reduced form I synthesis by at least tenfold. The effect was *cis*-acting and independent of the nucleotide sequence inserted. Thus, increasing the distance between *ori* core and the late gene side auxiliary sequence had an effect similar to deleting this auxiliary sequence. Since the 21-bp GC-rich repeats include sequences required for transcription of both early and late genes, and the 72-bp repeats contain enhancer elements required for efficient transcription of early genes, activation of SV40 *ori*-core does not appear to require transcription. Consistent with this hypothesis is the observation that initiation of DNA replication at the SV40 *ori* region *in vitro* was not inhibited by concentrations of α-amanitin that inhibit RNA polymerases II and III (Li and Kelly, 1984; Ariga, 1984b; Decker *et al.*, 1986).

Perhaps the simplest explanation for auxiliary sequences is that they facilitate binding of replication proteins to the *ori*-core. For example, the auxiliary sequence on the early gene side includes the strongest T-Ag binding site. However, binding of SV40 T-Ag to site II *in vitro* is independent of site I, and site II alone has a high-affinity binding region on its early gene side (Fig. 4, Section II.F.2; Tegtmeyer *et al.*, 1981, 1983; Tenen *et al.*, 1983b). Another explanation is that auxiliary sequences may affect the structure of chromatin, making the *ori* sequence more accessible to replication proteins. About 25% of SV40 chromosomes were hypersensitive to DNase I in two regions, the 72-bp repeats (i.e., the transcriptional enhancer elements) and the *ori* core (Gerard *et al.*, 1982; Fromm and Berg, 1983; Jongstra *et al.*, 1984; Innis and Scott, 1984). Together, these regions contained up to eight individual cleavage sites; the exact number of sites and their locations depended on whether the genome was replicating, nonreplicating, integrated into cellular chromosomes, or unintegrated (Cremisi, 1981; Cereghini and Yaniv, 1984). Furthermore, analysis of these hypersensitive sites in various insertion and deletion mutants revealed a direct correlation between hypersensitivity over the *ori* sequence and its proximity to the 21-bp GC-rich repeats (Innis and Scott, 1984; Jongstra *et al.*, 1984). Therefore, loss or relocation of the 21-bp repeat region could reduce access or *ori*-core to proteins required for DNA replication. Consistent with this hypothesis is the observation that the accessibility of the *Bgl*I restriction site in *ori*-core to cleavage by *Bgl*I decreased about 50% when 56–90 bp were inserted between the core and late side auxiliary sequences (Innis and Scott, 1984). The fraction of nonreplicating, mature SV40 chromosomes that can be cleaved by *Bgl*I has been reported as 30% (Shelton *et al.*, 1980), 50% (Tack *et al.*, 1981), and 100% (Varshavsky *et al.*, 1978), perhaps reflecting differences in chromatin structure resulting from differences in virus strains or methods of chromatin preparation. However, *Bgl*I sensitivity was the same for replicating as for nonreplicating chromosomes (Tack *et al.*, 1981), indicating that changes in chromatin structure over the core sequence did not occur during DNA replication.

2. PyV

The major difference between the *ori* sequences of PyV and SV40 is that PyV requires the promoter–enhancer sequences on the late gene side of *ori* core while SV40 does not. At least four elements can be identified in the PyV *ori* sequence by analysis of various deletion mutants, and are referred to as α, β, core, and auxiliary sequences. Deletions were constructed in PyV DNA that had been cloned into plasmid vectors, and then the PyV was either excised and transfected into 3T3 cells (Luthman *et al.*, 1982; Katinka and Yaniv, 1983), or the plasmids were used to transfect *ori*-defective PyV-transformed mouse cells (COP cells: Tyndall *et al.*, 1981; Katinka and Yaniv, 1983; MOP cells; Muller *et al.*, 1983a). Plasmids containing α-β-core, α-core or β-core all replicated when provided with PyV T-Ag either from the infecting viral DNA itself or from an integrated PyV genome; only α-β or core alone did not replicate (Table II). Thus, core sequence together with sequences encompassed by *either* the α- or β-element constituted the PyV *ori*. All of the deletions in Fig.

8 that inhibited DNA replication were *cis*-acting. The early gene side boundary of the *ori*-core sequence (nucleotide 42) was defined by deletion 75 which still allowed about 15% replication; deletion 77 did not replicate while deletions 1004, 76, pBBg(H), PP1(B)BG(H), PB503d1300 replicated as well as the parental sequence (Fig. 8). As with SV40, sequences adjacent to the early gene side of *ori*-core that include T-Ag binding site A facilitated replication. The borders of this auxiliary sequence, nucleotide 43 to 64, were defined by mutants 75 and 76.

Defining the *cis*-acting sequences required for replication on the late gene side or *ori* is more complex. The late gene side border of *ori* at first appears to be about the *Pvu*II site at position 5131. Deletions that ex-

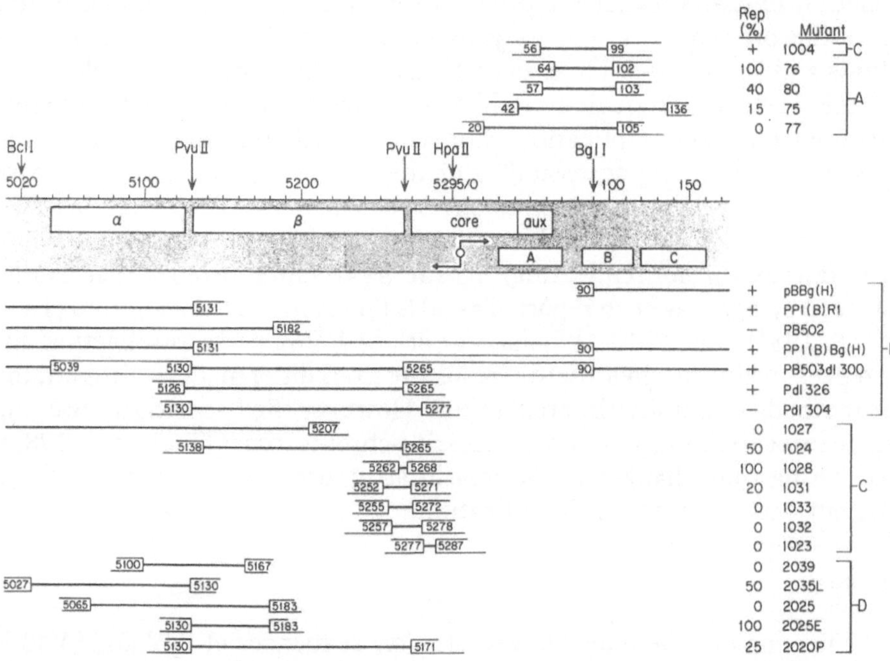

FIGURE 8. PyV mutants that define the PyV origin of replication. Deletion mutants were selected from the data of (A) Katinka and Yaniv (1983), (B) Muller *et al.* (1983a), (C) Luthman *et al.* (1982), and (D) Tyndall *et al.* (1981) that defined the PyV *cis*-acting sequences required to initiate viral DNA replication. The representations are as described in Fig. 7. The outer limits of four elements are defined and referred to as α, β, core, and auxiliary sequences. In addition, the three major T-Ag binding sites (A, B, and C) are indicated, along with the 34-bp palindrome ( ), and restriction sites *Bcl*I, *Pvu*II, *Hpa*II, and *Bgl*I. The numbering system in bp is from Griffin (1980) and runs from the *Hpa*II site toward the early genes. The sequences deleted from plasmids carrying the PyV *ori* region are indicated by bars with an open box (PyV sequence) at each end. The number in the box is the PyV nucleotide bordering the deletion. The numbers shown are those given by the individual investigators. No attempt was made to correct them for strain differences (Pomerantz *et al.*, 1983). Subtle differences other than the ones reported may also be present, and strain differences may not affect DNA replication in the various mutants. Consequently, the borders given for each element could vary by several nucleotides.

tended farther toward the *Hpa*II site prevented replication (Fig. 8; PB502, Pd1304, 1027, 2039, 2025). However, the effect of these deletions was eliminated if sequences between the *Pvu*II site at position 5130 and the *Bcl*I site were retained. This region from nucleotides 5039 to 5126 contains the α-element whose borders are defined by mutants PB503d1300, Pd1326, 1024, 2025E, and 2020P. The ability of the α-element to activate DNA replication depended on its orientation and/or position relative to the core-element (Muller *et al.*, 1983a). The β-element resides in the sequences between the two *Pvu*II sites, nucleotides 5131 to 5265. The late side border of β was defined by mutants PP1(B)R1, PP1(B)Bg(H), and 2035L, but the early gene side border of β could not be defined because mutants in this region retained the α-element. Either α-core or β-core was sufficient for T-Ag-dependent DNA replication in mouse somatic cells, but only β-core [mutants PP1(B)R1, PP1(B)Bg(H)] replicated when injected into the nuclei of one-cell or two-cell mouse embryos along with a plasmid that expressed PyV T-Ag (Wirak *et al.*, 1985). Thus, although α and β are interchangeable in mouse fibroblasts, they must have independent functions in that they can be distinguished by different cell types in the same organism. Mutants 2039 and 2025, which lacked part of both α and β, failed to replicate. The existence of a separate core-element was evident from deletions that left the α-element intact but eliminated β and extended beyond the *Pvu*II site at position 5265 toward the *Hpa*II site (Pd1304). These mutants did not replicate. Small deletions at this *Pvu*II site (mutants 1028, 1031, 1033, 1032, 1023) placed the late gene side border for the core-element at nucleotide 5271 (mutant 1031). Mutant 1031 which reduced replication by 80% indicated the presence of a late gene side auxiliary region between nucleotides 5272 and 5268 analogous to the SV40 *ori* region. As with SV40, most, but not all, single base-pair mutations within *ori*-core, and particularly within the 33-bp palindrome, blocked PyV DNA replication (Fig. 10; Triezenberg and Folk, 1984; Luthman *et al.*, 1984).

The function of α and β in DNA replication appears to be related to their role in transcription (Fig. 9). The α-element is a strong enhancer of PyV gene activity (de Villiers and Schaffner, 1981; Tyndall *et al.*, 1981; Mueller *et al.*, 1984; Herbomel *et al.*, 1984). The β-element is a relatively weak enhancer that can be converted to a strong enhancer by single base-pair mutations selected by their ability to allow PyV to replicate in mouse embryonal carcinoma cell line F9 (Linney and Donerly, 1983; Herbomel *et al.*, 1984). Mutations in the β-element that allowed viral replication in F9 cells were *cis*-acting for both viral gene expression and DNA replication, because PyV DNA did not replicate in F9 cells even when coinfected with an F9 host-range PyV mutant that did replicate and synthesize T-Ag (Fujimura and Linney, 1982). The DNase I hypersensitive sites normally associated with transcriptionally active regions of chromatin are found in the α and β regions of PyV chromosomes (Herbomel *et al.*, 1981). Substitution of both PyV α and β (*Pvu*II at position 5265 to *Bcl*I, Fig. 8) by a single copy of the SV40 72-bp enhancer element, in either orientation,

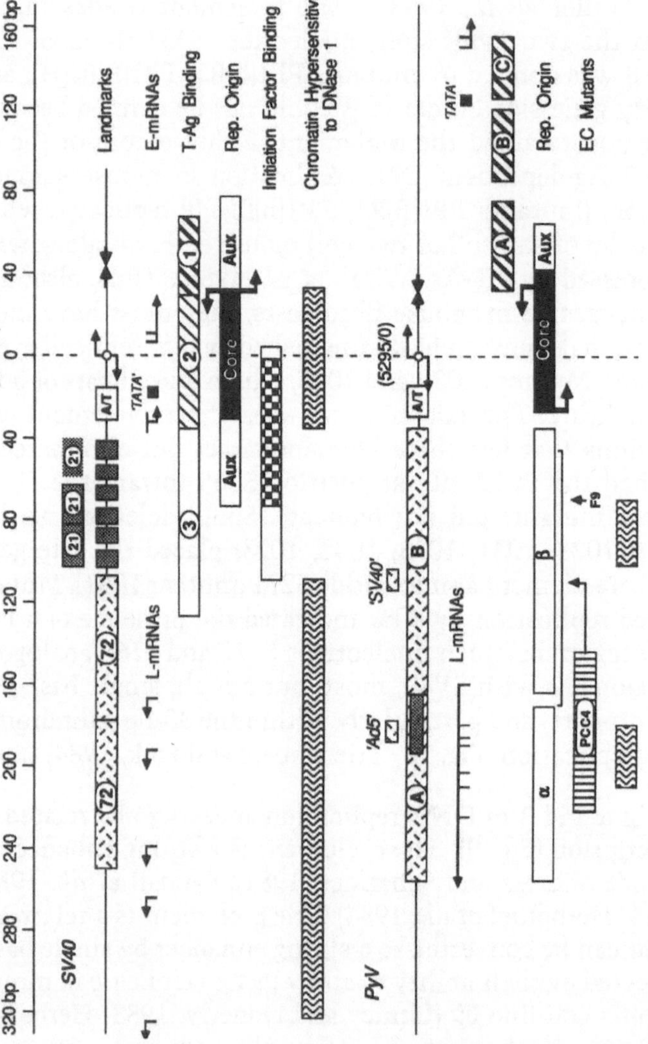
SV40
PyV
(0/5243)
(5295/0)
Landmarks
E-mRNAs
T-Ag Binding
Rep. Origin
Initiation Factor Binding
Chromatin Hypersensitivity to DNase 1
L-mRNAs
EC Mutants
'TATA'
A/T
Core
Aux
'SV40'
'Ad5'
PCC4
F9
α
β
320 bp
280
240
200
160
120
80
40
0
40
80
120
160 bp

FIGURE 9. SV40 and PyV *ori* regions. The center of the major palindrome found in SV40 and PyV *ori* was defined as "0," and the distance in nucleotides to the left or right indicated for simplicity in comparing the two genomes. The conventional numbering systems for SV40 (1 to 5243) and PyV (1 to 5295) are in opposite direction. *SV40* landmarks include two 72-bp perfect repeats that contain enhancer elements, two perfect 21-bp repeats and one imperfect 21-bp repeat that comprise the early promoter, six $G_{3-4}GC_2Pu_2$ repeats (shaded boxes), 17-bp A/T sequence, 27-bp palindrome (pinwheel), and 14-bp inverted repeat (><). Early mRNA initiates at two sites within the 27-bp palindrome before DNA replication begins and then at sites 40 bases upstream after DNA replication begins (Buchman *et al.*, 1984). DNA binding sites for wt SV40 T-Ag (Figs. 3, 4) are shaded according to binding affinity (1 > 2 > 3). The *cis*-acting sequences that function as the SV40 origin of DNA replication (*ori*) include sequences that are required for replication (Core) and auxiliary sequences (Aux) that facilitate replication three- to fivefold (Fig. 7). The transition points from discontinuous to continuous DNA synthesis define the origin of bidirectional replication (OBR) and are indicated by arrows protruding from *ori*-core that point in the direction of synthesis (Figs. 15, 16). A DNA binding site for a factor(s) required to initiate SV40 DNA replication is based on competition *in vitro* between plasmids containing *ori* and plasmids carrying deletions in and about *ori* (Yamaguchi and DePamphilis, 1986). Regions in SV40 chromatin that are hypersensitive to cleavage by DNase I are indicated by boxes with wavy lines; many specific cleavage sites exist within these regions (Jongstra *et al.*, 1984; Cereghini and Yaniv, 1984; Innes and Scott, 1984). *PyV* landmarks include major enhancer elements A and B that contain Ad5 and SV40 enhancer consensus sequences (mottled shading; Tyndall *et al.*, 1981; de Villiers and Schaffner, 1981; Linney and Donerly, 1983; Herbomel *et al.*, 1984; Mueller *et al.*, 1984; Veldman *et al.*, 1985). Shaded area in enhancer A designates minimum required sequence (Herbomel *et al.*, 1984; Veldman *et al.*, 1985). The 15-bp A/T sequence, 34-bp palindrome, and 14-bp inverted repeat are similar to those in SV40. DNA binding sites A, B, and C for wt PyV T-Ag have similar affinities; the affinity of PyV T-Ag for the PyV *ori*-core region is at least tenfold lower (Fig. 5). *cis*-Acting sequences that function as the PyV origin of replication (α, β, and Core) are defined by deletion and point mutations (Fig. 8). The transition points from discontinuous to continuous DNA synthesis that define the origin of bidirectional replication are based on unpublished data of E. A. Hendrickson, C. Fritze, and M. L. DePamphilis. Host-range mutations in PyV that allow replication in mouse embryonal carcinoma (EC) cell line PCC4 involve duplications of sequences in the α-element (range indicated by striped box) that replace sequences in the β-element (range indicated by bar, Melin *et al.*, 1985a,b). Single-point mutations can allow PyV replication in EC F9 cells (Sekikawa and Levine, 1981; Fujimura *et al.*, 1981a; deSimone *et al.*, 1985). DNase I hypersensitive sites in PyV chromomsomes are also indicated (Herbomel *et al.*, 1981).

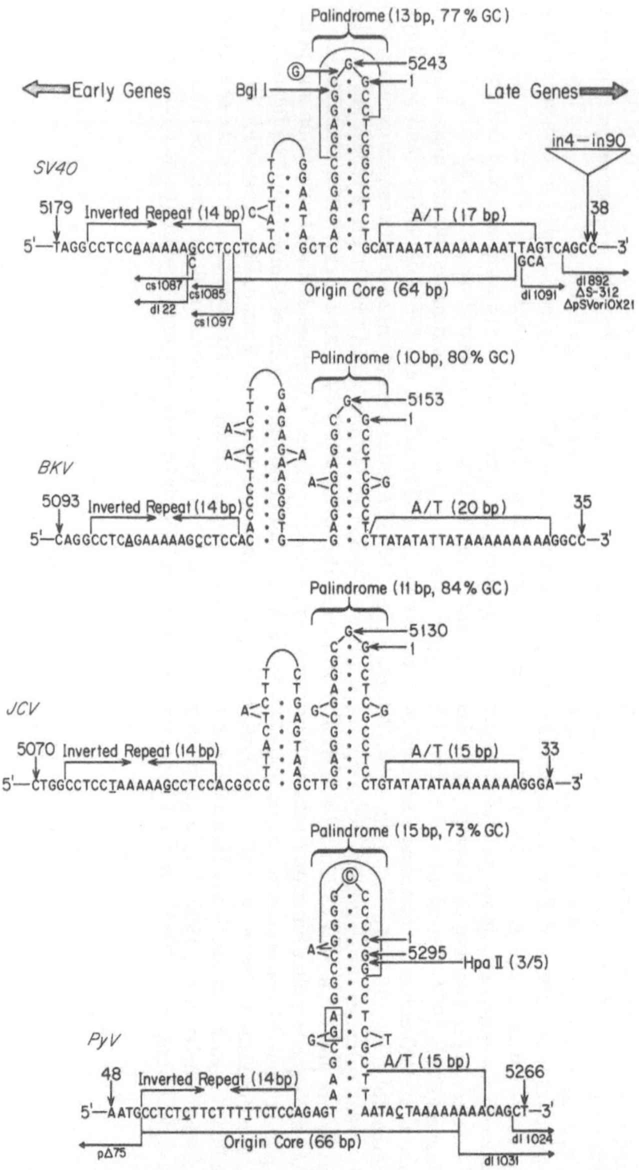

FIGURE 10. The core sequences from the origins of replication of SV40 (Fiers *et al.*, 1978; Reddy *et al.*, 1978; van Heuversywn and Fiers, 1979; Buchman *et al.*, 1980), BKV (Dhar *et al.*, 1978; Seif *et al.*, 1979, 1980a,b; Yang and Wu, 1979), JCV (Miyamura *et al.*, 1983; Frisque, 1983; Frisque *et al.*, 1984), and PyV. The sequence shown is the template for E-mRNA with potential secondary structure included only as an aid for making comparisons; it is unlikely that secondary structure exists *in vivo.* The SV40 numbering system is from Buchman *et al.* (1980), and that from PyV is from B. E. Griffin *et al.* (1980). Note that PyV is numbered in the opposite direction of SV40, BKV, and JCV, but the orientation of each sequence is the same with respect to their early and late genes. Griffin and co-workers (Soeda *et al.*, 1979; Griffin, 1980) reported a T instead of a C at position 5 (circled) that was missing from Deininger *et al.* (1980), while Deininger *et al.* (1980) reported an additional AG at positions 15, 16 (boxed) which was absent from Griffin (1980). However, later sequences for PyV

TABLE II. Viral Sequences That Allow DNA Replication in Mouse Embryonic and Differentiated Cells[a]

	Replication in mouse			
		Embryos		
Plasmid	Oocytes	Alone	+ PyV T-Ag gene	Differentiated cells (PyV T-Ag$^+$)
pML	–	–	–	–
pML-SV40(*ori*$^+$,T-Ag$^+$)	–	3% +	3% +	–
pML-PyV(*ori*$^+$,T-Ag$^+$)	–	+	+	+
PyV T-Ag$^-$ plasmids				
pML-PyV(α-β-core)		–	+	+
pML-PyV(β-core)		–	+	+
pML-PyV(α-core)		–	–	+
pML-PyV(α-β)		–	–	–
pML-PyV(core)				–

[a] Covalently closed, superhelical DNA plasmids were either injected into the nuclei of one-cell or two-cell mouse embryos, or transfected into mouse MOP cells which express PyV T-Ag from an integrated copy of PyV *ori*$^-$ DNA (Wirak *et al.*, 1985). In embryos, PyV T-Ag was provided by coinjecting pML-PyV(*ori*$^+$,T-Ag$^+$), a plasmid containing a complete copy of the PyV early gene region. +, maximum response; –, no response. *ori*$^+$ indicates complete origin of replication sequence (Figs. 7, 8), while α, β, and core refer to the individual *ori* components (Fig. 9).

was sufficient to restore both PyV DNA replication and T-Ag expression in mouse 3T6 or COP cells, but substitution with the SV40 21-bp repeats had no significant effect (de Villiers *et al.*, 1984; Campbell and Villarreal, 1985). Since the SV40 enhancer element functions in either orientation, it is not likely to function by initiating transcription through *ori*-core, but rather act as a binding site for a protein that modifies *ori*-core's interaction with the replication complex. The fact that only β-core functioned as *ori* in mouse embryos (Wirak *et al.*, 1985; Table II) suggests that only the β-element can function as an enhancer in mouse embryos. In fact, although the α-element provided a 3.5-fold higher enhancement of gene activity in mouse fibroblasts compared to mouse embryonal carcinoma cells, the β-element, which was a poorer enhancer than α by about 3-fold, nevertheless had the same enhancer activity in both cell lines (Herbomel *et al.*, 1984).

PyV DNA synthesis likely begins in or near the core sequence because this is the only common homology among papovavirus *ori* regions (Fig.

(Katinka and Yaniv, 1983; Triezenberg and Folk, 1984; Luthman *et al.*, 1984; Cowie and Kamen, 1984; Pomerantz and Hassell, 1984) confirmed the presence of the AG and revealed that position 5 was a G. The nucleotides deleted by mutants in Figs. 7 and 8 that define the outer limits for SV40 and PyV *ori*-core sequences are shown, and the regions in SV40 and PyV where small deletions and point mutations most frequently inactivate *ori* are indicated by a bracket around the palindromic sequence.

10) and the place where bidirectional DNA replication begins in SV40 (Hay and DePamphilis, 1982). Therefore it is interesting to note two differences in the functional organization of these regions. First, both the TATA box and early mRNA initiation sites are located within the SV40 *ori*-core (Figs. 3, 4), whereas both are located outside of the PyV *ori*-core on the early gene side (Figs. 5, 9). Second, SV40 T-Ag binds readily to the SV40 *ori*-core sequence at site II, but PyV T-Ag binds very poorly to the PyV *ori*-core sequence (Figs. 4, 5). Therefore, perhaps an enhancer element compensates for one or both of these differences by altering the chromatin structure of PyV *ori*-core to make it accessible to initiation proteins.

3. Common Features of *ori*-Core among Papovaviruses

The *ori*-core sequences for SV40, PyV, BKV, and JCV are remarkably similar (Fig. 10). They share at least three features in common: an AT sequence of 15 to 20 bp on the late gene side, a GC-rich palindrome of 23 to 34 bp in the center, and an inverted repeat of 14 bp on the early gene side. A second palindrome exists in SV40, BKV, and JCV, but it differs considerably in size, and it is absent from PyV. The location of a large, GC-rich palindromic sequence adjacent to an AT-rich region raises the possibility that the palindromic region may assume a cruciform structure in dsDNA or a hairpin structure in ssDNA. Some palindromic sequences have been shown to form cruciform structures *in vitro* under conditions that promote unwinding of the DNA such as high negative superhelical density and high temperature, and under conditions that stabilize base pairing in the duplex stem portion (Lilley, 1980; Panayotatos and Wells, 1981; Mizuuchi *et al.*, 1982; Gellert *et al.*, 1983; Singleton, 1983; Courey and Wang, 1983; Lilley and Kemper, 1984; Sinden and Pettijohn, 1984). Those palindromes that have GC-rich stem lengths of 10 bp or greater, flanked by AT-rich regions, and allow a loop of 3–6 bases are the ones most likely to form cruciforms *in vitro* under optimal conditions. Purified covalently closed, superhelical SV40 and PyV DNA contain enough free energy to denature localized regions of the genome, making them sensitive to ssDNA-specific endonucleases (Beard *et al.*, 1973; Kato *et al.*, 1973; Germond *et al.*, 1974), DNA binding proteins (Morrow and Berg, 1974; Reed *et al.*, 1975; Lescure *et al.*, 1978), and base modifications (Lebowitz *et al.*, 1976). However, none of these experiments revealed ssDNA in the *ori* region. One possibility is that *ori* may exhibit single-stranded character only when the DNA is organized into chromatin. However, the cruciform conformation of palindromes in dsDNA is not favored kinetically under physiological conditions (Courey and Wang, 1983; Gellert *et al.*, 1983; Singleton, 1983; Sinden and Pettijohn, 1984), and cruciforms were not detected *in vivo* by transforming *E. coli* with a stable palindrome (Mizuuchi *et al.*, 1982), by using psoralen to cross-link duplex DNA regions (Sinden *et al.*, 1983), or by using S1 nuclease to cut ssDNA regions in isolated chromosomes (Herman *et al.*, 1979). Nevertheless, since 8% to 20% of mature chromosomes contained

T-Ag (Reiser *et al.*, 1980; Segawa *et al.*, 1980; Tack and DePamphilis, 1983; Garcea and Benjamin, 1983a; Milthorp *et al.*, 1984), it is possible that only a fraction of the chromosomes expressed the cruciform structure prior to initiation of DNA replication, and therefore went undetected. Some palindromic sequences can, however, form a stable hairpin structure in an ssDNA template *in vitro* (Ricca *et al.*, 1982; Patton and Chae, 1982; Weaver and DePamphilis, 1984), and therefore may do so when replication forks pass through these regions. These hairpins, as well as some nonpalindromic sequences, arrested the progress of DNA polymerase α *in vitro* (DePamphilis *et al.*, 1980; Weaver and DePamphilis, 1982, 1984), but only nonpalindromic arrest sites were detected in *ori* or other genomic regions of SV40 during its replication *in vivo* (Weaver and DePamphilis, 1984). Thus, although palindromic sequences can, in principle, operate through the formation of unique secondary structures, there is no evidence that such structures are involved in DNA replication *in vivo*.

One advantage of a unique *ori* sequence in viral DNA replication is to provide a strong initiation signal that allows 5×10^3 bp of viral DNA to replicate preferentially in the presence of 5×10^9 bp of cellular DNA. Thus, one would expect that more than one *ori* per genome would be advantageous. In fact, variants of SV40 (Lee *et al.*, 1975), PyV (Griffin and Fried, 1975; Magnusson and Nilsson, 1982), and a human papovavirus (Khoury *et al.*, 1974; Martin *et al.*, 1974) that were selected by their ability to outgrow the parent virus during serial passage at a high multiplicity of infection, all contained multiple copies of the *ori* region. The rate of DNA replication in the variant relative to that of the helper virus in the variant stock was an exponential function of the number of *ori* segments per genome (Lee and Nathans, 1979). In an SV40 mutant with two *ori* regions, both were shown to be active, and one could be inactivated without affecting the other (Shenk, 1978). Therefore, it is the *ori* sequence itself and not its genomic location that is recognized during initiation of replication, and it is the rate of initiation that determines the overall rate of viral DNA replication. Most of the variants selected were defective and required a helper virus to produce virions, suggesting that wild-type virus has only one *ori* because one is sufficient and more would limit the amount of genetic information the virus could encapsidate.

C. Cellular Factors That Control Replication

Expression of SV40 or PyV T-Ag in either permissive or nonpermissive cells induced cellular DNA synthesis that resulted in one doubling of cellular DNA (Section II.C; Hiscott and Defendi, 1979a,b; Gershey, 1979), and SV40 (Pages *et al.*, 1973) and PyV (Loche, 1979) DNA replication took place only after synchronized permissive cells entered S phase. Thus, papovaviruses replicate their chromosomes only with the help of cellular replication factors. Superinfection of SV40-infected CV-1 cells that were already active in viral DNA replication revealed a 9-hr lag period before the superinfecting virus initiated its own DNA replication (Rinaldy *et al.*, 1982). This time may be required to uncoat and possibly modify the viral chromosome before it is receptive to initiation factors. Viral DNA can reenter the replication pool up to about 4 hr after replicating (Green and Brooks, 1978; Roman, 1979; Ozer *et al.*, 1981). During this time, a competition exists between reinitiation of DNA replication and conversion of newly replicated DNA into previrion complexes that appears to involve binding of viral proteins to a specific *cis*-acting viral DNA sequence (Wang and Roman, 1981; Wang *et al.*, 1985).

Of the three components that regulate papovavirus DNA replication (*ori*, T-Ag, and permissive cell factors), the cellular factors are the least well characterized. Permissive cells appear to provide factors required for replication rather than nonpermissive cells providing factors that repress replication. SV40 can be rescued from most SV40-transformed mouse and hamster cell lines upon fusion with monkey cells or SV40-transformed permissive cells (Jensen and Koprowski, 1969; Watkins, 1975; Botchan *et al.*, 1980). Viral DNA replication first appears in nonpermissive cell nucleus and then in the permissive cell nucleus, indicating that permissive cell factors are diffusible (Watkins and Dulbecco, 1967; Steplewski and Koprowski, 1969). Failure to rescue integrated SV40 from transformed permissive cells reflects either a defective *ori* or an altered T-Ag because some of these cell lines are superinfectable with SV40 DNA or virus (Swetly *et al.*, 1969; Gluzman *et al.*, 1977; Gluzman, 1981) or produce virus when tsA-transformed cells are grown at the permissive temperature (Noonan *et al.*, 1976; Folk and Bancuk, 1976). Permissive cell factors are encoded by one or more chromosomes. PyV growth in hybrid cell lines containing both mouse and hamster chromosomes was proportional to the number of mouse (permissive) chromosomes present (Basilico *et al.*, 1970), and SV40 replication in monkey–hamster hybrids corresponded to retention of a particular monkey (permissive) chromosome (Garver *et al.*, 1980). However, mouse–human cell hybrids supported replication of SV40 only when human chromosomes were in excess, and PyV only when mouse chromosomes were in excess (Huebner *et al.*, 1977). The reason for this apparent suppression by nonpermissive chromosomes is not clear.

Cells can be nonpermissive for papovavirus production for reasons other than failure to replicate viral DNA. For example, mouse embryonal carcinoma cells (EC cells) block both transcription and replication of PyV or SV40 DNA (Levine, 1982). Human fibroblasts are equivalent to established permissive monkey cell lines in their ability to replicate SV40 DNA although they yield little virus (Ozer *et al.*, 1981). Primary human embryonic kidney cells are permissive for BKV, but only BKV DNA replication was detected in an SV40-transformed HEK cell line (Major and Matsumura, 1984). However, there are three basic mechanisms by which permissive cell factors can control initiation of viral DNA replication. They can regulate the concentration of T-Ag at the level of its transcription or translation, they can posttranslationally modify T-Ag to an active or stable form, and they can directly participate in the initiation process by forming a T-Ag/initiation factor(s)/origin of DNA replication complex.

The higher level (about threefold) of T-Ag produced in SV40-infected monkey cells compared with SV40-infected mouse cells has been suggested as an explanation for the difference in the two cells' ability to replicate SV40 DNA (Graessmann *et al.*, 1981). This conclusion rested on the observation that high levels of SV40 T-Ag induced the appearance of SV40 capsid antigen in mouse nuclei (Graessmann *et al.*, 1978), and the expression of late viral genes was thought to require viral DNA replication, although viral DNA synthesis was not measured. However, it is now clear that SV40 T-Ag can induce late gene expression in the absence of viral DNA replication (Keller and Alwine, 1984; Brady *et al.*, 1984; Hartzell *et al.*, 1984). Furthermore, levels of nuclear T-Ag undetected by immunofluorescence were sufficient for SV40 DNA synthesis in permissive cells (Lanford and Butel, 1980a). Thus, there is no clear correlation between the concentration of T-Ag and the ability to initiate viral DNA replication.

Only a small fraction of T-Ag appears to be capable of initiation of viral DNA synthesis, and several posttranslational modifications have been identified that may be related to T-Ag activity (Section II.D). DNA damaging agents such as radiation, chemical carcinogens and mutagens can induce viral DNA replication in a variety of transformed cell lines (discussed in Seidman and Salzman, 1983, pp. 44–48), and DNA damage can induce enzymes such as ADP-ribose polymerase (Benjamin and Gill, 1980) that may modify T-Ag. Differences in the size of T-Ag isolated from permissive and nonpermissive cells have also been reported, but these appear to represent artifacts generated during the extraction procedure (Section II.E). Thus, it is not yet clear which, if any, of the known T-Ag modifications are involved in initiation of viral DNA replication. This situation is likely to change with the availability of subcellular systems capable of initiating viral DNA replication *in vitro* (Section III.E).

One cell-specific control over viral DNA replication is transcriptional regulation of T-Ag synthesis. Mouse EC cells, the stem cells of teratocarcinomas that can differentiate into a variety of somatic cell types, have many similarities to multipotent cells of 7- to 8-day mouse embryos (Martin, 1980). PyV and SV40 virions entered and uncoated in EC cells, but viral mRNA was substantially reduced and neither early nor late viral proteins were detected unless the EC cells were first induced to differentiate (Swartzendruber *et al.*, 1977; Boccara and Kelly, 1978; Segal and Khoury, 1979; Fujimura *et al.*, 1981b; Trevor and Lehman, 1982; Levine, 1982; Dandolo *et al.*, 1983). PyV mutants have been isolated that replicated efficiently in EC cells, and the mutations mapped in the viral enhancer–promoter region (Fig. 9; Sekikawa and Levine, 1981; Katinka *et al.*, 1980, 1981; Vasseur *et al.*, 1980; Fujimura *et al.*, 1981a; Melin *et al.*, 1985a,b). While these host-range mutants continued to replicate in differentiated mouse cells as well as the EC cell line on which they were selected, some of them did not replicate in other, independently isolated,

EC cell lines. Some of the mutations were simple alterations in the β-element (e.g., F9 cell mutants), while others (e.g., PCC4 cell mutants) duplicated a sequence overlapping the Ad5 E1A-like enhancer (nucleotides 5100 to 5122), deleted part of the β-element, and transposed the duplicated sequence into the β-element. The mutated PyV enhancer, in contrast to wild-type PyV or murine leukemia virus enhancers, allowed expression of nonpolyoma genes in EC cells (Linney and Donerly, 1983; Linney *et al.*, 1984).Thus, the production of PyV T-Ag is insufficient in some mouse cells to allow PyV DNA replication.

The activity of viral enhancers from SV40, PyV, and murine sarcoma virus (MSV) varied up to fivefold among different cell species; SV40 enhancer was more effective in primate cells while PyV and MSV enhancers were better in mouse cells (de Villiers *et al.*, 1982; Laimins *et al.*, 1982; Kriegler and Botchan, 1983). Similarly, the enhancer region of the immunoglobulin gene preferentially stimulated transcription of genes in immunoglobulin-producing B lymphocytes (Banerji *et al.*, 1983; Gilles *et al.*, 1983; Mercola *et al.*, 1983), and cell-specific expression of insulin and chymotrypsin genes was determined by the 5′-flanking regions of these genes (Walker *et al.*, 1983). Furthermore, the tissue tropism and virulence of murine leukemia virus-induced lymphomas was associated with the LTR enhancerlike sequences (Chatis *et al.*, 1983; der Groseillers *et al.*, 1983). It appears that one or more cellular proteins specifically interact with enhancer sequences to stimulate selectively the expression of specific genes according to cell type (Scholer and Gruss, 1984; Emerson and Felsenfeld, 1984).

A second control over viral DNA replication involves the ability of T-Ag to interact correctly with the *ori* region. PyV DNA failed to replicate in EC F9 cells even when coinfected with a host-range PyV mutant that produced normal T-Ag and replicated its own DNA (Fujimura and Linney, 1982). Therefore, the PyV region (β-element, Fig. 9) carrying this EC host-range mutation was required in *cis* to initiate DNA replication as well as to activate expression of early viral genes. The β-element also confered cell-specific recognition of the PyV origin of DNA replication. In the presence of T-Ag that was provided from an integrated PyV genome, both PyV α-core and β-core function as *ori* sequences in differentiated mouse cells, but only β-core functioned as an *ori* sequence when coinjected into the nuclei of mouse embryos together with another plasmid that provided T-Ag (Wirak *et al.*, 1985; Table II). deVilliers *et al.* (1982, 1984) observed that substitution of the 72-bp repeats (enhancer elements) from SV40 for the *Pvu*II(5265)–*Bcl*I region of PyV (Figs. 8, 9) allowed PyV DNA to replicate in both mouse and human (HeLa) cells, and that the mouse immunoglobulin gene enhancer allowed replication in mouse myeloma cells but not in mouse fibroblasts. These data suggest that the enhancer sequence alone could determine host range. However, Campbell and Villarreal (1985) and J. Hassell (personal communication) observed that SV40 enhancer elements allowed PyV *ori*-core to replicate in mouse cells but *not* in CV-1 monkey cells. Conversely, substitution of the PyV enhancer for the SV40 enhancer allowed efficient SV40 T-Ag production in mouse cells, but not SV40 DNA replication, indicating that an inadequate concentration of viral T-Ag was not responsible for failure of SV40 DNA replication in mouse cells (Weber *et al.*, 1984). SV40 virus with a PyV enhancer still replicated in permissive monkey cells. Plasmids containing either PyV or SV40 *ori*-core coupled with either PyV α–β region or SV40 72-bp + 21-bp repeat region (Fig. 9) were used to transfect CV-1 monkey cells expressing either SV40 (COS cells) or PyV (COP cells) T-Ag, or 3T3 mouse cells expressing either PyV (MOP cells) or SV40 (MOS cells) T-Ag (J. Hassell, personal communication). The result was that replication depended only on a unique interaction between *ori*-core, T-Ag, and permissive host cell factors. The SV40 *ori* functioned to initiate DNA replication only when present in COS cells, and the PyV *ori* functioned only in MOP cells, regardless of which viral gene enhancer region was present. Therefore, host-range specificity for cells of different animal species depends only on the interaction of *ori*, T-Ag, and cell factors. Although enhancer sequences are required to activate the PyV *ori*, they do not determine species-specific host range, but they may determine cell-specific host range within the same organism. Apparently, viral DNA replication occurs in a permissive cell species because those cell factors cause an active replication complex to form between T-Ag and its homologous *ori*-core region. In the case of PyV, the nature of the enhancer element required for *ori* function can also impart cell-type specificity, at least within the same organism.

The fact that mouse EC cells neither replicate nor transcribe PyV DNA suggests that permissive cell factors are not expressed during early mouse embryonic development (Kelly and Condamine, 1982; Levine, 1982). However, injection of PyV DNA sequences into the nuclei of mouse one-cell or two-cell embryos revealed that the β-core configuration of PyV *ori* allowed DNA replication, but only in the presence of PyV T-Ag (Table II; Wirak *et al.*, 1985). Therefore, mouse permissivity factors are expressed at the earliest stages in development. EC cell lines are either atypical of normal mouse development, or represent specific stages in embryonic development where permissivity factors are not expressed. A likely candidate is the inner cell mass of the blastocyst. The SV40 *ori* sequence allowed replication at about 3% the level seen with PyV. This may represent a low level of embryonic expression of permissive cell factors common to all mammalian cells. Alternatively, SV40 *ori* may be homologous to certain cellular *ori* sequences. A mouse chromosomal DNA sequence has been isolated that shares partial homology with SV40 *ori*, initiates replication in an SV40 *in vitro* replication system, and replicates when transfected into COS cells (Ariga *et al.*, 1985).

Cellular factors may also affect the mechanism of DNA replication as well as regulate its initiation. Infection of nonpermissive cells with SV40 (Chia and Rigby, 1981) or semipermissive cells with human papilloma virus (Wettstein and Stevens, 1982) resulted in a small amount of head-to-tail concatenated DNA instead of normal replicating intermediates, suggesting that replication occurred via a "rolling-circle" mechanism. Concatemer formation did not require SV40 T-Ag. In contrast, concatemers were not a product of either PyV or SV40 DNA replication in mouse embryos (Wirak *et al.*, 1985). PyV strain A2 DNA replicated at a low rate in EC cells via some T-Ag-independent mechanism, whereas T-Ag was required for PyV DNA replication after EC cells had differentiated *in vitro* (Dandolo *et al.*, 1984). Interpreting the role of T-Ag in these studies depended on inactivation of tsA mutants.

D. Role of *ori* in Excision and Integration of Viral DNA

In both PyV (Pellegrini *et al.*, 1984) and SV40 (Conrad *et al.*, 1982; Miller *et al.*, 1984), excision and subsequent amplification of integrated copies of viral DNA required the *ori* region, the homologous T-Ag, and permissive cell factors (reviewed in DePamphilis and Wassarman, 1982, pp. 54–55). In the absence of permissive cell factors (e.g., in nonpermissive transformed cells), some data suggested that *ori* was still activated by T-Ag, but, if so, it did so in the same controlled manner that cellular origins were activated during S phase. Otherwise, amplification and/or replication of the viral DNA would occur continuously in proliferating transformed cell lines containing an intact (i.e., rescuable) viral genome. A functional *ori* was not required for integration of viral DNA, although transformation efficiency was increased about six-fold when replication could occur (Muller *et al.*, 1983; Dailey *et al.*, 1984). The resulting cell lines contained tandem head-to-tail arrangements of integrated viral genomes, suggesting that viral DNA replication occurred at a low level in the absence of permissive cell factors using a rolling-circle mechanism (Kornberg, 1980) that initiated at *ori*. In fact, nonintegrated head-to-tail SV40 concatemers have been identified in SV40-infected mouse cells (Chia and Rigby, 1981).

E. Subcellular Systems for the Replication of Viral Chromosomes

The development and properties of subcellular systems that either continue replication (including chromatin assembly) of endogenous viral chromosomes already present in the infected cells, or that carry out DNA replication upon addition of purified DNA have been discussed in detail by DePamphilis and Wassarman (1982, pp. 59–63) and Richter and Otto (1983). These systems have contributed significantly to our understanding of the events at replication forks and the maturation of replicating intermediates, but little toward our knowledge of initiation of replication at *ori*. Despite several encouraging reports that initiation of SV40 or PyV DNA replication occurred *in vitro* (Magnusson and Nilsson, 1979; Waldeck *et al.*, 1979; Cuzin and Clertant, 1980; Oda *et al.*, 1980), no further progress with these systems has been reported.

By taking advantage of recombinant DNA and blotting-hybridization technology to increase the signal-to-noise ratio, three new systems have been developed that appear capable of addressing the problem of initiation. Ariga and Sugano (1983) reported that a high-salt extract of HeLa cell nuclei and the cytoplasm from SV40-infected COS-1 cells initiated bidirectional replication in SV40 DNA or plasmids containing SV40 *ori*. Li and Kelly (1984) used a low-salt extract of either SV40-infected COS-1 or BSC-40 cells or uninfected permissive cells supplemented with purified SV40 T-Ag (Li and Kelly, 1985) to achieve results similar to those of Ariga and Sugano. In both systems, DNA synthesis was dependent on T-Ag and an intact *ori*-core sequence; a 4- or 6-bp deletion at the *Bgl*I site prevented replication. DNA synthesis was inhibited by aphidicolin but not by ddTTP, indicating the involvement of α-polymerase. Neither system was sensitive to α-amanitin, demonstrating that RNA polymerases II and III were not required (Ariga and Sugano, 1983; Li and Kelly, 1984; Ariga, 1984b). Both systems allowed extensive semiconservative DNA replication, although only about 1% of the input DNA appeared to be utilized for replication (Ariga and Sugano, 1983) and a small fraction (< 0.2%) of the input DNA contained either one or two replication forks (Li and Kelly, 1984). Finally, Decker *et al.*, (1986) have developed conditions that specifically stimulate initiation of DNA synthesis at *ori* in endogenous viral chromosomes present in a hypotonic extract from SV40-infected CV-1 cells. This system was also insensitive to α-amanitin.

Most of the DNA synthesis did not appear to occur via the normal RI mechanism in the system described by Ariga (1984a). T-Ag appeared to catalyze the breaking of one strand in the *ori* region to produce form II DNA, the product most rapidly labeled, which was then converted into form I DNA with varying extents of superhelical density (Ariga, 1984a). These form II molecules did not represent topologically relaxed early RI since higher-molecular-weight DNA, equivalent to topologically relaxed late RI, did not appear concurrently in the agarose gel analysis. Therefore, these form II DNA intermediates appeared to contain multiple nicks, resulting in release of nascent DNA chains of varying lengths. The superhelical turns presumably resulted from nucleosome assembly and sealing of the strand breaks, and DNA synthesis presumably resulted from a rolling-circle mechanism by initiating synthesis at the 3′-OH end of the strand break, or by DNA primase–DNA polymerase α initiating synthesis *de novo*. In either event, extensive strand displacement implies the participation of additional proteins such as topoisomerases and helicases which may explain the requirement for high concentrations of ATP (Li and Kelly, 1984).

Ariga (1984b) also observed that the human Alu family sequence functioned as an *ori* sequence in his SV40 DNA replication system. The Alu segment contained a GAGGC consensus sequence arranged similarly to T-Ag binding site II in SV40 *ori*, which may explain its activity in the *in vitro* system. However, the same plasmid did not function as an origin when transfected into COS cells (R. Tjian, personal communication), or when added to the system of Li and Kelly (1985). The latter system appeared specific for the SV40 *ori* region and permissive cell factors from monkey (CV-1) or human cells (HeLa); Alu family repeats, BKV and JCV *ori* sequences were less than 10% as effective as SV40. Ariga *et al.* (1985) also identified mouse cell DNA sequences that acted as ARS sequences in yeast, and *ori* sequences in their SV40 *in vitro* replication system. One mouse ARS failed to replicate when transfected into COS cells, but another one did replicate. Low-stringency T-Ag/DNA interactions might account for induction of cellular DNA replication which occurs even with T-Ag mutants that are incapable of initiating viral DNA replication (Section II.C). SV40 T-Ag binds to GAGGC sequences in DNA from many sources (Wright *et al.*, 1984), including a polymer of *Xho*I restriction sites (Pomerantz and Hassell, 1984), indicating that mammalian chromosomes frequently contain sequences, such as the Alu repeat family, that share some functional homology with the *ori*-core sequence of papovaviruses.

Initiation of replication on purified DNA added to cell extracts has several potential pitfalls that should be considered. First, purified form I DNA (Hay and DePamphilis, 1982) and mature (nonreplicating) viral chromosomes (Krauss and Benbow, 1981; Tack and DePamphilis, 1983) can contain early RI molecules as well. Therefore, if only a small fraction of added DNA (or chromosomes) appears to initiate replication, it could be due to contaminating RI. Furthermore, since T-Ag may be associated with replication forks in RI (Section II.F.5), T-Ag might stimulate DNA synthesis at replication forks in contaminating RI as well as synthesis at *ori* in form I DNA. One important control for this artifact was the use of *E. coli* plasmid DNA containing the SV40 *ori*. Since only the plasmid containing a functional SV40 *ori* replicated, *de novo* initiation was strongly indicated. However, a second artifact was observed by Jong and Scott (1984) who reported that *E. coli* plasmids carrying yeast ARS elements (the putative yeast *ori* sequence) could contain RNA specifically associated with the ARS-plasmid, and not the control plasmid. This RNA was not removed by the usual purification procedures and therefore acted as a primer for DNA synthesis on these ARS-plasmids either with yeast cell extracts or with purified DNA polymerases. A third potential artifact is transcriptional activation. RNA polymerases can use covalently closed, superhelical DNA as a template for RNA synthesis which generates a primer-template for DNA polymerases. When the template was nicked at a low rate to allow DNA unwinding, the nicked molecules supported extensive DNA synthesis (Champoux and McConaughy, 1975). In this respect, it is interesting to note that RNase A, an ssRNA endonuclease that does not remove RNA primers from replication forks (Kaufmann, 1981), was a potent inhibitor of initiation of SV40 DNA replication *in vitro* (Ariga and Sugano, 1983), and that a high ATP concentration promoted initiation of RNA synthesis. It is possible, but not likely, that RNA polymerase I, which is resistant to α-amanitin and was implicated previously in the initiation of DNA synthesis in subcellular systems (Brun and Weissbach, 1978), promotes transcription at the SV40 *ori*, which includes both early and late gene promoters. In this regard, it is also important to recall that T-Ag promotes L-mRNA transcription even in the absence of DNA replication (Keller and Alwine, 1984; Brady *et al.*, 1984). Thus, a detailed characterization of any *in vitro* system will be necessary in order to interpret the biological significance of the data.

F. Replicating Intermediates

DNA replication was initiated at the *ori* of SV40(I) and PyV(I) DNA (Section III.B). Initiation occurred preferentially on those molecules with the highest negative superhelical density (Chen and Hsu, 1984), indicating that only chromosomes with a full complement of nucleosomes can initiate replication. Alternatively, the increased torsional strain in these chromosomes may have resulted from a specific DNA gyrase-like activity used to activate origins of replication and/or transcription. Such an activity seems to exist in *Xenopus* oocytes where transcription from injected plasmids occurred only on chromosomes whose topological strain was not stable, as is typical with chromosomes organized into classical nucleosomes, but could be relaxed by coinjection of nucleases, topoisomerase I, or novobiocin, a DNA gyrase inhibitor (Ryoji and Worcel, 1984). Once initiation occurs, papovavirus DNA replication proceeded semiconservatively (Magnusson *et al.*, 1972; Hirt, 1966) and bidirectionally with both forks advancing at similar rates (Danna and Nathans, 1972; Fareed *et al.*, 1972; Crawford *et al.*, 1973, 1975; Robberson *et al.*, 1975; Seidman *et al.*, 1978; Bjursell *et al.*, 1979; Tapper *et al.*, 1979; Tapper and DePamphilis, 1980; Martin and Setlow, 1980). The resulting RI (Fig. 11A) contained three loops of DNA, no visible ends, and two forks with some ssDNA character as judged by chromatography on BND-cellulose, electron microscopy, sensitivity to ssDNA-specific endonucleases, and superhelical density (Bourgaux *et al.*, 1969, 1971; Levine *et al.*, 1970; Bourgaux and Bourgaux-Ramoisy, 1971; Meinke and Goldstein, 1971; Sebring *et al.*, 1974; Herman *et al.*, 1979). RI DNA sedimented as a broad peak centered at 25 S and was a precursor to form I (21 S) and form II (16 S) DNA in pulse–chase experiments. The two loops of nascent dsDNA (the two sibling molecules) are topologically relaxed while the unreplicated DNA contains negative superhelical turns (Sebring *et al.*, 1971, 1974; Jaenisch *et al.*, 1971; Bourgaux and Bourgaux-Ramoisy, 1972; Salzman *et al.*, 1973, 1975; Roman *et al.*, 1974). Denaturation of RI DNA released all nascent DNA strands while the parental strands remained intertwined; daughter strands were not longer than one genome in length. Therefore, parental strands remained covalently sealed during replication,

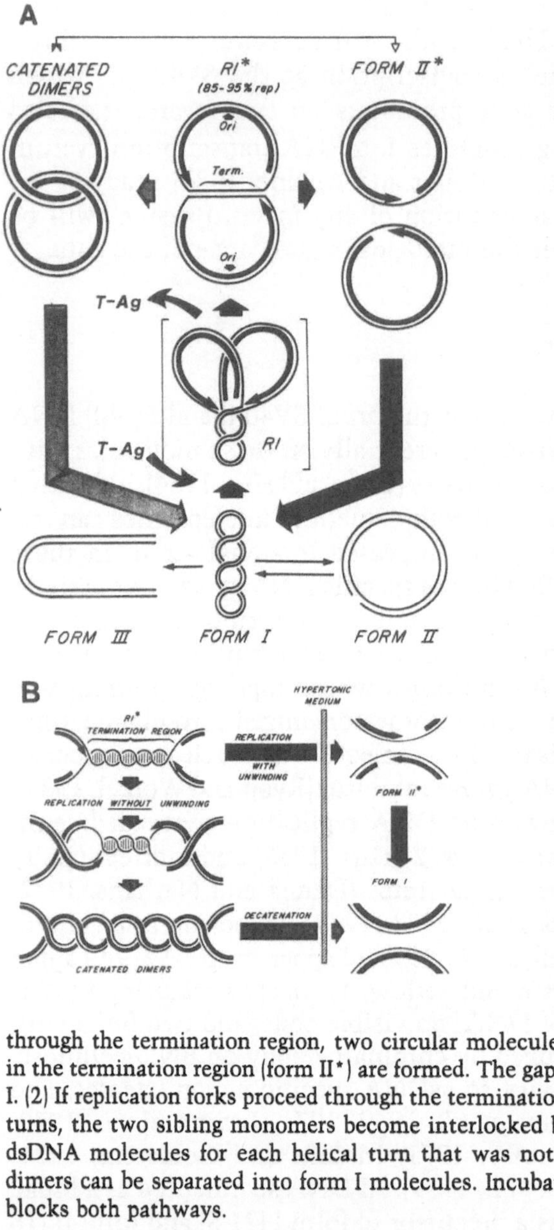

FIGURE 11. Major SV40 and PyV DNA replication intermediates. (A) Form I DNA is the circular, double-stranded, covalently closed, superhelical mature form of viral DNA. Bidirectional replication from the unique origin of replication generates RI DNA, the circular replicating intermediate form of DNA containing two replication forks and covalently closed, parental DNA strands. RI* DNA is replicating intermediate DNA about 90% completed. Form II* DNA is the circular form of viral DNA with a nick or gap in the newly synthesized DNA strand in the termination region. Form II DNA is the circular form of viral DNA containing a randomly placed phosphodiester bond interruption ("nick") or gap of one or more nucleotides in either of the two strands. Form III DNA is the linear, double-stranded form of viral DNA. Forms I, II, and II* DNA all contain one copy of the genome and are therefore monomers. Catenated dimer DNA contains two circular copies of the genome interlocked by one or more intertwines of duplex DNA. (B) Two alternative pathways for separation of sibling DNA molecules. The termination region is represented for RI*, form II*, catenated dimers, and form I DNA. (1) If the remaining Watson–Crick turns between the replication forks are unwound prior to fork movement through the termination region, two circular molecules with a gap in the nascent strand in the termination region (form II*) are formed. The gap is filled and sealed to produce form I. (2) If replication forks proceed through the termination region without unwinding helical turns, the two sibling monomers become interlocked by one intertwine between the two dsDNA molecules for each helical turn that was not removed. The resulting catenated dimers can be separated into form I molecules. Incubation of cells in hypertonic medium blocks both pathways.

and the major replication pathway was not a "rolling-circle" mechanism (Kornberg, 1980). This mechanism assumes that a topoisomerase activity releases the torsional strain that accumulates as the DNA helix is unwound, either by acting as a swivel in front of replication forks, or intertwining the two sibling molecules behind the forks (Champoux and Been, 1980). Superhelical turns are not expressed *in vivo* because both unreplicated and newly replicated DNA are organized into nucleosomes (Section V). As replication proceeded, the changes in superhelical density and conformation of RI DNA allowed these molecules to be fractionated according to the extent of their replication using either equilibrium centrifugation in CsCl–ethidium bromide (or propidium diiodide) gradients (Sebring *et al.*, 1971; Salzman *et al.*, 1973; Fareed *et al.*, 1973a; Grossman *et al.*, 1974; Salzman *et al.*, 1975) or electrophoresis in agarose gels (Tegtmeyer and Macasaet, 1972; Martin, 1977; Tapper and DePamphilis, 1978; Tapper *et al.*, 1982; Buckler-White *et al.*, 1982).

Electron microscopic analysis of SV40(RI) (Tapper and DePamphilis, 1980; Buckler-White *et al.*, 1982) and PyV(RI) (Crawford *et al.*, 1974; Bjursell *et al.*, 1979; Buckler-White *et al.*, 1982) DNA revealed that movement of the two replication forks in the same molecule was not synchronized. In SV40, the total population of replication forks advanced at the same average rate, resulting in termination of replication at a point about 180° from *ori*. However, the normal distribution about the mean revealed that only one-third of the molecules contained two forks at the same distance from *ori*; the average pair of replication forks were out of synchrony by 12% of the distance traveled. This degree of asymmetry was similar at all stages of replication, suggesting that forks move at different rates instead of, or in addition to, leaving *ori* at different times. Therefore, forks close to *ori* appear to be initiated concurrently and move out at similar rates (Martin and Setlow, 1980), but, as a result of nonuniform rates of travel, when the average replication fork traverses 50% of genome, the other fork will only have traversed 44% of the genome, leaving the two forks separated by an average of 315 bp (Tapper and DePamphilis, 1980). In PyV, the medium fork asymmetry was about 400 bp compared with 190 bp observed for SV40 (Buckler-White *et al.*, 1982), but again the degree of fork asymmetry appeared the same throughout replication (Bjursell *et al.*, 1979). The fraction of unidirectionally replicating SV40 or PyV RI *in vivo* (one active replication fork) has been reported to be 8–10% (Robberson *et al.*, 1975; Martin and Setlow, 1980) to 32–35% (Buckler-White *et al.*, 1982).

Later during infection, large circular and linear DNA concatemers were found containing 2 to 10 tandemly arranged, head-to-tail SV40 (Martin *et al.*, 1976; Rigby and Berg, 1978) or PyV (Ganz and Sheinin, 1983) genomes that constituted less than 0.1% of the viral DNA (Ganz and Sheinin, 1983). Mixed infections with two mutants showed that each concatemer contained a single genome and, therefore, must have resulted from DNA replication, perhaps via a rolling-circle mechanism (Rigby and Berg, 1978). Circular concatemers that consist of multiples of two may arise through replication of circular dimers that originated by intramolecular recombination either in RI or in catenated dimers (Jaenisch and Levine, 1971a,b; Martin *et al.*, 1976). Although DNA circles with a tail longer than one genome in length (rolling circles) have been detected by electron microscopy of SV40 (Martin and Setlow, 1980; Tapper *et al.*,

1982) and PyV (Bjursell *et al.*, 1979) DNA, RI molecules were the dominant replication form. As the lytic cycle progressed, the fraction of putative rolling circles among RI increased from 0 to 6% (Bjursell, 1978; Bjursell *et al.*, 1979). Addition of 2′-deoxy-2′-azidocytidine (Cz) inhibited replication 80–90% and increased the fraction of rolling-circle molecules in the RI pool to 33% (Bjursell, 1978). In the presence of Cz, the lengths of tails on putative rolling circles varied from 20% to 200% the length of one genome, and the size of these molecules was consistent with about 75% having initiated unidirectional replication at the normal *ori* (Bjursell, 1978).

Rolling-circle replicating molecules (DNA circles with a tail) could arise from RI molecules fractured at the nuclease-sensitive, ssDNA region in one of the two replication forks in replicating chromosomes (Herman *et al.*, 1979; Cusick *et al.*, 1981; Tapper *et al.*, 1982; Gourlie and Pigiet, 1983) while the remaining fork continued replication. In the absence of a termination signal, replication would then continue unidirectionally around the genome. For example, cleavage of one fork in the early RI that accumulate in Cz-treated cells (Bjursell *et al.*, 1977; Eliasson *et al.*, 1981) would generate the observed rolling-circle molecules that initiated replication at *ori*. Cleavage of RI DNA appears to occur in subcellular DNA replication systems as well. Both the extent of fork asymmetry and the fraction of unidirectionally replicating molecules increased significantly when viral DNA replication continued in isolated nuclei that were not supplemented with cytosol (Bjursell *et al.*, 1979; Buckler-White *et al.*, 1982). Similarly, DNA replication in nuclei from SV40-infected CV-1 cells (Tapper *et al.*, 1982) and in PyV chromosomes in a hypotonic nuclear extract (Gourlie and Pigiet, 1983) accumulated RI broken at one replication fork. These data allowed a reanalysis of other results and revealed that fractured RI were more evident in subcellular systems such as nuclei (Tapper and DePamphilis, 1978; Gourlie and Pigiet, 1983) and nuclear extracts (Su and DePamphilis, 1978) than with intact cells (Tapper and DePamphilis, 1978). The fraction of radiolabel appearing as fractured RI depended on incubation conditions and whether the fate of RI was followed by a pulse–chase or continuous labeling protocol. Addition of cytosol (Tapper *et al.*, 1982) or omission of $MgCl_2$ from the reaction mix (Gourlie and Pigiet, 1983) reduced dramatically the amount of fractured RI that accumulated. Therefore, it appears that divalent metal ions activate an endonuclease activity associated with replication forks unless a cytosol factor is present. Perhaps loss of this cellular factor as the viral lytic cycle progresses converts SV40(RI) DNA into the "rolling-circle" molecules that are most evident late during infection.

SV40(III) DNA, generally a minor (< 1%) species in lytic infections, has been observed at higher levels, particularly at late times during infection (Gruss and Sauer, 1977). However, it did not behave as a transient intermediate in viral DNA replication, and appeared to result from an unusual amount of recombination between viral and cellular DNA. Purified viral DNA contained cellular DNA of a discrete size, and SV40 (III) DNA consisted of defective genomes containing cellular sequences.

In summary, the normal mode of papovavirus DNA replication is via bidirectionally replicating RI molecules that predominate when the rate of viral DNA synthesis is greatest. Concatenated DNA forms, and rolling-circle intermediates are consistently associated with abnormal physiological conditions such as late during infection of permissive cells when the cells are dying, infection of nonpermissive cells incapable of efficient viral DNA replication, treatment of cells with drugs that inhibit DNA replication, and manipulation of subcellular viral DNA replication systems. The fact that unidirectional replicating DNA occurred more frequently with PyV RI than with SV40 RI may be related to the less efficient DNA replication normally observed with PyV-infected mouse cells compared with SV40-infected monkey cells.

G. Separation of Sibling DNA Molecules

1. Late Replicating Intermediates (RI*)

Elongation of nascent DNA chains on SV40(RI) DNA proceeds bidirectionally until replication is 85–95% completed. At this point, late replicating DNA intermediates (RI*, Fig. 11A) accumulate to a level 2–3 times greater than observed for an equivalent sample of RI at earlier stages in replication, indicating that separation of sibling molecules is a slow step in replication. This accumulation was demonstrated by both electrophoretic and electron microscopic analysis of purified SV40(RI) containing nascent DNA strands uniformly labeled in intact cells (Tapper and DePamphilis, 1978; Sundin and Varshavsky, 1980; Tapper *et al.*, 1982; Tack and DePamphilis, 1983). Viral DNA was purified by several different procedures, and the distribution of RI in agarose gels was standardized in different ways to eliminate artifacts. The accumulation of SV40(RI*) was also demonstrated by fractionation of SV40(RI) DNA that was pulse-labeled in isolated nuclei (Tapper and DePamphilis, 1978) and nuclear extracts (Seidman and Salzman, 1979; D. Tapper, unpublished data) supplemented with cytosol. However, the extent to which RI* was observed varied considerably. Earlier reports that SV40(RI*) DNA accumulated up to eightfold over earlier RI appear to have overestimated the extent of accumulation because of difficulty in identifying replicating molecules by electron microscopy (Levine *et al.*, 1970; Bourgaux *et al.*, 1971), and by not separating, in dye-density gradients, RI* from SV40(II*) and dimeric DNA which are labeled rapidly during replication (Mayer and Levine, 1972). In contrast, analysis of PyV(RI) DNA by electron microscopy (Bjursell and Magnusson, 1976; Bjursell *et al.*, 1977), distribution of pulse-labeled DNA (Magnusson and Nilsson, 1979), and gel electrophoresis (Martin, 1977) did not detect an accumulation of RI*. Part of this discrepancy may reflect differences in data analysis since electrophoretic fractionation of PyV(RI) DNA (Bjursell and Magnusson, 1976; Bjursell *et al.*, 1977, 1979; Martin, 1977) and SV40(RI) DNA (Tegtmeyer and Macasaet, 1972; Chou *et al.*, 1974; Tapper and DePamphilis, 1978; Sundin and Varshavsky, 1980) under comparable conditions generated remarkably similar patterns. Part of the observed variation among experiments may reflect the physiological conditions of the infected cells. Increasing the concentration of cell culture media by as little as 20% not only increased the fraction of nascent DNA in catenated dimers (Sundin and Varshavsky, 1981) but also the fraction in RI* (DePamphilis *et al.*, 1983a; Weaver *et al.*, 1985). A third problem is the sensitivity of RI structure to conditions that either relax the covalently closed parental strands or break one of the parental strands at a replication fork, dramatically altering the distribution of RI in various fractionation procedures. The previously reported accumulation of RI* at about 80% completion (Tapper and DePamphilis, 1978) most likely resulted from SV40(RI) DNA in which the ssDNA at one replication fork was broken (Tapper *et al.*, 1982; Gourlie and Pigiet, 1983).

SV40 replication forks are arrested at specific sites in the termination region. The 3′-ends of long nascent DNA chains on purified SV40(RI) DNA were labeled specifically, then cut with a restriction endonuclease, and the sizes of the released fragments determined by gel electrophoresis (Tapper and DePamphilis, 1980). The results showed that DNA synthesis on the forward arms of replication forks paused at a minimum of two major and ten minor sites that were separated one from another an average of 112 bp and were spread over a distance of about 1320 bp. The two major DNA replication arrest sites were separated by about 470 bp of unreplicated DNA centered at 2743 bp, about 52% of the genome from the *Bgl*I site at *ori* (nt 5256; wt800, Fig. 16). This region encompassed the normal termination site for DNA [*Hin*dII + III(B/G) junction] (Danna and Nathans, 1972; Lai and Nathans, 1975; Tapper *et al.*, 1979). Therefore, most replication forks pause when replication is 91% completed, consistent with the accumulation of RI*. Analysis of the locations of PyV replication forks based on the change in mobility of DNA restriction fragments during gel electrophoresis also revealed the accumulation of forks at specific sites flanking the termination region 180° from *ori* (Buckler-White and Pigiet, 1982). In addition, a smaller accumulation of forks was observed on the late gene side of *ori*, and the general pattern suggested that arrest sites occurred throughout the genome. This conclusion was also reached by Zannis-Hadjopoulos *et al.* (1983), based on the distribution of 3′-ends of cloned SV40 nascent DNA strands, and by Weaver and DePamphilis (1984) who analyzed the 5′-end locations of nascent SV40 DNA at the *Bgl*I, *Msp*I, and *Eco*RI restriction sites. Nevertheless, it is clear from all of these data that the most prominent site where replication forks accumulate is in the termination region.

The locations of about 80% of the prominent *in vivo* replication arrest sites corresponded to sites that also arrested α-polymerase on the same DNA template *in vitro* (Weaver and DePamphilis, 1984). However, changing the genomic location of the SV40 termination region through specific deletions essentially eliminated the accumulation of RI* without inhibiting replication (Weaver *et al.*, 1985). Therefore, it would appear that the accumulation of replication forks in the normal termination region was the consequence rather than the cause of some rate-limiting step unique to the normal termination site, such as difficulty in unwinding parental DNA strands (Section III.G.4). When replication forks are forced to pause, they may do so at natural obstacles such as nucleosomes and DNA synthesis arrest sites (Weaver and DePamphilis, 1982, 1984). Trifonov and Mengeritsky (1984) suggested that these accumulations of SV40 replication forks corresponded to the theoretically preferred locations of nucleosomes in SV40 chromosomes. Although this hypothesis is appealing (Section IV.G), the number and distribution of the pause sites analyzed were insufficient to be statistically significant (Zannis-Hadjopoulos *et al.*, 1983), and the absence of pause sites under some nucleosomes was not considered equally with the presence of sites under other nucleosomes.

Separation of sibling molecules does not require a unique termination site on the genome because the normal termination site could be moved to other locations relative to *ori* or removed altogether without preventing replication; termination continued to occur approximately 180° from *ori* (Brockman *et al.*, 1975; Lai and Nathans, 1975; Griffin and Fried, 1975; Weaver *et al.*, 1985). Therefore, separation of sibling molecules appeared to occur at whatever sequence the two oncoming replication forks happen to meet. Because all replication forks did not arrive simultaneously at the point 180° from *ori* (Bjursell *et al.*, 1979; Tapper and DePamphilis, 1980; Buckler-White *et al.*, 1982), the actual sites where termination occurred represent a distribution about the mean termination site that reflects the variation from one RI molecule to the next. This would explain why replication fork arrest sites were distributed about the mean termination site by ±450 bp (Tapper and DePamphilis, 1980), and the gap in the nascent DNA strand of SV40(II*) DNA was distributed over a 730-bp region (Chen *et al.*, 1976). These data define a "termination region" in which separation occurs. Although a specific sequence is not required for separation to occur, the termination site does affect the efficiency with which parental strands are unwound, and thus affects the proportion of RI* that produce catenated dimers instead of form II* (Weaver *et al.*, 1985; Section III.G.4).

2. Form II* DNA

Under conditions that produce the maximum rate of viral DNA replication, the two sibling molecules in RI* are released as form II* DNA (Fig. 11A). Denaturation of SV40(II*) DNA revealed that all of its nascent DNA was present as a linear molecule approximately one genome in length (Fareed *et al.*, 1973a; Tapper *et al.*, 1982; Weaver *et al.*, 1985) with its ends located in the genomic region where termination of replication occurred (Fareed *et al.*, 1973a; Laipis *et al.*, 1975; Chen *et al.*, 1976) and separated by a gap of about 50 nucleotides (Chen *et al.*, 1976). SV40(II*) DNA has been observed in virus-infected monkey cells (Fareed *et al.*, 1973a; Laipis *et al.*, 1975; Chen *et al.*, 1976; Weaver *et al.*, 1985) and in isolated nuclei incubated in the presence or absence of cytosol (Tapper *et al.*, 1979, 1982). In whole cells, SV40(II*) DNA appeared as rapidly as SV40(RI) DNA disappeared and SV40(I) was synthesized (Fareed *et al.*, 1973a; Weaver *et al.*, 1985). In nuclei incubated with cytosol, a physical map of the gradient of pulse-labeled nascent SV40 DNA was consistent with SV40(II*) DNA acting as a transient intermediate in the formation of SV40(I) DNA (Tapper *et al.*, 1979); the concentration of nascent DNA plateaued in the termination region of form II*, but peaked at the point where replication forks terminated in form I DNA. When isolated nuclei were incubated in the absence of cytosol, SV40(RI*) DNA containing 18–30% of a pulse-label carried out either *in vivo* or *in vitro* was converted during a 1-hr chase period into SV40(II*) DNA that contained 12–20% of the pulse-label (Tapper *et al.*, 1982). SV40(I) DNA was not synthesized in the absence of cytosol factors and the SV40(II) DNA that accumulated contained radiolabeled Okazaki fragments in addition to an essentially full-length nascent DNA strand (Tapper *et al.*, 1979, 1982). Therefore, separation of sibling molecules does not require covalently closed DNA. Separation can occur prior to the joining of Okazaki fragments to long nascent DNA chains, and the nascent strand is interrupted where DNA synthesis last occurred, the termination region.

3. Catenated Dimers

Covalently closed catenated dimers, circular dimers, and higher DNA oligomers have been identified in SV40 and PyV lytic infections (Cuzin *et al.*, 1970; Meinke and Goldstein, 1971; Jaenisch and Levine, 1971b; Crawford *et al.*, 1973; Martin *et al.*, 1976; Sundin and Varshavsky, 1980, 1981). Catenated dimers consist of two dsDNA monomers locked together by 1 to 7 intertwines between their respective genomes, and may contain two form I DNA molecules, two form II DNA molecules, or one form I and one form II molecule (Sundin and Varshavsky, 1980). Circular dimers consist of two dsDNA monomers tandemly linked in a head-to-tail arrangement (Goff and Berg, 1977; Rigby and Berg, 1978). Higher oligomers containing tandemly arranged head-to-tail genomes were also found late in infection and thought to arise by a rolling-circle mechanism (Section III.E.). *In vivo,* dimers resulted from replication rather than intermolecular recombination, since 95% of the dimers formed after infection with two SV40 mutants contained two copies of the same genome (Goff and Berg, 1977). Intramolecular recombination in either RI or catenated dimers could produce circular dimers which could again undergo intramolecular recombination to produce form I DNA (Wake and Wilson, 1980). However, this pathway for form I synthesis must be inefficient relative to replication because infection of cells with dimeric SV40 DNA yielded dimeric DNA as the predominant product (Jaenisch and Levine, 1971a), and similar amounts of both dimeric and monomeric PyV DNA were synthesized when tsA PyV-transformed mouse cells were incubated at the permissive temperature (Cuzin *et al.*, 1970).

Catenated dimers were labeled rapidly during the period of maximum DNA synthesis (Cuzin *et al.*, 1970; Jaenisch and Levine, 1973; Sundin and Varshavsky, 1980). About half of the pulse-labeled covalently closed dimers were catenated and half were circular, and the catenated dimers disappeared with a half-life of 3.7 hr, generating circular dimers and monomers (Jaenisch and Levine, 1973). Since the sedimentation properties of dimers are similar to RI, dimers could account for the observation that about 10% of the radiolabel in SV40(RI) or PyV(RI) DNA appeared to remain in RI during a pulse–chase experiment (Meinke and Goldstein, 1971; Tegtmeyer and Macasaet, 1972; De Pamphilis *et al.*, 1975). However, pulse-labeled covalently closed catenated dimers have also been observed to disappear completely in 1.3 hr, similar to the time required for a 90% turnover in RI DNA, and consistent with a role for catenated dimers as transient intermediates in replication (Sundin and and Varshavsky, 1980). One problem in comparing these data is that total viral DNA was extracted in the earlier experiments using the SDS–NaCl method of Hirt (1967) in which catenated dimers accounted for 1–2% of the DNA (Jaenisch and Levine, 1972, 1973; Goff and Berg, 1977), whereas the later experiments first extracted viral DNA in the form of chromosomes which contained 10–20% catenated dimers. Second, the chromosomal form of viral DNA changes with time during a pulse–chase period from replicating chromosomes (90 S) to mature chromosomes (70 S) to previrion complexes (200 S) to virions (250 S) (De Pamphilis and Wassarman, 1982, pp. 78–82). Thus, the contribution of any particular DNA species may have been a property of its structure as well as its ability to be extracted from nuclei as viral chromosomes rather than DNA. The role of catenated dimers in replication is discussed below.

4. Two Pathways for Separation of Sibling Chromosomes

Separation of the two sibling molecules in SV40(RI*) DNA occurs via two alternative pathways (Fig. 11A). If the parental DNA strands between the converging replication forks are completely unwound during replication, then two form II* DNA molecules are released when the forks collide. However, if all of the helical turns of the parental DNA are not removed before the two forks meet, then catenated dimers are formed with the two sibling chromosomes intertwined once for each parental

DNA helical turn that was not removed (Fig. 11B). The choice of termination pathway is affected by at least three parameters: the osmolarity of the cell culture medium, the multiplicity of infection (m.o.i.), and the DNA sequence at which termination occurs (Weaver *et al.*, 1985). Both hypertonic medium and high concentrations of intracellular replicating DNA promoted the formation of catenated dimers instead of form II* DNA, whereas cells infected with low amounts of virus and incubated in isotonic medium produced predominantly form II* DNA. Surprisingly, the effect of hypertonic medium, which was readily observed even at low multiplicities of infection, was eliminated when termination occurred at sequences other than wild-type. This was not a function of DNA size since the same result was obtained with genomes of 3.3 to 8.2 kb. Neither did it result from the presence of some unique termination sequence since the 232-bp SV40 *ori* region was the only sequence common to all the genomes. Therefore, the DNA sequence at which two replication forks meet must determine the pathway by which separation of sibling molecules occurs.

Sundin and Varshavsky (1981) had previously demonstrated that increasing the osmolarity of the culture medium in a variety of ways caused the reversible accumulation of newly synthesized catenated dimers. However, they did not observe a similar response with RI*. In contrast, Weaver *et al.* (1985) consistently observed that *both* RI* and catenated dimers accumulated concomitantly and reversibly in response to high osmolarity (DePamphilis *et al.*, 1983a). The difference between these results was not due to viral strain differences because small shifts in the termination region did not change the ratio of RI* to catenated dimers (Weaver *et al.*, 1985). One difference that may be relevant is that Sundin and Varshavsky (1981) first isolated viral chromosomes and then extracted viral DNA from the chromosomes whereas Weaver *et al.* (1985) extracted viral DNA directly from infected cells by the method of Hirt (1967). Since replicating chromosomes are more difficult to extract from nuclei than nonreplicating chromosomes (Su and DePamphilis, 1978; Seidman and Salzman, 1979), replicating DNA may have been preferentially excluded from Sundin and Varshavsky's analysis. Sundin and Varshavsky (1981) also observed about equal amounts of nascent DNA in catenated dimers and form II + II* in 1.0X medium, indicating that catenated dimers were readily formed under normal physiological conditions. In contrast, under apparently the same conditions, others found that catenated dimers were barely detectable (Weaver *et al.*, 1985; DePamphilis *et al.*, 1983a). The simplest explanation is that small differences (20%) in medium concentration can make significant differences in the fraction of catenated dimers present.

The action of hypertonic medium is twofold. First, it inhibits unwinding of parental DNA which results in accumulation of RI*, reduction in the rate of DNA synthesis, and formation of catenated dimers. The effect is specific for replication forks in the termination region, since only those RI in the final stages of replication were arrested (i.e., RI*; De Pamphilis *et al.*, 1983; Weaver *et al.*, 1985). Sundin and Varshavsky (1981) also observed the reduction in DNA synthesis and, in addition, demonstrated that the number of intertwines in catenated dimers increased with increasing osmolarity from 1–5 intertwines in 1.0X medium to 1–25 intertwines in 1.75X medium, suggesting that failure to unwind DNA in front of replication forks began when the forks were separated by as much as 250 bp. Second, hypertonic medium inhibited the decatenation process, resulting in the accumulation of newly formed catenated dimers which were rapidly decatenated when cells were shifted back to isotonic medium (Sundin and Varshavsky, 1981; Weaver *et al.*, 1985). The inhibition of decatenation cannot be the only effect of hypertonic medium as previously suggested (Sundin and Varshavsky, 1981) because it would be difficult to account for the accumulation of RI*, the increased number of intertwines in catenated dimers, or the failure to observe catenated dimers when termination occurs at sequences other than the normal SV40 termination region. The possibility that decatenation is significantly faster at other termination regions seems unlikely because catenated intertwines in SV40 appeared to be distributed over the entire chromosome (Varshavsky *et al.*, 1983) whereas the remaining duplex DNA turns will always be localized at the termination region. It is also possible that form II* DNA is generated by separation of catenated dimers consisting of two form II* molecules (Fig. 11A). However, this is a minor pathway since the synthesis of form I DNA is faster than the completion of decatenation. The average number of intertwines per dimer decreased in the order: form II/form II dimers > form II/form I dimers > form I/form I dimers (Sundin and Varshavsky, 1980).

A likely target of the hypertonic medium effect is topoisomerase II. It may be required to relax superhelical turns generated when helicases separate the parental DNA strands in the termination region, as well as separating the two interlocked monomers in catenated dimers (Liu *et al.*, 1980; Hsieh and Brutlag, 1980; Miller *et al.*, 1981; Goto and Wang, 1982). Consistent with this hypothesis is the observation that yeast temperature-sensitive mutants in topoisomerase II accumulate catenated 2-μm plasmids (DiNardo *et al.*, 1984) and interlocked nuclei (Uemura and Yanagida, 1984) at the restrictive temperature. Similarly, an *E. coli* temperature-sensitive mutant in DNA gyrase, a bacterial enzyme with properties similar to topoisomerase II, accumulates interlocked nucleoids (Steck and Drlica, 1984). However, Colwill and Sheinin (1983) observed that mouse cell line tsA1S9 was deficient in topoisomerase II at the restrictive temperature, but still appeared to replicate polyoma DNA normally (Sheinin, 1976; Ganz and Sheinin, 1983). Since topoisomerase II appears to interact efficiently with most, but not all, DNA sequences (Lui *et al.*, 1983), perhaps the wild-type SV40 termination sequence is a poor substrate for this enzyme so that modest inhibition by hypertonic medium accentuates the problem and thus promotes formation of catenated dimers. Topoisomerase II activity will copurify with SV40 chromosomes (Waldeck *et al.*, 1983; Liu *et al.*, 1983; Krauss *et al.*, 1984), but its function in SV40 DNA replication has not been demonstrated. Novobiocin, a specific inhibitor of bacterial DNA gyrase, inhibited SV40 DNA replication in nuclear extracts, but this effect could be accounted for by inhibition of DNA polymerase α (Edenberg, 1980).

The effect of hypertonic medium was most pronouncd at high m.o.i.; the fraction of catenated dimers increased sixfold as the m.o.i. increased from about 0.5 to 50 (Weaver *et al.*, 1985). Presumably, the rate of formation of catenated dimers is increased when the concentration of intracellular RI* is high and/or the rate of separation of catenated dimers is decreased because the protein(s) required for either step are in limited supply. Alternatively, high concentrations of DNA can facilitate catenation of form I and form II* DNA molecules that lie in close proximity because they were the recent product of RI* separation. For example, topoisomerase I can catenate and decatenate DNA circles if one of them has an ssDNA gap, and catenation is favored at high DNA concentration whereas decatenation is favored at low DNA concentration (Low *et al.*, 1984). Topoisomerase II performs the same functions with dsDNA. In fact, mammalian topoisomerase II rapidly interlocks SV40 chromosomes because of their aggregation in the reaction mixture (Liu *et al.*, 1983).

It seems clear from the results discussed above that neither catenated dimers nor form II* represent obligatory intermediates in replication of circular DNA molecules. Conditions have been found in which either form II* or catenated dimers predominate in the reaction. However, under normal physiological conditions, at low concentrations of replicating DNA and at most termination regions, complete unwinding of parental DNA to give form II* appears to be the major pathway. The alternative pathway (Fig. 11A) of decatenation prior to completing DNA replication to produce form II* is also possible, but there is no compelling reason to suggest that all replicating molecules must pass through a catenated intermediate. Other observations also indicate that formation of catenated dimers is a secondary pathway that appears under unfavorable conditions. Cycloheximide rapidly interferes with SV40 chromosome replication (Cusick *et al.*, 1984) and increases the fraction of catenated dimers by three- to fourfold (Jaenisch and Levine, 1972, 1973). Since addition of hypertonic medium can also inhibit initiation of protein synthesis (Saborio *et al.*, 1974), the two effects may be related. Similarly, dimeric and oligomeric viral DNA increased tenfold by 70 hr postinfection when the rate of DNA synthesis dropped and cells showed the cytopathic effects of virus infection (Martin *et al.*, 1976; Goff and Berg, 1977). Catenated dimers in other systems have been shown not to be transient intermediates in replication. Catenated dimers of mitochrondrial DNA did not consist of two newly replicated DNA molecules (Flory and Vinograd, 1973; Berk and Clayton, 1976; Bogenhagen and Clayton, 1976), suggesting that they were formed either by recombination or, more likely, by the action of topoisomerases on monomeric DNA. Sakakibara *et al.* (1976) demonstrated that catenated dimers of Col E1 DNA were produced by replication in a subcellular system, but that once formed they were stable and their conversion to monomeric DNA was insignificant, if it occurred at all.

Under normal physiological conditions, RI* accumulated two- to threefold and replication forks accumulated at specific DNA sites separated by about 470 bp (Tapper and DePamphilis, 1978, 1980; Tapper *et al.*, 1982). These *in vivo* arrest sites frequently coincided with arrest sites for DNA polymerase α on the same template *in vitro* (Weaver and DePamphilis, 1984). Variations reported in the fraction of SV40 and PyV RI* (DePamphilis and Wassarman, 1982) presumably reflected variation in the three parameters discussed above. Since RI* did not accumulate when replication terminated at other DNA regions, the arrest of replication forks at specific DNA sites is the consequence rather than the cause of termination. Replication forks are arrested at specific sites in the wild-type termination region because it is difficult to unwind, and DNA polymerase α pauses naturally at specific DNA sequences (Weaver and DePamphilis, 1982).

H. Termination of Replication

Replication terminates 180° from *ori* when the gap in form II* is filled in and the molecule sealed to produce form I DNA. This final gap-filling step, like the completion of Okazaki fragments (Anderson and DePamphilis, 1979), required gap-filling proteins found in the cytosol fraction (Tapper *et al.*, 1979, 1982). SV40(I) DNA purified from virions contained 26 ± 0.5 superhelical turns (Shure and Vinograd, 1976) that resulted from the presence of 24 ± 1 nucleosomes per genome (DePamphilis and Wassarman, 1982, p. 83). Thus, the purified DNA expressed approximately one superhelical turn for each nucleosome that was present in the viral chromosome (Germond *et al.*, 1975). Since nascent DNA on both arms of replication forks is assembled rapidly into nucleosomes (Section V.D.), superhelicity in the DNA will be expressed as soon as form II* DNA is sealed. Thus, when nucleosome assembly from newly synthesized histones was inhibited by addition of cycloheximide or puromycin to virus-infected cells (DePamphilis and Wassarman, 1982; Cusick *et al.*, 1984), the covalently closed DNA products of replication were deficient in superhelical turns (Bourgaux and Bourgaux-Ramoisy, 1972; White and Eason, 1973; Cheevers, 1973; Sen *et al.*, 1975; Yu and Cheevers, 1976a). Nucleosome assembly did not require concomitant DNA synthesis, simply the presence of nonnucleosomal dsDNA (White and Eason, 1973).

I. Comparison of Viral and Cellular Replicons

Papovavirus replicating DNA provides an appropriate model for the process of separation of sibling chromosomes and termination of DNA replication in mammalian replicons. The approach of two replication forks in SV40(RI) or PyV(RI) DNA is topologically equivalent to the final 5000 bp of replication between two adjacent cellular replicons, because neither "infinitely" long linear DNA molecules nor small circular, covalently closed molecules can unwind parental DNA strands by freely rotating one strand about the other (Fig. 12). In fact, the accumulation of replicon-sized DNA in eukaryotic cells implies that cellular replicons, like viral replicons, pause before undergoing separation of sibling molecules (DePamphilis and Wassarman, 1980). Therefore, the slower rate of fork movement in viral replicons may reflect the actual rate of fork movement in cellular replicons during the final 2% to 10% of replicon maturation, while the average rate of cellular fork movement is much greater. SV40 and PyV DNA had an unexpectedly long replication time of 15–25 min (Danna and Nathans, 1972; Fareed *et al.*, 1973a; Crawford *et al.*, 1973; Manteuil and Girard, 1974; Tapper and DePamphilis, 1978; Tapper *et al.*, 1979). Thus, replication fork movement in mammalian chromosomes averaged 1500 to 3000 bp/min per fork (Edenberg and Huberman, 1975), while replication forks in SV40(RI) DNA at 37°C proceeded at about 145 bp/min per fork (Tapper *et al.*, 1979).

Papovavirus replication does not require a specific termination sequence. Presumably, separation of newly replicated cellular chromosomes also does not require a specific termination sequence, because genetic rearrangements or deletions that remove a termination site would prevent separation of sibling chromosomes. In contrast, if an origin of replication

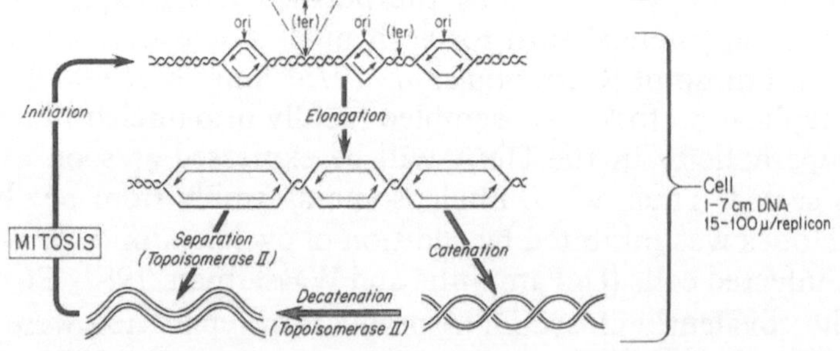

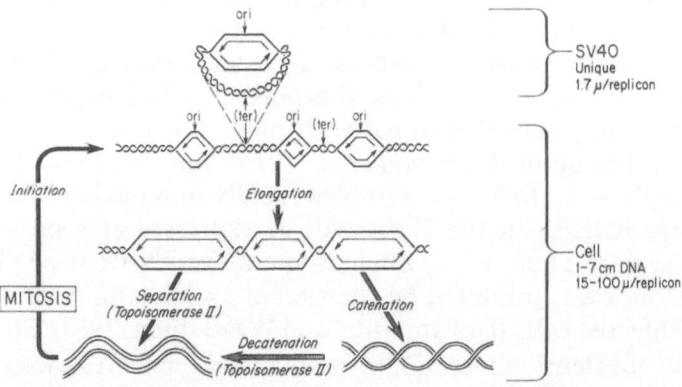

FIGURE 12. Topological problems encountered during maturation of viral replicons are equivalent to those in the final 2% to 10% of replication in converging cellular replicons. Nonreplicated DNA is indicated by helical regions and newly replicated DNA by parallel lines. Origins of replication (*ori*) and termination sites for replication (ter) are indicated.

was lost, that region of cellular DNA would still be replicated by initiation at origins upstream and downstream (Blumenthal *et al.*, 1974). Nevertheless, there is some evidence for cellular termination signals. Induction of PyV DNA replication by incubation of PyV-transformed rat cells with mitomycin C revealed that the PyV and flanking cellular DNA sequences were amplified at least tenfold in chromosomal DNA, and that replication forks were arrested within a 400-bp cellular DNA fragment located about 2000 bp to one side of the integrated PyV genome (Baran *et al.*, 1983). Sequences that block movement of replication forks have also been identified in the plasmid R6K (Germino and Bastia, 1981), and at the termination region for *E. coli* (Kuempel *et al.*, 1977; Louarn *et al.*, 1979) and *B. subtilis* (Weiss and Wake, 1984) DNA replication, where they may be involved in segregation of sibling chromosomes during cell division. However, there is no evidence that any of these sequences are required for replication.

It would be advantageous for the cell to avoid intertwines (catenation) between two sibling molecules since this could interfere with separation of sister chromatids during mitosis. For example, if catenation were an obligatory step in replication, then sister chromatids would be intertwined about once every 30–50 μm (average *ori*-to-*ori* distance), a *minimum* of about 2000 intertwines per chromosome. Thus, based on the results with SV40, DNA unwinding should be completed at most termination sites and the two sibling chromosomes separated, particularly when cells are under normal physiological conditions and the concentration of replication forks is low. Since about 8×10^3 replication forks exist at any time during S phase (Hand, 1979), they constitute only about 0.1% of the total nuclear DNA (600 bp DNA/fork). Nevertheless, when catenation does occur, topoisomerase II could act as a surveillance system to separate the chromosomes (Section III.G.4).

The question of whether or not initiation of papovavirus DNA replication is representative of initiation of cellular DNA replication is more difficult to answer. Eukaryotic cellular chromosomes, like papovavirus chromosomes, do appear to utilize *ori* sequences for initiation of replication. Cellular *ori* sequences are implicated strongly by the temporal order of gene replication (Epher *et al.*, 1981; Calza *et al.*, 1984; Heintz *et al.*, 1983), the selective amplification of specific genes (Osheim and Miller, 1983; Mariani and Schimke, 1984; Stark and Wahl, 1984), the existence of autonomously replicating sequences (Struhl, 1983; Fangman *et al.*, 1983), and the fact that all bacterial, viral plasmid, and organelle genomes require one or more *ori* sequences (Kornberg, 1980). The major evidence against a requirement for cellular *ori* sequences is twofold. First, the average distance between chromosomal replication bubbles in preblastomere *Drosophila* embryos was at least five times smaller than in differentiated *Drosophila* cells (Blumenthal *et al.*, 1974; Zakian, 1976; McKnight and Miller, 1977), suggesting that embryonic cells initiate DNA replication at many more sequences than do differentiated cells. Second, all dsDNA molecules injected into *Xenopus* eggs replicated equally well, regardless of their sequence (Harland and Laskey, 1980; McTiernan and Stambrook, 1984; Mechali and Kearsey, 1984). Some indication of sequence-specific replication was reported (Chambers *et al.*, 1982; Hines and Benbow, 1982), but the interpretation of these results has been questioned (Mechali and Kearsey, 1984). Thus, eukaryotic chromosome replication in embryonic cells may not require specific *ori* sequences. However, this was not case with dsDNA injected into the nuclei of mouse one-cell or two-cell embryos (Wirak *et al.*, 1985). Plasmid DNA failed to replicate unless it contained a papovavirus *ori* sequence, and PyV *ori*-dependent replication required PyV T-Ag. Furthermore, only the β-core configuration of PyV *ori* was active in both embryos and differentiated mouse cells; and α-core sequence was active only in differentiated cells although it did not interfere with β-core activity (Table II). These results are consistent with a requirement for *cis*-acting sequences to replicate DNA in mammalian nuclei throughout development, and demonstrate that *ori* sequences can be specifically activated by certain cell types. The reason that mammalian embryos respond so differently to injected DNA compared with amphibian embryos may be related to biological differences. Nonmammalian embryos undergo cell cleavage at rates 40 to 200 times greater than mammalian embryos (Blumenthal *et al.*, 1974; Kelly and Condamine, 1982; Newport and Kirschner, 1982), consistent with a large store of predisposed replication factors that may override normal control signals (Laskey *et al.*, 1983; Zierler *et al.*, 1985).

The question remains as to whether or nor cellular DNA sequences can be identified that will also allow replication of injected plasmid DNA. One important difference between cellular and papovavirus initiation mechanisms is the fact that cellular chromosomes, which can contain a single DNA molecule 10 cm long, initiate replication concurrently at many sites, and, although the number of sites per chromosome can change during early stages in development, initiation does not occur twice at the same site during the same S phase (Blumenthal *et al.*, 1974). Thus, the major function of T-Ag may be to ensure that viral genomes, which are analogous to individual mammalian replicons, are replicated many times within a single S phase; a process not unlike the amplification of specific cellular genes. In the absence of T-Ag, "permissive cell factors," which may be part of the normal cellular DNA replication machinery, may initiate replication only once per cellular S phase.

Another relationship between viral and cellular *ori* sequences may be in the use of enhancer elements. The fact that PyV *ori* has an enhancer but SV40 *ori* does not reveals that enhancers are not a special feature of viral *ori* sequences (Section III.B). Furthermore, the enhancer element in PyV *ori* imparts cell-specific activation of this *ori* sequence (Section III.C). Thus, enhancer elements could provide organisms with a cell-specific mechanism for amplifying certain genes at specific times during development when the cellular equivalent of T-Ag is present. Furthermore, the observation that transcriptionally active genes are replicated early in S phase while inactive DNA sequences are replicated late indicates a possible role for enhancer elements as active components of cellular *ori* sequences.

IV. DNA SYNTHESIS AT REPLICATION FORKS

A. Summary

The data available for SV40 and PyV DNA replication forks can be interpreted on the basis of a simple model (Fig. 13) which has been described previously (Anderson and DePamphilis, 1979; DePamphilis and Wassarman, 1980, 1982). DNA synthesis on the forward arm of replication forks occurs in the same direction as fork movement and is predominantly, if not exclusively, a continuous process carried out by DNA polymerase $\alpha[C_1C_2]$. However, DNA synthesis on the retrograde arm must occur in the opposite direction and is therefore a discontinuous process involving the repeated initiation of Okazaki fragments. Thus, as forks advance, the forward arm is maintained as dsDNA while a ssDNA region is exposed on the retrograde arm that acts as an "initiation zone" for the synthesis of Okazaki fragments. Assuming that nucleosome disassembly in front of the fork is the rate-limiting step in fork movement, the size of this initiation zone has been suggested to be 220 ± 73 nucleotides, the average distance from one nucleosome core to the next. Since nucleo-

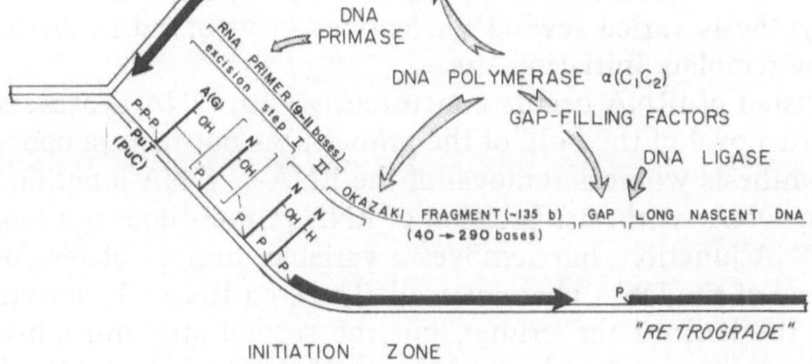

FIGURE 13. Events at an SV40 DNA repliation fork. Arrowheads represent the growing 3′-ends of nascent DNA strands (solid black lines) which rest on the DNA template (open line).

somes are arranged in a near-random fashion with respect to DNA sequence, the size and composition of initiation zones will vary extensively from one replicating chromosome to the next.

By some stochastic process, the enzyme complex DNA primase–DNA polymerase $\alpha[C_1C_2]$ selects one of many possible sites on the template within the initiation zone to initiate synthesis of a short RNA primer, referred to as "initiator RNA" (iRNA), on whose 3′-OH end DNA polymerase α rapidly initiates DNA synthesis. Initiation events occur 80% of the time at 3′-dPuT and 20% of the time at 3′-dPuC in the template, with the first ribonucleotide complementary to either dT or dC. Although these sites occur, on average, once every 7 nucleotides, initiation events occur only once every 135 bases (the average size of Okazaki fragments). Therefore, once synthesis of an RNA primer is initiated, elongation of the resulting nascent chain appears rapid enough to prevent additional initiation events downstream. Neither the template sequence encoding the RNA primer, nor the transition point in the template where RNA synthesis changes to DNA synthesis shows any sequence preference. The transition from RNA to DNA synthesis generally occurs 9–11 bases downstream, but it can vary from 2 to 12 bases downstream, depending on the template initiation site. DNA synthesis continues until 10–15 nucleotides remain to be incorporated whereupon one or more protein "gap-filling factors" are required to allow α-polymerase to complete DNA synthesis. DNA ligase then seals the 3′-end of the Okazaki fragment to the 5′-end of the long, nascent DNA strand. Okazaki fragments that have completed their synthesis range in size from 40 bases (those that initiated closest to the growing daughter strand) to 290 bases long (those that initiated closest to the replication fork). Thus, the structure of a typical iRNA-initiated Okazaki fragment is $(p)ppA/G(pN)_{7-9}$ $pN\text{–}p\text{–}dN(pN)_{40-290}$. Purified DNA primase–DNA polymerase α synthesized RNA primers on ssDNA templates *in vitro* that were similar to iRNA *in vivo,* except that they were initiated at different sites in the template and were about 3 bases shorter. The transition from RNA to DNA synthesis varied several nucleotides *in vivo* and *in vitro* even at the same template initiation site.

Excision of iRNA occurs concurrently with DNA synthesis and ligation. Removal of the bulk of the primer does not require concomitant DNA synthesis whereas removal of the RNA–p–DNA junction is facilitated by DNA synthesis. Excision of RNA primers does not stop at the RNA–DNA junction, but removes a variable number of residues from the 5′-end of the DNA chain as well. Thus, an RNase H activity could remove the bulk of the primer, but the second step must involve an exonuclease that degrades both RNA and DNA. The 5′-end of the degraded Okazaki fragment becomes the 5′-end of the long nascent DNA strand, the fork advances to another nucleosome, a new Okazaki fragment initiation zone is exposed, and the process begins again.

B. Okazaki Fragments

1. Okazaki Fragments Are Transient Intermediates in DNA Replication

During semiconservative DNA replication, newly synthesized DNA appears coincidentally on both arms of replication forks, yet all known DNA polymerases extend polynucleotide chains only at their 3′-OH termini (Kornberg, 1980). This paradox is resolved by the repeated initiation of short nascent DNA chains ("Okazaki fragments") on retrograde DNA templates where the direction of DNA synthesis is opposite to the direction of fork movement (Fig. 13). Purified SV40(RI) and PyV(RI) DNA pulsed-labeled *in vivo* (Fareed and Salzman, 1972; Perlman and Huberman, 1977; Hunter *et al.*, 1977) or in subcellular systems (Pigiet *et al.*, 1973, 1974; Magnusson *et al.*, 1973; Francke and Hunter, 1974; Qasba, 1974; Anderson *et al.*, 1977; Anderson and DePamphilis, 1979; Eliasson and Reichard, 1979) contained 50% or less of their nascent DNA in the form of short chains with an average length of 135 nucleotides (Flory, 1975; Anderson and DePamphilis, 1979) even when radiolabeled for very short time periods (Francke and Hunter, 1974; Perlman and Huberman, 1977; Anderson *et al.*, 1977) or terminally labeled at their 5′-ends (Pigiet *et al.*, 1974; Hay *et al.*, 1984). Continued incubation resulted in a rapid, quantitative joining of Okazaki fragments to long nascent DNA chains. The average number of Okazaki fragments per replication fork was equal to the ratio of short (Okazaki fragments) to long nascent DNA chains. The actual measurement required end-labeling only nascent DNA chains, and subtracting that portion of nascent chains which could represent the "long" daughter strands from young replicating DNA molecules (Tegtmeyer, 1972; Anderson and DePamphilis, 1979). Only about one-fourth of the replication forks appeared to have an Okazaki fragment (Hay *et al.*, 1984); calculations based on data from any experiments with PyV, SV40, and mammalian cells gave an upper limit of one Okazaki fragment per fork (Herman *et al.*, 1979; Anderson and DePamphilis, 1979; DePamphilis and Wassarman, 1980).

Two sources of short DNA fragments could be mistaken for true Okazaki fragments. First, mammalian cells contained short pieces of *nonnascent* DNA that were detected by end-labeling procedures (Kowalski and Denhardt, 1978). However, only nascent Okazaki fragments could be labeled completely with BrUdR or thymidine (Pigiet *et al.*, 1973; Francke and Hunter, 1974; Francke and Vogt, 1975). The second problem was incorporation and subsequent excision of dUMP into nascent DNA, a metabolic pathway that causes breaks in nascent DNA chains, generating false "Okazaki fragments" (DePamphilis and Wassarman, 1980). However, this artifact did *not* account for the Okazaki fragments observed during DNA replication because:

1. About 50% of the Okazaki fragments were covalently attached to a short oligoribonucleotide ("iRNA," Section IV.C); dUMP repair products were not.
2. SV40 Okazaki fragments originated almost exclusively from retrograde arms of replication forks (see below); dUMP repair should have occurred on both arms.
3. dUTPase in cytosol prevented incorporation of dUTP into DNA in either cell lysates or nuclei supplemented with cytosol unless excessive amounts of dUTP were added (Wist *et al.*, 1978; Grafstrom *et al.*, 1978; Gautschi *et al.*, 1978).
4. Variation of dTTP levels (Narkhammer-Meuth *et al.*, 1981a; Tapper *et al.*, 1982) or addition of uracil to suppress uracil-*N*-glycosidase (Wist *et al.*, 1978; Brynolf *et al.*, 1978) or addition of dUTPase (Wist, 1979) had little, if any, effect on the size or amount of Okazaki fragments.
5. PyV cell extracts contained only 0.4% as much dUTP as dTTP (Nilsson *et al.*, 1980) and only one dTTP pool served as a substrate for DNA synthesis (Nicander and Reichard, 1983).
6. Less than 1 fmole dUMP/μmole DNA was detected in human cells unless they were treated with methotrexate to inhibit dTTP synthesis, which raised the level of dUMP about 800-fold (Goulian *et al.*, 1980).

Although these artifacts were generally not a problem in DNA replication studies, they could become so when DNA was radiolabeled after it was purified and under conditions that depressed nucleotide pools. For example, incubation of cells in the presence of hydroxyurea resulted in inhibition of PyV DNA synthesis, accumulation of very short nascent DNA fragments that did not contain RNA primers, and disappearance of cellular dUTP (Magnusson, 1973; Magnusson *et al.*, 1973). Perhaps misincorporation and excision of another unusual base occurred in the presence of hydroxyurea.

2. Distribution of Okazaki Fragments between the Two Arms of a Replication Fork

Theoretically, Okazaki fragments *must* originate from retrograde templates, whereas reinitiation of DNA synthesis on forward templates is not required and may be prohibited. Once DNA synthesis begins at *ori*, synthesis in the same direction as fork movement could continue until termination of replication occurs. If replication forks advance (i.e., DNA templates are unwound) faster than DNA synthesis can keep pace, then Okazaki fragments may be initiated on forward as well as retrograde templates. However, if the enzyme complex responsible for initiation of Okazaki fragments in eukaryotes behaves like the *E. coli* primasome and travels along the DNA template in a 5′ to 3′ direction (Arai and Kornberg,

1981b), it could only operate on the retrograde template where the direction of its travel is *toward* the fork. On the forward template, the initiation complex would travel *away* from the fork and collide with DNA polymerase coming in the opposite direction. Furthermore, DNA primase–DNA polymerase α strongly favored initiating DNA synthesis on the 3′-end of a preformed RNA or DNA primer over *de novo* synthesis of its own RNA primer (Yamaguchi *et al.*, 1985a). This would strongly favor continuous synthesis of DNA on forward arms of forks. Unfortunately, conflicting data on the distribution of Okazaki fragments between the two arms of replication forks in prokaryotic (Ogawa and Okazaki, 1980) as well as eukaryotic systems (DePamphilis and Wassarman, 1980) preclude a simple picture applicable to all situations.

If Okazaki fragments from SV40 or PyV originate only from retrograde templates, they should not contain self-complementary sequences since replication is bidirectional from a unique origin. However, in some cases, 70% to 90% of Okazaki fragments self-annealed (Pigiet *et al.*, 1973; Fareed *et al.*, 1973b; Magnusson, 1973; Flory, 1975; Narkhammer-Meuth *et al.*, 1981a) while in others only 52% (Narkhammer-Meuth *et al.*, 1981b) or 20%–30% formed duplex DNA (Francke and Hunter, 1974; Francke and Vogt, 1975; Hunter *et al.*, 1977). Even if experiments that used hydroxyurea (Magnusson, 1973; Laipis and Levine, 1973) or that inadequately separated nascent SV40 DNA from contaminating cellular and viral DNA or RNA that might have annealed with nascent chains (Fareed *et al.*, 1973b) are omitted, the conflict remains: isolated PyV Okazaki fragments have been reported as 75% dsDNA (Pigiet *et al.*, 1973; Flory, 1973; Narkhammer-Meuth *et al.*, 1981a), 52% dsDNA (Narkhammer-Meuth *et al.*, 1981b); 30% dsDNA (Francke and Hunter, 1974; Francke and Vogt, 1975; Hunter *et al.*, 1977), and 20% dsDNA, when only shorter fragments were tested (Francke and Vogt, 1975), suggesting that from 63% to 90% of Okazaki fragments in different experiments came from the same side of replication forks.

The situation with SV40 was clarified when nascent DNA chains were annealed to separated strands of DNA restriction fragments taken from both sides of *ori.* Using short nascent DNA fragments radiolabeled under a variety of conditions and then isolated from purified SV40(RI) DNA, at least 80% of Okazaki fragment DNA labeled in whole cells annealed to retrograde templates while long nascent DNA strands annealed equally well to both templates (Perlman and Huberman, 1977). The same was true for 90% of the DNA fragments labeled by polynucleotide kinase (Cusick *et al.*, 1981), 96% under optimal hybridization conditions (Hay *et al.*, 1984), 95% labeled in isolated nuclei supplemented with cytosol (Kaufmann *et al.*, 1978; Kaufmann, 1981), and the bulk of Okazaki fragments labeled specifically in their 5′-RNA moiety (Kaufmann, 1981). Efforts to alter the relative rates of DNA synthesis and ligation by varying temperature had no effect on the distribution of Okazaki fragments (Perlman and Huberman, 1977). Finally, the frequency at which RNA-primed Okazaki fragments were initiated at specific nucleotide locations on each arm of a replication fork was compared, and 99% were found on retrograde templates (Hay *et al.*, 1984). The small number of initiation events that appeared to occur on forward arms may, in fact, have resulted from a fraction of DNA molecules that either replicated unidirectionally (Buckler-White *et al.*, 1982) or initiated replication at sites other than *ori* (Martin and Setlow, 1980). Thus, initiation of Okazaki fragments in SV40(RI) DNA may occur exclusively on retrograde templates.

Similar experiments with PyV(RI) DNA yielded a different result. Either 60% (Narkhammer-Meuth *et al.*, 1981a), 70% (Flory, 1975), or 80% (Hunter *et al.*, 1977) of Okazaki fragments labeled in isolated nuclei, and about 74% (Narkhammer-Meuth *et al.*, 1981b) of Okazaki fragments labeled in cells originated from retrograde arms. Okazaki fragments from both arms were essentially identical in size, and 25% to 40% of the forward arm fragments appeared to contain iRNA (Narkhammer-Meuth *et al.*, 1981a), eliminating the possibility that they arose from either radiolytic breaks or misincorporation and excision of unusual bases. *In vitro*, nascent PyV iRNA was labeled in isolated nuclei with either [β-^{32}P]-GTP or [^{3}H]rNTPs, and RNA–DNA linkages were identified by transfer of ^{32}P from labeled nascent DNA to rNMPs (Narkhammer-Meuth *et al.*, 1981a). Although the measured cpm were small, the results indicated that 36% to 89% of all Okazaki fragments had iRNA. Surprisingly, a larger fraction of forward arm fragments contained iRNA than did retrograde fragments. *In vivo*, PyV(RI) DNA was isolated from infected cells, the 5′-ends of nascent DNA chains labeled with ^{32}P and then annealed to separated strands of restriction fragments (Narkhammer-Meuth *et al.*, 1981b). Fragments annealed to either retrograde or forward templates were found to contain iRNA as evidenced by the release of [^{32}P]pNp upon alkaline hydrolysis. Furthermore, the extent of self-annealing among Okazaki fragments was not diminished significantly by selectively eliminating RNA-primed DNA chains from the population, indicating that the RNA-primed DNA chains came from both arms of the fork. Thus, all of the experiments with PyV DNA indicate that discontinuous DNA synthesis could occur on both arms of replication forks, although the fraction of initiation events on forward arms varied with experimental conditions.

Since it seems unlikely that SV40 and PyV DNA replication are fundamentally different, two explanations are possible. First, DNA synthesis may be discontinuous on forward arms of SV40 DNA, but either the frequency of initiation events or the rate of joining of Okazaki fragments to longer strands varies with the virus and physiological conditions. For example, Okazaki fragments were not detected on the forward template of phage P2 until a ts DNA ligase mutant of *E. coli* was incubated at the restrictive temperature (Ogawa and Okazaki, 1980). Second, the fraction of viral DNA that replicated unidirectionally (Robberson *et al.*, 1975; Martin and Setlow, 1980; Buckler-White *et al.*, 1982), or that contained viable deletions which displaced *ori* with respect to other viral sequences (Francke and Vogt, 1975) varied and appeared to be more prevalent with PyV than with SV40(RI) DNA. Thus, the same template sequence represents the retrograde template in one RI molecule but the forward template in another RI molecule. It is interesting to note that SV40 tsA209 synthesized Okazaki fragments on both arms of its forks at the *permissive* temperature (Dinter-Gottlieb and Kaufmann, 1982). Perhaps altered T-Ag can affect the direction of replication, the origin of replication, or the frequency of DNA synthesis initiation events at replication forks.

C. RNA Primers, DNA Primase, and DNA Synthesis

All known DNA polymerases require both a template and a 3′-OH nucleotide primer in order to initiate DNA synthesis (Kornberg, 1980). In eukaryotes, the template is usually DNA, although retrovirus reverse transcriptase and mammalian DNA polymerase γ can also use RNA templates (DePamphilis and Wassarman, 1980). The primer can be DNA (e.g., parvoviruses; Berns and Hauswirth, 1982) or a single dNMP covalently attached to a protein (e.g., adenovirus; Kelly, 1982) However, in the examples of SV40, PyV, and mammalian DNA replication, the primer is RNA. This RNA, referred to as initiator RNA (iRNA), consists of short, transient oligoribonucleotides, relatively uniform in length but variable in composition, whose 5′-ends are either pppA or pppG and whose 3′-ends are attached covalently to newly synthesized DNA (Fig. 13). iRNA is synthesized by the enzyme DNA primase as a primer for DNA polymerase α. Table III summarizes the properties of DNA primase *in vitro* and iRNA synthesis *in vivo.* The evidence for iRNA, its synthesis by DNA primase, and its excision by as yet unidentified enzymes is summarized below.

1. RNA–DNA Covalent Linkages

Perhaps the most critical test of iRNA is its covalent linkage to nascent DNA. RNA–p–DNA linkages have been measured quantitatively by release of [2′(3′)-^{32}P]rNMPs from [^{32}P]-DNA labeled *in vitro* and then incubated with alkali or RNase (Pigiet *et al.*, 1974; Hunter and Francke, 1974; Anderson *et al.*, 1977; Kaufmann *et al.*, 1977). The validity of this ^{32}P-transfer assay was demonstrated by showing that the soluble ^{32}P-labeled products copurified with rNMPs (Pigiet *et al.*, 1974; Anderson *et al.*, 1977) could be converted into cyclic [2′:3′-^{32}P]rNMPs, and then enzymatically cleaved into [3′-^{32}P]rNMPs (Anderson *et al.*, 1977). The amount of ^{32}P-transfer was proportional only to the specific activity of the [α-^{32}P]dNTP substrates (Anderson *et al.*, 1977) which were prepared from purified nucleosides and treated with periodate to destroy contaminating

TABLE III. Comparison of iRNA Synthesized on SV40 DNA Templates *in Vivo* with iRNA Synthesized by CV-1 DNA Primase–DNA Polymerase α on the Same Templates *in Vitro*

	iRNA properties							Initiation sites	
	5′-Terminal rN (%)				Size (bases)		RNA–DNA		
Measurements[a]	A	G	C	U	Range	Peak	junctions[b]	Template sequence[c]	Frequency (sites/bases)
In vivo[d]									
rN–$\overset{*}{p}$–dN							"Random"		
5′–$\overset{*}{p}$rNp	62	24	7	7	1,2–12	9–11			
G$\overset{*}{p}$pprN	71	29	0	0	1,2–12	9–11			
Sequence	78	10	4	8	3–12	7–8	"Random"	3′-dPu$\underline{T}$ (70%) 3′-dPu$\underline{C}$ (8%)	1/5–1/15
In vitro[a]									
rN–$\overset{*}{p}$–dN							"Random"		
G$\overset{*}{p}$pprN	72	28	0	0	1–9				
Sequence	55	45	0	0		6–8		3′-d(Py)$_n$C$\underline{T}$TT(Py)$_n$ (80%) 3′-d(Py)$_n$C$\underline{C}$C(Py)$_n$ (20%)	1/12–1/40

[a] Measurements: rN–$\overset{*}{p}$–dN—The relative frequencies of all 16 possible dinucleotides at RNA–DNA linkages were determined by transfer of ^{32}P from DNA radiolabeled with one of the four dNTP to each 2′(3′)-rNMP upon alkaline hydrolysis (Anderson *et al.*, 1977). 5′-$\overset{*}{p}$rNp—The 5′-terminal ribonucleotide composition was determined by DEAE-Sephadex chromatography of the nucleotides released upon alkaline hydrolysis of 5′-^{32}P-labeled nascent DNA chains (Hay *et al.*, 1984). G$\overset{*}{p}$pprN—The 5′-terminal ribonucleotide composition was determined by PEI-chromatography of nucleotides released upon nuclease P1 digestion of nascent DNA chains that had been "capped" with [α-^{32}P]-GTP (Hay *et al.*, 1984; Yamaguchi *et al.*, 1985a). Sequence—The 5′-terminal ribonucleotide composition was determined by locating either the 5′-end of $\overset{*}{p}$–RNA–DNA chains (*in vivo*) or the 5′-end of G$\overset{*}{p}$pp–RNA–DNA chains on the template sequence (Hay and DePamphilis, 1982; Hay *et al.*, 1984; Yamaguchi *et al.*, 1985b).

[b] All 16 possible RNA–DNA junctions were found at frequencies equivalent to those for the corresponding dinucleotides in the DNA template.

[c] Consensus DNA template sequence selected for iRNA initiation sites is shown with the position complementary to the 5′-terminal iRNA ribonucleotide underlined. The ratio of 5′-pppA/5′-pppG initiation events depended upon the relative affinity of DNA primase–DNA polymerase α for specific DNA template sequences (Tseng and Ahlem, 1984; Yamaguchi *et al.*, 1985b) and the ratio of ATP/GTP in the reaction mixture (Eliasson and Reichard, 1978; Yamaguchi *et al.*, 1985a,b).

[d] Data from Anderson *et al.* (1977), DePamphilis *et al.* (1979) Hay *et al.* (1984), and unpublished results of E. A. Hendrickson and M. L. DePamphilis.

[e] Data from Yamaguchi *et al.* (1985a,b).

ribonucleotides (Pigiet *et al.*, 1974; Anderson *et al.*, 1977). RNA–p–DNA linkages were found only with RI DNA. Misincorporation of dNTPs into RNA under the conditions used in these experiments was insignificant (Pigiet *et al.*, 1974; Eliasson and Reichard, 1979).

In both SV40 and PyV DNA, all 16 possible RNA–p–DNA linkages were found at frequencies suggesting a near-random distribution on the genome. The measured frequencies for SV40 DNA were similar to those predicted from a nearest-neighbor analysis of the genome (Anderson *et al.*, 1977; DePamphilis *et al.*, 1979), and the frequencies of RNA–p–DNA linkages in specific regions of the genome were the same as those from the total genome (Kaufmann *et al.*, 1977). The nucleotide locations of RNA–p–DNA linkages have been mapped to the single nucleotide level in three regions of the SV40 genome, and no preference for a particular linkage was observed (Hay *et al.*, 1984).

2. Location, Size, and Composition of iRNA

The RNA–p–DNA covalent linkages found in nascent DNA were accounted for by the presence of a short oligoribonucleotide at the 5′-end of the DNA that began with either ATP or GTP. Since alkaline hydrolysis of RNA–DNA linkages did not reduce the size of nascent DNA chains, the RNA moiety must reside near one of the termini (Pigiet *et al.*, 1974; Kaufmann *et al.*, 1977). About 40% of Okazaki fragments labeled in isolated nuclei were shortened by 7–8 residues when treated with either alkali or RNase T2 (Kaufmann *et al.*, 1977). Since at least 30% of the Okazaki fragments contained an RNA–p–DNA linkage, the RNA moiety must have been attached to the 5′-end of the DNA. Similarly, 30% of PyV Okazaki fragments contained intact iRNA (labeled in nuclei with [β-^{32}P]-ATP or GTP) which accounted for about 80% of the RNA–p–DNA linkages present (Eliasson and Reichard, 1978). The internal oligoribonucleotide sequence was nearly random (Reichard *et al.*, 1974).

About 50% of SV40 Okazaki fragments contained iRNA as judged by the alkaline sensitivity of 5′-end-labeled DNA chains from SV40(RI) DNA, and by the fraction of label recovered as oligoribonucleotides after digestion with T4 DNA polymerase 3′-to-5′ exonuclease (T4 exo) (Hay *et al.*, 1984). Since the 5′-terminal ribonucleotide composition of *all* SV40 iRNA–DNA chains contained 14% rC and rU, while the (p)ppRNA–DNA chains began exclusively with either ATP or GTP (Table III, "*in vivo*"), about 28% of the iRNA was partially degraded. These data are consistent with iRNA excision occurring concurrently with Okazaki fragment elongation (see Section IV.C.4).

DNase I or T4 exo digestion of purified nascent DNA chains from PyV (Reichard *et al.*, 1974; Eliasson and Reichard, 1978) or SV40 (Kaufmann, 1981; Hay and DePamphilis, 1982; Hay *et al.*, 1984) released oligoribonucleotides 7–12 residues long with 1–3 dNMPs at their 3′-end. More recently, T4 exo was shown to leave a single dNMP at the 3′-end of iRNA after digestion of nascent DNA (Yamaguchi *et al.*, 1985a). Under these conditions, 36% of SV40 iRNA–DNA chains contained 9–11 ribonucleotides and 17% consisted predominantly of 2–3 residues (Hay *et al.*, 1984). Therefore, to accommodate the ^{32}P-transfer experiments described above and the polarity of the T4 exo activity, iRNA must be attached to the 5′-ends of nascent DNA chains. This was confirmed using vaccinia virus guanylyltransferase ("capping enzyme") to specifically label the 5′-ends of pppRNA–DNA and ppRNA–DNA chains by addition of [α-^{32}P]-GTP (Hendrickson and DePamphilis, unpublished data), revealing the sizes of intact iRNA to vary from about 2 to 12 bases.

Mapping the locations of 5′-ends as well as RNA–p–DNA linkages of RNA-primed nascent DNA chains on the SV40 genome revealed that initiation events occur at specific nucleotide sites (Tapper and DePamphilis, 1980; Hay and DePamphilis, 1982; Hay *et al.*, 1984). These sites were found, on average, once every 7 bases, although the frequency varied from 1 in 4 to 1 in 14 bases, depending on the template sequence. If all of these sites were used in a single RI molecule, the average length of Okazaki fragments would have been only 8 residues! Therefore, once initiation occurred at one site, DNA synthesis must have been rapid enough to convert the remaining template into dsDNA before additional initiation events occurred. Synthesis of SV40 iRNA was initiated primarily on the second nucleotide (underlined) of the template sequence 3′-dPu$\underline{T}$ and secondarily at 3′-dPu$\underline{C}$ with a preference for 3′-dPu$\underline{T}$T or 3′-dPu$\underline{T}$A; purine-rich sites (PyPuPu, PuPuPu) were excluded. Based on the composition of initiation sites (Hay *et al.*, 1984), about 80% of SV40 RNA primers began with pppApPu, similar to RNA primers from *E. coli* (Kitani *et al.*, 1985) and *Drosophila* (Kitani *et al.*, 1984). Preference for two Pu residues at the 5′-ends of iRNA simply reflected the fact that there were more 3′-Pu$\underline{T}$Py sites available in the template than 3′-Pu$\underline{T}$Pu sites. Regions particularly active in initiation of Okazaki fragments (Tapper and DePamphilis, 1980) represented sequences containing a high percentage of 3′-dPu$\underline{T}$ sites. The mapping data indicated that 70–80% of iRNA would begin with ATP and the rest with GTP. Using vaccinia virus guanylyltransferase to label iRNA with [α-^{32}P]-GTP, 71% of initiation events were found to have begun with ATP and 29% with GTP; UTP and CTP were absent (Hendrickson and DePamphilis, unpublished data). The use of preferred initiation sites could explain the apparent anomalies observed in iRNA composition when one or more dNTPs were substituted for rNTPs (Eliasson and Reichard, 1979; Tseng and Goulian, 1980). Nearest-neighbor frequencies may change because, rather than simply substituting the dNTP analogue at the same initiation sites, the nucleotide composition in the reaction mixture may alter the preference for certain sites, as observed with DNA primase site selection *in vitro* (see Section IV.C.3). The size and composition of iRNA from SV40 and PyV are strikingly similar to iRNA from eukaryotic cells (Tseng *et al.*, 1979; Kitani *et al.*, 1984) and *E. coli* (Kitani *et al.*, 1985).

The length of RNA primers was not unique but varied with the initiation site used. Although analysis of the average size of isolated iRNAs indicated that all RNA primers were not the same size, part of this variation could have been due to the fact that heterogeneous sequences of the same length do not have the same mobility during gel electrophoresis, and part due to partial degradation of iRNAs. However, measuring the actual distance in nucleotides from the 5′-end of iRNA to the RNA–p–DNA junction demonstrated that the typical length of iRNA varied from 6 to 9 bases with some primers apparently as short as 3 and some as long as 12 bases (Hay *et al.*, 1984). However, this estimate treated each initiation site equally, and did not take into account the frequency at which each site was used. Thus, the fact that total end-labeled iRNA has a preferred length of 9–11 ribonucleotides reflects a preference for initiation at some sites over others. The transition from RNA to DNA synthesis, in contrast to the initiation of RNA synthesis, showed no preference for base sequence and often occurred at two or more adjacent nucleotides.

3. DNA Primase

iRNA is apparently synthesized by DNA primase from the template at replication forks rather than added on to the ends of DNA chains or synthesized as random-sequence primers prior to replication. The viral RI-associated oligoribonucleotide described above has been labeled in isolated nuclei with [β-^{32}P]-ATP, [β-^{32}P]-GTP, or [α-^{32}P]rNTPs (Narkhammer-Meuth *et al.*, 1981a), and a fraction of the label can be recovered as pppAp or pppGp (Reichard *et al.*, 1974; Eliasson and Reichard, 1978; Kaufmann, 1981). The labeled iRNA in SV40(RI) DNA survived treatment with RNase A, suggesting that it was entirely base-paired to the template (Kaufmann, 1981). ATP or GTP was used to initiate iRNA synthesis at a frequency that depended on their relative concentrations (Eliasson and Reichard, 1978), and iRNA synthesis was proportional to the concurrent rate of DNA synthesis (Eliasson and Reichard, 1979). However, addition of rNTPs to nuclei never stimulated DNA synthesis more than 20–30% (Anderson *et al.*, 1977; Eliasson and Reichard, 1979; Tapper *et al.*, 1982). This presumably occurred because endogenous rNTPs were present in the extracts, and because dNTPs could partially substitute for rNTPs when the dNTP/rNTP ratio was high (Eliasson and Reichard, 1979; Tseng and Goulian, 1980).

Synthesis of iRNA was not carried out by either RNA polymerase II or III because these enzymes were inhibited specifically by α-amanitin whereas SV40 (Kaufmann, 1981; Tapper *et al.*, 1982; Li and Kelly, 1984; Ariga, 1984b; Decker *et al.*, 1986), PyV (Hunter and Francke, 1974; Reichard and Eliasson, 1979), and HeLa cell (Krokan *et al.*, 1975; Brun and Weissbach, 1978) DNA synthesis were resistant to α-amanitin. RNA polymerase I had been implicated as the "primase" by virtue of the ability of an RNA polymerase I antiserum to inhibit Okazaki fragment synthesis in a subcellular system (Brun and Weissbach, 1978), but the evidence now favors a unique enzyme, DNA primase, which may or may not be a subunit of another RNA polymerase. DNA primase activity, the ability to synthesize an RNA primer on a DNA template for use by DNA polymerase, has been detected in extracts of monkey cells (Kaufmann and

Falk, 1982), human lymphocytes (Tseng and Ahlem, 1982), *Xenopus* eggs (Mechali and Harland, 1982; Riedel *et al.*, 1982), and *Drosophila* embryos (Yoda and Okazaki, 1983). DNA primase copurified with DNA polymerase α from monkey cells (Yamaguchi *et al.*, 1985a), mouse cells (Yagura *et al.*, 1982; Tseng and Ahlem, 1983), human cells (Wang *et al.*, 1984; Gronostajski *et al.*, 1984), calf thymus (Hubscher, 1983), *Drosophila* embryos (Conaway and Lehman, 1982; Kaguni *et al.*, 1983), *Xenopus* eggs (Shioda *et al.*, 1982) and with yeast DNA polymerase I (Plevani *et al.*, 1984; Singh and Dumas, 1984), the homologue of α-polymerase. Furthermore, DNA primase activity in mouse (Yagura *et al.*, 1983a) and human (Wang *et al.*, 1984) cells was immunoprecipitated with monoclonal antibodies against DNA polymerase α. When the DNA primase–DNA polymerase complex was provided with a DNA template and the complementary ribo- and deoxyribonucleoside triphosphates, it synthesized RNA-primed DNA in a fashion characteristic of DNA synthesis in *vivo:* synthesis was resistant to α-amanitin, and an oligoribonucleotide, 7–12 bases long, beginning with A or G was found covalently attached to the 5′-ends of nascent DNA chains. In fact, the sequence of one iRNA was shown to be complementary to its template initiation site (Yamaguchi *et al.*, 1985b). Thus, it appears that a DNA primase–DNA polymerase α complex is responsible for initiation of Okazaki fragments at eukaryotic replication forks. However, two properties of DNA primase–DNA polymerase α *in vitro* were different from iRNA synthesis *in vivo: in vitro*-synthesized iRNA was about 3 nucleotides shorter than iRNA synthesized *in vivo,* and DNA sequence-specific initiation sites *in vitro* were distinctly different from initiation sites selected *in vivo* (Table III, Fig. 16; Yamaguchi *et al.*, 1985a,b).

Three parameters affected the selection of iRNA initiation sites *in vitro* and the length of iRNA synthesized: (1) the sequence of the DNA template, (2) the ATP/GTP ratio in the reaction mixture, and (3) the secondary structure of the template. Initiation events by DNA primase–DNA polymerase α from CV-1 cells were found, on average, once every 16 bases, although the actual frequency varied from 1 in 13 to 1 in 35 bases among different DNA templates (about one-half the frequency observed *in vivo*), and some sites were highly preferred over others (Yamaguchi *et al.*, 1985b). The sites shared a consensus sequence of either 3′-dC$\underline{\text{T}}$TT or 3′-dC$\underline{\text{C}}$C centered within a pyrimidine-rich sequence that was perhaps about 7 to 25 residues in length. The initiating nucleotide of the RNA primer [identified by labeling (p)ppRNA–DNA chains with [α-^{32}P]-GTP using vaccinia guanylyltransferase] was complementary to the underlined base. Hence, the initiation sites selected by DNA primase–DNA polymerase α on SV40 DNA templates *in vitro* were distinctly different from the sites selected on the same DNA templates during SV40 replication *in vivo* (Fig. 16). Similar results were observed with purified DNA primase from mouse cells (Tseng and Ahlem, 1984), suggesting that site selection is carried out by the DNA primase component alone.

Preference for specific initiation sites and the extent of their utilization were modulated by the ATP/GTP ratio in the reaction mixture (Yamaguchi *et al.*, 1985a,b), similar to the synthesis of PyV iRNA–DNA chains in isolated nuclei (Eliasson and Reichard, 1978). High ATP concentrations similar to those found in mammalian cells (Cory *et al.*, 1976) promoted initiation at pyrimidine-rich sites containing 3′-dCTTT, while high GTP concentrations promoted initiation at pyrimidine-rich sites containing 3′-dCCC (Yamaguchi *et al.*, 1985b). At ATP/GTP ratios similar to those found in mammalian cells (20 : 1), ATP starts were preferred 4 : 1 over GTP starts; CTP and UTP starts were not detected. When the ATP/GTP ratio was changed to 0.5 : 1, ATP and GTP starts were equal. Nevertheless, changing the ATP/GTP ratio did not make the *in vitro* sites the same as in the *in vivo* sites; it simply gave preference to one subset of *in vitro* sites over another.

The influence of secondary structure (i.e., dsDNA hairpins) was inferred by Tseng and Ahlem (1984) who showed that a 6-bp deletion in the 27-bp palindrome located within the SV40 *ori* (Fig. 16) reduced the frequency of initiation at start sites adjacent to the palindrome by 80%. This palindromic sequence has been shown to form a hairpin when present as a ssDNA template under the conditions employed in DNA polymerase assays (Weaver and DePamphilis, 1984). It is possible that hairpin formation prevents initiation within a palindromic sequence as well, since initiation events *in vitro* were not detected within the SV40 27-bp palindrome on either template (Tseng and Ahlem, 1984; Yamaguchi *et al.*, 1985b). Folding of the DNA template also appears to affect selection of initiation sites for *E. coli* DNA primase (Sims and Benz, 1980).

4. iRNA Excision

iRNA must be completely excised from the 5′-end of the growing daughter strand before it is joined to the 3′-end of a new Okazaki fragment. Otherwise, the resulting DNA will contain one or more intermittent ribonucleotides. RNA–p–DNA linkages were shown to exist only transiently in PyV RI (Pigiet *et al.*, 1974), and, in SV40 RI, were shown to be excised at the same rate that Okazaki fragments are joined to nascent DNA (Anderson *et al.*, 1977), consistent with their role as an intermediate in replication. Excision of iRNA on replicating SV40 DNA occurred in two steps (Anderson and DePamphilis, 1979). The bulk of the iRNA (measured by ^{32}P-transfer to rNMPs) was removed even in the absence of concomitant DNA synthesis or complete elongation and ligation of Okazaki fragments. This step could be carried out by RNase H, an enzyme that endonucleolytically removes all ribonucleotides in RNA/DNA hybrids, but one that does not remove RNA–DNA linkages (Berkower *et al.*, 1973; Stavrianopoulos *et al.*, 1976; Ogawa and Okazaki, 1984). Excision of the remaining 5′-p-rN–(pdN)$_n$ linkage (measured by ^{3}; s2P-transfer to rNDPs) occurred 2.5 times more slowly when Okazaki fragment maturation was retarded by Ara-ATP or the absence of gap-filling factors from the cytosol fraction (Anderson and DePamphilis, 1979). Furthermore, comparison of the locations of 5′-ends of DNA chains carrying iRNA (i.e., the RNA–p–DNA junction) with the 5′-ends of DNA chains without a terminal rNMP residue, indicated that a variable number of 5′-terminal dNMPs had been excised *in vivo* from nascent DNA as well as the iRNA moiety (Hay *et al.*, 1984). Thus, this excision step appeared to involve a 5′–3′ exonuclease activity that may have been coupled with DNA polymerase α, analogous to the role proposed for *E. coli* DNA polymerase I (Kornberg, 1980; Ogawa and Okazaki, 1984). Consistent with this hypothesis is the fact that those Okazaki fragments whose 3′-ends were farthest from the 5′-ends of long nascent DNA chains most frequently had an iRNA–DNA chain downstream (Anderson and DePamphilis, 1979), indicating that complete iRNA excision was accelerated by close proximity with the 3′-end of a growing Okazaki fragment. Since the relative rates of the two excision steps could vary with the conditions of the assay, the second excision may sometimes involve two or more rNMPs (Anderson *et al.*, 1979; Hay and DePamphilis, 1982). The DNA remaining after excision of iRNA contains a 5′-PO_4 (Anderson and DePamphilis, 1979).

Essentially all of the Okazaki fragments on purified SV40(RI) DNA were present in one of three stages in their metabolism (Anderson and DePamphilis, 1979): (1) About 20% were separated from longer nascent DNA chains by a single phosphodiester bond interruption that could be sealed with DNA ligase alone. (2) About 30% were separated by a gap of ssDNA and required both DNA polymerase and DNA ligase for joining. (3) The remaining 50% faced an RNA primer and required 5′–3′ exonuclease, DNA polymerase, and DNA ligase activities in order to be joined to the long DNA strands. Therefore, at least 50% of the long nascent DNA chains do not contain iRNA, as well as 50% of the Okazaki fragments (Hay *et al.*, 1984). This distribution reflected the steady-state balance in initiation, elongation, and joining of Okazaki fragments with the concurrent excision of iRNA, as expected from a stochastic model for initiating Okazaki fragment synthesis within an initiation zone of variable size (Section IV.G).

5. iRNA *in Vivo*

Since many of the earlier studies on iRNA were carried out using a variety of subcellular systems, it was important to establish that iRNA was part of DNA replication *in vivo* as well as *in vitro*. The following results, some of which were mentioned above, fulfill this purpose. Hay and DePamphilis (1982) and Hay *et al.* (1984) demonstrated that purified

SV40(RI) DNA from virus-infected CV-1 cells contained Okazaki fragment-size DNA carrying a short oligoribonucleotide covalently attached to the 5′-end of the DNA. The size, sequence composition, and RNA–p–DNA linkages of these iRNA molecules were essentially the same as those obtained from nascent DNA chains radiolabeled in subcellular systems. Furthermore, about 50% of the Okazaki fragments on purified SV40(RI) DNA that had been pulse-labeled with [^{3}H]-Thd *in vivo* could be joined to longer nascent DNA *only* if treated by an enzyme such as *E. coli* DNA polymerase I that was capable of excising iRNA (Anderson and DePamphilis, 1979). Treatment of RI with DNA polymerase I was not required for ligation when RNA–p–DNA linkages were excised by allowing DNA replication to proceed in nuclei washed free of cytosol, and the time required for loss of this polymerase I requirement was the same as for loss of RNA–p–DNA linkages (Anderson and DePamphilis, 1979). Finally, PyV Okazaki fragments labeled *in vivo* were sensitive to 5′-OH-specific spleen exonuclease following treatment with alkali or RNase A plus RNase T1 which removed RNA and generated 5′-OH termini on those DNA chains that were covalently attached to RNA (Narkhammer-Meuth *et al.*, 1981b). Since alkaline hydrolysis of apyrimidinic DNA yielded 5′-PO_4 termini (Machida *et al.*, 1981), these results could not be explained by incorporation and excision of unusual bases.

D. Gap-Filling Step

Gap-filling represents a distinctive step in the maturation of Okazaki fragments that is carried out by DNA polymerase α in conjunction with one or more proteins referred to as gap-filling factors (Fig. 13). In the absence of cytosol, nuclei carried out the complete excision of iRNA (Anderson *et al.*, 1977), although joining of Okazaki fragments was 10–20 times slower (DePamphilis and Berg, 1975; Francke and Hunter, 1975; Otto and Reichard, 1975; Edenberg *et al.*, 1976, 1977; Su and DePamphilis, 1976, 1978; Anderson *et al.*, 1977). About 70% of these fragments were separated from long nascent strands by a gap of 10–15 nucleotides, while the remaining 30% were mature Okazaki fragments ready for the ligation step (Anderson and DePamphilis, 1979). DNA polymerase α was required since the gap-filling step was inhibited by aphidicolin (Wist and Prydz, 1979; Weaver *et al.*, 1980; Krokan, 1981b; Lonn and Lonn, 1983); those that did join had already completed DNA synthesis and only required DNA ligase activity, an aphidicolin-insensitive enzyme (Weaver *et al.*, 1980). However, α-polymerase, in contrast to T4 DNA polymerase, was not able to complete DNA synthesis in short gaps in purified DNA primer-templates (Fisher *et al.*, 1979; DePamphilis *et al.*, 1980), or at the ends of linear DNA molecules (Pritchard *et al.*, 1983). In fact, partially purified α-polymerase alone or together with mammalian DNA ligase did not stimulate DNA synthesis in nuclei at enzyme concentrations equivalent to those found in cytosol (Hershey *et al.*, 1973; DePamphilis and Berg, 1975; Francke and Hunter, 1975; Krokan *et al.*, 1977; Brun and Weissbach, 1978; Fraser and Huberman, 1978; Reinhard *et al.*, 1979). Thus, one or more of the heat-labile, cytosol proteins that could be purified free of DNA polymerase activity (DePamphilis, unpublished data) and stimulated Okazaki fragment ligation in subcellular systems appeared to represent DNA polymerase α cofactors that increased the enzyme's efficiency to synthesize DNA in short gaps.

Treatment of virus-infected cells or subcellular systems with FdUrd (Salzman and Thoren, 1973; Manteuil and Girard, 1974), hydroxyurea (Magnusson, 1973; Laipis and Levine, 1973), Ara-CTP (Hunter and Francke, 1975), or low concentrations of dNTPs (LeBlanc and Singer, 1976; Eliasson and Reichard, 1979) resulted in a preferential accumulation of Okazaki-like fragments separated by gaps of ssDNA (Salzman and Thoren, 1973). These data were interpreted in favor of two different DNA polymerases, one with a low dNTP K_m that elongated Okazaki fragments and one with a high dNTP K_m that completed the gap-filling step. However, several alternative explanations should be considered. First, FUrd and hydroxyurea decrease intracellular dNTP pools and thus might increase the rate of misincorporation and excision of dUTP and other abnormal bases, thereby increasing the fraction of "Okazaki-like" DNA fragments. Second, Ara-CTP was incorporated into nascent DNA, making the primer-template a poorer substrate for α-polymerase (Manteuil *et al.*, 1974; Fry, 1983). Third, apparent dNTP K_m values observed with α-polymerase varied with the DNA substrate from 0.04 μM in isolated nuclei to 0.6 μM on DNase I-activated DNA, to 6.5 μM on denatured DNA (Eliasson and Reichard, 1979; Pritchard *et al.*, 1983) as well as with the source of enzyme (Pritchard *et al.*, 1983). Therefore, the utilization of dNTPs was not a simple function of the enzyme used but was determined by the structure and composition of the replication complex.

E. DNA Polymerase α[C_1C_2]

The evidence is compelling that DNA polymerase α is solely responsible for DNA synthesis on both sides of SV40 and PyV replication forks. Aphidicolin, a specific inhibitor of α-polymerase (Spadari *et al.*, 1982), blocked all steps in SV40 DNA replication (Krokan *et al.*, 1979; Dinter-Gottlieb and Kaufmann, 1982, 1983; Waqar *et al.*, 1983), including the gap-filling step. Similarly, Ara-CTP, which inhibited α-polymerase 20–30 times more effectively than either β- or γ-polymerase (Krokan *et al.*, 1979; Fry, 1983), inhibited viral DNA synthesis in essentially the same manner as isolated α-polymerase (Krokan *et al.*, 1979). In contrast, dideoxy-TTP, a selective inhibitor of both β- and γ-polymerase, had little effect on any of the steps in DNA replication (van der Vliet and Kwant, 1978; Edenberg *et al.*, 1978; Krokan *et al.*, 1979; DePamphilis *et al.*, 1979;

Waqar *et al.*, 1983). α-Polymerase activity consistently increased three- to tenfold with the onset of DNA replication in infected cells (Wintersberger and Wintersberger, 1975; Narkhammer and Magnusson, 1976; Mechali *et al.*, 1977), and copurified with replicating SV40 (Edenberg *et al.*, 1978; Otto and Fanning, 1978; Tsubota *et al.*, 1979; Otto *et al.*, 1979) and PyV (Gourlie *et al.*, 1981) chromosomes. The ratio of α-polymerase molecules to PyV(RI) DNA was 4.8 ± 0.9 throughout the sedimentation gradient (Gourlie *et al.*, 1981). DNA polymerase β was never detected, although low amounts of DNA polymerase γ were found with SV40 chromosomes. A specific requirement for α-polymerase in SV40 DNA replication was demonstrated by the finding that α-polymerase, but not β- or γ-polymerase, could reconstitute viral DNA replication in *N*-ethylmaleimide-inactivated nuclear extracts (Krokan *et al.*, 1979). Furthermore, the sensitivity of α-polymerase to chloride and acetate anions corresponded to that of subcellular SV40 DNA replication systems (Richter *et al.*, 1980), and the extent of DNA synthesis on SV40 chromosomes was proportional to the fraction of α-polymerase present (Waqar *et al.*, 1983).

Essentially all of the α-polymerase could be extracted from monkey cells as a complex with at least two proteins, C_1 and C_2, that act in concert as stimulatory cofactors specific for α-polymerase from the same species (Pritchard and DePamphilis, 1983; Pritchard *et al.*, 1983). These stimulatory cofactors, originally recognized with HeLa DNA polymerase α (Lamothe *et al.*, 1981), could be dissociated from α-polymerase by phosphocellulose or hydrophobic chromatography, or by treatment with nonionic detergents. Since the responses of $\alpha[C_1C_2]$-polymerase to aphidicolin, dideoxy-TTP, and *N*-ethylmaleimide were the same as those of α-polymerase alone, the "holo" form of this enzyme, which also contained DNA primase activity (Yamaguchi *et al.*, 1985a; Vishwanatha and Baril, unpublished data), appeared to be the replicative enzyme.

The C_1C_2 complex functions in primer recognition (Pritchard *et al.*, 1983). Both α-polymerase alone ("core") and with C_1C_2 ("holo") required a minimum of about 12 bases of template per primer and exhibited the same activity on DNA substrates with low ratios of template to primer. However, at high template to primer ratios, C_1C_2 stimulated "core" enzyme from 180- to 1800-fold when DNA substrate concentrations were low [$v = V(S)/K$]. Maximum stimulation occurred at about 1000 bases per primer, the consequence of a dramatic reduction in the K_m of the DNA substrate, or more specifically, the primer itself. The additional template was not required for extensive elongation since stimulation was observed with the first nucleotide incorporated. C_1C_2 had no effect on the K_m for dNTP substrates, the frequency or intensity of DNA sequence signals that arrested α-polymerase (Weaver and DePamphilis, 1982, 1984), or the processivity (9–11 bases) of α-polymerase. Therefore, C_1C_2 specifically increased the ability of α-polymerase to find a primer and insert the first nucleotide. The fact that extensive ssDNA primer-templates were better substrates for $\alpha[C_1C_2]$-polymerase than optimally activated

DNase I-treated DNA was for α-polymerase, demonstrated that ssDNA participated in the reaction, presumably by allowing the "holo"-enzyme to slide along the template until it finds a primer. In contrast, ssDNA inhibited α-polymerase, demonstrating that binding is nonproductive, and that the enzyme must continually rebind in order to locate a primer.

What is the function of DNA polymerase $\alpha[C_1C_2]$ at replication forks? With mammalian DNA, this enzyme could utilize the low concentration of replication forks (about 0.04% of total DNA) present during S phase 100–1000 times more effectively than α-polymerase alone. Presumably, C_1C_2 also lowers the K_m for the ssDNA templates that are used as substrates for DNA primase–DNA polymerase α, and, in addition, may change site selection from the pyrimidine-rich sequences observed *in vitro* to the PuPy sites observed *in vivo* (Table III). In fact, DNA primase stimulatory cofactors described for mouse DNA primase–DNA polymerase α (Yagura *et al.*, 1983b), appeared very similar to C_1C_2. However, a paradox arises from the data. DNA polymerase α activity coupled to DNA primase was not inhibited by aphidicolin, while the same enzyme preparations using preformed primers were inhibited by this drug (Tseng and Ahlem, 1982; Wang *et al.*, 1984; Plevani *et al.*, 1984). Since DNA replication in all eukaryotic systems examined so far was inhibited by aphidicolin (DePamphilis and Wassarman, 1980, 1982; Spadari *et al.*, 1982; Section IV.D), Okazaki fragment synthesis must occur in two steps. First, DNA primase–DNA polymerase α may synthesize a short RNA–DNA fragment about 20–25 nucleotides long, an aphidicolin-insensitive process, and then disengage in order to travel with the replication fork and repeat the process at each initiation zone (Fig. 14). DNA polymerase $\alpha[C_1C_2]$ could then initiate synthesis on the 3′-end of the resulting RNA–DNA primer in an aphidicolin-sensitive step. Several pieces of data are consistent with this hypothesis. First, DNA primase–DNA polymerase α synthesized RNA–DNA chains 20–25 nucleotides long in the presence of excess template (Yamaguchi *et al.*, 1985a), revealing a processivity that was essentially the same as DNA polymerase $\alpha[C_1C_2]$ (Pritchard *et al.*, 1983). This could explain the accumulation of short (about 30 nucleotides) RNA-primed DNA chains in PyV(RI) DNA (Eliasson and Reichard, 1978). Furthermore, in the presence of aphidicolin, 30- to 35-nucleotide-long RNA-primed nascent DNA chains accumulated in SV40(RI) DNA (Dinter-Gottlieb and Kaufmann, 1983).

F. Replication Proteins Associated with Viral Chromosomes

SV40 chromosomes, isolated from a hypotonic nuclear extract by sedimentation through glycerol gradients, contained at least 30 major proteins that could be labeled with [^{35}S]methionine (Milavetz *et al.*, 1980). Some of these (e.g., the major capsid protein VP1) could be cross-linked to nascent DNA containing BrdUMP by UV irradiation (Tsutsui *et al.*,

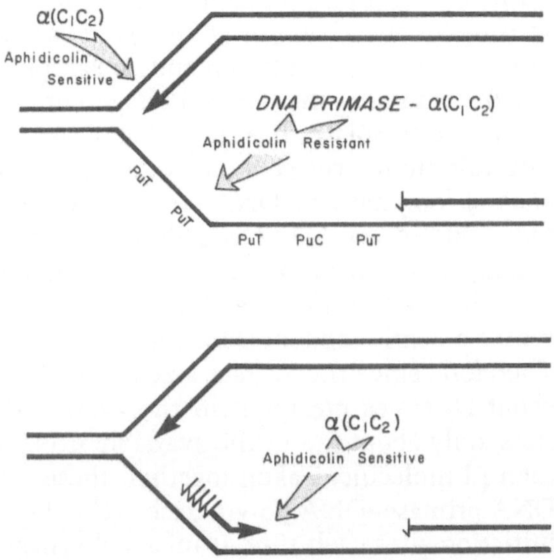

FIGURE 14. Proposed roles of DNA primase–DNA polymerase α and DNA polymerase α[C_1C_2] in the initiation and synthesis of DNA at replication forks. DNA primase–DNA polymerase α synthesizes a short RNA–DNA chain 25–35 nucleotides long in an aphidicolin-resistant reaction. This is used by DNA polymerase α[C_1C_2] to initiate DNA synthesis in an aphidicolin-sensitive reaction. Once this sequence has occurred on the forward DNA template at *ori*, DNA polymerase α[C_1C_2] continues DNA synthesis on the forward arm of the replication fork without the need for further initiation events by DNA primase–DNA polymerase α.

1983). Replicating SV40 and PyV chromosomes isolated by similar procedures also were associated with DNA polymerase α, perhaps DNA polymerase γ (Section IV.E), a ssDNA binding protein that stimulated α-polymerase activity (Otto *et al.*, 1979), a ssDNA-dependent ATPase that was also present in uninfected CV-1 cells (Brewer *et al.*, 1983), topoisomerase I (Sen and Levine, 1975; Keller *et al.*, 1977; Tsubota *et al.*, 1979; Hamelin and Yaniv, 1979; Gourlie *et al.*, 1981), topoisomerase II (Waldeck *et al.*, 1983; Liu *et al.*, 1983; Krauss *et al.*, 1984), DNA ligase (Krauss *et al.*, 1984), uracil-DNA glycosylase (Krokan, 1981a), poly(ADP-ribose) polymerase (Prieto-Soto *et al.*, 1983), and RNase H (Gourlie *et al.*, 1981b) activities. A tightly bound protein also has been reported near *ori* on SV40 DNA. Electron microscopy of viral chromosomes treated with 1 M NaCl (Griffith *et al.*, 1975) or virions treated with 5% SDS (Kasamatsu and Wu, 1976a,b) revealed SV40(II) DNA with protein remaining at map position 0.7. Similar experiments have shown that both VP1 (Brady *et al.*, 1981) and topoisomerase I (Hamelin and Yaniv, 1979) can form stable complexes with SV40 DNA, but their genomic locations were not determined. Stable complexes may actually result from the methods of purification since topoisomerase I did not always copurify with SV40 chromosomes (Young and Champoux, 1978).

G. Initiation of DNA Synthesis at Replication Forks

Two key observations support a stochastic model for initiation of Okazaki fragments at SV40 replication forks (Anderson and DePamphilis, 1979; DePamphilis *et al.*, 1979, 1980, 1983a,b; DePamphilis and Wassarman, 1980, 1982; DePamphilis, 1984). First, mature Okazaki fragments waiting for the action of DNA ligase vary in length from 40 to 290 nucleotides (Fig. 13; Anderson and DePamphilis, 1979; DePamphilis *et al.*, 1979). Therefore, since Okazaki fragments originated predominantly from retrograde templates (Section IV.B.2), they were not initiated at uniform intervals around the genome but at many different distances from the 5′-ends of long nascent DNA strands. Second, iRNA initiation sites were found once every 4–14 bases along the template, and the only consensus among these initiation sites was 3′-dPuT or 3′-dPuC in the template; both the internal iRNA sequence and the RNA–p–DNA linkages represented a nearly random selection of template sequences (Section IV.C.2). Therefore, since the average size of an Okazaki fragment (135 bases) was about 20 times greater than the average distance between iRNA start sites, only about 5% of the available iRNA initiation sites were used on each RI molecule. Taken together, these two observations revealed that DNA primase–DNA polymerase $\alpha[C_1C_2]$ selected one of several possible initiation sites each time it initiated synthesis at a replication fork (Fig. 14). Since the sites chosen in any one template region varied from one RI to the next, the selection process was stochastic.

A stochastic model explains the observation that the fraction of Okazaki fragments containing RNA at their 5′-ends was essentially the same for long fragments as for shorter ones, because some initiation events occurred close to the 5′-end of the growing strand and some occurred farther away (Kaufmann *et al.*, 1977; Eliasson and Reichard, 1978). The size of an Okazaki fragment does not necessarily correspond to its "age." Naturally, the shortest Okazaki fragments (10–30 bases) more frequently carried an iRNA because they were enriched for newly initiated DNA chains. Assuming that the probability of excising iRNA increases with time after its synthesis, a fraction of the long nascent DNA strands would be expected to carry iRNA, as observed (Pigiet *et al.*, 1974; Kaufmann *et al.*, 1977; Eliasson and Reichard, 1978; Kaufmann, 1981), because some Okazaki fragments would be initiated close enough to the growing daughter strand's 5′-end to allow ligation to precede excision of the iRNA from the Okazaki fragment.

The question remains: why is an Okazaki fragment initiated an average of once every 135 bases? The simplest explanation is that the frequency of Okazaki fragment initiation is determined by the periodicity of nucleosomes in front of replication forks (Section V.A). If nucleosome disassembly is the overall rate-limiting step in replication, then as replication forks move from one nucleosome core to the next, the forward arm will be maintained as dsDNA by continuous DNA synthesis whereas the length of ssDNA on retrograde arms will be determined by the average distance between nucleosome core particles (220 ± 73 bases). Thus, the maximum length of Okazaki fragments would be about 290 bases and the average length about 110 bases, in excellent agreement with actual measurements. The fact that only 25% of SV40 replication forks had an

Okazaki fragment is consistent with fork movement (creation of an initiation zone) being the rate-limiting step in replication (Hay *et al.*, 1984).

H. Initiation of DNA Synthesis in the *ori* Region

In order to compare the events at replication forks with DNA synthesis events in and around the *ori* sequence, the nucleotide locations and relative frequencies of RNA–p–DNA linkages and 5′-ends of RNA–DNA chains were mapped in three different regions of the SV40 genome, and the size distribution of iRNA in each region was determined (Hay and DePamphilis, 1982; Hay *et al.*, 1984). The first region was a 311-bp DNA fragment that contained *ori.* The second was a 263-bp fragment containing the *Msp*I restriction site (about 350 bp from *ori*). The third region was a 360-bp fragment containing the *Eco*RI site (about 1780 bp from *ori*). No significant differences were observed in the initiation events that occurred in each of the three regions. The size distribution of iRNA, the sequence composition of iRNA initiation sites, and the distribution of initiation sites between the two arms of replication forks were essentially the same in the *ori* region (Figs. 15, 16A) as they were

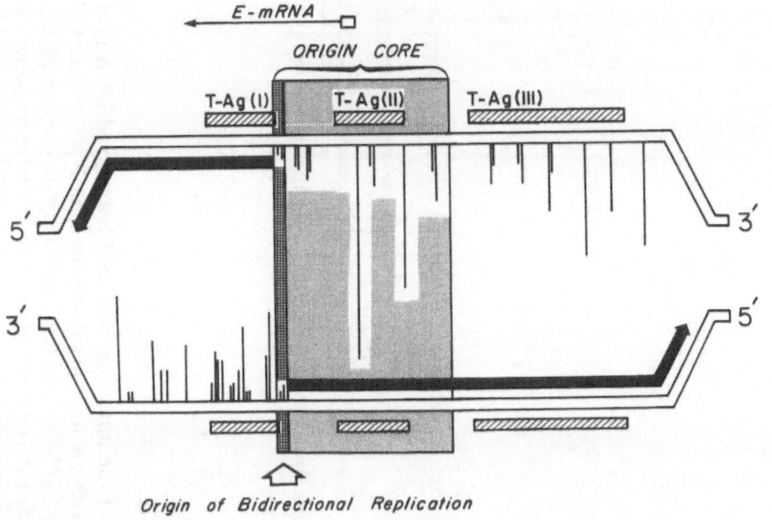

FIGURE 15. SV40 replication bubble in the *ori* region. Continuous synthesis of nascent DNA (solid bold lines with arrowheads) on the forward arms of replication forks began at one of the vertical bars indicating the positions and relative frequencies of RNA–p–DNA covalent linkages (Hay and DePamphilis, 1982; Hay *et al.*, 1984). The 64-bp *ori* sequence (Fig. 10), the three SV40 T-Ag DNA binding sites defined by methylation experiments (Fig. 4), and the direction and template for early E-mRNA (Fig. 3) are indicated. The transition between the presence and absence of RNA–p–DNA linkages on each strand of *ori* defines the origin of bidirectional DNA replication (OBR).

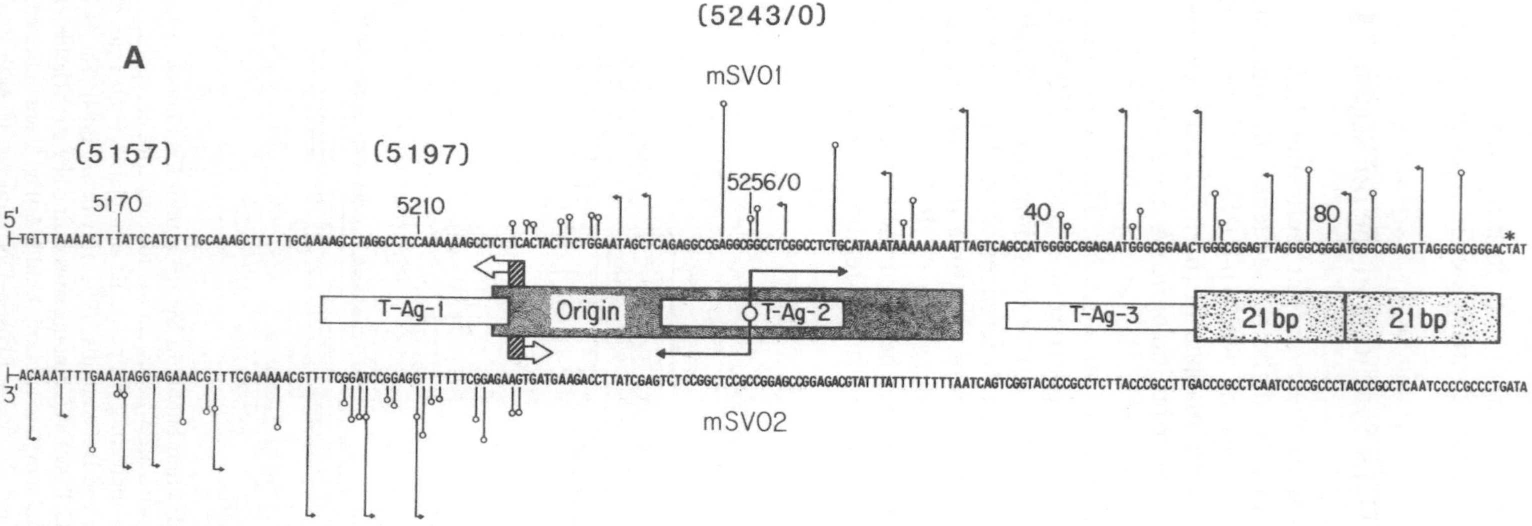

FIGURE 16A. Maps of the initiation sites for RNA-primed DNA synthesis in the SV40 *ori* region. DNA template initiation sites are shown for iRNA (vertical bars with an arrowhead) and the dNMP position in RNA–p–DNA covalent linkages (vertical bars with a circle) for nascent DNA chains synthesized by SV40-infected CV-1 cells (Hay and DePamphilis, 1982; Hay *et al.*, 1984). The arrowheads indicate the direction of synthesis, and the length of each bar is proportional to the relative frequency at which the site was used. However, the relative intensities of 5′-ends of iRNA cannot be compared with those of RNA–p–DNA junctions. Note that an iRNA initiation event is always about 10 bases upstream from one or more RNA–p–DNA linkages. The transition between the appearance of initiation sites and their absence on each side of *ori* defines the origin of bidirectional replication (Hay and DePamphilis, 1982; Hay *et al.*, 1984; also shown in Fig. 15).

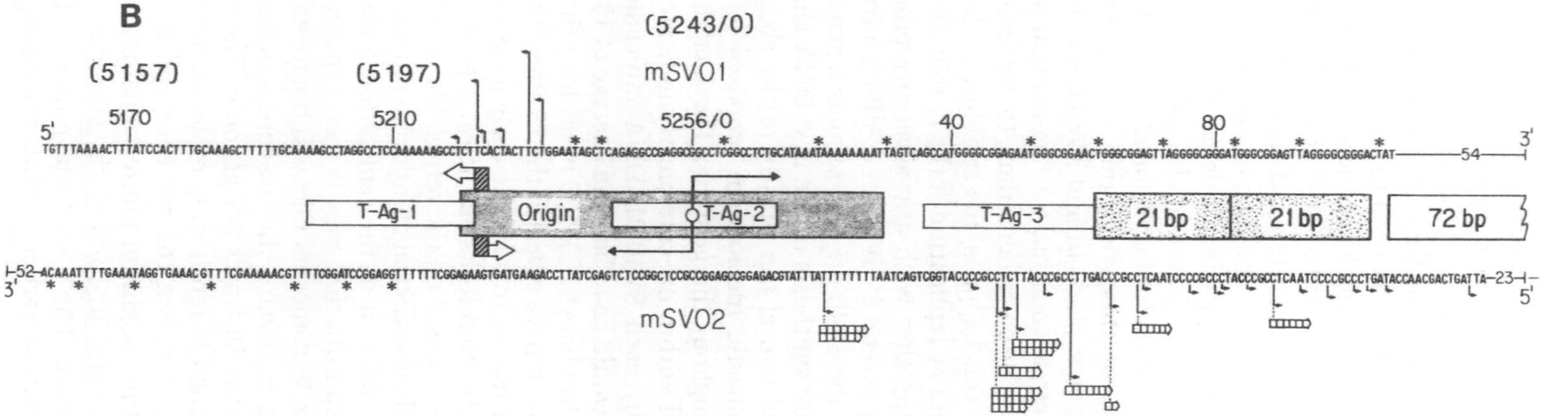

FIGURE 16B. DNA template initiation sites are shown for purified CV-1 cell DNA primase–DNA polymerase α (Yamaguchi *et al.*, 1985b). The ATP/GTP ratio in the reaction mixture was 20 : 1. Vertical bars with an arrowhead designate the template nucleotide complementary to the 5′-terminal ribonucleotide in the RNA primer. The arrowhead indicates the direction of synthesis. An asterisk indicates the position of 5′-terminal ribonucleotides of iRNA synthesized *in vivo* (vertical bars with arrowhead in panel A). Nucleotides are numbered with respect to the SV40 wt800 sequence which contains an A in place of a G at position 5222, and an additional 13 bp in the enhancer sequence (Hay *et al.*, 1984). The numbers for SV40 strain 776 are given above in parentheses. mSV01 and mSV02 are the two strands of the *Bst*N1 311-bp DNA fragment containing the SV40 *ori* region cloned into phage M13mp7 (Hay and DePamphilis, 1982). Also indicated in this figure are the primary DNA binding sites for T-Ag (D2 protein) (T-Ag-1, 2, and 3; Fig. 4), the *ori*-core sequence (Figs. 7, 10), a 27-bp palindrome (⮌), and the 21-bp and 72-bp repeated sequences (promoter–enhancer elements).

in the *Msp*I and *Eco*RI regions (Section IV.C2). Only the frequency of initiation sites in the three regions varied, and this was correlated with the frequency of 3′-dPuT Okazaki fragment initiation sites in the template. Therefore, synthesis of the first nascent DNA chain appeared to be initiated by the same mechanism used to initiate Okazaki fragment synthesis at replication forks throughout the genome. For comparison, the initiation sites selected *in vitro* by CV-1 cell DNA primase–DNA polymerase α on the SV40 *ori* region templates are also shown (Fig. 16B).

These data, together with the average length of iRNA on nascent DNA chains in each region, provided a detailed picture of the DNA synthesis initiation events in and around the *ori* sequence (Figs. 15, 16A), and suggested a simple model for initiation of DNA replication at *ori* (Fig. 17). Presumably, T-Ag together with other cellular components acts as an initiation complex that binds in and around *ori*-core to separate the two parental DNA strands and create a "replication bubble" that offers two "initiation zones," one on either strand, for initiation of RNA-primed DNA synthesis. Once this happens, the simplest explanation that would account for all the data at replication forks is that DNA primase–DNA polymerase $\alpha[C_1C_2]$, together with any accessory proteins that may be needed, selects one of several 3′-dPuT or 3′-dPuC initiation sites on the E-mRNA template in the SV40 *ori*-core sequence (data are not yet available for PyV) to initiate synthesis of the first DNA chain. The selection process clearly favored two of the sites in *ori* by about tenfold, and reflected the same stochastic mechanism proposed for site selection at replication forks throughout the genome. The fact that 13 RNA–DNA linkages were mapped within *ori*-core clearly indicates that all of these sites were not used by each SV40(RI) DNA molecule. Otherwise, the nascent DNA chains would have been an average of 15 bases long in this region and therefore the chains 100–200 bases long that were mapped in this experiment would not have revealed the 13 sites. The first RNA–DNA chain continues its synthesis toward the early gene region and becomes the forward arm of replication forks on the early gene side of *ori*. As soon as an initiation zone is exposed on the retrograde template of these forks, initiation of the second Okazaki fragment occurs in the opposite direction and becomes the forward arm of the replication forks traveling toward the late gene region. An initiation zone is then exposed on the retrograde template of these forks, and another Okazaki fragment is initiated whose direction of synthesis is toward the *ori*-core sequence. The initiation process is repeated about 38 times per genome. The net result of these activities is a bidirectionally replicating molecule whose forks originate at the transition point between the presence and absence of initiation sites on each strand [the "origin for bidirectional replication" (OBR)]. In SV40, this point lies at the junction of the strongest T-Ag binding site and the *ori*-core sequence. The 5′-end mapping data (Fig. 16A) were in excellent agreement with an earlier electron microscopic analysis of rep-

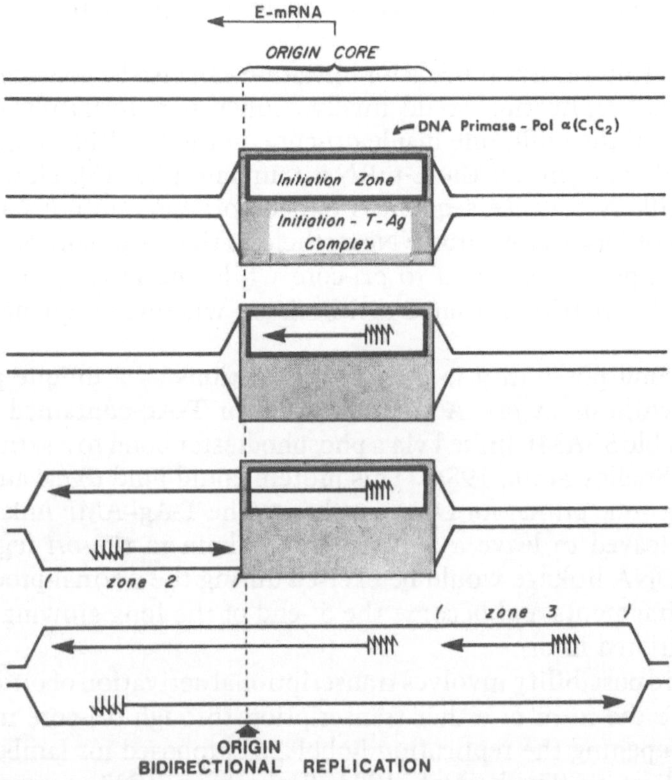

FIGURE 17. Proposed sequence of events in the initiation of SV40 DNA replication. The origin of replication is opened by an initiation complex directed to *ori* by T-Ag. DNA primase–DNA polymerase $\alpha[C_1C_2]$ initiates synthesis of RNA-primed DNA at one of several possible initiation sites on the E-mRNA template strand within *ori*-core. Successive initiation zones then develop as replication forks first advance toward the early gene side and then the late gene side of *ori*. The events at the first initiation zone within *ori* are essentially the same as the events at replication forks throughout SV40.

lication bubbles that placed the OBR about 35 bp to one side of the *Bgl*I site (Martin and Setlow, 1980). Later data from the same laboratory placed it 69 ± 40 bp from the *Bgl*I site (Zannis-Hadjopoulos *et al.*, 1983).

If this asymmetrical model for initiation of SV40 DNA replication is correct, then something must physically block initiation events on the L-mRNA side of *ori*-core. This follows because 3′-dPuT and 3′-dPuC sites are available on the L-mRNA template, and because the two major initiation sites on the E-mRNA template lie within a palindromic sequence, ensuring that at least one of the initiation sites is represented on the opposite strand and yet these sites were not utilized. The most likely candidate for blocking initiation of DNA synthesis on the L-mRNA template is the T-Ag initiation complex itself. Converting the L-mRNA template into an RNA/DNA duplex is unlikely because none of the L-mRNAs that have been mapped originated upstream of *ori* (DePamphilis and Wassarman, 1982).

Several alternative models for initiation can also be considered. First, bidirectional replication could involve concurrent initiation events on either side of the OBR, one inside *ori*-core on the E-mRNA template and one outside *ori*-core on the L-mRNA template (Fig. 15). However, this seems unlikely because sequences outside *ori*-core are not required for initiation of DNA replication. Nevertheless, the T-Ag portion of an initiation complex could bind to *ori*-core while the DNA primase–DNA polymerase α portion initiated synthesis on whatever sequence was adjacent.

A second possibility is that T-Ag functions as a unique primer to initiate synthesis at *ori*. A small fraction of T-Ag contained an easily exchangeable 5′-AMP linked via a phosphodiester bond to a serine residue on T-Ag (Bradley *et al.*, 1984). This protein could bind to *ori* and provide a ribonucleotide primer for DNA synthesis. The T-Ag–AMP linkage would then be cleaved to leave a 5′-AMP–DNA chain in the *ori* region. This RNA–p–DNA linkage would be excised during the normal processing of Okazaki fragments and become the 5′-end of the long growing daughter strand initiated in *ori*.

A third possibility involves transcriptional activation of *ori*-core. This could take the form of either transcription through *ori*-core in order to facilitate opening the replication bubble, as proposed for lambda and T7 DNA replication (Furth *et al.*, 1982; Richardson, 1983), or providing an mRNA primer to initiate replication specifically within *ori*-core, as proposed for plasmid col E1 (Itoh and Tomizawa, 1980). However, active transcription into *ori*-core seems unlikely in SV40 because neither the promoters nor the enhancer regions are required for replication, and initiation of SV40 DNA replication *in vitro* was resistant to α-amanitin (Ariga, 1984b; Li and Kelly, 1984; Decker *et al.*, 1986). Once replication begins, E-mRNa start sites shift from nucleotide positions 5232–5237 to positions 27–33 (Figs. 3, 4). These late lytic E-mRNA starts could provide RNA primers for initiation, but attempts to identify RNA primers originating from this point have failed (Hendrickson and DePamphilis, unpublished data). As discussed in Sections III.B and III.C, a transcriptional enhancer element (but not necessarily transcription) is necessary in addition to *ori*-core for initiation of PyV DNA replication.

Finally, the dinucleotide, adenosine-5′-tetraphospho-5′-adenosine (Ap_4A), has been implicated in the initiation of cellular DNA replication (Weinmann-Dorsch *et al.*, 1984). Ap_4A functioned as a primer for DNA polymerase α on a poly(dT) template (Rapaport *et al.*, 1981). The reaction was greatly stimulated by C_1C_2 (Rapaport *et al.*, 1981) which allowed the enzyme to use RNA primers as short as a dinucleotide (Pritchard *et al.*, 1983). Ap_4A also functioned as a primer on a dsDNA segment containing a region of oligo(dT):oligo(dA) (Zamecnik *et al.*, 1982). In fact, Ap_4A was about 30 times better at initiating synthesis on the oligo(dT) region in dsDNA than in the corresponding ssDNA template. However, only dATP was incorporated; synthesis stopped when it tried to enter a GC region. Zamecnik *et al.* (1982) suggested that Ap_4A might be used to initiate DNA synthesis at SV40 *ori*-core. This appears unlikely because *in vivo* initiation events at the 17-bp AT-rich region, the most likely target for Ap_4A melting of dsDNA, are all on the A-rich template rather than the T-rich template (Fig. 10). Furthermore, Ap_4A neither stimulated nor inhibited initiation of SV40 DNA synthesis on endogenous viral chromosomes (Decker *et al.*, 1986).

DNA synthesis in the *ori* region appeared to respond to inhibitors differently than DNA synthesis at other regions of the genome. Inhibition of viral DNA synthesis by hydroxyurea (Bjursell and Magnusson, 1976), 2′-deoxy-2′-azidocytidine (Bjursell *et al.*, 1977; Eliasson *et al.*, 1981), or aphidicolin (Dinter-Gottlieb and Kaufmann, 1982; Decker *et al.*, 1986) resulted in a pool of RI enriched for young replicating DNA molecules, suggesting that initiation of DNA replication was less sensitive to reduced levels of dNTPs, misincorporation of nucleotide analogues, and inhibition of DNA polymerase α than replication forks in other regions of the genome (see Section IV.E; Fig. 14). In contrast, initiation of DNA replication appears more sensitive to Miracil D, chloroquine (Manteuil and Girard, 1974), butyrate (Wawra *et al.*, 1981; Daniell *et al.*, 1982), and cycloheximide (Yu *et al.*, 1975; Yu and Cheevers, 1976a,b; Cusick *et al.*, 1984) than does maturation of RI.

V. CHROMATIN STRUCTURE AND ASSEMBLY AT REPLICATION FORKS

A. Summary

The structure of chromatin at SV40 DNA replication forks (and presumably cellular and PyV forks as well) can be divided into at least four distinct domains (Fig. 18). The *prereplicative chromatin* in front of replication forks is equivalent to mature (unreplicated) chromatin whose histone composition and nucleosome structure are indistinguishable from the host cell. The actual sites of DNA synthesis, consisting of Okazaki fragments and an average of 125 bp of the newly replicated daughter molecules, are free of nucleosomes and therefore referred to as *prenucleosomal DNA.* Thus, the enzymes responsible for DNA synthesis utilize nonnucleosomal DNA templates. Nucleosome assembly occurs rapidly on both arms of the fork but the initial structure is *immature chromatin* because it is hypersensitive to nonspecific endonucleases. *Postreplicative chromatin* consists of newly assembled nucleosomes that cannot be distinguished from mature chromatin. The structure, spacing, and phasing with respect to DNA sequence of nucleosomes on replicating viral chro-

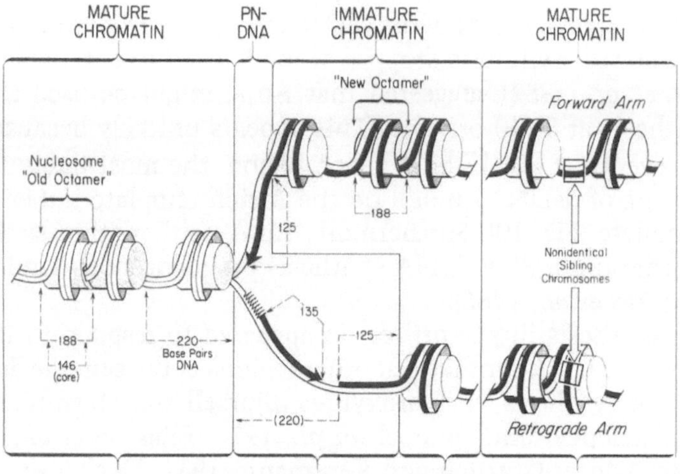

FIGURE 18. Chromatin structure at SV40 replication forks. Nascent DNA is represented by solid bold lines with arrowheads at their 3′-ends. A nucleosome is a 110 × 55-Å cylinder consisting of about 1¾ turns of duplex DNA (20 Å in diameter) coiled around a histone octamer [2(H2A, H2B, H3, H4)] (Finch *et al.*, 1977). "Old" histone octamers are speckled cylinders, "new" histone octamers are white cylinders. Numbers give the average distance in nucleotides. PN-DNA, prenucleosomal DNA.

mosomes are indistinguishable from mature viral chromosomes. "Old" prefork histone octamers appear to be distributed randomly to both arms of the fork, new histone octamers are assembled on both arms, and chromatin maturation occurs on both arms. Nucleosomes do not occupy the same DNA sites on all chromosomes, but many different preferred sites. In fact, the distribution of nucleosomes on one arm of a single replication fork is not the same as that on the opposite arm of the same fork. Creation of "nucleosome-free regions" associated with transcriptional enhancer regions is a post-DNA replication event.

B. Mature Viral Chromosomes

The preparation, composition, and structure of papovavirus chromosomes and previrion complexes have been reviewed previously [Cremisi, 1979; DePamphilis and Wassarman, 1980, 1982 (pp. 78–88); Das and Niyogi, 1981]. New isolation techniques have been introduced (Boyce *et al.*, 1982; Garcea and Benjamin, 1983a), but the basic picture remains the same: nascent DNA first appears in replicating chromosomes (about 90 S) that are converted into mature chromosomes (about 70 S) which are assembled into previrion complexes (about 200 S) and then into virions (about 250 S). The histone composition of viral chromosomes was shown to be essentially the same as that of cellular chromatin (Shelton *et al.*,

1980). Mature (nonreplicating) SV40 and PyV chromosomes contained an average of 24 ± 1 nucleosomes per genome (Saragosti *et al.*, 1980; Moyne *et al.*, 1981) that were indistinguishable from the nucleosomes of their host cell (Fig. 1A,B). Each nucleosomal core particle contained 146 ± 3 bp of DNA with a helix repeat of 10.4 bases (Shelton *et al.*, 1980). Digestion of chromosomes with micrococcal nuclease (Bellard *et al.*, 1976; Shelton *et al.*, 1978b, 1980), and analysis of chromosomes by electron microscopy (Saragosti *et al.*, 1980; Moyne *et al.*, 1981) revealed that 21–25% of the viral DNA was not associated with nucleosomes, but was distributed throughout the genome as highly variable regions of internucleosomal DNA. The amount of DNA associated with each nucleosome ranged from 188 to 197 bp by nuclease digestion, and 170 to 200 bp by electron microscopy, depending to some extent on the method of chromosome preparation.

Analyses of the locations of nucleosomes with respect to DNA sequence ("nucleosome phasing") revealed a large number of preferred sites (Ponder and Crawford, 1977; Mengeritsky and Trifonov, 1984; Scott *et al.*, 1984) which together constitute a nearly random distribution of nucleosomes throughout the genome. This concept is consistent with nucleosome phasing studies on cellular genes as well (Igo-Kemenes *et al.*, 1982; Wu *et al.*, 1983; Smith and Lieberman, 1984). Various restriction endonucleases cleaved only a fraction of viral chromosomes regardless of the concentration of enzyme, the time of incubation, or the number of enzymes used, demonstrating the absence of a unique phase relationship (Cremisi *et al.*, 1976b; Liggens *et al.*, 1979; Das *et al.*, 1979; Shelton *et al.*, 1980; Tack *et al.*, 1981; Glotov *et al.*, 1982). Typically, about 25–30% of the chromosomes were cleaved under conditions in which 100% of bare viral DNA was cleaved, although this varied from 15 to 50% depending on the enzyme chosen and the preparation of chromatin. None of the restriction enzymes tested, whose sites mapped throughout the genome, cut all of its sites in all of the chromosomes, ruling out a unique phase relationship of nucleosomes to DNA sequence. Results with replicating chromosomes were indistinguishable from those with mature chromosomes (Tack *et al.*, 1981), or chromosomes assembled *in vitro* (Simpson and Stein, 1980). Furthermore, the partial accessibility of restriction sites to restriction enzymes *in vitro* was consistent with their partial accessibility to trimethylpsoralen cross-linking *in vivo* (Carlson *et al.*, 1982). Taken together, the data suggest that nucleosomes occupy a large number of possible sites to produce a nearly random phase relationship with the DNA. Assuming 22 to 24 randomly phased nucleosomes, and 5243 bp per SV40 genome, protection of all 146 bp of nucleosomal core DNA would predict cleavage of 33–39% of the chromosomes, protection of all 165 bp of chromatosome DNA (Simpson, 1978) would predict 25–31% cleavage, and protection of all 188 bp of nucleosomal core plus linker DNA would predict 15–22% cleavage. Thus, mature SV40 and PyV chromosomes contained fewer nucleosomes than could be accommodated by the genome (about 28), and they had an average center-to-center spacing between neighboring core particles of 220 ± 74 bp (Fig. 18).

The major exception to a random distribution of nucleosomes occurred over the promoter–enhancer region adjacent to the late gene side of either the SV40 or PyV *ori*-core sequence in about 25% of mature viral chromosomes (DePamphilis and Wassarman, 1982). This region promoted the formation of chromatin that was hypersensitive to nonspecific endonucleases (Section III.B.1) and to exonuclease digestion beginning at a cleaved *Bgl*I site within *ori* (Scott *et al.*, 1984), and appeared as nucleosome-free regions in the electron microscope (Saragosti *et al.*, 1980; Jakobovits *et al.*, 1980, 1982). However, since the DNase I cleavage sites were interdigitated with DNase I-resistant sites (Cremisi, 1981; Cereghini *et al.*, 1984; Cereghini and Yaniv, 1984), the DNA in this region remained complexed with protein, perhaps even with a nucleosome that was lost under the spreading conditions for electron microscopy (Muller *et al.*, 1978). This region was also hyperaccessible to chemical reactions carried out *in vivo* (Beard *et al.*, 1981; Robinson and Hallick, 1982), although the extent varied with the reagent used (Seidman *et al.*, 1983). The *ori*-enhancer region was not hyperaccessible in previrion or virion particles (Hartmann and Scott, 1981, 1983; Kondoleon *et al.*, 1983). Other nuclease-hypersensitive sites were found in SV40 chromosomes, particularly at the termination region (Shakhov *et al.*, 1982).

The hypersensitive region near the SV40 *ori* was not formed during nucleosome assembly at replication forks since it appeared in only about 25% of the viral chromosomes, and was absent from newly replicated chromatin (Weintraub, 1979; Varshavsky *et al.*, 1983). Furthermore, newly assembled SV40 chromatin bore little or no phase relationship with the sequence of nascent DNA (Fig. 18; Tack *et al.*, 1981). Thus hypersensitivity appears to result from the interaction of a limited supply of sequence-specific proteins with promoter–enhancer elements (Scholer and Gruss, 1984; Emerson and Felsenfeld, 1984). However, the extent of chromatin hypersensitivity in the *ori*-enhancer region was greater in SV40 chromosomes that underwent replication than in those that did not (Cereghini and Yaniv, 1984). This indicated that the presence of immature viral chromatin produced by replication (see below) may have facilitated subsequent alteration of chromatin structure.

C. Segregation of "Old" Histone Octamers

Newly replicated SV40 DNA, like that of its host, is rapidly assembled into nucleosomes both in whole cells (DePamphilis and Wassarman, 1980; Klempnauer *et al.*, 1980; Herman *et al.*, 1981) and in subcellular systems (Shelton *et al.*, 1978a, 1980; Cusick *et al.*, 1981), resulting in a doubling of the number of nucleosomes during each round of chromosome replication. The deposition of newly synthesized histones onto nascent DNA follows a complex pathway (Worcel *et al.*, 1978; Cremisi and Yaniv, 1980; Jackson and Chalkley, 1981a,b; Jackson *et al.*, 1981) that resembles the assembly of nucleohistone complexes *in vitro* under physiological conditions (Ellison and Pulleyblank, 1983). However, histone octamers, once formed, are relatively stable; mixing of preexisting histones with newly synthesized histones occurs rarely, it at all (Leffak *et al.*, 1977; Prior *et al.*, 1980; Manser *et al.*, 1980; Russev and Hancock, 1982; Leffak, 1983), implying that newly replicated chromatin contains both "old" histone octamers that had been present on the DNA prior to replication and "new" histone octamers that were assembled entirely from nascent histones. Thus, as DNA replication forks advance, old histone octamers could be distributed *conservatively* with all of them going to *either* the forward *or* the retrograde arm and the newly synthesized histones going exclusively to the opposite arm, or *dispersively* with old histone octamers distributed randomly between the two arms of the fork and intermingled with new histone octamers.

Support for *conservative* histone segregation comes from four types of experiments. (1) Nucleosomes assembled from old histone octamers in the absence of protein synthesis had an internucleosomal spacing similar to normal nascent nucleosomes (Weintraub, 1976; Seale, 1976, 1978). (2) Long stretches of nucleosome-free DNA had been observed in the electron microscope with chromatin from cycloheximide-treated cells (Riley and Weintraub, 1979). (3) Old histone octamers tended to segregate in groups, even in the absence of cycloheximide (Leffak *et al.*, 1977; Leffak, 1983). (4) Nascent, nucleosomal DNA assembled in the absence of protein synthesis originated predominantly from forward arms of SV40 replication forks, and, in cells, from only one of the two sibling chromosomes (Seidman *et al.*, 1979; Roufa and Marchionni, 1982).

On the other hand, similar experiments carried out in other laboratories were interpreted to support *dispersive* histone segregation. (1) Chromatin replicated in the absence of protein synthesis did not contain long stretches of nonnucleosomal DNA (Jackson and Chalkley, 1981a; Pospelov *et al.*, 1982). (2) Chromatin replicated in the presence of density-labeled proteins, in cells actively synthesizing proteins, contained oligonucleosomes (5–9 monomers) of hybrid density associated with radiolabeled, nascent DNA (Russev and Hancock, 1982). (3) Electron microscopy of polyoma viral DNA replicated in the absence of protein synthesis revealed one-half of the normal complement of nucleosomes distributed throughout the genome (Cremisi *et al.*, 1978). (4) Histones labeled in one cell cycle were distributed equally to both nascent DNA strands during the subsequent cell cycle (Jackson and Chalkley, 1981b; Fowler *et al.*, 1982).

One of the strongest pieces of evidence in support of conservative histone segregation came from an analysis of the distribution of newly assembled nucleosomes between the two arms of DNA replication forks after SV40 tsB11 (a mutant in virion assembly)-infected monkey cells had been treated with cycloheximide to inhibit assembly of nucleosomes from newly synthesized histones (Seidman *et al.*, 1979). In these experiments, radiolabeled, nascent DNA was isolated from nucleosomes released by micrococcal nuclease (MNase) and hybridized to separated strands of DNA restriction fragments. At least 80–90% of the old histone octamers were found to protect nascent DNA that came from the forward arm of replication forks. However, when the same experiment was carried out by Cusick *et al.* (1984), using both tsB11 and wild-type virus, as well as several different protocols for radiolabeling and isolating nascent nucleosomal DNA, the old histone octamers were found to protect nascent DNA equally on *both* arms of replication forks, consistent with a dispersive model of histone segregation. In addition, Cusick *et al.* (1984) digested isolated replicating SV40 chromosomes with two strand-specific exonucleases that excised nascent DNA from either the forward or the retrograde side of replication forks to determine the ratio of prenucleosomal DNA. Pretreatment of cells with cycloheximide did not result in an excess of prenucleosomal DNA on either side of replication forks, but did increase the amount of internucleosomal DNA, consistent with an equal deficiency of nucleosomes on both arms of replication forks. A dispersive model of histone segregation is also consistent with the fact that newly assembled nucleosomal oligomers from both arms of replication forks were hypersensitive to MNase, indicating the same process of chromatin assembly occurs on both arms (Cusick *et al.*, 1983). Although it is difficult to rule out the possibility that both arms of the replication fork were protected from MNase by some artifact such as a redistribution of histone octamers over the newly replicated regions as a result of nucleosome sliding, or the presence of proteins other than histone octamers (e.g., virion proteins), these problems do not appear to explain the difference between the two reports. Cusick *et al.* (1984) pointed out one unexpected artifact that can occur with blotting-hybridization experiments in which the ssDNA fragments were transferred to nitrocellulose paper (Seidman *et al.*, 1979) rather than covalently bound to a DBN-type paper (Cusick *et al.*, 1984). The faster-migrating ssDNA strand of each restriction fragment had a tendency to bind less tightly to nitrocellulose than did its complementary sequence. Thus, subsequent hybridization of nascent nucleosomal DNA would, under these circumstances, appear to hybridize preferentially to DNA fragments representing the *forward* template at replication forks.

Seidman *et al.* (1979) also reported that long stretches of nonnucleosomal nascent DNA could be released by digestion of cellular chromatin from cycloheximide-treated cells with *Hae*III, consistent with conservative histone segregation. However, Annunziato and Seale (1984) observed that all of the nascent DNA released by *Hae*III contained nucleosomes, although the internucleosomal DNA regions were more heterogeneous in length than observed with normal chromatin, consistent with a dispersive mechanism of histone segregation.

The model for histone segregation at replication forks (Fig. 18) that best fits all of the available data is one that is neither exclusively conservative nor exclusively dispersive, but one in which the extent of either mode depends on the biology of the system examined. Nucleosomes may be dispersed in groups (Pospelov *et al.*, 1982), with the size of the group determined by the rate of fork movement and the arrangement of nucleosomes in a particular part of the genome. The side of the replication fork to which clusters of old octamers are passed may also be related to gene expression (Seidman *et al.*, 1979) rather than DNA replication. Furthermore, DNA replication is quite sensitive to physiological conditions. Changing the osmolarity of the cell culture medium changes the rate of DNA replication, the extent of unwinding of parental DNA strands, and the fraction of interlocked progeny DNA molecules (Varshavsky *et al.*, 1983; DePamphilis *et al.*, 1983; Weaver *et al.*, 1985). Perhaps cell culture conditions also affect the segregation of histone octamers.

D. Structure of Chromatin at Replication Forks

Electron microscopic analyses of replicating cellular (McKnight *et al.*, 1978) and viral (Cremisi *et al.*, 1978; Seidman *et al.*, 1978) chromosomes revealed a beaded appearance for prereplicative and newly replicated chromatin that was virtually indistinguishable with respect to bead diameter and periodicity from that of nonreplicating chromatin, even directly proximal to DNA replication forks. Such studies strongly suggested that extended regions of bare DNA did not accumulate at replication forks under normal conditions. Therefore, assembly of newly synthesized DNA into nucleosomes (or nucleosomelike particles) was a relatively rapid process, as expected, since the ratio of protein to DNA remained close to 1 : 1 during replication of cellular and viral chromatin (Cremisi *et al.*, 1976a; Levy and Jakob, 1978; Hancock, 1978; Cremisi, 1979; DePampilis and Wassarman, 1980). However, stretches of bare DNA as long as 500 nucleotides at the sites of DNA synthesis would have gone undetected, and longer regions of bare DNA may have been lost as a result of redistribution of nucleosomes during sample preparation (Muller *et al.*, 1978). Therefore, the distance of newly assembled nucleosomes from the sites of DNA synthesis in replicating SV40 chromosomes was measured by excising nascent DNA strands with either *E. coli* exonuclease III (Exo III degrades one strand of duplex DNA in the 3′–5′ direction), or phage T7 gene 6 exonuclease (T7 Exo degrades one strand of duplex DNA in the 5′–3′ direction) (DePamphilis *et al.*, 1979, 1980; Herman *et al.*, 1981). About 200–600 bases of nascent DNA were radiolabeled *in vivo* so that only one or two nucleosomes on each arm of the fork contained a significant portion of [^{3}H]-DNA. Neither exonuclease removed nascent DNA from nucleosomes at replication forks that had been stripped of most nonhistone proteins. The number of nucleotides incorporated into replicating chromosomes (i.e., length of nascent DNA) was determined by comparing the fraction of nascent [^{3}H]-DNA in purified SV40(RI) DNA released by Exo III with that released from uniformly labeled SV40 [^{32}P]-DNA restriction fragments present in the same reaction. Since the fraction of radiolabel released by exonuclease was proportional to the distance the enzyme traveled, the fraction of [^{3}H]-DNA released could be converted into nucleotides excised. Accordingly, the average distance from either 3′- or 5′-ends of long nascent chains to the first nucleosome on either arm of replication forks was found to be 125 ± 20 nucleotides. Since the average size of Okazaki fragments is 135 nucleotides and they were separated from the 5′-ends of longer DNA chains by about 40 nucleotides (Anderson and DePamphilis, 1979), the first nucleosome on retrograde arms was an average of 300 bases from the 5′-ends of Okazaki fragments (Fig. 18).

Despite the rapid appearance of nucleosomes (or nucleosomelike particles) on nascent chromatin, newly assembled cellular and viral chromatin is hypersensitive to nonspecific endonucleases such as MNase, DNase I and DNase II, suggesting that a maturation process follows nucleosome assembly (Klempnauer *et al.*, 1980; Cusick *et al.*, 1981, 1983). Nascent SV40 chromosomal DNA was digested about fivefold faster and about 25% more extensively than DNA in nonreplicating viral chromosomes, releasing both nucleosomal monomers and essentially bare fragments of duplex DNA (3–7 S) that had an average length of 120 bp and subsequently were degraded during the course of the digestion. On the other hand, MNase digestion of uniformly labeled mature SV40 chromosomes released only nucleosomal monomers and oligomers. The characteristics of the 3–7 S DNA released from replicating chromosomes were assessed by examining its sensitivity to MNase and ssDNA-specific S1 endonuclease, sedimentation behavior before and after deproteinization, buoyant density in CsCl after formaldehyde treatment, and size on agarose gels (Cusick *et al.*, 1981). In addition, it was observed that MNase digestion of purified SV40 DNA also resulted in the release of short DNA fragments that subsequently were degraded. Therefore, a DNA–protein complex was *not* required to account for the appearance of nonnucleosomal DNA fragments from replicating chromosomes. The fraction of newly synthesized DNA released from replication forks as 3–7 S fragments depended upon both the length of the radiolabeling period (i.e., number of labeled nucleotides per fork) and the extent of MNase digestion. The longer the radiolabeling period and the longer the time of digestion with MNase, the less 3–7 S DNA was observed. Cellular replication forks, which proceed about 20 times faster than SV40 replication forks, required significantly shorter radiolabeling periods than SV40 in order to observe 3–7 S DNA (Klempnauer *et al.*, 1980). In experiments with replicating cellular chromatin, the 3–7 S DNA was shown to cosediment with nucleosomal monomers at very low salt concentrations but not at high salt concentrations, suggesting that structurally altered, MNase-hypersensitive nucleosomes present at replication forks dissociated in high salt to release subnucleosomal lengths of DNA (Schlaeger, 1982). However, it seems more likely that under the exceptionally low salt conditions employed, short DNA fragments adventitiously associated with either nucleosomes or core histones released during MNase digestion (Cotton and Hamkalo, 1981).

The 3–7 S DNA was dsDNA fragments that were released from prenucleosomal DNA on both sides of replication forks (Cusick *et al.*, 1981). Nascent radiolabeled DNA in the 3–7 S fragments hybridized to both forward and retrograde DNA templates. Predigestion of replicating SV40 chromosomes with *either* Exo III or T7 Exo *partially* eliminated the release of 3–7 S DNA upon subsequent digestion of chromosomes with MNase, whereas predigestion with *both* Exo III and T7 Exo *completely* eliminated release of 3–7 S DNA. The 3–7 S DNA did not consist solely of Okazaki fragment DNA because 90% of the Okazaki fragment DNA annealed only to retrograde templates. Furthermore, 3–7 S DNA represented at least twice as much nascent DNA as found in Okazaki fragments, and was observed even when Okazaki fragments represented an insignificant fraction of the radiolabeled DNA in replicating chromosomes. Okazaki fragments are also part of the prenucleosomal DNA because either Exo III or T7 Exo removed at least 80% of the radiolabel in Okazaki fragments, with a corresponding decrease in the fraction of prenucleosomal DNA (Cusick *et al.*, 1981). Furthermore, 50–60% of the Okazaki fragments in either replicating chromosomes or purified replicating DNA were released by ssDNA-specific endonucleases; the rest remained with the cleaved molecules because of ssDNA regions too small to be recognized under the conditions employed (Herman *et al.*, 1979). At least 90% of the Okazaki fragments released from formaldehyde-fixed SV40 chromosomes were not contained in nucleosomes because they had the same buoyant density as purified DNA, they had the same sensitivity to MNase as purified DNA, and they sedimented at 4–6 S before and after treatments to remove proteins (Herman *et al.*, 1979). These results strongly suggest that synthesis of Okazaki fragments was completed prior to assembly of nascent DNA into nucleosomes.

Prenucleosomal DNA alone does not account for the nuclease hypersensitivity of immature chromatin. Pulse–chase experiments *in vitro* revealed a time-dependent chromatin maturation process that involved two distinct steps: (1) conversion of prenucleosomal DNA into immature nucleosomal oligomers and (2) maturation of newly assembled chromatin into a structure with increased nuclease resistance (Cusick *et al.*, 1983). When nascent prenucleosomal DNA was excised by the combined action of Exo III and T7 Exo, subsequent digestion of the remaining chromatin with MNase revealed the same degree of hypersensitivity observed prior to exonuclease treatment. Furthermore, newly assembled nucleosomal oligomers, isolated after a brief MNase digestion of replicating viral chromosomes, were also hypersensitive to MNase relative to oligomers isolated from mature chromosomes. Nascent DNA isolated from immature oligomers hybridized to both forward and retrograde templates, indicating that immature nucleosomes were present on *both* arms of replication forks. Inhibition of DNA polymerase α by aphidicolin inhibited conversion of prenucleosomal DNA into nucleosomes but did not inhibit chromatin maturation (i.e., decreasing MNase hypersensitivity). Unfortunately, analysis of the nucleoprotein products from MNase digestion of replicating and mature SV40 chromosomes failed to detect a change in nucleosome structure that corresponded to the loss of nuclease hypersensitivity. One likely step in chromatin maturation is the addition of histone H1 which protects linker DNA from nuclease attack and prevents sliding of nucleosomes during nuclease digestion (reviewed in Cusick *et al.*, 1983; D'Anna and Tobey, 1984).

Comparison of replicating and mature SV40 chromosomes by analysis with MNase (Klempnauer *et al.*, 1980; Cusick *et al.*, 1981, 1983), restriction endonucleases (Tack *et al.*, 1981), and electron microscopy (Cremisi *et al.*, 1978; Seidman *et al.*, 1978) demonstrated that the size and arrangement of nucleosomes on newly replicated chromatin were essentially the same as those on mature chromatin. The nearly random phase relationship between DNA sequence and chromatin structure was established within 400 bp of replication forks and was present on both arms of the same fork (Tack *et al.*, 1981), consistent with the presence of nucleosomes an average of 125 bp from ends of long nascent DNA strands (Herman *et al.*, 1981). Recent nuclease digestion studies on newly synthesized cellular chromatin (Annunziato and Seale, 1984; Jakob *et al.*, 1984; Smith *et al.*, 1984) led to the conclusion that newly synthesized DNA initially lacks the regular periodicity of nucleosomal organization, although nucleosome core particles were present. When nucleosomal periodicity appeared, the chromatin was still hypersensitive and a shorter nucleosomal repeat was commonly observed. However, this shortening appeared to occur as a result of sliding of nucleosomes during MNase digestion (Jackson *et al.*, 1981; Vaury *et al.*,, 1983; Smith *et al.*, 1984). In fact, if nucleosomal spacing at replication forks was closer than in mature chromatin, some nucleosomes in each replication bubble would have to be ejected in order to achieve their normal spacing! Thus, the simplest model to explain the available data is that newly replicated chromatin first appears as prenucleosomal DNA which is then assembled into nucleosomes with little, if any, phase relationship to the DNA. As the density of nucleosomes increases, a periodic structure is formed whose spacing is eventually locked into place by histone H1 to produce mature chromatin.

Two aspects of chromatin structure that relate to replication or transcription are worth noting. First, the fact that both immature chromatin and actively transcribed chromatin are hypersensitive to nonspecific endonucleases suggests similarities in their structures. Weintraub (1979) has proposed that immature chromatin is more accessible to those gene regulation proteins that eventually determine whether or not chromatin structure remains in an active or inactive form. Second, the nucleoprotein structure of nucleosomal monomers, released by MNase from replicating SV40 chromosomes digested in nuclear extracts, were dramatically different from those from replicating SV40 chromosomes that had been purified by sedimentation through sucrose gradients (Cusick *et al.*, 1983). Since SV40(RI) DNA in nuclear extracts supplemented with cytosol could be converted into SV40(I) DNA while SV40(RI) DNA in chromosomes from sucrose gradients could only complete Okazaki fragment synthesis (Su and DePamphilis, 1978), this difference in chromatin structure may account for the difference in DNA replication.

VI. CONCLUDING PERSPECTIVE

The future for papovavirus chromosome replication probably lies in answering those questions dealing with the mechanism for initiating viral DNA replication, the regulation of viral DNA replication, the relationship between DNA replication and gene expression, and the specific interactions between cellular and viral components that determine host-range specificity. Furthermore, the technical and conceptual knowledge developed with papovavirus chromosomes should be applicable to studying cloned cellular sequences. Specific areas of research would include the mechanism by which T-Ag activates replication at a viral *ori* sequence. T-Ag appears responsible for capturing the cellular replication machinery and bringing it to the viral genome, but the actual mechanism remains a mystery. The recent availability of *in vitro* systems that initiate SV40 DNA replication should open new avenues to this problem. A second question concerns the nature of permissive cell factors and their normal biological function in cells. The strong specificity involved in forming an active T-Ag/permissive cell factor/*ori* complex suggests that permissive cell factors normally interact with specific DNA sequences and proteins within the cell. Permissive cell factors may normally activate certain cellular *ori* sequences once per S phase, but, in the presence of T-Ag, they may activate the viral *ori* multiple times per S phase. Do cells have *ori* sequences similar to those found in viruses? The fact that DNA injected into mammalian embryonic nuclei did not replicate unless a known *ori* sequence was present suggests that cellular *ori* sequences could be isolated using this approach. Finally, the fact that transcriptional enhancer elements are required to activate PyV *ori*-core but not SV40 *ori*-core indicates a fundamental difference in *ori* sequences and their recognition by cellular components. Enhancers may be part of some or all cellular *ori* sequences, and the cell's mechanisms for regulation of gene expression may thus be involved in the temporal and cell-specific regulation of DNA replication.

REFERENCES

Acheson, N. H., 1980, Lytic cycle of SV40 and polyoma virus, in: *DNA Tumor Viruses: The Molecular Biology of Tumor Viruses, Part 2* (J. Tooze, ed.), Cold Spring Harbor Laboratory, Cold Spring Harbor, N.Y.

Adams, G., Steiner, U., and Seuwen, K., 1983, Proliferative activity and ribosomal RNA content of 3T3 and SV40-3T3 cells, *Cell Biol. Int. Rep.* **7**:955–962.

Anderson, S., and DePamphilis, M. L., 1979, Metabolism of Okazaki fragments during simian virus 40 DNA replication, *J. Biol. Chem.* **254**:11495.

Anderson, S., Kaufmann, G., and DePamphilis, M. L., 1977, RNA primers in simian virus 40 DNA replication: Identification of transient RNA–DNA covalent linkages in replicating DNA, *Biochemistry* **16**:4990.

Annunziato, A. T., and Seale, R. L., 1984, Presence of nucleosomes within irregularly cleaved fragments of newly replicated chromatin, *Nucleic Acids Res.* **12**:6179.

Antman, K. H., and Livingston, D. M., 1980, Intracellular neutralization of SV40 tumor antigens following microinjection of specific antibody, *Cell* **19**:627.

Arai, K., and Kornberg, A., 1981a, Mechanism of dnaB protein action. III. Allosteric role of ATP in the alteration of DNA structure by dnaB protein in priming replication, *J. Biol. Chem.* **256**:5260–5266.

Arai, K., and Kornberg, A., 1981b, Unique primed start of phage ϕX174 DNA replication and mobility of the primosome in a direction opposite chain synthesis, *Proc. Natl. Acad. Sci. USA* **78**:69–73.

Ariga, H., 1984a, Identification of the replicative intermediates in SV40 replication *in vitro*, *Nucleic Acids Res.* **12**:6053.

Ariga, H., 1984b, Replication of cloned DNA containing the Alu family sequence during cell extract-promoting SV40 DNA synthesis, *Mol. Cell. Biol.* **4**:1476.

Ariga, H., and Sugano, S., 1983, Initiation of SV40 DNA replication *in vitro*, *J. Virol.* **48**:481.

Ariga, H., Tsuchihashi, Z., Naruto, M., and Yamada, M., 1985, Cloned mouse DNA fragments can replicate in an SV40 T-antigen system in vivo and in vitro, *Mol. Cell. Biol.* **5**:563–568.

Ball, R. K., Siegel, B., Quellhorst, S., Brandner, G., and Braun, D. G., 1984, Monoclonal antibodies against simian virus 40 nuclear large T tumor antigen: Epitope mapping, papova virus cross-reaction and cell surface staining, *EMBO J.* **3**:1485–1491.

Banerji, J., Olson, L., and Schaffner, W., 1983, A lymphocyte-specific cellular enhancer is located downstream of the joining region in immunoglobulin heavy chain genes, *Cell* **33**:729–740.

Baran, N., Neer, A., and Manor, H., 1983, "Onion skin" replication of integrated polyoma virus DNA and flanking sequences in polyoma-transformed rat cells: Termination within a specific cellular DNA segment, *Proc. Natl. Acad. Sci. USA* **80**:105.

Basilico, C., Matsuya, Y., and Green, H., 1970, The interaction of polyoma virus with mouse/hamster somatic cell hybrids, *Virology* **41**:295.

Baumann, E., and Hand, R., 1979, Protein kinase activity associated with D2 hybrid protein related to SV40 T-antigen: Some characteristics of the reaction products, *Proc. Natl. Acad. Sci. USA* **76**:3688–3692.

Baumann, E., and Hand, R., 1982, Phosphorylation and dephosphorylation alters the structure of D^2 hybrid T antigen, *J. Virol.* **44**:78–87.

Beard, P., Morrow, J. F., and Berg, P., 1973, Cleavage of circular superhelical SV40 DNA to a linear duplex by S1 nuclease, *J. Virol.* **12**:1303.

Beard, P., Kaneko, M., and Cerutti, P., 1981, N-Acetoxyacetylaminofluorene reacts preferentially with a control region of intracellular SV40 chromosome, *Nature* **291**:84.

Bellard, M., Oudet, P., Germond, J., and Chambon, P., 1976, Subunit structure of simian virus 40 minichromosome, *Eur. J. Biochem.* **70**:543.

Benjamin, R. I., and Gill, D. M., 1980, ADP-ribosylation in mammalian cell ghosts: Dependence of poly(ADP-ribose) synthesis on strand breakage in DNA, *J. Biol. Chem.* **255**:10493.

Bergsma, D. J., Olive, D. M., Hartzell, S. W., and Subramanian, K. N., 1982, Territorial limits and functional anatomy of the simian virus 40 replication origin, *Proc. Natl. Acad. Sci. USA* **79**:381–385.

Berk, A. J., and Clayton, D. A., 1976, Mechanism of mitochondrial DNA replication in mouse L-cells: Topology of circular daughter molecules and dynamics of catenated oligomer formation, *J. Mol. Biol.* **100**:85–102.

Berkower, I., Leis, J., and Hurwitz, J., 1973, Isolation and characterization of an endonuclease from *E. coli* specific for RNA in RNA–DNA hybrid structure, *J. Biol. Chem.* **248**:5914–5921.

Berns, K. I., and Hauswirth, W. W., 1982, Organization and replication of parvovirus DNA, in: *Organization and Replication of Viral DNA* (A. S. Kaplan, ed.), pp. 4–30, CRC Press, Boca Raton, Fla.

Bjursell, G., 1978, Effects of 2′-deoxy-2′-azidocytidine on polyoma virus DNA replication: Evidence for rolling circle type mechanism, *J. Virol.* **26**:136.

Bjursell, G., and Magnusson, G., 1976, Replication of early replicative intermediates during hydroxy-urea inhibition, *Virology* **74**:249.

Bjursell, G., Skoog, L., Thelander, L., and Soderman, G., 1977, 2′-Deoxy-2′-azidocytidine inhibits the initiation of polyoma DNA synthesis, *Proc. Natl. Acad. Sci. USA* **74**:5310.

Bjursell, G., Munck, V., and Therkelsen, A. J., 1979, Polyoma DNA synthesis in isolated nuclei: Evidence for defective replication forks, *J. Virol.* **30**:929.

Blumenthal, A. B., Kriegstein, H. J., and Hogness, D. S., 1974, The units of DNA replication in *Drosophila melanogaster* chromosomes, *Cold Spring Harbor Symp. Quant. Biol.* **38**:205.

Boccara, M., and Kelly, F., 1978, Expression of polyoma virus in heterokaryons between embryonal carcinoma cells and differentiated cells, *Virology* **90**:147–150.

Bogenhagen, D., and Clayton, D. A., 1976, Thymidylate nucleotide supply for mitochondrial DNA synthesis in mouse L-cells, *J. Biol. Chem.* **251**:2938.

Botchan, M., Stringer, J., Mitchison, T., and Sambrook, J., 1980, Integration and excision of SV40 DNA from the chromosome of a transformed cell, *Cell* **20**:143–152.

Bourgaux, P., and Bourgaux-Ramoisy, D., 1971, A symmetrical model for polyoma virus DNA replication, *J. Mol. Biol.* **62**:513.

Bourgaux, P., and Bourgaux-Ramoisy, D., 1972, Unwinding of replicating polyoma virus DNA, *J. Mol. Biol.* **70**:399.

Bourgaux, P., Bourgaux-Ramoisy, D., and Dulbecco, R., 1969, The replication of the ring-shaped DNA of polyoma virus. I. Identification of the replicative intermediate, *Proc. Natl. Acad. Sci. USA* **64**:701.

Bourgaux, P., Bourgaux-Ramoisy, D., and Seiler, P., 1971, Replication of the ring-shaped DNA of polyoma virus. II. Identification of molecules at various stages of replication, *J. Mol. Biol.* **59**:195.

Bourre, F., and Sarasin, A., 1983, Targeted mutagenesis of SV40 DNA induced by UV light, *Nature* **305**:68–70.

Boyce, F. M., Sundin, O., Barsoum, J., and Varshavsky, A., 1982, New way to isolate SV40 nucleoprotein complexes from infected cells: Use of a thiol-specific reagent, *J. Virol.* **42**:292–296.

Bradley, M. K., Griffin, J. D., and Livingston, D. M., 1981, Phosphotransferase activities associated with large T-antigen, in: *Eighth Cold Spring Harbor Conference on Cell Proliferation*, pp. 1263–1271, Cold Spring Harbor Laboratory, Cold Spring Harbor, N. Y.

Bradley, M. K., Griffin, J. D., and Livingston, D. M., 1982, Relationship of oligomerization to enzymatic and DNA-binding properties of the SV40 large T antigen, *Cell* **28:**125–134.

Bradley, M. K., Hudson, J., Villaneuva, M. S., and Livingston, D. M., 1984, Specific *in vitro* adenylation of the simian virus 40 large tumor antigen, *Proc. Natl. Acad. Sci. USA* **81:**6574–6578.

Brady, J., Lavialle, C., Radonovich, M. F., and Salzman, N. P., 1981, Stable association of viral protein VP_1 with SV40 DNA, *J. Virol.* **39:**432.

Brady, J., Bolen, J. B., Radonovich, M., Salzman, N., and Khoury, G., 1984, Stimulation of simian virus 40 late gene expression by simian virus 40 tumor antigen, *Proc. Natl. Acad. Sci. USA* **81:**2040–2044.

Brewer, B. J., Martin, S. R., and Champoux, J. J., 1983, A cellular single-stranded DNA dependent ATPase associated with simian virus 40 chromatin, *J. Biol. Chem.* **258:**4496–4502.

Brockman, W. W., Gutai, M. W., and Nathans, D., 1975, Evolutionary variants of SV40: Characterization of cloned complementary variants, *Virology* **66:**115.

Brun, G., and Weissbach, A., 1978, Initiation of HeLa cell DNA synthesis in a subnuclear system, *Proc. Natl. Acad. Sci. USA* **75:**5931.

Brynolf, K., Eliasson, R., and Reichard, P., 1978, Formation of Okazaki fragments in polyoma DNA synthesis caused by misincorporation of uracil, *Cell* **13:**573.

Buchman, A. R., Burnett, L., and Berg, P., 1980, The SV40 nucleotide sequence, in: *DNA Tumor Viruses: The Molecular Biology of Tumor Viruses, Part 2* (J. Tooze, ed.), pp. 799–829, Cold Spring Harbor Laboratory, Cold Spring Harbor, N.Y.

Buchman, A. R., Fromm, M., and Berg, P., 1984, Complex regulation of SV40 early-region transcription from different overlapping promoters, *Mol. Cell. Biol.* **4:**1900.

Buckler-White, A. J., and Pigiet, V., 1982, Isolation and characterization of replication forks from discrete regions of the polyoma genome, *J. Virol.* **44:**499.

Buckler-White, A. J., Humphrey, G. W., and Pigiet, V., 1980, Association of polyoma T-antigen and DNA with the nuclear matrix from lytically infected 3T6 cells, *Cell* **22:**37–46.

Buckler-White, A. J., Krauss, M. R., Pigiet, V., and Benbow, R. M., 1982, Asynchronous bidirectional replication of polyoma virus DNA, *J. Virol.* **43:**885–895.

Burger, C., and Fanning, E., 1983, Specific DNA binding activity of T antigen subclasses varies among different SV40-transformed cell lines, *Virology* **126:**19–31.

Calza, R. E., Eckhardt, L. A., DelGiudice, T., and Schildkraut, C., 1984, Changes in gene position are accompanied by a change in time of replication, *Cell* **36:**689–696.

Campbell, B. A., and Villarreal, L. P., 1985, Host species specificity of polyomavirus DNA replication is not altered by the SV40 72 base-pair repeats, *Mol. Cell. Biol.* **5:**1534–1537.

Carlson, J. O., Pfenninger, O., Sinden, R., Lehman, J. M., and Pettijohn, D. E., 1982, New procedure using a psoralen derivative for analysis of nucleosome associated DNA sequences in chromatin of living cells, *Nucleic Acids. Res.* **10:**2043.

Carmichael, G., Schaffhausen, B. S., Mandel, G., Liang, T. J., and Benjamin, T. L., 1984, Transformation by polyoma virus is drastically reduced by substitution of phenylalanine for tyrosine at residue 315 of middle-sized tumor antigen, *Proc. Natl. Acad. Sci. USA* **81:**679.

Carroll, R. B., and Gurney, E. G., 1982, Time-dependent maturation of the SV-40 large T-antigen–p53 complex studied by using monoclonal antibodies, *J. Virol.* **44:**565.

Cereghini, S., and Yaniv, M., 1984, Assembly of transfected DNA into chromatin: Structural changes in the origin–promoter–enhancer region upon replication, *EMBO J.* **3:**1243.

Cereghini, S., Saragosti, S., Yaniv, M., and Hamer, D. H., 1984, SV40-α-globulin hybrid minichromosomes, differences in DNase I hypersensitivity of promoter and enhancer sequences, *Eur. J. Biochem.* **144:**545.

Chalberg, M. D., and Kelly, T. J., Jr., 1979, Adenovirus DNA replication *in vitro:* Origin and direction of daughter strand synthesis, *J. Mol. Biol.* **135:**999–1012.

Chambers, J. C., Watanabe, S., and Taylor, J. H., 1982, Dissection of a replication origin of *Xenopus* DNA, *Proc. Natl. Acad. Sci. USA* **79:**5572–5576.

Champoux, J. J., and Been, M. D., 1980, Topoisomerase and the swivel problem, in: *Mechanistic Studies of DNA Replication and Genetic Recombination* (B. Alberts, ed.), p. 809, Academic Press, New York.

Champoux, J. J., and McConaughy, B. L., 1975, Priming of superhelical SV40 DNA by *E. coli* RNA polymerase for *in vitro* DNA synthesis, *Biochemistry* **14:**307.

Chatis, P. A., Holland, C. A., Hartley, H. W., Rowe, W. P., and Hopkins, N., 1983, Role for the 3′-end of the genome in determining disease specificity of friend and moloney murine leukemia viruses, *Proc. Natl. Acad. Sci. USA* **80:**4408–4411.

Chaudry, F., Belsham, G. J., and Smith, A. E., 1982a, Biochemical properties of the 145,000-dalton super-T antigen from simian virus 40-transformed BALB/c 3T3 clone 20 cells, *J. Virol.* **45:**1098–1106.

Chaudry, F., Harvey, R., and Smith, A. E., 1982b, Structure and biochemical functions of four simian virus 40 truncated large T-antigens, *J. Virol.* **44:**54–66.

Cheevers, W. P., 1973, Protein and mRNA requirements for superhelicity of polyoma virus DNA, *Nature New Biol.* **242:**202.

Chen, M. C. Y., Birkenmeier, E., and Salzman, N. P., 1976, SV40 DNA replication: Characterization of gaps in the termination region, *J. Virol.* **17:**614.

Chen, S. S., and Hsu, M.-T., 1983, Intracellular forms of simian virus 40 nucleoprotein complexes. V. Enrichment for "active" simian virus 40 chromatin by differential precipitation with Mg^{2+}, *J. Virol.* **46:**808–817.

Chen, S. S., and Hsu, M.-T., 1984, Evidence for variation of supercoiled densities among SV40 nucleoprotein complexes and for higher supercoil density in replicating complexes, *J. Virol.* **51:**14.

Chen, S., Grass, D. S., Blanck, G., Hoganson, N., Manley, J. L., and Pollack, R. E., 1983, A functional simian virus 40 origin of replication is required for the generation of a super T antigen with a molecular weight of 100,000 in transformed mouse cells, *J. Virol.* **48:**492–502.

Chia, W., and Rigby, P. W. J., 1981, Fate of viral DNA in nonpermissive cells infected with SV40, *Proc. Natl. Acad. Sci. USA* **78:**6638.

Chou, J. Y., and Martin, R. G., 1975, DNA infectivity and the induction of host DNA synthesis with temperature-sensitive mutants of SV40, *J. Virol.* **15:**145.

Chou, J. Y., Avila, J., and Martin, R. G., 1974, Viral DNA synthesis in cells infected by temperature-sensitive mutants of SV40, *J. Virol.* **14:**116–122.

Chou, P. Y., and Fasman, G., 1978, Empirical predictions of protein conformation, *Annu. Rev. Biochem.* **47:**251–276.

Christiansen, J. B., and Brockman, W. W., 1982, Effects of large and small T-antigens on DNA synthesis and cell division in SV40-transformed BALB/c 3T3 cells, *J. Virol.* **44:**574.

Clark, R., Lane, D. P., and Tjian, R., 1981, Use of monoclonal antibodies as probes of simian virus 40 T antigen ATPase activity, *J. Biol. Chem.* **256:**11854–11858.

Clark, R., Peden, K., Pipas, J. M., Nathans, D., and Tjian, R., 1983, Biochemical activities of T-antigen proteins encoded by simian virus 40 A gene deletion mutations, *Mol. Cell. Biol.* **3:**220–228.

Clayton, C. E., Lovett, M., and Rigby, P. W. J., 1982, Functional analysis of a SV40 super T-antigen, *J. Virol.* **44:**974.

Clertant, P., and Cuzin, F., 1982, Covalent affinity labeling by periodate-oxidized [^{32}P]ATP of the large-T proteins of polyoma and SV40 viruses, *J. Biol. Chem.* **257:**6300–6305.

Clertant, P., and Seif, I., 1984, A common function for polyoma virus large-T and papillomavirus E1 proteins?, *Nature* **311:**276–279.

Clertant, P., Gaudray, P., and Cuzin, F., 1984a, Salt-stable binding of the large T protein to DNA in polyoma virus chromatin, *EMBO J.* **3:**303–307.

Clertant, P., Gaudray, P., May, E., and Cuzin, F., 1984b, The nucleotide binding site detected by affinity labeling in the large T proteins of polyoma and SV40 viruses is distinct from their ATPase catalytic site, *J. Biol. Chem.* **259:**15196–15203.

Coca-Prados, M., Vidali, G., and Hsu, M.-T., 1980, Intracellular forms of SV40 nucleoprotein complexes. III. Study of histone modifications, *J. Virol.* **36:**353.

Cohen, G. L., Wright, P. J., DeLucia, A. L., Lewton, B. A., Anderson, M. E., and Tegtmeyer, P., 1984, Critical spatial requirement within the origin of simian virus 40 DNA replication, *J. Virol.* **51**:91–96.

Colby, W. W., and Shenk, T., 1982, Fragments of the SV40 transforming gene facilitate transformation of rat embryo cells, *Proc. Natl. Acad. Sci. USA* **79**:5189.

Cole, C. N., Landers, T., Goff, S. P., Manteuil-Brutlag, S., and Berg, P., 1977, Physical and genetic characterization of deletion mutants of SV40 constructed *in vitro*, *J. Virol.* **24**:277.

Colwill, R. W., and Sheinin, R., 1983, tsA1S9 locus in mouse L cells may encode a novobiocin binding protein that is required for DNA topoisomerase II activity, *Proc. Natl. Acad. Sci. USA* **80**:4644.

Conaway, R. C., and Lehman, I. R., 1982, A DNA primase activity associated with DNA polymerase α from *Drosophila melanogaster* embryos, *Proc. Natl. Acad. Sci. USA* **79**:2523.

Conrad, S. E., and Botchan, M. R., 1982, Isolation and characterization of human DNA fragments with nucleotide sequence homologies with the simian virus 40 regulatory region, *Mol. Cell. Biol.* **2**:949–965.

Conrad, S. E., Liu, C., and Botchan, M. R., 1982, Fragment spanning the SV40 replication origin is the only DNA sequence required in *cis* for viral excision, *Science* **218**:1223.

Cory, J. G., Mansell, M. M., and Whiteford, T. W., Jr., 1976, Control of ribonucleotide reductase in mammalian cells, *Adv. Enzyme Regul.* **14**:45.

Cosman, D. J., and Tevethia, M. J., 1981, Characterization of a temperature-sensitive, DNA-positive, nontransforming mutant of simian virus 40, *Virology* **112**:605–624.

Cotton, R. W., and Hamkalo, B. A., 1981, Nucleosome dissociation at physiological strengths, *Nucleic Acids Res.* **9**:445.

Courey, A. J., and Wang, J. C., 1983, Cruciform formation in a negatively supercoiled DNA may be kinetically forbidden under physiological conditions, *Cell* **33**:817.

Covey, L, Choi, Y., and Prives, C., 1984, Association of simian virus 40 T antigen with the nuclear matrix of infected and transformed cells, *Mol. Cell. Biol.* **4**:1384–1392.

Cowie, A., and Kamen, R., 1984, Multiple binding sites for polyomavirus large T antigen within regulatory sequences of polyomavirus DNA, *J. Virol.* **52**:750–760.

Crawford, L. V., Syrett, C., and Wilde, A., 1973, The replication of polyoma DNA, *J. Gen. Virol.* **21**:515–521.

Crawford, L. V., Robbins, A. K., and Nicklin, P. M., 1974, Location of the origin and terminus of replication in polyoma virus DNA, *J. Gen. Virol.* **25**:133–142.

Crawford, L. V., Robbins, A. K., Nicklin, P. M., and Osborn, K., 1975, Polyoma DNA replication: Location of the origin in different virus strains, *Cold Spring Harbor Symp. Quant. Biol.* **39**:219.

Cremisi, C., 1979, Chromatin replication revealed by studies of animal cells and papovaviruses (SV40 and polyoma virus), *Microbiol. Rev.* **43**:297.

Cremisi, C., 1981, The appearance of DNase I hypersensitive sites at the 5′-end of the late SV40 genes is correlated with the transcriptional switch, *Nucleic Acids Res.* **9**:5949.

Cremisi, C., and Yaniv, M., 1980, Sequential assembly of newly synthesized histones in replicating SV40 DNA, *Biochem, Biophys. Res. Commun.* **92**:1117.

Cremisi, C., Pignatti, P. F., Croissant, O., and Yaniv, M., 1976d, Chromatin-like structures in polyoma virus and simian virus 40 lytic cycle, *J. Virol.* **17**:204.

Cremisi, C., Pignatti, P. F., and Yaniv, M., 1976b, Random location and absence of movement of nucleosomes in vivo on SV40 nucleoprotein complex isolated from infected cells, *Biochem. Biophys. Res. Commun.* **73**:548.

Cremisi, C., Chestier, A., and Yaniv, M., 1978, Assembly of SV40 and polyoma minichromosomes during replication, *Cold Spring Harbor Symp. Quant. Biol.* **42**:409–416.

Cusick, M. E., Herman, T. M., DePamphilis, M. L., and Wassarman, P. M., 1981, Structure of chromatin at DNA replication forks: Prenucleosomal DNA is rapidly excised from replicating simian virus 40 chromosomes by micrococcal nuclease, *Biochemistry* **20**:6648–6658.

Cusick, M. E., Lee, K.-S., DePamphilis, M. L., and Wassarman, P. M., 1983, Structure of chromatin at DNA replication forks: Nuclease hypersensitivity results from both pre-nucleosomal-DNA and an immature chromatin structure, *Biochemistry* **22**:3873–3884.

Cusick, M. E., DePamphilis, M. L., and Wassarman, P. M., 1984, Dispersive nucleosome segregation during replication of SV40 chromosomes, *J. Mol. Biol.* **178**:249.

Cuzin, F., and Clertant, P., 1980, Initiation of polyoma virus DNA replication in vitro and its dependence on the viral gene A protein, *Nucleic Acids Res.* **8**:4377.

Cuzin, F., Vogt, M., Dieckman, M., and Berg, P., 1970, Induction of virus multiplication in 3T3 cells transformed by a thermosensitive mutant of polyoma virus. II. Formation of oligomeric polyoma DNA molecules, *J. Mol. Biol.* **47**:317.

Dailey, L., Pellegrini, S., and Basilico, C., 1984, Deletion of the origin of replication impairs the ability of polyomavirus DNA to transform cells and to form tandem insertions, *J. Virol.* **49**:984–987.

Dandolo, L., Blangy, D., and Kamen, R., 1983, Regulation of polyoma virus transcription in murine embryonal carcinoma cells, *J. Virol.* **47**:55–64.

Dandolo, L., Aghion, J., and Blangy, D., 1984, T-antigen-independent replication of polyomavirus DNA in murine embryonal carcinoma cells, *Mol. Cell. Biol.* **4**:317–323.

Daniell, E., Burg, J. L., and Fedor, M. J., 1982, DNA and histone synthesis in butyrate-inhibited BSC-1 cells infected with SV40, *Virology* **116**:196.

D'Anna, J. A., and Tobey, R. A., 1984, Changes in histone H1 content and chromatin structure of cells blocked in early S phase by 5-fluorodeoxyuridine and aphidicolin, *Biochemistry* **23**:5024.

Danna, K. J., and Nathans, D., 1972, Studies of SV40 DNA. IV. Bidirectional replication of SV40 DNA, *Proc. Natl. Acad. Sci. USA* **69**:3097.

Darlix, J.-L., Khandjian, E. W., and Weil, R., 1984, Nature and origin of the RNA associated with SV40 large tumor antigen, *Proc. Natl. Acad. Sci. USA* **81**:5425–5429.

Das, G. C., and Niyogi, S. K., 1981, Structure, replication and transcription of the SV40 genome, *Prog. Nucleic Acid Res. Mol. Biol.* **25**:187–240.

Das, G. C., Allison, D. P., and Niyogi, S. K., 1979, Sites including those of origin and termination of replication are not freely available to single-cut restriction endonucleases in the supercompact form of simian virus 40 minichromsome, *Biochem. Biophys. Res. Commun.* **89**:17.

Decker, R. S., Yamaguchi, M., Bradley, M., and DePamphilis, M. L., 1986, Initiation of DNA replication in SV40 chromosomes and in SV40 recombinant DNA plasmid DNA, manuscript in preparation.

Deininger, P. L., Esty, A., LaPorte, P., Hsu, H., and Friedmann, T., 1980, The nucleotide sequence and restriction enzyme sites of the polyoma genome, *Nucl. Acids Res.* **8**:855–860.

DeLeo, A., Jay, G., Appella, E., Dubois, G., Law, L., and Old, L., 1979, Detection of a transformation-related antigen in chemically induced sarcomas and other transformed cells of the mouse, *Proc. Natl. Acad. Sci. USA* **76**:2420.

DeLucia, A. L., Lewton, B. A., Tjian, R., and Tegtmeyer, P., 1983, Topography of simian virus 40 A protein–DNA complexes: Arrangement of pentanucleotide interaction sites at the origin of replication, *J. Virol.* **46**:143–150.

DePamphilis, M. L., 1984, Papovavirus chromosomes as a model for the replication of mammalian chromosomes, in: *Genetics: New Frontiers*, Proceedings of the XV International Congress of Genetics, New Delhi, 1984, pp. 369–384.

DePamphilis, M. L., and Berg, P., 1975, Requirement of a cytoplasmic fraction for synthesis of SV40 DNA in isolated nuclei, *J. Biol. Chem.* **250**:4348.

DePamphilis, M. L., and Wassarman, P. M., 1980, Replication of eukaryotic chromosomes: A close-up of the replication fork, *Annu. Rev. Biochem.* **49**:627.

DePamphilis, M. L., and Wassarman, P. M., 1982, Organization and replication of papovavirus DNA, in: *Organization and Replication of Viral DNA* (A. S. Kaplan, ed.), pp. 37–114, CRC Press, Boca Raton, Fla.

DePamphilis, M. L., Beard, P., and Berg, P., 1975, Synthesis of superhelical SV40 DNA in cell lysates, *J. Biol. Chem.* **250**:4340.

DePamphilis, M. L., Anderson, S., Bar-Shavit, R., Collins, E., Edenberg, H., Herman, T., Karas, B., Kaufmann, G., Krokan, H., Shelton, E., Su, R., Tapper, D., and Wassarman, P. M., 1979, The replication and structure of simian virus 40 chromosomes, *Cold Spring Harbor Symp. Quant. Biol.* **43:**679.

DePamphilis, M. L., Anderson, S., Cusick, M., Hay, R., Herman, T., Krokan, H., Shelton, E., Tack, L, Tapper, D., Weaver, D., and Wassarman, P. M., 1980, The interdependence of DNA replication, DNA sequence and chromatin structure in simian virus 40 chromosomes, in: *Mechanistic Studies of DNA Replication and Genetic Recombination* (B. Alberts, ed.), pp. 55–78, ICN–UCLA Symposia on Molecular and Cellular Biology, Volume 19, Academic Press, New York.

DePamphilis, M. L., Chalifour, L. E., Charette, M. F., Cusick, M. E., Hay, R. T., Hendrickson, E. A., Pritchard, C. G., Tack, L. C., Wassarman, P. M., Weaver, D. T., and Wirak, D. O., 1983a, Papovavirus chromosomes as a model for mammalian DNA replication, in: *Mechanisms of DNA Replication and Recombination,* (N. R. Cozzarelli, ed.), pp. 423–447, UCLA Symposia on Molecular and Cellular Biology, New Series, Volume 10, Liss, New York.

DePamphilis, M. L., Cusick, M. E., Hay, R. T., Pritchard, C. G., Tack, L. C., Wassarman, P. M., and Weaver, D. T., 1983b, Chromatin structure, DNA sequences and replication proteins: Searching for the principles of eukaryotic chromosome replication, in: *New Approaches in Eukaryotic DNA Replication* (A. M. deRecondo, ed.), pp. 203–233, Plenum Press, New York.

Deppert, W., 1978, Simain virus 40 (SV40)-specific proteins associated with the nuclear matrix isolated from adenovirus type 2–SV40 hybrid virus-infected HeLa cells carry SV40 U-antigen determinants, *J. Virol.* **26:**165–178.

Deppert, W., and Walter, G., 1976, Simian virus 40 (SV40) tumor-specific proteins in nucleus and plasma membrane of HeLa cells infected by adenovirus 2–SV40 hybrid virus $Ad2^{+}ND2$, *Proc. Natl. Acad. Sci. USA* **73:**2505–2509.

Deppert, W., Gurney, E. G., and Harrison, R. O., 1981, Monoclonal antibodies against simian virus 40 tumor antigens: Analysis of antigenic binding sites, using adenovirus type 2–simian virus 40 hybrid viruses, *J. Virol.* **37:**478–482.

der Groseillers, L., Rassant, E., and Jolicoeur, P., 1983, Thymotropism of murine leukemia virus is conferred by its long terminal repeat, *Proc. Natl. Acad. Sci. USA* **80:**4203–4207.

Desiderio, S. V., and Kelly, T. J., Jr., 1981, Structure of the linkage between adenovirus DNA and the 55,000 molecular weight terminal protein, *J. Mol. Biol.* **145:**319–337.

de Simone, V., LaMantia, G., Lania, L., and Amati, P., 1985, Polyomavirus mutation that confers a cell-specific *cis* advantage for viral DNA replication, *Mol. Cell. Biol.* **5:**2142–2146.

de Villiers, J., and Schaffner, W., 1981, A small segment of polyoma virus DNA enhances the expression of a cloned β-globin gene over a distance of 1400 base pairs, *Nucleic Acids Res.* **9:**6251.

de Villiers, J., Olson, L., Tyndall, C., and Schaffner, W., 1982, Transcriptional "enhancers" from SV40 and polyoma virus show a cell type preference, *Nucleic Acids Res.* **10:**7965–7976.

de Villiers, J. Schaffner, W., Tyndall, C., Lupton, S., and Kamen, R., 1984, Polyoma virus DNA replication requires an enhancer, *Nature* **312:**242–246.

Dhar, R., Lai, C.-J., and Khoury, G., 1978, Nucleotide sequence of the DNA replication origin for human papovavirus BKV: Sequence and structural homology with SV40, *Cell* **13:**345.

Dilworth, S. M., and Griffin, B. E., 1982, Monoclonal antibodies against polyoma tumor antigens, *Proc. Natl. Acad. Sci. USA* **79:**1059.

Dilworth, S. M., Cowie, A., Kamen, R. I., and Griffin, B. E., 1984, DNA binding activity of polyoma virus large tumor antigen, *Proc. Natl. Acad. Sci. USA* **81:**1941–1945.

DiMaio, D., and Nathans, D., 1980, Cold-sensitive regulatory mutants of simian virus 40, *J. Mol. Biol.* **140:**129–142.

DiMaio, D., and Nathans, D., 1982, Regulatory mutants of SV40: Effect of mutations at a T-antigen binding site on DNA replication and expression of viral genes, *J. Mol. Biol.* **156:**531.

DiNardo, S., Voelkel, K., and Sternglanz, R., 1984, DNA topoisomerase II mutant of *Saccharomyces cerevisiae*: Topoisomerase II is required for segregation of daughter molecules at the termination of DNA replication, *Proc. Natl. Acad. Sci. USA* **81**:2626.

Dinter-Gottlieb, G., and Kaufmann, G., 1982, Uncoupling of SV40 tsA replicon activation from DNA chain elongation by temperature shifts and aphidicolin arrest, *Nucleic Acids Res.* **10**:763–773.

Dinter-Gottlieb, G., and Kaufmann, G., 1983, Aphidicolin arrest irreversibly impairs replicating SV40 chromosomes, *J. Biol. Chem.* **258**:3809.

Dixon, R. A., and Nathans, D., 1985, Purification of simian virus 40 large T antigen by immunoaffinity chromatography, *J. Virol.* **53**:1001–1004.

Dorn, A., Brauer, D., Fanning, E., and Knippers, R., 1982, Subclasses of simian virus 40 large tumor antigen: Partial purification and DNA binding properties of two subclasses of tumor antigen from productivity infected cells, *Eur. J. Biochem* **128**:53–62.

Dubensky, T. W., and Villarreal, L. D., 1984, The primary site of replication alters the eventual site of persistent infection by polyomavirus in mice, *J. Virol.* **50**:541.

Eckhart, W., 1983, Role of polyoma T-antigens in malignant cell transformation, *Prog. Nucleic Acid Res. Mol. Biol.* **29**:119.

Edenberg, H. J., 1980, Novobiocin inhibition of SV40 DNA replication, *Nature* **286**:529.

Edenberg, H. J., and Huberman, J. A., 1975, Eucaryotic chromosome replication, *Annu. Rev. Genet.* **9**:245.

Edenberg, H. J., Waqar, M. A., and Huberman, J. A., 1976, Subcellular systems for synthesis of SV40 DNA *in vitro*, *Proc. Natl. Acad. Sci. USA* **73**:4392–4396.

Edenberg, H. J., Waqar, M. A., and Huberman, J. A., 1977, DNA synthesis by partially purified SV40 chromosomes, *Nucleic Acids Res.* **4**:3083–3095.

Edenberg, H. J., Anderson, S., and DePamphilis, M. L. 1978, Involvement of DNA polymerase α in simian virus 40 DNA replication, *J. Biol. Chem.* **253**:3273.

Edwards, C. A. F., Khoury, G., and Martin, R. G., 1979, Phosphorylation of T-antigen and control of T-antigen expresison in cells transformed by wild-type and tsA mutants of SV40, *J. Virol.* **29**:753.

Eliasson, R., and Reichard, P., 1978, Replication of polyoma DNA in isolated nuclei, synthesis and distribution in initiator RNA, *J. Biol. Chem.* **253**:7469.

Eliasson, R., and Reichard, P., 1979, Replication of polyoma DNA in isolated nuclei. VI. Initiation RNA synthesis during nucleotide depletion, *J. Mol. Biol.* **129**:393.

Eliasson, R., Pontis, E., Reichard, P., and Eckstein, F., 1981, Replication of polyoma DNA in nuclei isolated from azidocytidine-inhibited fibroblasts, *J. Biol. Chem.*, **256**:9044.

Ellison, M. J., and Pulleyblank, D. E., 1983, Pathways of assembly of nucleohistone complexes formed *in vitro* under physiological conditions, *J. Biol. Chem.* **258**:13321–13327.

Ellman, M., Bikel, I., Figge, J., Roberts, T. M., Schlossman, R., and Livingston, D. M., 1984, Localization of the SV40 small t antigen in the nucleus and cytoplasm of monkey and mouse cells, *J. Virol.* **50**:623–628.

Emerson, B. M., and Felsenfeld, G., 1984, Specific factor conferring nuclease hypersensitivity at the 5′-end of the chicken adult β-globin gene, *Proc. Natl. Acad. Sci. USA* **81**:95–99.

Epher, E., Rifkind, R. A., and Marks, P. A., 1981, Replication of a and b globin DNA sequences occurs during early S phase in murine erythroleukemia cells, *Proc. Natl. Acad. Sci. USA* **78**:3058.

Fangman, W. L., Hice, R. H., and Chlebowicz-Siedziewska, E., 1983, ARS replication during the yeast S-phase, *Cell* **32**:831–838.

Fanning, E., Nowak, B., and Burger, C., 1981, Detection and characterization of multiple forms of simian virus 40 large T-antigen, *J. Virol.* **37**:92–102.

Fanning, E., Westphal, K.-H., Brauer, D., and Corlin, D., 1982, Subclasses of simian virus large T antigen: Differential binding of two subclasses of T antigen from productively infected cells to viral and cellular DNA, *EMBO J.* **1**:1023–1028.

Fareed, G. C., and Salzman, N. P., 1972, Intermediate in SV40 DNA chain growth, *Nature New Biol.* **238**:274.

Fareed, G. C., Garon, C. F., and Salzman, N. P., 1972, Origin and direction of SV40 DNA replication, *J. Virol.* **10**:484.

Fareed, G. C., Kerlie, M. L., and Salzman, N. P., 1973a, Characterization of SV40 DNA component II during viral DNA replication, *J. Mol. Biol.* **74:**95.

Fareed, G. C., Khoury, G., and Salzman, N. P., 1973b, Self-annealing of 4S strands from replicating SV40 DNA, *J. Mol. Biol.* **77:**457.

Feunteun, J., Sompayrac, L., Fluck, M., and Benjamin, T., 1976, Localization of gene functions in polyoma virus DNA, *Proc. Natl. Acad. Sci. USA* **73:**4169–4173.

Fiers, W., Contreras, R., Haegeman, G., Rogiers, R., van de Voorde, A., van Heuversywn, H., van Herreweghe, J., Volckaert, G., and Ysebaert, M., 1978, The complete nucleotide sequence of SV40 DNA, *Nature* **273:**113.

Finch, J. T., Lutter, L. C., Rhodes, D., Brown, R. S., Rushton, B., Levitt, M., and Klug, A., 1977, Structure of nucleosome core particles of chromatin, *Nature* **269:**29.

Fisher, P. A., Wang, T. S.-F., and Korn, D., 1979, Enzymological characterization of DNA polymerase α, basic catalytic properties, processivity and gap utilization of the homogeneous enzyme from human KB cells, *J. Biol. Chem.* **254:**6128.

Floros, J., Jonak, G., Galanti, N., and Baserga, R., 1981, Induction of cell DNA replication in G1-specific ts mutants by microinjection of SV40 DNA, *Exp. Cell Res.* **132:** 215–223.

Flory, P. J., Jr., 1975, Strandedness of newly synthesized short pieces of polyoma DNA from isolated nuclei, *Nucleic Acids Res.* **4:**1449.

Flory, P. J., Jr., and Vinograd, J., 1973, 5-Bromodeoxyuridine labeling of monomeric and catenated circular mitochondrial DNA in HeLa cells, *J. Mol. Biol.* **74:**81–94.

Folk, W. R., and Bancuk, J., 1976, Polyoma genome in hamster BHK-21-C13 cells: Integration into cellular DNA and induction of the viral replicon, *J. Virol.* **20:**133.

Fowler, E., Farb, R., and El-Saidy, S., 1982, Distribution of the core histones H2A, H2B, H3 and H4 during cell replication, *Nucleic Acids Res.* **10:**735–748.

Francke, B., and Eckhart, W., 1973, Polyoma gene functions required for viral DNA synthesis, *Virology* **55:**127–137.

Francke, B., and Hunter, T., 1974, *In vitro* polyoma DNA synthesis: Discontinuous chain growth, *J. Mol. Biol.* **83:**99.

Francke, B., and Hunter, T., 1975, In vitro polyoma DNA synthesis: Requirement for cytoplasmic factors, *J. Virol.* **15:**97–107.

Francke, B., and Vogt, M., 1975, In vitro polyoma DNA synthesis: Self annealing properties of short DNA chains, *Cell* **5:**205.

Fraser, J. M. K., and Huberman, J. A., 1978, In vitro HeLa cell DNA synthesis. II. Partial characterization of soluble factors stimulating nuclear DNA synthesis, *Biochim. Biophys. Acta* **520:**271.

Frisque, R. J., 1983, Nucleotide sequences of the region encompassing the JC virus origin of DNA replication, *J. Virol.* **46:**170–176.

Frisque, R. J., Bream, G. L., and Cannella, M. T., 1984, Human polyomavirus JC virus genome, *J. Virol.* **51:**458.

Fromm, M., and Berg, P., 1982, Deletion mapping of DNA regions required for SV40 early region promoter function *in vivo*, *J. Mol. Appl. Genet.* **1:**457.

Fromm, M., and Berg, P., 1983, SV40 early- and late-region promoter functions are enhanced by the 72-bp repeat inserted at distant locations and inverted orientations, *Mol. Cell. Biol.* **3:**991.

Fry, M., 1983, Eukaryotic DNA polymerases, in: *Enzymes of Nucleic Acid Synthesis and Modification* (S. T. Jacob, ed.), Volume I, pp. 39–92, CRC Press, Boca Raton, Fla.

Fujimura, F. K., and Linney, E., 1982, Polyoma mutants that productively infect F9 embryonal carcinoma cells do not rescue wild-type polyoma in F9 cells, *Proc. Natl. Acad. Sci. USA* **79:**1479–1483.

Fujimura, F. K., Deininger, P. L., Friedmann, T., and Linney, E., 1981a, Mutation near the polyoma DNA replication origin permits productive infection of F9 embryonal carcinoma cells, *Cell* **23:**809–814.

Fujimura, F. K., Silbert, P. E., Eckhart, W., and Linney, E., 1981b, Polyoma virus infection of retinoic acid-induced differentiated teratocarcinoma cells, *J. Virol.* **39:**306–312.

Furth, M. E., Dove, W. F., and Meyer, B. J., 1982, Specificity determinants for bacteriophage lambda DNA replication. III. Activation of replication in mutants by transcription outside of *ori*, *J. Mol. Biol.* **154**:65.

Galanti, N., Jonak, G. J., Soprano, K. J., Floros, J., Kaczmarek, L., Weissman, S., Reddy, V. B., Tilghman, S. M., and Baserga, R., 1981, Characterization and biological activity of cloned SV40 DNA fragments, *J. Biol. Chem.* **256**:6469.

Ganz, P. R., and Sheinin, R., 1983, Synthesis of multimeric polyoma virus DNA in mouse L-cells: Role of the tsA1S9 gene product, *J. Virol.* **46**:768.

Garcea, R. L., and Benjamin, T. L., 1983a, Isolation and characterization of polyoma nucleoprotein complexes, *Virology* **130**:65–75.

Garcea, R. L., and Benjamin, T. L., 1983b, Host range transforming gene of polyoma virus plays a role in virus assembly, *Proc. Natl. Acad. Sci. USA* **80**:3613–3617.

Garver, J. J., Pearson, P. L., Abrahams, P. J., and van der Eb, A. J., 1980, Control of SV40 replication by a single chromosome in monkey–hamster cell hybrids, *Somatic Cell Genet.* **6**:443–453.

Gaudray, P., Clertant, P., and Cuzin, F., 1980, ATP phosphohydrolase (ATPase) activity of a polyoma virus T-antigen, *Eur. J. Biochem.* **109**:553–560.

Gaudray, P., Tyndall, C., Kamen, R., and Cuzin, F., 1981, The high affinity binding site on polyoma virus DNA for the viral large-T protein, *Nucleic Acids Res.* **9**:5697.

Gautschi, J.R., Burkhalter, M., and Baumann, E. A., 1978, Comparative utilization of bromodeoxyuridine and iododeoxyuridine triphosphates for mammalian DNA replication *in vitro*, *Biochim. Biophys. Acta* **518**:31.

Gellert, M., O'Dea, M. H., and Mizuuchi, K., 1983, Slow cruciform transitions in palindromic DNA, *Proc. Natl. Acad. Sci. USA* **80**:5545.

Gerard, R. D., Woodworth-Gutai, M., and Scott, W. A., 1982, Deletion mutants which affect the nuclease-sensitive site in SV40 chromatin, *Mol. Cell. Biol.* **2**:782.

Germino, J., and Bastia, D., 1981, Termination of DNA replication in vitro at a sequence-specific replication terminus, *Cell* **23**:681.

Germond, J., Vogt, V., and Hirt, B., 1974, Characterization of the single-strand specific nuclease S1 on double stranded supercoiled polyoma DNA, *Eur. J. Biochem.* **43**:591.

Germond, J. E., Hirt, B., Oudet, P., Gross-Bellard, M., and Chambon, P., 1975, Folding of the DNA double helix in chromatin-like structures from SV40, *Proc. Natl. Acad. Sci. USA* **72**:1843.

Gersehey, E., 1979, SV40–host cell interactions during lytic infection, *J. Virol.* **30**:76.

Ghosh, P., and Lebowitz, P., 1981, SV40 early mRNA's contain multiple 5′-termini upstream and downstream from a Hogness–Goldberg sequence: A shift in 5′-termini during the lytic cycle is mediated by large "T" antigen, *J. Virol.* **40**:224–240.

Giacherio, D., and Hager, L. P., 1979, A poly(dT) stimulated ATPase activity associated with simian virus 40 large T-antigen, *J. Biol. Chem.* **254**:8113–8120.

Giacherio, D., and Hager, L., 1980, A specific DNA unwinding activity associated with SV40 T-antigens, *J. Biol. Chem.* **255**:8963.

Gidoni, D., Scheller, A., Barnet, B., Hantzopoulos, P., Oren, M., and Prives, C., 1982, Different forms of simian virus 40 large tumor antigens varying in their affinities for DNA, *J. Virol.* **42**:456–466.

Gilles, S. D., Morrison, S. L., Oi, V. T., and Tonegawa, S., 1983, A tissue-specific transcription enhancer element is located in the major intron of a rearranged immunoglobulin heavy chain gene, *Cell* **33**:717–728.

Glotov, B. O., Rudin, A. V., and Severin, E. S., 1982, Conditions for sliding of nucleosomes along DNA: SV40 minichromosomes, *Biochim. Biophys. Acta.* **696**:275–284.

Gluzman, Y., 1981, SV40-transformed simian cells support the replication of early SV40 mutants, *Cell* **23**:175–182.

Gluzman, Y., and Ahrens, B., 1982, SV40 early mutants that are defective for viral synthesis but competent for transformation of cultured rat and simian cells, *Virology* **123**:78–92.

Gluzman, Y. Davidson, J., Oren, M., and Winocour, E., 1977, Properties of permissive monkey cells transformed by UV-irradiated SV40, *J. Virol.* **22**:256.

Gluzman, Y., Frisque, R. J., and Sambrook, J., 1980, Origin-defective mutants of SV40, *Cold Spring Harbor Symp. Quant. Biol.* **44**:293.

Gluzman, Y., Reichl, H., and Solnick, D., 1982, Helper-free adenovirus type-5 vectors, in: *Eukaryotic Viral Vectors* (Y. Gluzman, ed.), pp. 187–192, Cold Spring Harbor Laboratory, Cold Spring Harbor, New York.

Goff, S. P., and Berg, P., 1977, Structure and formation of circular dimers of SV40 DNA, *J. Virol.* **24**:295.

Goldman, N. D., Brown, M., and Khoury, G., 1981a, Modification of SV40 T-antigen by poly ADP-ribosylation, *Cell* **24**:567–572.

Goldman, N., Howley, P., and Khoury, G., 1981b, Functional interaction between the early viral proteins of SV40 and adenovirus, *Virology* **109**:303.

Gooding, L. R., and O'Connell, K. A., 1983, Recognition by cytotoxic T lymphocytes of cells expressing fragments of the SV40 tumor antigen, *J. Immunol.* **131**:2580–2586.

Goto, T., and Wang, J. C., 1982, Yeast topoisomerase II: An ATP-dependent type II topoisomerase that catalyzes the catenation, decatenation, unknotting, and relaxation of dsDNA rings, *J. Biol. Chem.* **257**:5866.

Goulian, M., Bleile, B., and Tseng, B. Y., 1980, Methotrexate-induced misincorporation of uracil into DNA, *Proc. Natl. Acad. Sci. USA* **77**:1956.

Gourlie, B. B., and Pigiet, V. P., 1983, Polyoma virus minichromosomes: Characterization of the products of in vitro DNA synthesis, *J. Virol.* **45**:585.

Gourlie, B. B., Pigiet, V., Breaux, C. B., Krauss, M. R., King, C. R., and Benbow, R. M., 1981, Polyoma virus minichromosomes: Associated enzyme activities, *J. Virol.* **38**:862.

Graessmann, A., Graessmann, M., Guhl, E., and Mueller, C., 1978, Quantitative correlation between SV40 T-Ag synthesis and late viral gene expression in permissive and nonpermissive cells, *J. Cell Biol.* **77**:1–8.

Graessmann, A., Graessmann, M., and Mueller, C., 1979, SV40 and polyoma virus gene expression explored by the microinjection technique, *Curr. Top. Microbiol. Immun.* **87**:1–21.

Graessmann, A., Graessmann, M., and Mueller, C., 1981, Regulation of SV40 gene expression, *Adv. Cancer Res.* **35**:111–149.

Graessmann, M., and Graessmann, A., 1976, Early SV40 specific RNA contains information for T-antigen formation and chromatin replication, *Proc. Natl. Acad. Sci. USA* **73**:366.

Grastrom, R. H., Tseng, B. Y., and Goulian, M., 1978, The incorporation of uracil into animal cell DNA in vitro, *Cell* **15**:131.

Green, M. H., and Brooks, T. L., 1978, Recently replicated SV40 DNA is a preferential template for transcription and replication, *J. Virol.* **26**:235.

Greenspan, D. S., and Carroll, R. B., 1981, Complex of simian virus 40 large tumor antigen and 48,000-dalton host tumor antigen, *Proc. Natl. Acad. Sci. USA* **78**:105–109.

Griffin, B. E., 1980, Structure and genomic organization of SV40 and polyoma virus, in: *DNA Tumor Viruses: The Molecular Biology of Tumor Viruses, Part 2* (J. Tooze, ed.), pp. 61–124, Cold Spring Harbor Laboratory, Cold Spring Harbor, N.Y.

Griffin, B. E., and Dilworth, S. M., 1983, Polyomavirus: An overview of its unique properties, *Adv. Cancer Res.* **39**:183–268.

Griffin, B., and Fried, M., 1975, Amplification of a specific region of the polyoma virus genome, *Nature* **256**:175.

Griffin, B. E., Fried, M., and Cowie, A., 1974, Polyoma DNA: A physical map, *Proc. Natl. Acad. Sci. USA* **71**:2077–2081.

Griffin, B. E., Soeda, E., Barrell, B. G., and Staden, R., 1980, Sequence and analysis of polyoma virus DNA, in: *DNA Tumor Viruses: The Molecular Biology of Tumor Viruses, Part 2* (J. Tooze, ed.), pp. 831–896, Cold Spring Harbor Laboratory, Cold Spring Harbor, N.Y.

Griffin, J. D., Light, S., and Livingston, D. M., 1978, Measurements of the molecular size of simian virus 40 T antigen, *J. Virol.* **27**:218–226.

Griffin, J. D., Spangler, G. J., and Livingston, D. M., 1979, Protein kinase activity associated with simian virus 40 T antigen, *Proc. Natl. Acad. Sci. USA* **76**:2610–2614.

Griffin, J. D., Spangler, G. J., and Livingston, D., 1980, Enzymatic activities associated with the SV40 large T antigen, *Cold Spring Harbor Symp. Quant. Biol.* **44**:113–122.

Griffith, J., Dieckmann, M., and Berg, P., 1975, Electron microscope localization of a protein bound near the origin of SV40 DNA replication, *J. Virol.* **15**:167.

Gronostajski, R. M., Field, J., and Hurwitz, J., 1984, Purification of a primase activity associated with DNA polymerase α from HeLa cells, *J. Biol. Chem.* **259**:9479.

Grossman, L. I., Watson, R., and Vinograd, J., 1974, Restricted uptake of ethidium bromide and propidium diodide by denatured closed circular DNA in buoyant cesium chloride, *J. Mol. Biol.* **86**:271.

Gruss, C., Baumann, E., and Knippers, R., 1984, DNA binding properties of a mutant T-antigen from the SV40-transformed human cell line SV80, *J. Virology* **50**:943.

Gruss, P., and Sauer, G., 1977, Infectious linear DNA sequences replicating in SV40 infected cells, *J. Virol.* **21**:565.

Gurney, E. G., Harrison, R. O., and Fenno, J., 1980, Monoclonal antibodies against simian virus 40 T-antigens: Evidence for distinct subclasses of large T antigen and for similarities among nonviral T-antigens, *J. Virol.* **34**:752–763.

Gutai, M. W., and Nathans, D., 1978, Evolutionary variants of SV40: Nucleotide sequence of a conserved SV40 DNA segment containing the origin of viral DNA replication as an inverted repetition, *J. Mol. Biol.* **126**:259.

Hamelin, C., and Yaniv, M., 1979, Nicking-closing enzyme is associated with SV40 *in vivo* as a sodium dodecyl sulfate-resistant complex, *Nucleic Acids Res.* **7**:679.

Hancock, R., 1978, Assembly of new nucleosomal histones and new DNA into chromatin, *Proc. Natl. Acad. Sci. USA* **75**:2130.

Hand, R., 1979, *Cell Biology: A Comprehensive Treatise,* Vol. 2, pp. 389–437, Academic Press, New York.

Hansen, U., Tenen, D. G., Livingston, D. M., and Sharp, P. A., 1981, T antigen repression of SV40 early transcription from two promoters, *Cell* **27**:603–612.

Harland, R. M., and Laskey, R. A., 1980, Regulation of replication of DNA microinjected into eggs of *Xenopus laevis, Cell* **21**:761–771.

Harlow, E., Crawford, L. V., Pim, D. C., and Williamson, N. M., 1981a, Monoclonal antibodies specific for simian virus 40 tumor antigens, *J. Virol.* **39**:861–869.

Harlow, E., Pim, D., and Crawford, L. V., 1981b, Complex of SV40 large T-antigen and host 53,000 MW protein in monkey cells, *J. Virol.* **37**:564.

Harper, F., Florentin, Y., and Puvion, E., 1984, Localization of T antigen on simian virus 40 minichromosomes by immunoelectron microscopy, *EMBO J.* **3**:1235–1241.

Hartmann, J. P., and Scott, W. A., 1981, Distribution of DNase I-sensitive sites in SV40 nucleoprotein complexes from disrupted virus particles, *J. Virol.* **37**:908.

Hartmann, J. P., and Scott, W. A., 1983, Nuclease-sensitive sites in the two major intracellular SV40 nucleoproteins, *J. Virol.* **46**:1034–1038.

Hartzell, S. W., Byrne, B. J., and Subramanian, K. N., 1984, The SV40 minimal origin and the 72-bp repeat are required simultaneously for efficient induction of late gene expression with large tumor antigen, *Proc. Natl. Acad. Sci. USA,* **81**:6335–6339.

Hartzell, S. W., Yamaguchi, J., and Subramanian, K. N., 1983, SV40 deletion mutants lacking the 21-bp repeated sequences are viable, but have non-complementable deficiencies, *Nucleic Acids Res* **11**:1601.

Hassell, J. A., Lukanidin, E., Fey, G., and Sambrook, J., 1978, The structure and expression of two defective adenovirus 2/simian virus 40 hybrids, *J. Mol. Biol.* **120**:209–249.

Hassell, J. A., Muller, W. J., and Mueller, C. R., 1986, The dual role of the polyomavirus enhancer in transcription and DNA replication, *Cancer Cells,* Vol. 4, Cold Spring Harbor Press, in press.

Hay, R. T., and DePamphilis, M. L., 1982, Initiation of simian virus 40 DNA replication *in vivo:* Location and structure of 5′-ends of DNA synthesized in the *ori* region, *Cell* **28**:767–779.

Hay, R. T., Hendrickson, E. A., and DePamphilis, M. L., 1984, Sequence specificity for the initiation of RNA primed-SV40 DNA synthesis *in vivo, J. Mol. Biol.* **175**:131.

Hayaishi, O., and Ueda, K., 1977, Poly(ADP-ribose) and ADP-ribosylation of proteins, *Annu. Rev. Biochem.* **46**:95–116.

Hayday, A. C., Chaudry, F., and Fried, M., 1983, Loss of polyoma virus infectivity as a result of a single amino acid change in a region of polyoma virus large T-antigen which has extensive amino acid homology with simian virus 40 large T-antigen, *J. Virol.* **45**: 693–699.

Heintz, N. H., Milbrandt, J. D., Greisen, K. S., and Hamlin, J. L., 1983, Cloning of the initiation region of a mammalian chromosomal replicon, *Nature* **302**:439–441.

Henning, R., and Lange-Mütschler, J., 1983, Tightly associated lipids may anchor SV40 large T antigen in plasma membrane, *Nature* **305**:736–738.

Herbomel, P., Saragosti, S., Blangy, D., and Yaniv, M., 1981, Fine structure of the origin-proximal DNase I-hypersensitive region in wild-type and EC mutant polyoma, *Cell* **25**:651.

Herbomel, P., Bourachot, B., and Yaniv, M., 1984, Two distinct enhancers with different cell specificities coexist in the regulatory region of polyoma, *Cell* **39**:653.

Herman, T. M., DePamphilis, M. L., and Wassarman, P. M., 1979, Structure of chromatin at DNA replication forks: Okazaki fragments released from replicating simian virus 40 chromosomes by single-strand specific endonucleases are not in nucleosomes, *Biochemistry* **18**:4563.

Herman, T. M., DePamphilis, M. L., and Wassarman, P. M., 1981, Structure of chromatin at DNA replication forks: Location of the first nucleosomes on newly synthesized simian virus 40 DNA, *Biochemistry* **20**:621.

Hershey, H. V., Stieber, J. F., and Mueller, G. C., 1973, DNA synthesis in isolated nuclei: System for continuation of replication *in vitro*, *Eur. J. Biochem.* **34**:383.

Hines, P. J., and Benbow, R. M., 1982, Initiation of replication at specific origins in DNA molecules microinjected into unfertilized eggs of the frog *Xenopus laevis*, *Cell* **30**:459–468.

Hirt, B., 1966, Evidence for semiconservative replication of circular polyoma DNA, *Proc. Natl. Acad. Sci. USA* **55**:997.

Hirt, B., 1967, Selective extraction of polyoma DNA from infected mouse cell cultures, *J. Mol. Biol.* **26**:365–369.

Hiscott, J. B., and Defendi, V., 1979a, SV40 gene A regulation of cellular DNA synthesis. I. In permissive cells, *J. Virol.* **30**:590.

Hiscott, J., and Defendi, V., 1979b, SV40 gene A regulation of cellular DNA synthesis, II. In nonpermissive cells, *J. Virol.* **30**:802.

Hsieh, T.-S., and Brutlag, D., 1980, ATP-dependent DNA topoisomerase from *D. melanogaster* reversibly catenates duplex DNA rings, *Cell* **21**:115.

Hubscher, U., 1983, The mammalian primase is part of a high molecular weight DNA polymerase α polypeptide, *EMBO J.* **2**:133–136.

Huebner, K., Shander, M., and Croce, C. M., 1977, Suppression of replication of SV40 and polyoma virus in mouse–human hybrids, *Cell* **11**:25.

Hunter, T., and Francke, B., 1974, In vitro polyoma DNA synthesis: Involvement of RNA in discontinuous chain growth, *J. Mol. Biol.* **83**:123.

Hunter, T., and Francke, B., 1975, In-vitro polyoma DNA synthesis: Inhibition by 1-β-D-arabinofuranosyl CTP, *J. Virol.* **15**:759.

Hunter, T., Francke, B., and Bacheler, L, 1977, *In vitro* polyoma DNA synthesis: Asymmetry of short DNA chains, *Cell* **12**:1021.

Hunter, T., Hutchinson, M. A., and Eckhart, W., 1978, Translation of polyoma virus T antigen *in vitro*, *Proc. Natl. Acad. Sci. USA* **75**:5917–5921.

Hutchinson, N. I., Chang, L.-S., Pater, M. M., Bouck, N., Shenk, T. E., and di Mayorca, G., 1985, Characterization of a new simian virus 40 mutant, tsA 3900, isolated from deletion mutant tsA 1499, *J. Virol.* **53**:814–821.

Igo-Kemenes, T., Horz, W., and Zachau, H., 1982, Chromatin, *Annu. Rev. Biochem.* **51**:89.

Ikeda, J.-E., Enomoto, T., and Hurwitz, J., 1981, Replication of adenovirus DNA protein complex with purified proteins, *Proc. Natl. Acad. Sci. USA* **78**:884–888.

Innis, J. W., and Scott, W. A., 1984, DNA replication and chromatin structure of simian virus 40 insertion mutants, *Mol. Cell. Biol.* **4**:1499–1507.

Ito, Y., Brocklehurst, J. R., and Dulbecoo, R., 1977a, Virus-specific proteins in the plasma membrane of cells lytically infected or transformed by polyoma virus, *Proc. Natl. Acad. Sci. USA* **74**:4666–4670.

Ito, Y., Spurr, N., and Dulbecco, R., 1977b, Characterization of polyoma virus T-antigen, *Proc. Natl. Acad. Sci. USA* **74**:1259.

Itoh, T., and Tomizawa, J., 1980, Formation of an RNA primer for initiation of replication of Col E1 DNA by ribonuclease H, *Proc. Natl. Acad. Sci. USA* **77**:2450.

Jackson, V., and Chalkley, R., 1981a, A new method for the isolation of replicative chromatin: Selective deposition of histone on both new and old DNA, *Cell* **23**:121–134.

Jackson, V., and Chalkley, R., 1981b, A reevaluation of new histone deposition on replicating chromatin, *J. Biol. Chem.* **256**:5095–5103.

Jackson, V., Marshall, S., and Chalkley, R., 1981, The sites of deposition of newly synthesized histone, *Nucleic Acids Res.* **9**:4563–4581.

Jaenisch, R., and Levine, A. J., 1971a, Infection of primary African green monkey cells with SV40 monomeric and dimeric DNA, *J. Mol. Biol.* **61**:735.

Jaenisch, R., and Levine, A. J., 1971b, DNA replication in SV40 infected cells. V. Circular and catenated oligomes of SV40 DNA, *Virology* **44**:480.

Jaenisch, R., and Levine, A. J., 1972, DNA replication in SV40 infected cells. VI. The effect of cycloheximide on the formation of SV40 oligomeric DNA, *Virology* **48**:373.

Jaenisch, R., and Levine, A. J., 1973, DNA replication of SV40 infected cells. VII. Formation of SV40 catenated and circular dimers, *J. Mol. Biol.* **73**:199.

Jaenisch, R., Mayer, A., and Levine, A., 1971, Replicating SV40 molecules containing closed circular template DNA strands, *Nature New Biol.* **233**:72.

Jakob, K. M., Ben Yosef, S., and Tal, I., 1984, Reduced repeat length of nascent nucleosomal DNA is generated by replicating chromatin *in vivo, Nucleic Acids Res.* **12**:5015.

Jakobovits, E. B., Bratosin, S., and Aloni, Y., 1980, A nucleosome-free region in SV40 minichromosomes, *Nature* **285**:263.

Jakobovits, E. B., Bratosin, S., and Aloni, Y., 1982, Formation of a nucleosome-free region in SV40 minichromosomes is dependent upon a restricted segment of DNA, *Virology* **120**:340.

Jarvis, D. L., and Butel, J. S., 1985, Modification of simian virus 40 large tumor antigen by glycosylation, *Virology* **141**:173–189.

Jarvis, D. L., Lanford, R. E., and Butel, J. S., 1984, Structural comparisons of wild-type nuclear transport-defective simian virus 40 large tumor antigens, *Virology* **134**:168–176.

Jay, G., Khoury, G., DeLeo, A. B., Dippold, W. G., and Old, J. F., 1981a, p53 transformation-related protein: Detection of an associated phosphotransferase activity, *Proc. Natl. Acad. Sci. USA* **78**:2932–2936.

Jay, G., Nomura, S., Anderson, C. W., and Khoury, G., 1981b, Identification of the SV40 agnogene product: A DNA binding protein, *Nature* **291**:346–349.

Jensen, F. C., and Koprowski, H., 1969, Absence of a repressor in SV40-transformed cells, *Virology* **37**:687.

Jessel, D., Landau, T., Hudson, J., Lalor, T., Tenen, D. G., and Livingston, D. M., 1976, Identification of regions of the SV40 genome which contain preferred SV40 T antigen-binding sites, *Cell* **8**:535–545.

Jones, K. A., and Tjian, R., 1984, Essential contact residues within SV40 large T antigen binding sites I and II identified by alkylation-interference, *Cell* **36**:155–162.

Jones, K. A., Myers, R. M., and Tjian, R., 1984, Mutational analysis of simian virus 40 large T antigen DNA binding sites, *EMBO J.* **3**:3247–3255.

Jong, A. Y. S., and Scott, J. F., 1984, DNA synthesis in yeast cell-free extracts dependent on recombinant DNA plasmids purified from *E. coli, Nucleic Acids Res.* **12**:2943–2958.

Jongstra, J., Reudelhuber, T. L., Oudet, P., Benoist, C., Chae, C.-B., Jeltsch, J.-M., Mathis, D. J., and Chambon, P., 1984, Induction of altered chromatin structures by SV40 enhancer and promoter elements, *Nature* **307**:708.

Kaguni, L. S., Rossignol, J.-M., Conaway, R. C., and Lehman, I. R., 1983, Isolation of an intact DNA polymerase-primase from embryos of *Drosophila melanogaster, Proc. Natl. Acad. Sci. USA* **80**:2221.

Kalderon, D., and Smith, A. E., 1984, *In vitro* mutagenesis of a putative DNA binding domain of SV40 large-T, *Virology* **139**:109.

Kalderon, D., Richardson, W. D., Markham, A. F., and Smith, A. E., 1984a, Sequence requirements for nuclear localization of SV40 large T antigen, *Nature* **311**:33.

Kalderon, D., Roberts, B., Richardson, W. D., and Smith, A. E., 1984b, A short amino acid sequence able to specify nuclear localization, *Cell* **39**:499.

Kasamatsu, H., and Wu, M., 1976a, Protein–SV40 DNA complex stable in high salt and sodium dodecyl sulfate, *Biochem. Biophys. Res. Commun.* **68**:927.

Kasamatsu, H., and Wu, M., 1976b, Structure of a nickel DNA–protein complex isolated from SV40: Covalent attachment of the protein to DNA and nick specificity, *Proc. Natl. Acad. Sci. USA* **73**:1945.

Katinka, M., and Yaniv, M., 1983, DNA replication origin of polyoma virus: Early proximal boundary, *J. Virol.* **47**:244.

Katinka, M., Yaniv, M., Vasseur, M., and Blangy, D., 1980, Expression of polyoma early functions in mouse embryonal carcinoma cells depends on sequence rearrangements in the beginning of the late region, *Cell* **20**:393.

Katinka, M., Vasseur, M., Montreau, N., Yaniv, M., and Blangy, D., 1981, Polyoma DNA sequences involved in control of viral gene expression in murine embryonal carcinoma cells, *Nature* **290**:720.

Kato, A. C., Bartok, K., Fraser, M. J., and Denhardt, D. T., 1973, Sensitivity of superhelical DNA to a single-strand specific endonuclease, *Biochim. Biophys. Acta* **308**:68.

Kaufmann, G., 1981, Characterization of initiator RNA from replicating SV40 DNA synthesized in isolated nuclei, *J. Mol. Biol.* **147**:25.

Kaufmann, G., and Falk, H. H., 1982, An oligoribonucleotide polymerase from SV40-infected cells with properties of a primase, *Nucleic Acids. Res.* **10**:2309.

Kaufmann, G., Anderson, S., and DePamphilis, M. L., 1977, RNA primers in simian virus 40 DNA replication: Distribution of 5′-terminal oligoribonucleotides in nascent DNA, *J. Mol. Biol.* **116**:549.

Kaufmann, G., Bar-Shavit, R., and DePamphilis, M. L., 1978, Okazaki pieces grow opposite to the replication fork direction during SV40 DNA replication, *Nucleic Acids Res.* **5**:2535.

Keller, J. M., and Alwine, J. C., 1984, Activation of the SV40 late promoter: Direct effects of T-antigen in the absence of viral DNA replication, *Cell* **36**:381.

Keller, W., Muller, V., Eicken, I., Wendel, I., and Zentgraf, H., 1977, Biochemical and ultrastructural analysis of SV40 chromatin, *Cold Spring Harbor Symp. Quant. Biol.* **42**:227.

Kelly, F., and Condamine, H., 1982, Tumor viruses and early mouse embryos, *Biochim. Biophys. Acta* **651**:105.

Kelly, T. J., Jr., 1982, Organization and replication of adenovirus DNA, in: *Organization and Replication of Viral DNA* (A. S. Kaplan, ed.), pp. 116–146, CRC Press, Boca Raton, Fla.

Kelly, T. J., Jr., and Lewis, A. M., 1973, Use of nondefective adenovirus–simian virus 40 hybrids for mapping the simian virus 40 genome, *J. Virol.* **12**:643.

Khandjian, E. W., and Türler, H., 1983, Simian virus 40 and polyoma virus induce synthesis of heat shock proteins in permissive cells, *Mol. Cell. Biol.* **3**:1.

Khandjian, E. W., Matter, J.-M., Leonard, N., and Weil, R., 1980, Simian virus 40 and polyoma virus stimulate overall cellular RNA and protein synthesis, *Proc. Natl. Acad. Sci. USA* **77**:1476.

Khandjian, E. W., Loche, M., Darlix, J.-L., Cramer, R., Türler, H., and Weil, R., 1982, SV40 large tumor antigen: An "RNA binding protein?," *Proc. Natl. Acad. Sci. USA* **79**:1139.

Khoury, G., Fareed, G. C., Berry, K., and Martin, M. A., 1974, Characterization of a rearrangement in viral DNA: Mapping of the circular SV40-like DNA containing a triplication of a specific one-third of the viral genome, *J. Mol. Biol.* **87**:289.

Kitani, T., Yoda, K.-Y., and Okazaki, T., 1984, Discontinuous DNA replication of *Drosophila melanogaster* is primed by octaribonucleotide primer, *Mol. Cell. Biol.* **4**:1591.

Kitani, T., Ogawa, T., Yoda, K.-Y., and Okazaki, T., 1985, Evidence that discontinuous DNA replication in *E. coli* cells is primed by approximately 10–12 residues of RNA starting with purine, *J. Mol. Biol.* **184:**45–52.

Klempnauer, K.-H., Fanning, E., Otto, B., and Knippers, R., 1980, Maturation of newly replicated chromatin of SV40 and its host cell, *J. Mol. Biol.* **136:**359.

Klockmann, U., and Deppert, W., 1983, Acylated simian virus 40 large T-antigen: A new subclass associated with a detergent-resistant lamina of the plasma membrane, *EMBO J.* **2:**1151.

Klockmann, U., Staufenbiel, M., and Deppert, W., 1984, Membrane interactions of simian virus 40 large T-antigen: Influence of protein sequences and fatty acid acylation, *Mol. Cell. Biol.* **4:**1542.

Kondoleon, S. K., Robinson, G. W., and Hallick, L. M., 1983, SV40 virus particles lack a psoralen-accessible origin and contain an altered nucleoprotein structure, *Virology* **129:**261.

Kornberg, A., 1980, *DNA Replication*, Freeman, San Francisco.

Kowalski, J., and Denhardt, D. T., 1978, Most short DNA molecules isolated from 3T3 cells are not nascent, *Nucleic Acids Res.* **5:**4355.

Krauss, M. R., and Benbow, R. M., 1981, Polyoma virus minichromosomes: Associated DNA molecules, *J. Virol.* **38:**815.

Krauss, M. R., Gourlie, B. B., Bayne, M. L., and Benbow, R. M., 1984, Polyomavirus minichromosomes: Associated DNA topoisomerase II and DNA ligase activities, *J. Virol.* **49:**333–342.

Kress, M., May, E., Cassinger, R., and May, P., 1979, SV40 transformed cells express new species of proteins precipitable by SV40 anti-tumor serum, *J. Virol.* **31:**472–483.

Kriegler, M., and Botchan, M., 1983, Enhanced transformation by a simian virus 40 recombinant virus containing a Harvey murine sarcoma virus long terminal repeat, *Mol. Cell. Biol.* **3:**325–339.

Kriegler, M. P., Griffin, J. D., and Livingston, D. M., 1978, Phenotypic complementation of the SV40 tsA mutant defect in viral DNA synthesis following microinjection of SV40 T antigen, *Cell* **14:**983–994.

Krokan, H., 1981a, Preferential association of uracil-DNA glycosylase activity with replicating SV40 minichromosomes, *FEBS Lett.* **133:**89.

Krokan, H., 1981b, Role of mammalian DNA polymerases in replication and repair, in: *Chromosome Damage and Repair* (E. Seeberg and K. Kleppe, eds.), pp. 395–406, Plenum Press, New York.

Krokan, H., Bjorklid, E., and Prydz, H., 1975, DNA synthesis in isolated HeLa cell nuclei: Optimalization of the system and characterization of the product, *Biochemistry* **14:**4227.

Krokan, H., Schaffer, P., and DePamphilis, M. L., 1979, The involvement of eukaryotic DNA polymerases α and γ in the replication of cellular and viral DNA, *Biochemistry* **18:**4431–4443.

Kronkan, H., Wist, E., and Prydz, H., 1977, Effect of cytosol on DNA synthesis in isolated HeLa cell nuclei, *Biochem Biophys. Res. Commun.* **75:**414.

Kuempel, P. L., Duerr, S. A., and Seeley, N. R., 1977, Terminus region of the chromosome in *E. coli* inhibits replication forks, *Proc. Natl. Acad. Sci. USA* **74:**3927.

Lai, C.-J., and Nathans, D., 1974, Mapping temperature sensitive mutants of simian virus 40: Rescue of mutants by fragments of DNA, *Virology* **60:**466–474.

Lai, C.-J., and Nathans, D., 1975, Non-specific termination of SV40 DNA replication, *J. Mol. Biol.* **97:**113.

Lai, C.-J., Goldman, N. D., and Khoury, G., 1979, Functional similarity between the early antigens of simian virus 40 and human papovavirus BK, *J. Virol.* **30:**141–148.

Laimins, L. A., Khoury, G., Gorman, C., Howard, B., and Gruss, P., 1982, Host-specific activation of transcription by tandem repeats from SV40 and Moloney murine sarcoma virus, *Proc. Natl. Acad. Sci. USA* **79:**6453–6457.

Laipis, P. J., and Levine, A. J., 1973, DNA replication in SV40 infected cells. IX. The inhibition of a gap-filling step during discontinuous synthesis of SV40 DNA, *Virology* **56:**580.

Laipis, P., Sen, A. J., Levine, A., and Mulder, C., 1975, DNA replication in SV40 infected cells. X. The structure of the 16S gap circle intermediate in SV40 DNA synthesis, *Virology* **68:**115.

Lamothe, P., Baril, B., Chi, A., Lee, L., and Baril, E., 1981, Accessory proteins for DNA polymerase δ activity with single-strand DNA templates, *Proc. Natl. Acad. Sci. USA* **78:**4723.

Lane, D. P., and Crawford, L. V., 1979, T-antigen is bound to a host protein in SV40-transformed cells, *Nature* **278:**261–263.

Lane, D. P., and Hoeffler, W. K., 1980, SV40 large T shares an antigenic determinant with a cellular protein of molecular weight 68,000, *Nature* **288:**167–170.

Lanford, R. E., and Butel, J. S., 1980a, Inhibition of nuclear migration of wild-type SV40 tumor antigen by a transport-defective mutant of SV40–adenovirus 7 hybrid virus, *Virology* **105:**303–313.

Lanford, R. E., and Butel, J. S., 1980b, Biochemical characterization of nuclear and cytoplasmic forms of SV40 tumor antigens encoded by parental and transport-defective mutant SV40–adenovirus 7 hybrid viruses, *Virology* **105:**314–327.

Lanford, R. E., and Butel, J. S., 1984, Construction and characterization of an SV40 mutant defective in nuclear transport of T antigen, *Cell* **37:**801–813.

Lange-Mütschler, J., and Henning, R., 1983, A subclass of simian virus 40 T antigen with a high cell surface binding affinity, *Virology* **127:**333–344.

Lange-Mütschler, J., and Henning, R., 1984, Cell surface binding simian virus 40 large T antigen becomes anchored and stably linked to lipid of the target cells, *Virology* **136:**404–413.

Lania, L., Gundini-Attardi, D., Griffiths, N., Cooke, B., DeCicco, D., and Fried, M., 1980, The polyoma virus 100K large T-antigen is not required for the maintenance of transformation, *Virology* **101:**217–232.

Laskey, R. A., Harland, R. M., and Mechali, M., 1983, *Molecular Biology of Egg Maturation, Ciba Found. Symp.* **98:**25–43.

Learned, R. M., Myers, R. M., and Tjian, R., 1981, Replication in monkey cells of plasmid DNA containing the minimal SV40 origin, in: *The initiation of DNA Replication* (D. Ray, ed.), p. 555–566, Academic Press, New York.

Learned, R. M., Smale, S. T., Haltiner, M. M., and Tjian, R., 1983, Regulation of human ribosomal RNA transcription, *Proc. Natl. Acad. Sci. USA* **80:**3558–3562.

LeBlanc, D. J., and Singer, M. F., 1976, SV40 DNA replication in nuclear monolayers, *J. Virol.* **20:**78–85.

Lebowitz, J., Garon, C. C., Chen, C. Y., and Salzman, N. P., 1976, Chemical modifications of SV40 DNA by reaction with a water-soluble carbodiimide, *J. Virol.* **18:**205.

Lee, T. N. H., and Nathans, D., 1979, Evolutionary variants of SV40: Replication and encapsidation of variant DNA, *Virology* **92:**291.

Lee, T. N. H., Brockman, W. W., and Nathans, D., 1975, Evolutionary variants of SV40: Cloned substituted variants containing multiple initiation sites for DNA replication, *Virology* **66:**53.

Leffak, I. M., 1983, Stability of the conservative mode of nucleosome assembly, *Nucleic Acids Res.* **11:**2717–2732.

Leffak, I. M., Grainger, R., and Weintraub, H., 1977, Conservative assembly and segregation of nucleosomal histones, *Cell* **12:**837–845.

Leppard, K. N., and Crawford, L. V., 1984, An oligomeric form of simian virus 40 large T-antigen is immunologically related to the cellular tumor antigen p53, *J. Virol.* **50:**457–464.

Lescure, B., Chestier, A., and Yaniv, M., 1978, Transcription of polyoma virus DNA *in vitro.* II. Transcription of superhelical and linear polyoma DNA by RNA polymerase II, *J. Mol. Biol.* **124:**73.

Levine, A. J., 1982, The nature of the host range restriction of SV40 and polyoma viruses in embryonal carcinoma cells, *Curr. Top. Microbiol. Immunol.* **101:**1–27.

Levine, A. J., Kang, H. S., and Billheimer, F. E., 1970, DNA replication in SV40 infected cells. I. Analysis of replicating SV40 DNA, *J. Mol. Biol.* **50:**549.

Levy, A., and Jakob, K. M., 1978, Nascent DNA in nucleosome-like structures from chromatin, *Cell* **14**:259–267.

Lewis, A. M., Jr., Levine, A. S., Crumpacker, C. S., Levin, M. J., Samaha, R. J., and Henry, P. H., 1973, Studies of nondefective adenovirus 2–simian virus 40 hybrid viruses. V. Isolation of additional hybrids which differ in their simian virus 40-specific biological properties, *J. Virol.* **11**:655.

Lewton, B. A., DeLucia, A. L., and Tegtmeyer, P., 1984, Binding of simian virus 40 A protein to DNA with deletions at the origin of replication, *J. Virol.* **49**:9–13.

Li, J. J., and Kelly, T. J., 1984, SV40 DNA replication *in vitro*, *Proc. Natl. Acad. Sci. USA* **81**:6973.

Li, J. J., and Kelly, T. J., 1985, SV40 DNA replication *in vitro:* Specificity of initiation and evidence for bidirectional replication, *Mol. Cell. Biol.* **5**:1238.

Liggens, G., English, M., and Goldstein, D., 1979, Structural changes in SV40 chromatin as probed by restriction endonucleases, *J. Virol.* **31**:718.

Lilley, D. M. J., 1980, The inverted repeat as a recognizable structural feature in supercoiled DNA molecules, *Proc. Natl. Acad. Sci. USA* **77**:6468.

Lilley, D. M. J., and Kemper, B., 1984, Cruciform–resolvase interactions in supercoiled DNA, *Cell* **36**:413.

Linney, E., and Donerly, S., 1983, DNA fragments from F9 PyEC mutants increase expression of heterologous genes in transfected F9 cells, *Cell* **35**:693.

Linney, E., Davis, B., Overhauser, J., Chao, E., and Fan, H., 1984, Non-function of a Moloney murine leukaemia virus regulatory sequence in F9 embryonal carcinoma cells, *Nature* **308**:470.

Linzer, D. I. H., and Levine, A. J., 1979, Characterization of a 54K dalton cellular SV40 tumor antigen present in SV40-transformed cells and uninfected embryonal carcinoma cells, *Cell* **17**:43–52.

Liu, L. F., Liu, C.-C., and Alberts, B. M., 1980, Type II DNA topoisomerases: Enzymes that can unknot a topologically knotted DNA molecule via a reversible double-strand break, *Cell* **19**:697.

Liu, L. F., Halligan, B. D., Nelson, E. M., Rowe, T. C., Chen, G. L., and Tewey, K. M., 1983, Breakage and reunion of DNA helix by mammalian DNA topoisomerase II, in: *Mechanisms of DNA Replication and Recombination* (N. R. Cozzarelli, ed.), pp. 43–54, Liss, New York.

Loche, M. P., 1979, Studies on polyoma virus DNA replication in synchronized C3H2K cells, *J. Gen. Virol.* **42**:429.

Lonn, U., and Lonn, S., 1983, Aphidicolin inhibits the synthesis and joining of short DNA fragments but not the union of 10-Kb DNA replication intermediates, *Proc. Natl. Acad. Sci. USA* **80**:3996.

Louarn, J., Patte, J., and Louarn, J. M., 1979, Map position of the replication terminus on the *E. coli* chromosome, *Mol. Gen. Genet.* **172**:7.

Lovett, M., Clayton, C. E., Murphy, D., Rigby, P. W. J., Smith, A. E., and Chaudry, F., 1982, Structure and synthesis of a simian virus 40 super T-antigen, *J. Virol.* **44**:963–973.

Low, R. L., Kaguni, J. M., and Kornberg, A., 1984, Potent catenation of supercoiled and gapped DNA circles by topisomerase I in the presence of a hydrophilic polymer, *J. Biol. Chem.* **259**:4576.

Luthman, H., Nilsson, M.-G., and Magnusson, G., 1982, Non-contiguous segments of the polyoma genome required in cis for DNA replication, *J. Mol. Biol.* **161**:533.

Luthman, H., Osterlund, M., and Magnusson, G., 1984, Inhibition of polyoma DNA synthesis by base pair substitutions at the replication origin, *Nucleic Acids Res.* **12**:7503.

Machida, Y., Okazaki, T., Miyake, T., Ohtsuka, E., and Ikehara, M., 1981, Characterization of nascent DNA fragments produced by excision of uracil residues in DNA, *Nucleic Acids Res.* **9**:4755.

Magnusson, G., 1973, Hydroxyurea-induced accumulation of short fragments during polyoma DNA replication, I. Characterization of fragments, *J. Virol.* **12**:600.

Magnusson, G., and Nilsson, M.-G., 1979, Replication of polyoma DNA in isolated nuclei: Analysis of replication fork movement, *J. Virol.* **32**:386.

Magnusson, G., and Nilsson, M.-G., 1982, Viable polyoma virus variant with two origins of DNA replication, *Virology* **119**:12.

Magnusson, G., Winnacker, E. L., Eliasson, R., and Reichard, P., 1972, Replication of polyoma DNA in isolated nuclei. II. Evidence for semiconservative replication, *J. Mol. Biol.* **72**:539.

Magnusson, G., Pigiet, V., Winnacker, E. L., Abrams, R., and Reichard, P., 1973, RNA-linked short DNA fragments during polyoma replication, *Proc. Natl. Acad. Sci. USA* **70**:412.

Magnusson, G., Nilsson, M.-G., Dilworth, S. M., and Smolar, N., 1981, Characterization of polyoma mutants with altered middle and large T-antigens, *J. Virol.* **39**:673–683.

Major, E. O., and Matsumura, P., 1984, Human embryonic kidney cells: Stable transformation with an origin-defective SV40 DNA and use as hosts for human papovavirus replication, *Mol. Cell. Biol.* **4**:379–382.

Mann, K., and Hunter, T., 1979, Association of simian virus 40 T antigen with simian virus 40 nucleoprotein complexes, *J. Virol.* **29**:232–241.

Mann, K., and Hunter, T., 1980, Phosphorylation of SV40 large T antigen in SV40 nucleoprotein complexes, *Virology* **107**:526–532.

Manos, M. M., and Gluzman, Y., 1984, Simian virus 40 large T-antigen point mutants that are defective in viral DNA replication but competent in oncogenic transformation. *Mol. Cell. Biol.* **4**:1125–1133.

Manos, M. M., and Gluzman, Y., 1985, Genetic and biochemical analysis of transformation-competent, replication-defective simian virus 40 large T antigen mutants, *J. Virol.* **53**:120–127.

Manser, T., Thacher, T., and Rechsteiner, M., 1980, Arginine-rich histones do not exchange between human and mouse chromosomes in hybrid cells, *Cell* **19**:993–1003.

Manteuil, S., and Girard, M., 1974, Inhibitors of DNA synthesis: Their influence on replication and transcription of SV40 DNA, *Virology* **60**:438.

Manteuil, S., Kopecka, H., Caraux, J., Prunell, A., and Girard, M., 1974, In vivo incorporation of cytosine arabinoside into SV40 DNA, *J. Mol. Biol.* **90**:751.

Margolskee, R. F., and Nathans, D., 1984, Simian virus 40 mutant T antigens with relaxed specificity for the nucleotide sequence at the viral DNA origin of replication, *J. Virol.* **49**:386–393.

Mariani, B. D., and Schimke, R. T., 1984, Gene amplification in a single cell cycle in CHO cells, *J. Biol. Chem.* **259**:1901–1910.

Martin, G. R., 1980, Teratocarcinomas and mammalian embryogenesis, *Science* **209**:768–776.

Martin, M. A., Khoury, G., and Fareed, G. C., 1974, Specific reiteration of viral DNA sequences in mammalian cells, *Cold Spring Harbor Symp. Quant. Biol.* **39**:129.

Martin, M. A., Howley, P. M., Pyrne, J. C., and Garon, C. F., 1976, Characterization of supercoiled SV40 molecules in productively infected cells, *Virology* **71**:28.

Martin, R. F., 1977, Analysis of polyoma virus DNA replicative intermediates by agarose gel electrophoresis, *J. Virol.* **23**:827.

Martin, R., 1981, The transformation of cell growth and transmogrification of DNA synthesis by SV40, *Adv. Cancer Res.* **34**:1–68.

Martin, R. G., and Setlow, V. P., 1980, Initiation of SV40 DNA synthesis is not unique to the replication origin, *Cell* **20**:381.

Mason, D. H., Jr., and Takemoto, K. K., 1976, Complementation between BK human papovavirus and a simian virus 40 tsA mutant, *J. Virol.* **17**:1060–1068.

May, E., Lasne, C., Prives, C., Borde, J., and May, P., 1983, Study of the functional activities concomitantly retained by the 115,000 Mr super T antigen, an evolutionary variant of simian virus 40 large T antigen expressed in transformed rat cells, *J. Virol.* **45**:901–913.

Mayer, A., and Levine, A. J., 1972, DNA replication in SV40 infected cells. VIII. Distribution of replicating molecules at different stages of replication in SV40 infected cells, *Virology* **50**:328.

McCormick, F., and Harlow, E., 1980, Association of a murine 53,000 dalton phosphoprotein with SV40 large T antigen in transformed cells, *J. Virol.* **34**:213–224.

McCormick, F., Clark, R., Harlow, E., and Tjian, R., 1981, SV40 T-antigen binds specifically to a cellular 53K protein in vitro, *Nature* **292**:63.

McKay, R. D. G., 1981, Binding of a simian virus 40 T antigen-related protein to DNA, *J. Mol. Biol.* **145**:471–488.

McKay, R. D. G., 1983, Immunoassay for sequence-specific DNA–protein interaction, *Methods Enzymol.* **92**:138–146.

McKay, R. D. G., and DiMaio, D., 1981, Binding of an SV40 T antigen-related protein to the DNA of SV40 regulatory mutants, *Nature* **289**:810–813.

McKnight, S. L., and Miller, O. L., 1977, Electron microscope analysis of chromatin replication in the cellular blastoderm *Drosophila melanogaster* embryo, *Cell* **12**:795.

McKnight, S. L., Bustin, M., and Miller, O. L., 1978, Electron microscope analysis of chromosome metabolism in the *Drosophila melanogaster* embryo, *Cold Spring Harbor Symp. Quant. Biol.* **42**:741.

McTiernan, C. F., and Stambrook, P. J., 1984, Initiation of SV40 DNA replication after microinjection into *Xenopus* eggs, *Biochim. Biophys. Acta* **782**:295–303.

Mechali, M., and Harland, R. M., 1982, DNA synthesis in a cell-free system from *Xenopus* eggs: Priming and elongation on single-stranded DNA in vitro, *Cell* **30**:93.

Mechali, M., and Kearsey, S., 1984, Lack of specific sequence requirements for DNA replication in *Xenopus* eggs compared with high sequence specificity in yeast, *Cell* **38**:55–64.

Mechali, M., Girard, M., and de Recondo, A. M., 1977, DNA polymerase activities in growing cells infected with SV40, *J. Virol.* **23**:117.

Meinke, W., and Goldstein, D. A., 1971, Studies on the structure and formation of polyoma DNA replicative intermediates, *J. Mol. Biol.* **61**:543.

Melero, J. A., Stitt, D. T., Mangel, W. F., and Carroll, R. B., 1979, Identification of new polypeptide species (48–55k) immunoprecipitable by antiserum to purified large T antigen and present in SV40-infected and transformed cells, *Virology* **93**:466–480.

Melin, F., Pinon, H., Kress, C., and Blangy, D., 1985a, Isolation of polyomavirus mutants multiadapted to murine embryonal carcinoma cells, *J. Virol.* **53**:862.

Melin, F., Pinon, H., Reiss, C., Kress, C., Montreau, N., and Blangy, D., 1985b, Common features of polyomavirus mutants selected on PCC4 embryonal carcinoma cells, *EMBO J.* **4**:1799.

Mengeritsky, G., and Trifonov, E. N., 1984, Nucleotide sequence-directed mapping of the nucleosomes of SV40 chromatin, *Cell Biophys.* **6**:1–8.

Mercer, W. E., Nelson, D., DeLeo, A. B., Old, L. J., and Baserga, R., 1982, Microinjection of monoclonal antibody to protein p53 inhibits serum-induced DNA synthesis in 3T3 cells, *Proc. Natl. Acad. Sci. USA* **79**:6309–6312.

Mercer, W. E., Nelson, D., Hyland, J. K., Croce, C. M., and Baserga, R., 1983, Inhibition of SV40-induced cellular DNA synthesis by microinjection of monoclonal antibodies, *Virology* **127**:149–158.

Mercola, M., Wang, X.-F., Olsen, J., and Calammi, K., 1983, Transcriptional enhancer elements in the mouse immunoglobulin heavy chain locus, *Science* **221**:663–665.

Michel, M. R., and Schwyzer, M., 1982, Messenger ribonucleoproteins of cells infected by simian virus 40 contain large T antigen, *Eur. J. Biochem.* **129**:25–32.

Milavetz, B. I., Spotila, L. D., Thomas, R., and Huberman, J. A., 1980, Two-dimension analysis of proteins sedimenting with SV40 chromosomes, *J. Virol.* **35**:854.

Miller, J., Bullock, P., and Botchan, M., 1984, SV40 T-antigen is required for viral excision from chromosomes, *Proc. Natl. Acad. Sci. USA* **81**:7534–7538.

Miller, K. G., Liu, L. F., and Englund, P. T., 1981, A homogeneous type II DNA topoisomerase from HeLa cell nuclei, *J. Biol. Chem.* **256**:9334.

Milthorp, P., Baumann, E., and Hand, R., 1984, The distribution of large T antigen in simian virus 40 nucleoprotein complexes, *Can. J. Microbiol.* **30**:622–631.

Miyamura, T., Jikuya, H., Soeda, E., and Yoshiike, K., 1983, Genomic structure of human polyoma virus JC: Nucleotide sequence of the region containing replication origin and small-T-antigen gene, *J. Virol.* **45**:73.

Mizuuchi, K., Mizuuchi, M., and Gellert, M., 1982, Cruciform structures in palindromic DNA are favored by DNA supercoiling, *J. Mol. Biol.* **156**:229.

Montenarh, M., and Henning, R., 1983, Self-assembly of simian virus 40 large T antigen oligomers by divalent cations, *J. Virol.* **45**:531–538.

Montenarh, M., Kohler, M., and Henning, R., 1984, Oligomerization of simian virus 40 large T antigen is not necessarily repressed by temperature-sensitive A gene lesions, *J. Virol.* **49**:658–664.

Moore, G. R., and Williams, R. J. P., 1984, Protein antigenicity and protein mobility, *Nature* **312**:706.

Morrison, B., Kress, M., Khoury, G., and Jay, G., 1983, Simian virus 40 tumor antigen: Isolation of the origin-specific DNA-binding domain, *J. Virol.* **47**:106–114.

Morrow, J. F., and Berg, P., 1974, Location of the T4 gene 32 protein binding site on SV40 DNA, *J. Virol.* **12**:1631.

Moyne, G., Freeman, K., Saragosti, S., and Yaniv, M., 1981, A high resolution electron microscope study of nucleosomes from SV40 chromatin, *J. Mol. Biol.* **149**:735.

Mueller, C., Graessmann, A., and Graessmann, M., 1978, Mapping of early SV40-specific functions by microinjection of different early viral DNA fragments, *Cell* **15**:579–585.

Mueller, C. R., Mes-Masson, A.-M., Bouvier, M., and Hassell, J. A., 1984, Location of sequences in polyomavirus DNA that are required for early gene expression in vivo and in vitro, *Mol. Cell. Biol.* **4**:2594.

Muller, U., Zentgraf, H., Eicken, I., and Keller, W., 1978, Higher order structure of SV40 chromatin, *Science* **201**:406–415.

Muller, W. J., Mueller, C. R., Mes, A.-M., and Hassell, J. A., 1983a, Polyomavirus origin for DNA replication comprises multiple genetic elements, *J. Virol.* **47**:586–599.

Muller, W. J., Naujokas, M. A., and Hassell, J. A., 1983b, Polyomavirus–plasmid recombinants capable of replicating have an enhanced transforming potential, *Mol. Cell. Biol.* **3**:1670–1674.

Myers, R. M., and Tjian, R., 1980, Construction and analysis of simian virus 40 origins defective in tumor antigen binding and DNA replication, *Proc. Natl. Acad. Sci. USA* **77**:6491–6495.

Myers, R. M., Rio, D. C., Robbins, A. K., and Tjian, R., 1981a, SV40 gene expression is modulated by the cooperative binding of T antigen to DNA, *Cell* **25**:376–384.

Myers, R. M., Kligman, M., and Tjian, R., 1981b, Does simian virus 40 T antigen unwind DNA? *J. Biol. Chem.* **256**:10156–10160.

Myers, R. M., Williams, R. C., and Tjian, R., 1981c, Oligomeric structure of a simian virus 40 T antigen in free form and bound to DNA, *J. Mol. Biol.* **148**:347–353.

Narkhammer, M., and Magnusson, G., 1976, DNA polymerase activities induced by polyoma virus infection of 3T3 mouse fibroblasts, *J. Virol.* **18**:1.

Narkhammer-Meuth, M., Eliasson, R., and Magnusson, G., 1981a, Discontinuous synthesis of both strands at the growing fork during polyoma DNA replication *in vitro*, *J. Virol.* **39**:11.

Narkhammer-Meuth, M., Kowalski, J., and Denhardt, D. T., 1981b, Both strands of polyoma DNA are replicated discontinuously with RNA primers *in vivo*, *J. Virol.* **39**:21.

Nevins, J. R., 1982, Induction of the synthesis of a 70,000 dalton mammalian heat shock protein by the adenovirus E1A gene product, *Cell* **29**:913–919.

Newport, J., and Kirschner, M., 1982, A major developmental transition in early *Xenopus* embryos, *Cell* **30**:675–696.

Nicander, B., and Reichard, P., 1983, Dynamics of pyrimidine deoxynucleoside triphosphate pools in relationship to DNA synthesis in 3T6 mouse fibroblasts, *Proc. Natl. Acad. Sci. USA* **80**:1347.

Nilsson, S. V., and Magnusson, G., 1984, Activities of polyomavirus large T antigen proteins expressed by mutant genes, *J. Virol.* **51**:768–775.

Nilsson, S., Reichard, P., and Skoog, L., 1980, Deoxyuridine triphosphate pools after polyoma virus infection, *J. Biol. Chem.* **255**:9552.

Noonan, C. A., Bruggs, J. S., and Butel, J. S., 1976, Characterization of simian cells transformed by temperature-sensitive mutants of simian virus 40, *J. Virol.* **18**:1106–1119.

Oda, T., Watanabe, S., Hanakawa, S., and Nakamura, T., 1980, Complete in vitro DNA replication of SV40 chromatin in digitonin-treated permeable cells, *Acta Med. Okayama* **34**:409.

Ogawa, T., and Okazaki, T., 1980, Discontinuous DNA replication, *Annu. Rev. Biochem.* **49**:421.

Ogawa, T., and Okazaki, T., 1984, Function of RNase H in DNA replication revealed by RNase H defective mutants of *E. coli*, *Mol Gen. Genet.* **193**:231.

Okuda, A., Shimura, H., and Kimura, G., 1984, Abortive transformation of rat 3YI cells by simian virus 40: Viral functions overcoming inhibition of cellular proliferation under various conditions of culture, *Virology* **133**:35–45.

Oren, M., Winocour, E., and Prives, C., 1980, Differential affinities of SV40 large tumor antigen for DNA, *Proc. Natl. Acad. Sci. USA* **77**:220–224.

Oren, M., Maltzman, W., and Levine, A. J., 1981, Post-translational regulation of the 54K cellular tumor antigen in normal and transformed cells, *Mol. Cell. Biol.* **1**:101.

Osheim, Y. N., and Miller, O. L., 1983, Novel amplification and transcriptional activity of chorion genes in *Drosophila melanogaster* follicle cells, *Cell* **33**:543.

Otto, B., and Fanning, E., 1978, DNA polymerase alpha is associated with replicating SV40 nucleoprotein complexes, *Nucleic Acids Res.* **5**:1715.

Otto, B., and Reichard, P., 1975, Replication of polyoma DNA in isolated nuclei. V. Complementation on *in vitro* DNA replication, *J. Virol.* **15**:259.

Otto, B., Fanning, E., and Richter, A., 1979, DNA polymerases and a single-strand-specific DNA binding protein associated with SV40 nucleoprotein complexes, *Cold Spring Harbor Symp. Quant. Biol.* **43**:705.

Ozer, H. L., Slater, M. L., Dermody, J. J., and Mandel, M., 1981, Replication of SV40 DNA in normal human fibroblasts and in fibroblasts from *Xeroderma pigmentosum*, *J. Virol.* **39**:481–489.

Pabo, C. O., and Sauer, R. T., 1984, Protein–DNA recognition, *Annu. Rev. Biochem.* **53**:293–321.

Pages, J., Manteuil, S., Stehelin, D., Fiszman, M., Marx, M., and Girard, M., 1973, Relationship between replication of SV40 DNA and specific events of the host cell cycle, *J. Virol.* **12**:99.

Pan, S., and Knowles, B. B., 1983, Monoclonal antibody to SV40 T-antigen blocks lysis of cloned cytotoxic T-cell line specific for SV40 TSTA, *Virology* **125**:1–7.

Panayotatos, N., and Wells, R. D., 1981, Cruciform structures in supercoiled DNA, *Nature* **289**:466.

Patton, J., and Chae, C., 1982, A method for isolation of a large amount of a single-stranded DNA fragment, *Anal. Biochem.* **126**:231.

Paucha, E., Mellor, A., Harvey, R., Smith, A. E., Hewick, R. M., and Waterfield, M. D., 1978, Large and small tumor antigens from SV40 have identical amino termini mapping at 0.65 map units, *Proc. Natl. Acad. Sci. USA* **75**:2165–2169.

Pellegrini, S., Dailey, L., and Basilico, C., 1984, Amplification and excision of integrated polyoma DNA sequences require a functional origin of replication, *Cell* **36**:943.

Perlman, D., and Huberman, J. A., 1977, Asymmetric Okazaki piece synthesis during replication of SV40 DNA *in vitro*, *Cell* **12**:1029.

Pigiet, V., Winnacker, E. L., Eliasson, R., and Reichard, P., 1973, Discontinuous elongation of both strands at the replication forks in polyoma DNA replication, *Nature New Biol.* **245**:203.

Pigiet, V., Eliasson, R., and Reichard, P., 1974, Replication of polyoma DNA in isolated nuclei. III. The nucleotide sequence at the RNA–DNA junction of nascent strands, *J. Mol. Biol.* **84**:197.

Pintel, D., Bouck, N., and di Majorca, G., 1981, Separation of lytic and transforming functions of the simian virus 40 A region: Two mutants which are temperature sensitive for lytic functions have opposite effects on transformation, *J. Virol.* **38**:518–528.

Pipas, J. M., 1985, Mutations near the carboxyl terminus of the simian virus 40 large tumor antigen alter viral host range, *J. Virol* **54**:569–575.

Pipas, J. M., Peden, K. W., and Nathans, D., 1983, Mutational analysis of simian virus 40 T antigen: Isolation and characterization of mutants with deletions in the T-antigen gene, *Mol. Cell. Biol.* **3**:203–213.

Plevani, P., Badaracco, G., Augl, C., and Chang, L. M. S., 1984, DNA polymerase I and DNA primase complex in yeast, *J. Biol. Chem.* **259**:7532.

Polvino-Bodnar, M., and Cole, C. N., 1982, Construction and characterization of viable deletion mutants of simian virus 40 lacking sequences near the 3′ end of the early region, *J. Virol.* **43**:489–502.

Pomerantz, B. J., and Hassell, J. A., 1984, Polyomavirus and simian virus 40 large T antigens bind to common DNA sequences, *J. Virol.* **49**:925.

Pomerantz, B. J., Mueller, C. R., and Hassell, J. A., 1983, Polyomavirus large T antigen binds independently to multiple, unique regions on the viral genome, *J. Virol.* **47**:600–610.

Ponder, B. A. J., and Crawford, L. V., 1977, The arrangement of nucleosomes in nucleoprotein complexes from polyoma virus and SV40, *Cell* **11**:35.

Pospelov, V., Russev, G., Vassilev, L, and Tsanev, R., 1982, Nucleosome segregation in chromatin replicated in the presence of cycloheximide, *J. Mol. Biol.* **156**:79.

Postel, E. H., and Levine, A. J., 1976, The requirement of simian virus 40 gene A product for the stimulation of cellular thymidine kinase activity after viral infection, *Virology* **73**:206–216.

Prieto-Soto, A., Gourlie, B., Miwa, M., Pigiet, V., Sugimura, T., Malik, N., and Smulson, M., 1983, Polyoma virus minichromosomes: Poly ADP-ribosylation of associated chromatin proteins, *J. Virol.* **45**:600–606.

Prior, C. P., Cantor, C. R., Johnson, E. M., and Allfrey, V. G., 1980, Incorporation of exogenous pyrene-labeled histone into *Physarum* chromatin: A system for studying changes in nucleosomes assembled in vivo, *Cell* **20**:597–608.

Pritchard, C. G., and DePamphilis, M. L. 1983, Preparation of DNA polymerase α[C1C2] by reconstituting α-polymerase with its specific stimulatory cofactors, C1C2, *J. Biol. Chem.* **258**:9801–9809.

Pritchard, C. G., Weaver, D. T., Baril, E. F., and DePamphilis, M. L., 1983, DNA polymerase α cofactors C1C2 function as primer recognition proteins, *J. Biol Chem.* **258**:9810–9819.

Prives, C., Gilboa, E., Revel, M., and Winocour, E., 1977, Cell-free translation of simian virus 40 early messenger RNA coding for viral T-antigen, *Proc. Natl. Acad. Sci. USA* **74**:457–461.

Prives, C., Gluzman, Y., and Winocour, E., 1978, Cellular and cell-free synthesis of SV40 T antigen in permissive and transformed cells, *J. Virol.* **25**:587.

Prives, C., Beck, Y., and Shure, H., 1980, DNA binding properties of simian virus 40 T-antigens synthesized *in vivo* and *in vitro*, *J. Virol.* **33**:689–696.

Prives, C., Barnet, B., Scheller, A., Khoury, G., and Jay, G., 1982, Discrete regions of simian virus 40 large T antigen are required for nonspecific and viral origin-specific DNA binding, *J. Virol.* **43**:73–82.

Prives, C., Covey, L., Scheller, A., and Gluzman, Y., 1983, DNA-binding properties of simian virus 40 T-antigen mutants defective in viral DNA replication, *Mol. Cell. Biol.* **3**:1958–1966.

Qasba, P. K., 1974, Synthesis of SV40 DNA in isolated nuclei, *Biochem. Biophys. Res. Commun.* **60**:1338.

Rabek, J. P., Zakian, V. A., and Levine, A. J., 1981, The SV40 A gene product suppresses the adenovirus H5ts125 defect in DNA replication, *Virology* **109**:290.

Rapaport, E., Zamecnik, P. C., and Baril, E. F., 1981, Association of diadenosine 5′,5‴-P^1,P^4-tetraphosphate binding protein with HeLa cell DNA polymerase α, *J. Biol. Chem.* **256**:12148.

Rassoulzadegan, M., Gaudray, P., Canning, M., Trejo-Avila, L., and Cuzin, F., 1981, Two polyoma virus gene functions involved in the expression of the transformed phenotype in FR 373 rat cells, *Virology* **114**:489–500.

Rassoulzadegan, M., Cowie, A., Carr, A., Glaichenhaus, N., Kamen, R., and Cuzin, F., 1982, The roles of individual polyoma virus early proteins in oncogenic transformation, *Nature* **300**:713–718.

Rawlins, D. R., Collis, P., and Muzyczka, N., 1983, Characterization of am404, an amber mutation in the simian virus 40 T antigen gene, *J. Virol.* **47**:202–216.

Reddy, V. B., Thimmappaya, B., Dhar, R., Subramanian, K. N., Zain, B. S., Pan, J., Ghosh, P. K., Celma, M. L., and Weissman, S. M., 1978, The genome of SV40, *Science* **200**:494.

Reed, S. I., Ferguson, J., Davis, R. W., and Stark, G. R., 1975, T-antigen binds to SV40 DNA at the origin of replication, *Proc. Natl. Acad. Sci. USA* **72**:1605.

Reich, N. C., and Levine, A. J., 1982, Specific interaction of the SV40 T antigen–cellular p53 protein complex with SV40 DNA, *Virology* **117**:286–290.

Reich, N. C., Oren, M., and Levine, A. J., 1983, Two distinct mechanisms regulate the levels of a cellular tumor antigen, p53, *Mol. Cell. Biol.* **3**:2143–2150.

Reichard, P., and Eliasson, R., 1979, Synthesis and function of polyoma initiator RNA, *Cold Spring Harbor Symp. Quant. Biol.* **43**:271.

Reichard, P., Eliasson, R., and Soderman, G., 1974, Initiator RNA in discontinuous polyoma DNA synthesis, *Proc. Natl. Acad. Sci. USA* **71**:4901.

Reinhard, P., Maillart, P., Schluchter, M. Gautschi, J. R. and Schindler, R., 1979, An assay system for factors involved in mammalian DNA replication, *Biochim. Biophys. Acta.* **564**:141.

Reiser, J., Renart, J., Crawford, L. V., and Stark, G. R., 1980, Specific association of simian virus 40 tumor antigen with simian virus 40 chromatin. *J. Virol.* **33**:78–87.

Revie, D., Tseng, B. Y., Grafstrom, R. H., and Goulian, M., 1979, Covalent association of protein with replicative form DNA of parvovirus H-1, *Proc. Natl. Acad. Sci. USA* **76**:5539.

Ricca, G., Taylor, J., and Kalinyak, J., 1982, Simple rapid method for the synthesis of radioactively labeled cDNA hybridization probes utilizing bacteriophage M13mp7, *Proc. Natl. Acad. Sci. USA* **79**:724.

Rice, S. A., and Klessig, D. F., 1984, The function(s) provided by the adenovirus-specified, DNA-binding protein required for viral late gene expression is independent of the role of the protein in viral DNA replication, *J. Virol.* **49**:35–49.

Richardson, C. C., 1983, Bacteriophage T7: Minimal requirements for the replication of a duplex DNA molecule, *Cell* **33**:315.

Richardson, J., 1981, The anatomy and taxonomy of protein structure, *Adv. Protein Chem.* **34**:168–340.

Richter, A., and Otto, B., 1983, SV40 chromatin replication *in vitro*, in: *Replication of Viral and Cellular Genomes* (Y. Becker, ed.), p. 53–68, Nijhoff, The Hague.

Richter, A., Scheu, R., and Otto, B., 1980, Replication of SV40 chromatin *in vitro* depends on the amount of DNA polymerase α associated with replicating chromatin *Eur. J. Biochem.* **109**:67.

Riedel, H.-D., Konig, H., Stahl, H., and Knippers, R., 1982, Circular single-stranded M13-DNA as a template for DNA synthesis in protein extracts from *Xenopus laevis* eggs: Evidence for a eukaryotic DNA priming activity, *Nucleic Acids Res.* **10**:5621.

Rigby, P. W. J., and Berg, P., 1978, Does SV40 DNA integrate into cellular DNA during productive infection? *J. Virol.* **28**:475.

Rigby, P. W. J., and Lane, D. P., 1983, Structure and function of the simian virus 40 large T-antigen, in: *Advances in Virology and Oncology* (G. Klein, ed.), pp. 31–58, Raven Press, New York.

Riley, D., and Weintraub, H., 1979, Conservative segregation of parental histones during replication in the presence of cycloheximide, *Proc. Natl. Acad. Sci. USA* **76**:328.

Rinaldy, A., Feuteun, J., and Rosenberg, B. H., 1982, Prereplicative events involving SV40 DNA in permissive cells, *J. Virol.* **41**:237–243.

Rio, D., Robbins, A., Myers, R., and Tjian, R., 1980, Regulation of simian virus 40 early transcription *in vitro* by a purified tumor antigen, *Proc. Natl. Acad. Sci. USA* **77**:5706–5710.

Robberson, D. L., Crawford, L. V., Syrett, C., and James, A. W., 1975, Unidirectional replication of a minority of polyoma virus and SV40 DNAs, *J. Gen. Virol.* **26**:59–69.

Robinson, G. W., and Hallick, L, 1982, Mapping the *in vivo* arrangement of nucleosomes on SV40 chromatin by the photoaddition of radioactive hydroxymethyltrimethylpsoralen, *J. Virol.* **41**:78.

Roman, A., 1979, Kinetics or reentry of polyoma form I DNA into replication as a function of time postinfection, *Virology* **96**:660–663.

Roman, A., 1982, Alteration in the simian virus 40 maturation pathway after butyrate-induced hyperacetylation of histones, *J. Virol.* **44**:958–962.

Roman, A., Champoux, J. J., and Dulbecco, R., 1974, Characterization of replicative intermediates of polyoma virus, *Virology* **57**:147.

Rothwell, V. M., and Folk, W. R., 1983, Comparison of the DNA sequence of the Crawford small-plaque variant of polyomavirus with those of polyomaviruses A2 and strain 3, *J. Virol.* **48**:472.

Rotter, V., 1983, P53, a transformation-related cellular-encoded protein, can be used as biochemical marker for the detection of primary mouse tumor cells, *Proc. Natl. Acad. Sci. USA* **80**:2613.

Roufa, D. J., and Marchionni, M. A., 1982, Nucleosome segregation at a defined mammalian chromosomal site, *Proc. Natl. Acad. Sci. USA* **79**:1810.

Rudolph, K., and Mann, K., 1983, Salt-resistant association of simian virus 40 T antigen with simian virus 40 DNA in nucleoprotein complexes, *J. Virol.* **47**:276–286.

Ruley, H. E., and Fried, M., 1983, Sequence repeats in a polyoma virus DNA region important for gene expression, *J. Virol.* **47**:233.

Rundell, K., and Cox, J., 1979, Simian virus 40 T antigen affects the sensitivity of cellular DNA synthesis to theophylline, *J. Virol.* **30**:394–396.

Russev, G., and Hancock, R., 1982, Assembly of new histones into nucleosomes and their distribution in replicating chromatin, *Proc. Natl. Acad. Sci. USA* **79**:3143.

Ryder, K., DeLucia, A. L., and Tegtmeyer, P., 1983, Binding of SV40 A protein to the BK virus origin of DNA replication, *Virology* **129**:239–245.

Ryoji, M., and Worcel, A., 1984, Chromatin assembly in *Xenopus* oocytes: In vivo studies, *Cell* **37**:21.

Saborio, J. L., Pong, S. S., and Kock, G., 1974, Selective and reversible inhibition of initiation of protein synthesis in mammalian cells, *J. Mol. Biol.* **85**:195.

Sakakibara, Y., Suzuki, K., and Tomizawa, J., 1976, Formation of catenated molecules by replication of colicin E1 plasmid DNA in cell extracts, *J. Mol. Biol.* **108**:569.

Salzman, N. P., and Thoren, M. M., 1973, Inhibition in the joining of DNA intermediates to growing SV40 chains, *J. Virol.* **11**:721.

Salzman, N. P., Sebring, E. D., and Radonovich, M., 1973, Unwinding of parental strands during SV40 DNA replication, *J. Virol.* **12**:669.

Salzman, N. P., Lebowitz, J., Chen, M., Sebring, E., and Garon, C. F., 1975, Properties of replicating SV40 DNA molecules and mapping unpaired regions in SV40 DNA I,*Cold Spring Harbor Symp. Quant. Biol.* **39**:209.

Santos, M., and Butel, J. S., 1984, Antigenic structure of simian virus 40 large tumor antigen and association with cellular protein p53 on the surface of simian virus 40-infected and -transformed cells, *J. Virol.* **51**:376.

Saragosti, S., Moyne, G., and Yaniv, M., 1980, Absence of nucleosomes in a fraction of SV40 chromatin between the origin of replication and the region coding for the late leader RNA, *Cell* **20**:65.

Sauer, R. T., Pabo, C. O., Meyer, B. J., Ptashne, M., and Backman, K., 1979, Regulatory functions of the lambda repressor reside in the amino-terminal domain, *Nature* **279**:396.

Sauer, R. T., Yocum, R. R., Doolittle, R. F., Lewis, M., and Pabo, C. O., 1982, Homology among DNA binding proteins suggests use of a conserved super-secondary structure, *Nature* **298**:447.

Schaffhausen, B., 1983, Transforming genes and gene products of polyoma and SV40, *CRC Crit. Rev. Biochem.* **13**:215.

Schaffhausen, B. S., and Benjamin, T. L., 1976, Deficiency in histone acetylation in non-transforming host range mutants of polyoma virus, *Proc. Natl. Acad. Sci. USA* **73**:1092.

Schaffhausen, B. S., Silver, J. E., and Benjamin, T. L., 1978, Tumor antigen(s) in cells productivity infected by wild-type polyoma virus and mutant NG-18, *Proc. Natl. Acad. Sci. USA* **75**:79.

Scheidtmann, K.-H., Echle, B., and Walter, G., 1982, Simian virus 40 large T antigen is phosphorylated at multiple sites clustered in two separate regions, *J. Virol.* **44**:116.

Scheidtmann, K.-H., Hardung, M., Echle, B., and Walter, G., 1984a, DNA-binding activity of simian virus 40 large T antigen correlates with a distinct phosphorylation state, *J. Virol.* **50**:1–12.

Scheidtmann, K.-H., Schickedanz, J., Walter, G., Lanford, R., and Butel, J. S., 1984b, Differential phosphorylation of cytoplasmic and nuclear variants of simian virus 40 large T encoded by a simian virus 40–adneovirus 7 hybrid viruses, *J. Virol.* **50**:636–640.

Scheller, A., and Prives, C., 1985, Simian virus 40 and polomavirus large tumor antigens have different requirements for high-affinity sequence-specific DNA binding, *J. Virol.* **54**:532.

Scheller, A., Covey, L., Branet, B., and Prives, C., 1982, A small subclass of SV40 T antigen binds to the viral origin of replication, *Cell* **29:375.**

Schlaeger, E.-J., 1982, Replicative conformation of parental nucleosomes: Salt sensitivity of DNA–histone interaction and alteration of histone H1 binding, *Biochemistry* **21**:3167.

Schlegel, R., and Benjamin, T. L., 1978, Cellular alterations dependent upon the polyoma virus hr-t function: Separation of mitogen from transforming capacities, *Cell* **14**:587.

Scholer, H. R., and Gruss, P., 1984, Specific interaction between enhancer-containing molecules and cellular components, *Cell* **36**:403.

Schutzbank, T., Robinson, R., Oren, M., and Levine, A. J., 1982, SV40 large T antigen can regulate some cellular transcripts in a positive fashion, *Cell* **30**:481–490.

Schwyzer, M., Weil, R., Frank, G., and Zubex, H., 1980, Amino acid sequence analysis of fragments generated by partial proteolysis from large simian virus 40 tumor antigen, *J. Biol. Chem.* **255**:5627–5634.

Schwyzer, M., Tai, Y., Studer, E., and Michel, M. R., 1983, Binding sites for monoclonal antibodies and for mRNP's on SV40 large T-antigen determined with a cleavage map, *Eur. J. Biochem.* **137**:303–309.

Scott, W. A., Walter, C. F., and Cryer, B. L., 1984, Barriers to nuclease *Bal*31 digestion across specific sites in SV40 chromatin, *Mol. Cell. Biol.* **4**:604.

Seale, R. L., 1976, Studies on the mode of segregation of histone nu bodies during replication in HeLa cells, *Cell* **9**:423.

Seale, R. L., 1978, Nucleosomes associated with newly replicated DNA have an altered conformation, *Proc. Natl. Acad. Sci. USA* **75**:2717.

Sebring, E. D., Kelly, T. J., Jr., Thoren, M. M., and Salzman, N. P., 1971, Structure of replicating SV40 DNA molecules, *J. Virol.* **8**:478.

Sebring, E. D., Garon, C. F., and Salzman, N. P., 1974, Superhelical density of replicating SV40 DNA molecules, *J. Mol. Biol.* **90**:371.

Segal, S., and Khoury, G., 1979, Differentiation as a requirement for SV40 gene expression in F-9 embryonal carcinoma cells, *Proc. Natl. Acad. Sci. USA* **76**:5611–5615.

Segal, S., Levine, A. J., and Khoury, G., 1979, Evidence for non-spliced SV40 RNA in undifferentiated murine teratocarcinoma stem cells, *Nature* **280**:335–338.

Segawa, M., Sugano, S., and Yamaguchi, N., 1980, Association of simian virus 40 T antigen with replicating nucleoprotein complexes of simian virus 40, *J. Virol.* **35**:320–330.

Seidman, M., and Salzman, N. P., 1979, Late replicative intermediates are accumulated during SV40 DNA replication *in vivo* and *in vitro*, *J. Virol.* **30**:600.

Seidman, M., and Salzman, N. P., 1983, DNA replication of papovaviruses: *In vivo* studies, in: *Replication of Viral and Cellular Genomes* (Y. Becker, ed.), p. 29–52, Nijoff, The Hague.

Seidman, M. M., Garon, C. F., and Salzman, N. P., 1978, The relationship of SV40 replicating chromosomes to two forms of the nonreplicating SV40 chromosome, *Nucleic Acids Res.* **5**:2877.

Seidman, M. M., Levine, A. J., and Weintraub, H., 1979, The asymmetric segregation of parental nucleosomes during chromosome replication, *Cell* **18**:439.

Seidman, M., Slor, H., and Bustin, M., 1983, The binding of a carcinogen to the nucleosomal and non-nucleosomal regions of the SV40 chromosome *in vivo*, *J. Biol. Chem.* **258**:5215.

Seif, I., Khoury, G., and Dhar, R., 1979, The genome of human papovavirus BKV, *Cell* **18**:963. (See also *Cell* **19**, No. 2, 1980, Errata.)

Seif, I., Khoury, G., and Dhar, R., 1980a, A rapid enzymatic DNA sequence technique: Determination of sequence alterations in early SV40 temperature sensitive and deletion mutants, *Nucleic Acids Res.* **8**:225.

Seif, I., Khoury, G., and Dhar, R., 1980b, Sequence and analysis of the genome of human papovavirus BKV, in: *DNA Tumor Viruses: The Molecular Biology of Tumor Viruses, Part 2* (J. Tooze, ed.), pp. 897–922, Cold Spring Harbor Laboratory, Cold Spring Harbor, N.Y.

Sekikawa, K., and Levine, A. J., 1981, Isolation and characterization of polyoma host range mutants that replicate in nullipotential embryonal carcinoma cells, *Proc. Natl. Acad. Sci. USA* **78**:1100.

Sen, A., and Levine, A. J., 1975, SV40 nucleoprotein complex activity unwinds superhelical turns in SV40 DNA, *Nature* **249**:343.

Sen, A., Laipis, P., and Levine, A. J., 1975, DNA replication in SV40-infected cells. XI. The properties of SV40 DNA and nucleoprotein complex synthesized in the presence of cycloheximide, *Intervirology* **5**:122.

Setlow, V. P., Persico-DiLauro, M., Edwards, C. A. F., and Martin, R. G., 1980, The isolation of SV40 tsA/deletion, double mutants, and the induction of host DNA synthesis, *Virology* **101**:250–260.

Shakhov, A. N., Nedospasov, S. A., and Georgiev, G. P., 1982, Deoxyribonuclease II as a probe to sequence-specific chromatin organization: Preferential cleavage in the 72-bp modulator sequence of SV40 minichromosome, *Nucleic Acids Res.* **10**:3951.

Shalloway, D., Kleinberger, T., and Livingston, D. M., 1980, Mapping of SV40 DNA replication origin region binding sites for the SV40 T antigen by protection against exonuclease III digestion, *Cell* **20**:411.

Shaw, S. B., and Tegtmeyer, P., 1981, Binding of dephosphorylated A protein to SV40 DNA, *Virology* **115**:88–96.

Sheinin, R., 1976, Polyoma and cell DNA synthesis in mouse L-cells temperature sensitive for the replication of cell DNA, *J. Virol.* **17**:692.

Shelton, E. R., Kang, J., Wassarman, P. M., and DePamphilis, M. L., 1978a, Chromatin assembly in isolated mammalian nuclei, *Nucleic Acids Res.* **5**:349–362.

Shelton, E. R., Wassarman, P. M., and DePamphilis, M. L., 1978b, The structure of simian virus 40 chromosomes in nuclei from infected monkey cells, *J. Mol. Biol.* **125**:491–514.

Shelton, E. R., Wassarman, P. M., and DePamphilis, M. L., 1980, Structure, spacing and phasing of nucleosomes on isolated forms of mature simian virus 40 chromosomes, *J. Biol. Chem.* **255**:771–782.

Shenk, T., 1978, Construction of a viable SV40 variant containing two functional origins of DNA replication, *Cell* **13**:791.

Shenk, T. E., Carbon, J., and Berg, P., 1976, Construction and analysis of viable deletion mutants of simian virus 40, *J. Virol.* **18**:664–672.

Shioda, M., Nelson, E. M., Bayne, M. L., and Benbow, R. M., 1982, DNA primase activity associated with DNA polymerase α from *Xenopus laevis* ovaries, *Proc. Natl. Acad. Sci. USA* **79**:7209.

Shortle, D. R., and Nathans, D., 1979, Regulatory mutants of simian virus 40: Constructed mutants with base substitutions at the origin of replication, *J. Mol. Biol.* **131**:801–817.

Shortle, D. R., Margolskee, R. F., and Nathans, D., 1979, Mutational analysis of the simian virus 40 replicon: Pseudorevertants of mutants with a defective replication origin, *Proc. Natl. Acad. Sci. USA* **76**:6128–6131.

Shure, M., and Vinogard, J., 1976, The number of superhelical turns in native virion SV40 DNA determined by the band counting method, *Cell* **8**:215.

Silver, T., Schaffhausen, B., and Benjamin, T. L., 1978, Tumor antigens induced by nontransforming mutants of polyoma virus, *Cell* **15**:485.

Simanis, V., and Lane, D. P., 1985, An immunoaffinity purification procedure for SV40 large T antigen, *Virology* **144**:88–100.

Simmons, D. T., 1984, Stepwise phosphorylation of the NH_2-terminal region of the simian virus 40 large T antigen, *J. Biol. Chem.* **259**:8633–8640.

Simmons, D. T., Chang, C., and Martin, M. A., 1979, Multiple forms of polyoma virus tumor antigens from infected and transformed cells, *J. Virol.* **29**:881.

Simpson, R. T., and Stein, A., 1980, Random protection of single-cut restriction endonuclease sites in SV40 mini-chromosomes assembled in vitro, *FEBS Lett.* **111**:337.

Sims, J., and Benz, E. M., Jr., 1980, Initiation of DNA replication by the *E. coli* dnaG protein: Evidence that tertiary structure is involved, *Proc. Natl. Acad. Sci. USA*, **77**:900.

Simpson, R. T., 1978, Structure of the chromatosome, a chromatin particle containing 160 base pairs of DNA and all the histones, *Biochemistry* **17**:5524.

Sinden, R. R., and Pettijohn, D. E., 1984, Cruciform transitions in DNA, *J. Biol. Chem.* **259**:6593.

Sinden, R. R., Broyles, S. S., and Pettijohn, D. E., 1983, Perfect palindromic *lac* operator DNA sequence exists as a stable cruciform structure in supercoiled DNA *in vitro* but not *in vivo*, *Proc. Natl. Acad. Sci. USA* **80**:1797.

Singh, H., and Dumas, L. B., 1984, A DNA primase that copurifies with the major DNA polymerase from the yeast *Saccharomyces cerevisiae*, *J. Biol. Chem.* **259**:7936.

Singleton, C. K., 1983, Effects of salts, temperature, and stem length on supercoil-induced formation of cruciforms, *J. Biol. Chem.* **258**:7661.

Smith, A. E., Smith, R., and Paucha, E., 1978, Extraction and fingerprint analysis of SV40 large and small T-antigens, *J. Virol.* **28**:140–153.

Smith, M. R., and Lieberman, M. W., 1984, Nucleosome arrangement in α-satellite chromatin of African green monkey cells, *Nucleic Acids Res.* **21**:6493.

Smith, P. A., Jackson, V., and Chalkley, R., 1984, Two-stage maturation process for newly replicated chromatin, *Biochemistry* **23**:1576.

Soeda, E., Arrand, J. R., Smolar, N., and Griffin, B. E., 1979, Sequence from early region of polyoma virus DNA containing viral replication origin and encoding small, middle, and (part of) large T-antigens, *Cell* **17**:357.

Soprano, K. J., Galanti, N., Jonak, G. T., McKercher, S., Pipas, J. M., Peden, K. W. C., and Baserga, R., 1983, Mutational analysis of simian virus 40 T antigen: Stimulation of cellular DNA synthesis and activation of rRNA genes by mutants with deletions in the T-antigen gene, *Mol. Cell. Biol.* **3**:214–219.

Soule, H. R., and Butel, J. S., 1979, Subcellular localization of simian virus 40 large tumor antigen, *J. Virol.* **30**:523–532.

Spadari, S., Sala, F., and Pedrali-Noy, G., 1982, Aphidicolin: A specific inhibitor of nuclear DNA replication in eukaryotes, *Trends Biochem. Sci.* **7**:29.

Spillman, T., Giacherio, D., and Hager, L. P., 1979, Single strand DNA binding of SV40 tumor antigen, *J. Biol. Chem.* **254**:3100–3104.

Stahl, H., and Knippers, R., 1983, Simian virus 40 large tumor antigen on replicating viral chromatin: Tight binding and localization on the viral genome, *J. Virol.* **47**:65–76.

Stahl, H., Bauer, M., and Knippers, R., 1983, The simian-virus-40 large-tumor antigen in replicating viral chromatin: A salt-resistant protein–DNA interaction, *Eur. J. Biochem.* **134**:55–61.

Stahl, H., Droge, P., Zentgraf, H., and Knippers, R., 1985, A large-tumor-antigen-specific monoclonal antibody inhibits DNA replication of simian virus 40 minichromosomes in an *in vitro* elongation system, *J. Virol.* **54**:473–482.

Stark, G. R., and Wahl, G. M., 1984, Gene amplification, *Annu. Rev. Biochem.* **53**:447–491.

Staufenbiel, M., and Deppert, W., 1983, Different structure systems in the nucleus are targets for SV40 large T antigens, *Cell* **33**:173–181.

Stavrianopoulos, J. G., Gambino-Giuffrida, A., and Chargaff, E., 1976, RNase H of calf thymus: Substrate specificity, activation, inhibition, *Proc. Natl. Acad. Sci. USA* **73**:1087.

Steck, T. R., and Drlica, K., 1984, Bacterial chromosome segregation: Evidence for DNA gyrase involvement in decatenation, *Cell* **36**:1081.

Steplewski, Z., and Koprowski, H., 1969, Development of SV40 coat protein antigen in nonpermissive nuclei in heterokaryocytes, *Exp. Cell Res.* **57**:433.

Stillman, B. W., Lewis, J. B., Chou, L. T., Matthews, M. B., and Smart, J. E., 1981, Identification of the gene and mRNA for the adenovirus terminal protein precursor, *Cell* **23**:497–508.

Stringer, J. R., 1982, Mutant of simian virus 40 large T-antigen that is defective for viral DNA synthesis, but competent for transformation of cultured rat cells, *J. Virol.* **42**:854–864.

Struhl, K., 1983, The new yeast genetics, *Nature* **305**:391–396.

Su, R. T., and DePamphilis, M. L., 1976, In vitro replication of SV40 DNA in a nucleoprotein complex, *Proc. Natl. Acad. Sci. USA* **73**:3466.

Su, R. T., and DePamhilis, M. L., 1978, Simian virus 40 DNA replication in isolated replicating viral chromosomes, *J. Virol.* **28**:53–65.

Subramanian, K. N., and Shenk, T., 1978, Definition of the boundaries of the origin of SV40 DNA replication, *Nucleic Acids Res.* **5**:3635.

Sundin, O., and Varshavsky, A., 1980, Terminal stages of SV40 DNA replication proceed via multiple intertwined catenated dimers, *Cell*, **21**:103.

Sundin, O., and Varshavsky, A., 1981, Arrest of segregation leads to accumulation of highly intertwined catenated dimers: Dissection of the final stages of SV40 DNA replication, *Cell* **25**:659–669.

Swartzendruber, D. E., Friedrich, T. D., and Lehman, J. M., 1977, Resistance of teratocarcinoma stem cells to infection with SV40: Early events, *J. Cell. Physiol.* **93**:25.

Swetly, P., Brodano, G. B., Knowles, B., and Koprowski, H., 1969, Response of SV40-transformed cell lines and cell hybrids to superinfection with SV40 and its DNA, *J. Virol.* **4**:348.

Tack, L. C., Wassarman, P. M., and DePamphilis, M. L., 1981, Chromatin assembly, relationship of chromatin structure to DNA sequence during SV40 replication, *J. Biol. Chem.* **256**:8821.

Tapper, D. P., and DePamphilis, M. L., 1978, Discontinuous DNA replication: Accumulation of simian virus 40 DNA at specific stages in its replication, *J. Mol. Biol.* **120**:401.

Tapper, D. P., and DePamphilis, M. L., 1980, Preferred DNA sites are involved in the arrest and initiation of DNA synthesis during replication of simian virus 40 DNA, *Cell* **22**:97–108.

Tapper, D., Anderson, S., and DePamphilis, M. L., 1979, Maturation of replicating simian virus 40 DNA molecules in isolated nuclei by continued bidirectional replication to the normal termination region, *Biochim. Biophys. Acta* **565**:84–97.

Tapper, D. P., Anderson, S., and DePamphilis, M. L., 1982, Distribution of replicating simian virus 40 DNA in intact cells and its maturation in isolated nuclei, *J. Virol.* **41**:877–892.

Tegtmeyer, P., 1972, Simian virus 40 deoxyribonucleic acid synthesis: The viral replicon, *J. Virol* **10**:591–602.

Tegtmeyer, P., and Anderson, B., 1981, Partial purification of SV40 A protein and a related cellular protein from permissive cells, *Virology* **115**:67–74.

Tegtmeyer, P., and Macasaet, F., 1972, SV40 DNA synthesis: Analysis by gel electrophoresis, *J. Virol.* **10**:599.

Tegtmeyer, P., Rundell, K., and Collins, J. K., 1977, Modification of SV40 protein A, *J. Virol.* **21**:647.

Tegtmeyer, P., Anderson, B., Shaw, S. B., and Wilson, V. G., 1981, Alternative interactions of the SV40 A protein with DNA, *Virology* **115**:75–87.

Tegtmeyer P., Lewton, B. A., DeLucia, A. L., Wilson, V. G., and Ryder, K., 1983, Topography of simian virus 40 A protein–DNA complexes: Arrangement of protein bound to the origin of replication, *J. Virol.* **46**:151–161.

Templeton, D., and Eckhart, W., 1984, Characterization of viable mutants of polyomavirus cold sensitive for maintenance of cell transformation, *J. Virology*, **49**:799.

Tenen, D. G., Haines, L. L., and Livingston, D. M., 1982, Binding of an analog of the SV40 T antigen to wild-type and mutant viral replication origins, *J. Mol. Biol.* **157**:473–492.

Tenen, D. G., Livingston, D. M., Wang, S. S., and Martin, R. G., 1983a, Effects of a stem-loop structure within the SV40 replication origin upon SV40 T antigen binding to origin region sequences, *Cell* **34**:629–639.

Tenen, D. G., Taylor, T. S., Haines, L. L., Bradley, M. K., Martin, R. G., and Livingston, D. M., 1983b, Binding of simian virus 40 large T antigen from virus-infected monkey cells to wild-type and mutant viral replication origins, *J. Mol. Biol.* **168:**791–808.

Tevethia, M. J., 1984, Immortalization of primary mouse embryo fibroblasts with SV40 virions, viral DNA, and a subgenomic DNA fragment in quantitative assay, *Virology* **137:**414.

Tevethia, M. J., Slippey, A. E., and Cosman, D. J., 1981, Mapping of additional temperature-sensitive mutations (1600 series) on the genome of simian virus 40 by marker rescue, *Virology* **112:**789–794.

Tevethia, S. S., Lewis, A. J., Campbell, A. E., Tevethia, M. J., and Rigby, P. W., 1984, Simian virus 40 specific cytotoxic lymphocyte clones localize two distinct TSTA sites on cells synthesizing a 48 KD SV40 T antigen, *Virology* **133:**443–447.

Thummel, C., Tjian, R., Hu, S.-L., and Grodzicker, T., 1983, Translational control of SV40 T antigen expression from the adenovirus late promoter, *Cell* **33:**455–464.

Tilly, K., McKittrick, N., Zylicz, M., and Georgopoulos, C., 1983, The dnaK protein modulates the heat-shock response of *Escherichia coli*, *Cell* **34:**641–646.

Tjian, R., 1978, The binding site on SV40 DNA for a T antigen-related protein, *Cell* **13:**165–179.

Tjian, R., and Robbins, A., 1979, Enzymatic activities associated with a purified SV40 T-antigen related protein, *Proc. Natl. Acad. Sci. USA* **76:**610.

Tjian, R., Fey, G., and Graessmann, A., 1978, Biological activity of purified SV40 T-antigen proteins, *Proc. Natl. Acad. Sci. USA* **75:**1279.

Tooze, J. (ed.), 1981, *DNA Tumor Viruses: The Molecular Biology of Tumor Viruses, Part 2*, pp. 61–370a, Cold Spring Harbor Laboratory, Cold Spring Harbor, N.Y.

Tornow, J., and Cole, C. N., 1983a, Intracistronic complementation in the simian virus 40 A gene, *Proc. Natl. Acad. Sci. USA* **80:**6312–6316.

Tornow, J., and Cole, C. N., 1983b, Nonviable mutants of simian virus 40 with deletions near the 3′ end of gene A define a function for large T antigen required after onset of viral DNA replication, *J. Virol.* **47:**487–494.

Tornow, J., Polvino-Bodnar, M., Santangelo, G., and Cole, C. N., 1985, Two separable functional domains of SV40 large T-antigen: Carboxyl-terminal region of SV40 large T-antigen is required for efficient capsid protein synthesis, *J. Virol.* **53:**415.

Trevor, K., and Lehman, I. M., 1982, The interaction of polyoma virus with F9 embryonal carcinoma cells and chemically induced differentiated progeny: Fate of the viral DNA and expression of viral antigens, *J. Cell. Physiol.* (Suppl.) **2:**69.

Triezenberg, S. J., and Folk, W. R., 1984, Essential nucleotides in the polyomavirus origin region, *J. Virol.* **51:**437.

Trifonov, E. N., and Mengeritsky, G., 1984, SV40 replication pause sites map with the nucleosomes, *J. Virol.* **52:**1011.

Tseng, B. Y., and Ahlem, C. N., 1982, DNA primase activity from human lymphocytes, *J. Biol. Chem.* **257:**7280.

Tseng, B. Y., and Ahlem, C. N., 1983, A DNA primase from mouse cells, purification and partial characterization, *J. Biol. Chem.* **258:**9845.

Tseng, B. Y., and Ahlem, C. N., 1984, Mouse primase initiation sites in the origin region of simian virus 40, *Proc. Natl. Acad. Sci. USA* **81:**2342–2346.

Tseng, B. Y., and Goulian, M., 1980, Initiator RNA synthesis upon ribonucleotide depletion: Evidence for base substitution, *J. Biol. Chem.* **255:**2062.

Tseng, B. Y., Erickson, J. M., and Goulian, M., 1979, Initiator RNA of nascent DNA from animal cells, *J. Mol. Biol.* **129:**531.

Tsernoglou, D., Tucker, A. D., and van der Vliet, P. C., 1984, Crystallization of a fragment of the adenovirus DNA binding protein, *J. Mol. Biol.* **172:**237–239.

Tsubota, Y., Waqar, M. A., Burk, J. F., Milavetz, B. I., Evans, M. J., Kowalski, D., and Huberman, J. A., 1979, Association of enzymes with replicating and nonreplicating SV40 chromosomes, *Cold Spring Harbor Symp. Quant. Biol.* **43:**693.

Tsutsui, K., Watanabe, S., Katagiri, M., and Oda, T., 1983, Identification of proteins interacting with newly replicated DNA in SV40-infected cells by UV-induced DNA–protein crosslinking, *Nucleic Acids Res.* **11:**4793.

Türler, H., 1980, The tumor antigens and the early functions of polyoma virus, *Mol. Cell. Biochem.* **32**:63.

Tyndall, C., La Mantia, G., Thacker, C. M., Favaloro, J., and Kamen, R., 1981, A region of the polyoma virus genome between the replication origin and late protein coding sequences is required in *cis* for both early gene expression and viral DNA replication, *Nucleic Acids Res.* **9**:6231.

Uemura, T., and Yanagida, M., 1984, Isolation of type I and II DNA topoisomerase mutants from fission yeast: Single and double mutants show different phenotypes in cell growth and chromatin organization, *EMBO J.* **3**:1737.

van der Vliet, P. C., and Kwant, M. M., 1978, Role of DNA polymerase γ in adenovirus DNA replication, *Nature* **276**:532.

van Heuverswyn, H., and Fiers, W., 1979, Nucleotide sequence of the Hind-C fragment of SV40 DNA, *Eur. J. Biochem.* **100**:51–60.

van Roy, F., Fransen, L., and Fiers, W., 1983, Improved localization of phosphorylation sites in simian virus 40 large T antigen, *J. Virol.* **45**:315–331.

van Roy, F., Fransen, L., and Fiers, W., 1984, Protein kinase activities in immune complexes of simian virus 40 large T-antigen and transformation-associated cellular p53 protein, *Mol. Cell. Biol.* **4**:232–239.

Varshavsky, A., 1981, On the possibility of metabolic control of replicon "misfiring": Relationship to emergence of malignant phenotypes in mammalian cell lineages, *Proc. Natl. Acad. Sci. USA* **78**:3673.

Varshavsky, A. J., Sundin, O. H., and Bohn, M. J., 1978, SV40 viral minichromosome: Preferential exposure of the origin of replication as probed by restriction endonucleases, *Nucleic Acids Res.* **5**:3469.

Varshavsky, A., Sundin, O., Ozkaynak, E., Pan, R., Solomon, M., and Snapka, R., 1983, Final stages of DNA replication: Multiply intertwined catenated dimers as SV40 segregation intermediates, in: *Mechanisms of DNA Replication and Recombination* (N. Cozzarelli ed.), pp. 463–494, Liss, New York.

Vasseur, M., Kress, C., Montreau, N., and Blangy, D., 1980, Isolation and characterization of polyoma virus mutants able to develop in embryonal carcinoma cells, *Proc. Natl. Acad. Sci. USA* **77**:1068.

Vaury, C., Gilly, C., Alix, D., and Lawrence, J. J., 1983, Assembly kinetics of replicating chromatin: Isolation and characterization of prenucleosomal and nucleosomal DNA, *Biochem. Biophys. Res. Commun.* **110**:811–818.

Veldman, G. M., Lupton, S., and Kamen, R., 1985, Polyomavirus enhancer contains multiple redundant sequence elements that activate both DNA replication and gene expression, *Mol. Cell. Biol.* **5**:649.

Venaskatsen, S., and Moss, B., 1980, Donor and acceptor specificities of HeLa cell mRNA guanylyltransferase, *J. Biol. Chem.* **255**:2835.

Verderame, M. F., Kohtz, D. S., and Pollack, R. E., 1983, 94,000- and 100,000-molecular weight simian virus 40 T antigens are associated with the nuclear matrix in transformed and revertant mouse cells, *J. Virol.* **46**:575–583.

Wake, C. T., and Wilson, J. H., 1980, Defined oligomeric SV40 DNA: A sensitive probe of general recombination in somtic cells, *Cell* **21**:141.

Waldeck, W., Spaeren, U., Mastromei, G., Eliasson, R., and Reichard, P., 1979, Replication of polyoma DNA in nuclear extracts and nucleoprotein complexes, *J. Mol. Biol.* **135**:675.

Waldeck, W., Theobald, M., and Zentgraf, H., 1983, Catenation of DNA by eucaryotic topoisomerase II associated with SV40 minichromosomes, *EMBO J.* **2**:1255.

Walker, M. D., Edlund, T., Boulet, A. M., and Rutter, W. J., 1983, Cell-specific expression controlled by the 5′-flanking region of insulin and chymotrypsin genes, *Nature* **305**:557–561.

Wang, H.-T., and Roman, A., 1981, Cessation of reentry of SV40 DNA into replication and its simultaneous appearance in nucleoprotein complexes of the maturation pathway, *J. Virol.* **39**:255–262.

Wang, H.-T., Larsen, S. H., and Roman, A., 1985, A cis-acting sequence promotes removal of SV40 DNA from the replication pool, *J. Virol.* **53**:410–414.

Wang, T. S.-F., Hu, S.-Z., and Korn, D., 1984, DNA primase from KB cells, characterization of a primase activity tightly associated with immunoaffinity purified DNA polymerase-α, *J. Biol. Chem.* **259**:1854.

Waqar, M. A., Evans, M. J., Burke, J. F., Tsubota, Y., Plummer, M. J., and Huberman, J. A., 1983, *In vitro* DNA synthesis by an α-like DNA polymerase bound to replicating SV40 chromosomes, *J. Virol.* **48**:304.

Watkins, J. F., 1975, The SV40 rescue problem, *Cold Spring Harbor Symp. Quant. Biol.* **39**:355.

Watkins, J., and Dulbecco, R., 1967, Production of SV40 virus in heterokaryons of transformed and susceptible cells, *Proc. Natl. Acad. Sci. USA* **58**:1396.

Wawra, E., Pockl, E., Mullner, E., and Wintersberger, E., 1981, Effect of sodium butyrate on induction of cellular and viral DNA synthesis in polyoma virus-infected mouse kidney cells, *J. Virol.* **38**:973.

Weaver, D. T., and DePamphilis, M. L., 1982, Specific sequences in native DNA that arrest progress of DNA polymerase alpha, *J. Biol. Chem.* **257**:2075.

Weaver, D. T., and DePamphilis, M. L., 1984, The role of palindromic and nonpalindromic sequences in arresting DNA synthesis *in vitro* and *in vivo*, *J. Mol. Biol* **180**: 961–986.

Weaver, D., Krokan, H., and DePamphilis, M. L., 1980, Gap-filling: A unique step in the metabolism of Okazaki fragments that require both DNA polymerase α and protein cofactors, *J. Supramol. Struct. Suppl.* **4**:895.

Weaver, D. T., Fields-Berry, S., and DePamphilis, M. L., 1985, The termination region for SV40 DNA replication directs the mode of separation for the two sibling molecules, *Cell* **41**:565.

Weber, F., de Villiers, J., and Schaffner, W., 1984, An SV40 "enhancer trap" incorporates exogenous enhancers or generates enhancers from its own sequences, *Cell* **36**:983.

Weinmann-Dorsch, C., Hedi, A., Grummt, I., Albert, W., Ferdinand, F.-J., Friis, R. R., Pierron, G., Moll, W., and Grummt, F., 1984, Drastic rise of intracellular adenosine(5') tetraphospho(5')adenosine correlates with onset of DNA synthesis in eukaryotic cells, *Eur. J. Biochem.* **138**:179.

Weintraub, H., 1976, Cooperative alignment of nu bodies during chromosome replication in the presence of cycloheximide, *Cell* **9**:419.

Weintraub, H., 1979, Assembly of an active chromatin structure during replication, *Nucleic Acids Res.* **7**:781.

Weiss, A. S., and Wake, R. G., 1984, A unique DNA intermediate associated with termination of chromosome replication in *B. subtilis*, *Cell* **39**:683.

Wettstein, F. O., and Stevens, J. G., 1982, Variable-sized free episomes of Shope papilloma virus DNA are present in all non-virus-producing neoplasms and integrated episomes are detected in some, *Proc. Natl. Acad. Sci. USA* **79**:790.

White, M., and Eason, R., 1973, Supercoiling of SV40 DNA can occur independently of replication, *Nature New Biol.* **241**:46.

Wilson, V. G., Tevethia, M. J., Lewton, B. A., and Tegtmeyer, P., 1982, DNA binding properties of simian virus 40 temperature-sensitive A proteins, *J. Virol.* **44**:458–466.

Wintersberger, V., and Wintersberger, E., 1975, DNA polymerase in polyoma infected mouse kidney cells, *J. Virol.* **16**:1095.

Wirak, D. O., Chalifour, L. C. Wassarman, P. M., Muller, W. J., Hassell, J. A., and DePamphilis, M. L., 1985, Sequence-specific DNA replication in preimplantation mouse embryos, *Mol. Cell. Biol.* **5**:2924–2935.

Wist, E., 1979, Partial purification of deoxyuridine triphosphate nucleotideohydrolase and its effect on DNA synthesis in isolated HeLa cell nuclei, *Biochim. Biophys. Acta* **565**:98.

Wist, E., and Prydz, H., 1979, The effect of aphidicolin on DNA synthesis in isolated HeLa cell nuclei, *Nucleic Acids Res.* **6**:1583.

Wist, E., Unhjem, O., and Krokan, H., 1978, Accumulation of small fragments of DNA in isolated HeLa cell nuclei due to transient incorporation of dUMP, *Biochim. Biophys. Acta* **520**:253.

Woodworth-Gutai, M., Celeste, A., Sheflin, L., and Sclair, M., 1983, Naturally arising recombinants that are missing portions of the SV40 regulatory region, *Mol. Cell. Biol.* **3**:1930.

Worcel, A., Han, S., and Wong, M. L., 1978, Assembly of newly replicated chromatin, *Cell* **15**:969.

Wright, P. J., DeLucia, A. L., and Tegtmeyer, P., 1984, Sequence-specific binding of SV40 A protein to nonorigin and cellular DNA, *Mol. Cell. Biol.* **4**:2631–2638.

Wu, K. C., Strauss, F., and Varshavsky, A., 1983, Nucleosome arrangement in green monkey α-satellite chromatin, superimposition of non-random and apparently random patterns, *J. Mol. Biol.* **170**:93.

Yagura, T., Kozu, T., and Seno, T., 1982, Mouse DNA replicase, DNA polymerase associated with a novel RNA polymerase activity to synthesize initiator RNA of strict size, *J. Biol. Chem.* **257**:11121–11127.

Yagura, T., Tanaka, S., Kozu, T., Seno, T., and Korn, D., 1983a, Tight association of DNA primase with a subspecies of mouse DNA polymerase α, *J. Biol. Chem.* **258**:6698–6700.

Yagura, T., Kozu, T., and Seno, T., 1983b, Mechanism of stimulation by a specific protein factor of *de novo* DNA synthesis by mouse DNA replicase with fd phage single-stranded circular DNA, *Nucleic Acids Res.* **11**:6369.

Yamaguchi, M., and DePamphilis, M. L., 1986, DNA binding site for a factor(s) required to initiate SV40 DNA replication, *Proc. Natl. Acad. Sci. USA*, **83**:1646–1650.

Yamaguchi, M., Hendrickson, E. A., and DePamphilis, M. L., 1985a, DNA primase–DNA polymerase α from simian cells: Modulation of RNA primer synthesis by ribonucleoside triphosphates, *J. Biol. Chem.* **260**:6254–6263.

Yamaguchi, M., Hendrickson, E. A., and DePamphilis, M. L., 1985b, DNA primase–DNA polymerase α from simian cells: Sequence specificity of initiation sites on simian virus 40 DNA, *Mol. Cell. Biol.* **5**:1170–1183.

Yang, R. C. A., and Wu, R., 1979, BK virus DNA: Complete nucleotide sequence of a human tumor virus, *Science* **206**:456.

Yoda, K., and Okazaki, T., 1983, Primer RNA for DNA synthesis on single-stranded DNA template in a cell free system from *Drosophila melanogaster* embryos, *Nucleic Acids Res.* **11**:3433.

Young, L. S., and Champoux, J. J., 1978, Interaction of the DNA untwisting enzyme with the SV40 nucleoprotein complex, *Nucleic Acids Res.* **5**:623.

Yu, K., and Cheevers, W. P., 1976a, DNA synthesis in polyoma virus. IV. Mechanism of formation of closed-circular viral DNA deficient in superhelical turns, *J. Virol.* **17**:402.

Yu, K., and Cheevers, W. P., 1976b, DNA synthesis in polyoma virus infection. V. Kinetic evidence for two requirements for protein synthesis during viral DNA replication, *J. Virol.* **17**:415.

Yu, K., Kowalski, J., and Cheevers, W., 1975, DNA synthesis in polyoma virus infection. III. Mechanism of inhibition of viral DNA replication by cycloheximide, *J. Virol.* **15**:1409.

Zakian, V. A., 1976, Electron microscopic analysis of DNA replication in main band and satellite DNAs of *Drosophila virilis*, *J. Mol. Biol.* **108**:305.

Zamecnik, P. C., Rapaport, E., and Baril, E. F., 1982, Priming of DNA synthesis by diadenosine 5′,5‴-P^1,P^4-tetraphosphate with a double-stranded octadecamer as a template and DNA polymerase α, *Proc. Natl. Acad. Sci. USA* **79**:1791.

Zannis-Hadjopoulos, M., Chepelinsky, A. B., and Martin, R. G., 1983, Mapping of the 3′-end positions of SV40 nascent strands, *J. Mol. Biol.* **165**:599.

Zierler, M. K., Marini, N. J., Stowers, D. J., and Benbow, R. M., 1985, Stockpiling of DNA polymerases during cogenesis and embryogenesis in the frog, *Xenopus laevis*, *J. Biol. Chem.* **260**:974–981.

Zylicz, M., LeBowitz, J., McMacken, R., and Georgopoulos, G., 1983, The dnaK protein of *Escherichia coli* possesses an ATPase and autophosphorylating activity and is essential in *in vitro* DNA replication, *Proc. Natl. Acad. Sci. USA* **80**:6431–6435.

CHAPTER 4

Transformation by SV40 and Polyoma

ROGER MONIER

I. INTRODUCTION

The polyomaviruses, polyoma and simian virus 40 (SV40), are tumorigenic when injected into appropriate hosts, and can also alter the morphology and growth properties of cells in culture. Because they are relatively easy to produce and because their genomes are small DNA molecules, they were the first oncogenic viruses which could be studied by modern molecular genetics. Since their discovery in the late 1950s and early 1960s, they have attracted the attention of many oncologists who realized that their molecular biology could be studied in great detail.

Experiments using thermosensitive mutants (Dulbecco and Eckhart, 1970; Tegtmeyer, 1975) as well as DNA transfection of cultured cells (Van der Eb *et al.*, 1979; Novack *et al.*, 1980) or animal experiments (Moore *et al.*, 1980) with specified genome fragments have clearly established that early viral functions are sufficient to induce tumors *in vivo* and to transform cells *in vitro*, restricting further the area of interest in the viral genome.

Thermosensitive mutants, affecting viral DNA replication, were also very useful in introducing the idea that early viral functions are required not only for initiating but also for maintaining cell transformation (Osborn and Weber, 1975; Kimura and Itagaki, 1975; Tegtmeyer, 1975; Brugge and Butel, 1975; Martin and Chou, 1975; Seif and Cuzin, 1977). Although this point has been the subject of many discussions, it is now generally agreed that in the vast majority of transformed cells, some, if not all, of

ROGER MONIER • Laboratoire d'Oncologie Moléculaire, Institut Gustave-Roussy, Pavillon de Recherche, 49805 Villejuif Cédex, France.

the properties associated with the transformed phenotype are under the control of at least one viral function, which must be continuously expressed to ensure the maintenance of transformation.

Further experiments with appropriate DNA constructs (Treisman *et al.*, 1981; Zhu *et al.*, 1984) have made it possible to attribute distinct effects on cellular properties to the two main products of the polyoma virus early region, large T and middle T. As previously observed with region E1a of adenoviruses (Van der Eb *et al.*, 1977; Shiroki *et al.*, 1979), cloned DNA sequences that can only produce polyoma large T are sufficient to convert primary rodent cells into continuous cell lines, a process usually referred to as "immortalization" (Rassoulzadegan *et al.*, 1983). On the other hand, DNA sequences coding for middle T alter the morphological properties and anchorage dependence of transfected continuous cell lines (Treisman *et al.*, 1981) and complement large T-coding sequences for full transformation and tumorigenicity of rodent cells in primary cultures (Rassoulzadegan *et al.*, 1982). These observations are reminiscent of the cooperation between E1a and E1b regions of adenoviruses (Van der Eb *et al.*, 1977; Shiroki *et al.*, 1979) and between cellular oncogenes of the *myc* and *ras* types (Land *et al.*, 1983; Ruley, 1983). The SV40 genome does not code for a middle T-like product. SV40 large T probably carries both activities on the same molecule.

Transforming and tumorigenic viral functions are thus encoded by at least two early genes in the polyoma virus genome. Both SV40 and polyoma early regions code for another early protein, small t. Although small t is probably not essential for tumorigenicity of SV40 (Lewis and Martin, 1979), it nevertheless plays a role in cell transformation under some conditions (Shenk *et al.*, 1976; Sleigh *et al.*, 1978; Feunteun *et al.*, 1978).

Therefore, contrary to what was previously believed to be the case, the control of tumorigenesis and transformation by polyomaviruses is not ensured by a single viral gene. Moreover, immortalizing and transforming genes of DNA tumor viruses are interchangeable with cellular oncogenes in cotransfection experiments (Land *et al.*, 1983; Ruley, 1983). These recent observations have renewed interest in the study of oncogenesis by polyoma virus and SV40, since comparisons between these easily manipulated genomes and cellular oncogenes should be possible in molecular terms.

II. TUMORIGENICITY BY SV40 AND POLYOMA

A. SV40

SV40 and polyoma are not tumorigenic in their host species under normal circumstances. SV40 can be found in the kidneys of its natural host (rhesus monkeys) without apparent unfavorable consequences (Sweet and Hilleman, 1960).

SV40 was inadvertently injected into millions of children and adults as a contaminant of poliovirus vaccine (Sweet and Hilleman, 1960). So far no tumors have been recognized that could have been induced by SV40. It is unlikely, therefore, that SV40 is oncogenic in humans. SV40 DNA has been detected in an episomal form in 8 of 35 human brain tumors, but no evidence of an etiological role is available (Krieg *et al.*, 1981).

Early observations on experimental animals gave positive results only with hamsters, in which subcutaneous injection into newborn animals produced fibrosarcomas at the site of injection (Eddy *et al.*, 1962). In adult animals, intravenous injections of high doses of the virus were followed by the rapid appearance of leukemias, lymphomas, osteosarcomas, and reticulum cell sarcomas (Diamandopoulos, 1978).

SV40 was considered nontumorigenic in mice, until specified inbred strains were compared (Hargis and Malkiel, 1979). After injection into mice, SV40 was found to persist several months (Abramczuk *et al.*, 1984). It can induce tumors after long latency periods in congenic strains which have a low cytotoxic T-lymphocyte response, such as B10.D2 (H-2d), but not in B10.BR (H-2k) which has a high response (Abramczuk *et al.*, 1984).

Transgenic mice harboring the SV40 T-antigen genes have been obtained by microinjecting, in the male pronucleus of either C57 or F_2 hybrid eggs (obtained by mating C57 × SJL hybrid adults), a linearized plasmid containing the SV40 early region and a metallothionein fusion gene. A high percentage of transgenic adults developed choroid plexus tumors (papillomas or carcinomas) (Brinster *et al.*, 1984). Previously, Jaenisch and Mintz (1974) had not observed an increased incidence of tumors in mice developing from SV40-injected blastocycts and carrying SV40 DNA sequences in the cellular chromosomes of all tissues tested. In the transgenic mice studied by Brinster *et al.* (1984), it appears that some unknown mechanism inactivates the injected SV40 genes during early development and that an infrequent event activates them in a few cells during later development.

The expression of SV40 large T protein is all that is required for tumorigenicity. Mutants producing normal large T and no or altered small t (dl54-59 mutants) are able to induce tumors in newborn hamsters, although with an increased latency period as compared to the wild-type virus (Lewis and Martin, 1979). The dl54-59 mutant, dl884, was even found to produce tumors able to metastasize, a behavior never observed with wild-type SV40-induced tumors (Dixon *et al.*, 1982).

B. Polyoma

Polyoma virus is ubiquitous in many laboratory mouse colonies and populations of wild mice (Yabe *et al.*, 1961; Rowe, 1961) without an apparent role in the etiology of naturally occurring tumors. When injected into newborn animals of various laboratory strains, it induces tumors in a large number of organs [salivary glands, thymus, ovaries, mammary glands, and adrenals (Gross, 1953; Eddy *et al.*, 1958)]. Polyoma virus can also induce tumors after long latency at the site of injection in adult animals (Defendi, 1960).

Injection of large amounts of viral DNA into newborn hamsters (Israel *et al.*, 1979) or into newborn rats (Gelinas *et al.*, 1981) induces tumors. The effectiveness of the injected DNA is increased when the continuity of the sequences specifically coding for the C-terminal portion of large T is interrupted by *Eco*RI digestion (Israel *et al.*, 1979). A complete large T protein is therefore not required for tumorigenicity of polyoma DNA. This observation parallels that of Lania *et al.* (1981) on polyoma-transformed cells, which lose their functional large T after passage *in vivo* as tumor cells. According to Asselin *et al.* (1984), large T could even be entirely eliminated, since injection into newborn hamsters of a plasmid encoding only the middle T protein is enough to produce tumors. Nevertheless, polyoma middle T is far more tumorigenic in the presence of small t. As a matter of fact, when the experimental animal is the rat, cooperation between middle T and small t is absolutely required for tumorigenicity. A recombinant encoding both small t and middle T but not large T induced tumors as efficiently as did the wild-type DNA in newborn rats (Asselin *et al.*, 1984).

III. *IN VITRO* CELL TRANSFORMATION

Cells, freshly explanted from animals, do not grow continuously in tissue culture. After the period when the cells actively divide, the cultures enter a "crisis," and usually die. In some cases, the crisis is surmounted and a continuous cell line emerges. This "transformation" phenomenon never occurs spontaneously with chicken or human cells. It is more frequent with rodent cells, although the frequency of spontaneous transformation is variable according to species. For example, under similar conditions, Chinese hamster embryo cells are known to give rise to postcrisis continuous cell lines ten times more frequently than Syrian hamster embryo cells (Barrett, 1980; Kraemer *et al.*, 1983). At the same time the transformed cells acquire new growth control properties, they display characteristic morphological changes as compared to the original cultures, and frequently acquire enhanced tumorigenicity in the syngeneic host (Earle *et al.*, 1943a,b).

The rate of spontaneous transformation is usually very slow. It can be accelerated by various agents, such as the chemical carcinogen 20-methylcholanthrene in the original experiments of Earle *et al.* (1943a). Polyoma virus was the first DNA virus shown to produce transformation of cultured cells (Dawe and Law, 1959; Vogt and Dulbecco, 1960), a phenomenon that subsequently was interpreted as composed of two steps: a primary transformation caused by the virus, followed by further changes which resulted in increased transplantability (Vogt and Dulbecco, 1963).

The first experiments performed *in vitro* with DNA viruses used freshly explanted cultures, mainly derived from embryos. The heterogeneity of such cultures rendered the quantitative interpretation of these experiments hazardous. The establishment of continuous cell lines, devoid of tumorigenicity and displaying a tight regulation of their rate of division by the availability of serum growth factors and by the culture density, like the Syrian hamster BHK-21 (MacPherson and Stoker, 1962) or the mouse Swiss 3T3 (Todaro and Green, 1963), enabled investigators to use homogeneous well-characterized populations of cells. Nevertheless, continuous cell lines are already at least one step removed from completely normal cells. In recent years, it proved necessary to use both primary and continuous cells in order to analyze properly the transforming functions of oncogenic DNA viruses.

A. Properties of Transformed Cells Which Can Be Used for Selection

When cells are fully transformed by SV40 or polyoma virus, they usually possess the following properties, which are not displayed by normal freshly explanted cells:

1. Ability to grow continuously, without growth crisis, upon successive prolonged passages in culture. This property is loosely referred to as "immortalization." It should be realized that the "immortal" character applies to the population of cells in culture and not necessarily to every cell in the population.
2. Ability to form clones when seeded on a solid surface at low cell densities (clonability).
3. Low requirements for serum growth factors.
4. Ability to overgrow a continuous layer of cells, a property that enables the transformed cells to form multilayered colonies (dense foci), which are usually easy to distinguish from the background of untransformed cells.
5. Anchorage independence of growth.

If primary cell cultures are used in transformation experiments, any one of these five properties can be used as a basis for direct selection of transformants, which should be able to grow and form colonies or foci, while untransformed cells would not. With established cell lines, only the last three properties can be exploited for selection. Selection based on anchorage independence is the most stringent technique which can be used for obtaining transformed cells which possess all five properties. Using a nonselective analysis of SV40 transformation of either mouse 3T3 or primary rat embryonic cells, Risser and Pollack (1974) and Risser *et al.* (1974) distinguished two main types of transformants, which they designated as minimal and maximal. Minimal transformants resemble continuous cell lines of the 3T3 type and maximal transformants are able to overgrow a dense layer of cells and to form colonies in suspension. In similar experiments, where polyoma virus and SV40 were compared with respect to their respective ability to transform rat FR3T3 cells, Perbal and Rassoulzadegan (1980) observed that SV40 produced a wider range of phenotypes than polyoma. By and large, SV40 produced more minimal transformants than polyoma under similar conditions. The choice of the selection procedure is probably more important with SV40 than with polyoma virus in order to get fully transformed cells.

Using the same virus and the same type of cells, the precise phenotype of the transformants which can be obtained in a transformation experiment will depend therefore on the type of selection. Even using one particular type of selection, and one type of cell, different phenotypes can still be obtained, depending upon the precise conditions under which the transformation is performed. The most important factors which can influence the outcome of a transformation experiment, apparently include the state of growth of the cell at the time of infection and the number of cell divisions the cells are allowed to perform after infection and before selection (e.g., Rassoulzadegan *et al.*, 1978; Sompayrac and Danna, 1983b).

In the case of SV40-transformed cells, a strong correlation between anchorage independence of growth and tumorigenicity in nude mice is usually observed (Shin *et al.*, 1975). This correlation does not necessarily hold for other oncogenic DNA viruses of the papovavirus family (Grisoni *et al.*, 1984).

B. Other Properties of Transformed Cells

Besides the main properties displayed by transformed cells which can be used for selecting transformants, a set of other properties are usually acquired after transformation.

1. Enhanced Agglutination by Lectins

Several plant lectins, each of which binds to specific glycosyl groups, are able to produce the agglutination of cultured cells (Burger, 1973). Tumor cells are more susceptible to agglutination than are normal cells (Aub *et al.*, 1963) and this observation has been extended to cells transformed by SV40 and polyoma (Burger, 1969; Inbar *et al.*, 1969). The enhanced agglutinability of transformed versus normal cells appears to be due to an increased fluidity of the transformed cell membrane rather than to an increased number of lectin receptors (Nicholson, 1974).

2. Loss of Surface Fibronectin

In many cells transformed by viruses, the amount of the glycoprotein fibronectin associated with the cell surface is decreased (Hynes, 1973).

This decrease, which is not correlated with a decrease in biosynthesis, is not always observed with SV40-transformed highly tumorigenic cells (Steinberg *et al.*, 1979).

3. Changes in the Cytoskeleton

Disruption of the cytoskeleton is a very frequent consequence of viral transformation. Many SV40- or polyoma-transformed rodent cells do not display well-organized actin cables (McNutt *et al.*, 1973; Pollack *et al.*, 1975; Treisman *et al.*, 1981). The disappearance of actin cables correlates well with anchorage independence (Shin *et al.*, 1975). Revertants from transformed cells (McNutt *et al.*, 1973; Pollack *et al.*, 1975) regain a normal distribution of actin cables. Cells transformed by tsA mutant viruses, which produce a thermolabile T antigen, also regain a normal cytoskeleton when incubated at the restrictive temperature (Osborn and Weber, 1975; Pollack *et al.*, 1975).

4. Enhanced Uptake of Nutrients

Transformation of cells by polyoma and SV40 stimulated the rate of uptake of nutrients, compared to control untransformed cells (Cunningham and Pardee, 1969; Foster and Pardee, 1969; Isselbacher, 1972; Quinlan *et al.*, 1976; Inui *et al.*, 1979). This stimulated transport, which should not be interpreted as an unspecific increase in permeability, because it only affects specific nutrients, could be related to the ability displayed by transformed cells to grow in low serum concentrations. The stimulation of nutrient uptake is dependent upon an active T antigen in cells transformed by tsA mutants (Brugge and Butel, 1975; Seif and Cuzin, 1977).

5. Production of Plasminogen Activator

Transformation of rodent cells by SV40 and polyoma is accompanied by a greatly increased production of plasminogen activator (Ossowski *et al.*, 1973; Pollack *et al.*, 1974; Seif and Cuzin, 1977). A correlation between plasminogen activator production and anchorage independence of growth has been described (Pollack *et al.*, 1974). The causal role played by plasminogen activator in the phenotypic modifications which are characteristic of viral transformation is indicated by the observation that hamster fibroblasts transformed by SV40 or polyoma do not display all the morphological changes induced by transformation when the serum added to the culture medium is previously depleted of plasminogen (Ossowski *et al.*, 1973).

In polyoma tsa transformants, which possess a thermosensitive transformed phenotype, the production of plasminogen activator returns to normal levels upon incubation at the restrictive temperature (Seif and Cuzin, 1977).

C. Permissivity and Transformation

Many cell types can be infected by polyoma and SV40. Mouse cells for polyoma and African green monkey cells for SV40 are fully permissive, i.e., viral infection leads to viral multiplication and finally to cell lysis. Permissive cells thus cannot be transformed by a fully infectious virus.

Rodent cells, particularly rat or mouse cells, are totally nonpermissive for SV40. When they are infected, no virus production is observed and most of the cells survive infection. Nevertheless, the early viral functions are efficiently expressed in a fraction of the infected population which is a function of the multiplicity of infection. Among the many survivors, a small fraction of the infected population is eventually transformed. The process is very inefficient: no more than 0.1% of the infected population can be selected as transformed when 10–100 plaque-forming units (PFU) (100–10,000 physical particles) per cell is used. The shape of the dose–response curve nevertheless indicates that one virion is enough to transform a cell (Todaro and Green, 1966).

In order to circumvent the difficulties introduced by permissivity in transforming monkey cells by SV40, two approaches have been followed. UV-irradiated SV40 was used to infect CV-1 cells and transformants were selected by suspension in soft agar (Gluzman *et al.*, 1977). The isolation of transformed cells which could not support the replication of SV40 early mutants suggested that UV irradiation can inactivate SV40 replication functions without abolishing transforming functions. This hypothesis was actually proved correct (Gluzman and Ahrens, 1982). In a more recent approach, Gluzman *et al.* (1980) have used origin-defective mutants of SV40, which lack specific sequences required to initiate replication, to transform monkey CV-1 cells (Gluzman, 1981). Contrary to CV-1 cells transformed with UV-irradiated SV40, *ori*⁻ SV40-transformed cells (COS cells) can support the replication of SV40 early mutants (Gluzman, 1981).

The transformation of human cells, which are semipermissive for SV40, has also been facilitated by the use of *ori*⁻ SV40 mutants (Small *et al.*, 1982; Nagata *et al.*, 1983; Boast *et al.*, 1983; Major and Matsumura, 1984). Nevertheless, permissivity problems are not the only difficulties encountered in human cell transformation by SV40. Although SV-40 transformed human cells display an expanded life span, as compared to untransformed counterparts, they usually undergo senescence. Continuously growing human cell lines, which have been obtained after SV40 infection, have usually been passaged for prolonged periods of time without immediate selection and the role played by SV40 in the control of their final phenotype is unclear (see the review by Sack, 1981).

Contrary to SV40, there is no known cell type which can be infected by polyoma and which does not display some permissivity. Semipermissive clonable cell lines, like the Syrian hamster BHK-21 (MacPherson and Stoker, 1962) or the Fischer rat FR3T3 (Seif and Cuzin, 1977), have been developed and can be used to obtain polyoma transformants. It is likely that all the cells in such semipermissive populations do not continuously produce the cellular factors which are needed to ensure successful replication of the virus and that viable nonproducer transformants derive from this fraction of the population. Some clones of polyoma-transformed BHK cells can be induced to produce progeny virus when chilled from 39°C to 31°C (Folk, 1973). Similarly, some polyoma-transformed rat cells start synthesizing viral DNA after exposure to mitomycin C (Neer *et al.*, 1977).

In recent years, the use of appropriate techniques which enable the uptake of DNA by cells (e.g., Graham and Van der Eb, 1973) as well as the molecular cloning of viral DNA fragments has greatly extended the possibilities of studying the effects of selected viral genes on the transfected cell phenotype. The transfection techniques circumvent the limitations imposed by cellular sensitivity to infection and bypass the requirements for viral DNA uncoating, while molecular cloning techniques provide unlimited amounts of viral DNA sequences which do not contain some of the viral information required for replication.

D. Expression of Viral Functions in Transformed Cells

All transformed cell lines, established from SV40- or polyoma-infected cell cultures, contain virus-specific RNAs (Benjamin, 1966) and proteins, which were first detected by immunological techniques, using antisera derived from animals bearing virus-induced tumors (Black *et al.*, 1963; Habel, 1965). These "tumor antigens" can also be immunoprecipitated by a technique introduced by Tegtmeyer *et al.* (1975). Thus, continuous expression of viral information is observed, as a rule, in transformed cells.

RNA sequences which originate from the early part of the viral genomes are always detected (Khoury *et al.*, 1973; Kamen *et al.*, 1974). In most SV40-transformed cells, the two early proteins, large T and small t, can be immunoprecipitated from cell extracts (Crawford *et al.*, 1978). However in some SV40-transformed mouse or rat cell lines, the antitumor antisera precipitate proteins which are of much larger apparent molecular weight than large T (super-T) (May *et al.*, 1981; Lovett *et al.*, 1982; Chen *et al.*, 1983b). Upon further recloning, super-T-producing rat or mouse cells frequently segregate clones which produce large T of normal size (May *et al.*, 1984; Chen *et al.*, 1984). The transition between normal large T- and super-T-producing cells is due to a tandem duplication occurring inside the viral DNA early region (May *et al.*, 1981; Lovett *et al.*, 1982), a process which may require a functional replication origin (Chen *et al.*, 1983b). Super-T antigens, although they specifically recognize the SV40 *ori* sequences, are totally inefficient in viral DNA replication but retain some of the transforming ability of normal large T (May *et al.*, 1983).

In polyoma-transformed cells, mRNAs which correspond at least to the 5′-terminal half of the early region have been identified in all cell lines examined (Kamen *et al.*, 1974, 1980). Accordingly, middle T and small t are also detected (Ito and Spurr, 1980). On the contrary, mouse cells transformed by polyoma usually do not make full-sized large T (Ito and Spurr, 1980). This is usually observed with all rodent cell lines in which spontaneous or artificial induction of viral DNA does not occur (Kamen *et al.*, 1980). However, in polyoma transformants that do contain under some circumstances free viral DNA, the viral mRNAs are identical to those found in lytically infected cells and large T is also present.

E. Subcellular Localization of Viral Proteins in Transformed Cells

The cellular localization of large T in cells transformed by SV40 is mainly nuclear as shown by indirect immunofluorescence techniques. Moreover, a small fraction (about 5%) of the total large T is strongly associated with the nuclear matrix (Verderame *et al.*, 1983). Similarly, in polyoma-transformed cell lines, which produce large T, the localization of large T is mainly nuclear (Ito *et al.*, 1977a; Zhu *et al.*, 1984). The nuclear localization of SV40 large T is probably correlated with its DNA binding activity. In this respect, differences in behavior between SV40 large T from lytically infected cells (Scheidtmann *et al.*, 1984) and SV40 large T from the SV40-transformed human cell line SV80 (Bradley *et al.*, 1982) have been reported. In both types of cells, SV40 large T can be detected in subcellular extracts in multiple forms, either monomeric or polymeric. In lytically infected cells, the monomeric form, which apparently corresponds to newly synthesized hypophosphorylated molecules, has a greater affinity for DNA than do the polymeric forms (Scheidtmann *et al.*, 1984). In SV80 cells, on the contrary, the dimeric and tetrameric forms of large T are more active than the monomeric form (Bradley *et al.*, 1982). This interesting observation should be appreciated, nevertheless, in reference to the observation by Gruss *et al.* (1984) that large T purified from this particular cell line, SV80, has lost the high specificity and affinity of wild-type SV40 large T for wild-type SV40 DNA and does not bind specifically to SV40 DNA cloned from the SV80 cell line.

The transport of SV40 large T to the nucleus of infected or transformed cells shows a strong requirement for a particular amino acid sequence, rich in basic amino acids, namely the sequence located between residues 127 and 133: Lys-Lys-Lys-Arg-Lys-Val-Glu (Lanford and Butel, 1984; Kalderon *et al.*, 1984). Mutations occurring in this sequence alter the transport to the nucleus. In particular, the replacement of Lys^{128} by Thr (Kalderon *et al.*, 1984) or by Asn, as occurs in the SV40–adenovirus 7 hybrid PARA (cT) (Lanford and Butel, 1981), completely blocks the transfer of large T to the nucleus, as measured by indirect immunofluorescence. Lanford and Butel (1984) have constructed an SV40 mutant,

SV40 (cT)-3, linking the SV40 early sequences from PARA (cT) to late sequences from wild-type SV40. This mutant is defective on CV-1 cells, but can be replicated on COS-1 cells, although, in these cells, the defect in nuclear transport is *trans* dominant: SV40 (cT)-3-infected COS-1 cells do not show the nuclear indirect immunofluorescence specific of SV40 large T. This observation draws attention to the interpretation of experiments relying on the immunofluorescent-detection of viral antigens: it is clear, in this particular instance, that the level of intranuclear large T, sufficient to ensure the replication of the SV40 (cT)-3 mutant in COS-1 cells, is below the limit of sensitivity of the immunofluorescence technique.

Mutants affected in nuclear transport are still able to induce the formation of dense foci on monolayers of the established Rat-1 cell line (Kalderon *et al.*, 1984), but according to preliminary results reported by Lanford and Butel (1984) the cT mutation affects the efficiency of viral transformation of primary rodent cells.

In both SV40- and polyoma-transformed cells, protein molecules which behave immunologically as large T, are also detected in the plasma membrane (Ito *et al.*, 1977a,b; Segawa and Ito, 1982; Henning *et al.*, 1981). In all cases, these large T-like molecules represent a minor fraction of the total large T. Most of the SV40 large T-like molecules associated with the plasma membrane appear to be associated with a detergent-resistant lamina of the membrane from which it can be solubilized with the zwitterionic detergent, Empigen BB (Klockmann and Deppert, 1983). The transport to the membrane of these large T-like molecules is not decreased, but rather increased, in the case of the PARA (cT) SV40–adeno 7 hybrid (Lanford and Butel, 1982). The significance and origin of the large T-like membrane-located proteins will be further discussed in Section III.F.

Studies on the subcellular localization of small t of SV40 or polyoma have been greatly simplified by the use of mutants or recombinants, from which small t is preferentially or exclusively synthesized (Ellman *et al.*, 1984; Zhu *et al.*, 1984). Monoclonal anibodies directed against the amino acid sequences specific for small t have also been used (Montano and Lane, 1984). These different approaches have led to a similar conclusion for both SV40 and polyoma small t: this viral antigen is present at the same time in the cytoplasm and in the nucleus, and this conclusion was reached, in the case of SV40, both by immunofluorescence and by subcellular fractionation followed by immunoprecipitation (Ellman *et al.*, 1984). The precise localization of SV40 small t inside the nucleus appears to be distinct from that of large T (Montano and Lane 1984).

The polyoma-specific middle T antigen has been detected in the plasma membrane of polyoma-transformed cells (Ito *et al.*, 1977b). More recent experiments have indicated a complex distribution, which includes the perinuclear region, the cytoplasm, and the plasma membrane (Zhu *et al.*, 1984). A relationship between the localization of middle T and the phosphorylation level has been observed: while middle T is distributed approximately equally between the plasma membrane and the rest of the cell, 80–90% of the *in vivo* labeled ^{32}P-middle T is in the membrane fraction, mostly phophorylated on serine and threonine residues (Segawa and Ito, 1982). Most of the middle T-associated protein kinase activity, measured *in vitro*, is also found in the plasma membrane fraction (Segawa and Ito, 1982).

All mutants in which the C-terminus of middle T is missing are totally defective in transformation of continuous rodent cell lines (Novack and Griffin, 1981; Mes and Hassell, 1982; Templeton and Eckhart, 1982; Templeton *et al.*, 1984). The transformation-defective Py-1387-T mutant (Carmichael *et al.*, 1982), in which a G-C base pair at nucleotide position 1387 is replaced by a T-A base pair, produces normal-sized small t and large T but a truncated middle T which lacks the last 37 amino acid residues at the C-terminus including a 20-amino acid-long hydrophobic "tail." The truncated middle T is exclusively located in the cytoplasm of cells infected with the Py-1387-T mutant, and is totally inactive in an *in vitro* protein kinase assay. The membrane insertion of middle T thus appears to depend on the hydrophobic C-terminal tail and is essential for cell transformation.

F. Tumor-Specific Transplantation Antigens

Animals can be immunized against a further challenge with SV40- or polyoma-induced transplantable tumors. This immunological tolerance can be induced by injecting the virus itself, cells derived from virus-induced tumors, or *in vitro* transformed cells. In tumor-rejection tests, no cross-reactivity between SV40 and polyoma is observed, and cells of different species transformed by the same virus apparently carry the same transplantation antigen (Habel, 1961; Sjogren *et al.*, 1961; Khera *et al.*, 1963; Habel and Eddy, 1963; Koch and Sabin, 1963; Defendi, 1963; Sjogren, 1965). Thus, it appears likely that tumor-specific transplantation antigens (TSTA) are coded by the virus genome in SV40- or polyoma-transformed or tumor cells.

Immunization against tumor transplantation can also be obtained after injection into animals of cell membrane preparations from SV40-induced tumor cells (Tevethia and Rapp, 1966; Coggin *et al.*, 1967), suggesting that the TSTA is located on the cell surface.

The tumor rejection mechanism appears to be mediated by sensitized T lymphocytes and *in vitro* the presence of TSTA on tumor or transformed cells can be characterized by T-lymphocyte-mediated cell lysis (Tevethia *et al.*, 1974; Trinchieri *et al.*, 1976; Gooding, 1977; Knoles *et al.*, 1979).

The relationship between SV40 early proteins and TSTA has been established through the following observations:

1. The presence of large T antigen (or closely related antigens) at the cell surface can be demonstrated serologically with antisera directed against purified large T (Deppert and Henning, 1979; Soule and Butel, 1979; Deppert *et al.*, 1980; Henning *et al.*, 1981; Lange-Mütschler *et al.*, 1981).
2. Lactoperoxidase-catalyzed ^{125}I iodination of cell surface (Deppert and Henning, 1979; Deppert *et al.*, 1980) and of plasma membrane preparations from SV40-transformed cells (Soule *et al.*, 1980) detected a protein with the same apparent molecular weight as large T.
3. Analysis of SV40 tsA mutant behavior suggested that large T and TSTA are coded by the same viral gene (Anderson *et al.*, 1977; Tevethia and Tevethia, 1977; Deppert, 1980).
4. Purified large T can be used to immunize mice against an SV40 tumor cell challenge (Chang *et al.*, 1979). Similarly, the purified D2 protein, a large T-related protein encoded by the Ad2^{+}D2 adeno 2–SV40 hybrid virus, can be used to induce SV40-tumor immunity in mice and the generation of cytotoxic T lymphocytes (Tevethia *et al.*, 1980). Monoclonal antibodies directed against SV40 large T are moreover able to block *in vitro* sensitized T-lymphocyte cytotoxicity (Pan and Knowles, 1983).

There is little doubt therefore that TSTA is carried by large T itself or large T-related molecules located at the cell surface. Although some reports (Santos and Butel, 1984b) suggest that both N- and C-termini of large T are exposed on the cell surface of SV40-transformed cells, the antigenic sites recognized by cytotoxic T lymphocytes are specified by DNA sequences which correspond to the N-terminal half of large T (Tevethia *et al.*, 1983, 1984).

Large T from extracts of SV40-transformed cells binds to the surface of cells in culture (Lange-Mütschler and Henning, 1982) by a process which involves the formation of a covalent ester bond with cell surface lipids (Henning and Lange-Mütschler, 1983; Lange-Mütschler and Henning, 1984). Schmidt-Ullrich *et al.* (1982) have shown that, among the four large T-specific tryptic peptides which can be labeled after lactoperoxidase-catalyzed radioiodination of the outer surface of the plasma membrane, one can also be labeled with [^{14}C] glucosamine. Therefore, the large T-related molecules located at the cell surface and responsible for TSTA activity could be posttranslational modifications, by acylation and glycosylation, of nuclear large T. At the present time, nevertheless, it is not possible to exclude other alterations in the structure of membrane-located large T as compared to nuclear large T.

Large T appears to be constantly shed from and replaced at the surface of transformed cells (Santos and Butel, 1984a). These observations might be related to the mechanisms of the immunological responses in animals exposed to SV40-induced tumors or immunized with purified large T or D2 proteins.

G. Association of Cellular Proteins with Viral Proteins

With the exception of a report on the *in vitro* binding of SV40 small t to several cell proteins, including tubulin (Yang *et al.*, 1979), most of the experimental observations in this field were concerned with the formation of complexes between SV40 large T and the cellular protein p53 on the one hand, and between polyoma middle T and $pp60^{c\text{-}src}$ on the other.

The existence of a complex between SV40 large T and a cellular protein was recognized in all cell lines transformed by the wild-type virus (Lane and Crawford, 1979; Linzer and Levine, 1979; McCormick and Harlow, 1980). Because of its apparent molecular weight on SDS-polyacrylamide gels, the cellular protein is usually referred to as the p53 protein.

SV40 large T and the cellular p53 occur in equimolar amounts in the complex (Freed *et al.*, 1983) which sediments on sucrose density gradients in the 23–26 S fraction in extracts from both human SV40-transformed SV80 cells (Bradley *et al.*, 1982) and mouse SV40-transformed 3T3 cells (McCormick and Harlow, 1980). No p53 was found in the large T dimer 7 S and tetramer 15 S fraction (Bradley *et al.*, 1982).

In SV40-transformed cells, all of the available p53 appears to be complexed to large T (Greenspan and Carroll, 1981). A similar observation was made in adenovirus 5-transformed mouse 3T3 cells, where a complex between the adenovirus E1b protein and p53 has been demonstrated (Sarnow *et al.*, 1982). The large T–p53 complex has the same *in vitro* DNA binding properties, including specific binding to SV40 *ori* sequences, as free large T (Reich and Levine, 1982; Scheller *et al.*, 1982).

p53 is present in higher amounts in SV40-transformed mouse 3T3 cells than in nontransformed control cells. But increased levels of cellular p53 are not restricted to SV40-transformed cells. They are also found in many transformed cell lines, including rodent cell lines transformed by chemical carcinogens (DeLeo *et al.*, 1979; Jörnvall *et al.*, 1982) and many cell lines derived from human tumors (Crawford *et al.*, 1981). Some human tumor cell lines nevertheless do not contain increased amounts of p53 (e.g., HeLa cells; the EJ bladder carcinoma cell line) (Crawford *et al.*, 1981). High levels of p53 are also observed in embryo cells (Chandrasekaran *et al.*, 1981.)

p53 is a phosphoprotein (Linzer and Levine, 1979) localized in the nucleus (Dippold *et al.*, 1981). The presence of a protein kinase activity associated with p53 in large T–53 complex has been described by Jay *et al.* (1981), but this observation was not confirmed by van Roy *et al.* (1984).

In nontransformed mouse 3T3 cells, p53 appears to be a cell cycle protein, which could be involved in cell cycle regulation. When quiescent 3T3 cells are stimulated to divide, an increase in p53 gene transcription precedes an increased steady-state level of the protein, which peaks in the late G1 phase of the cell cycle (Reich and Levine, 1984). These observations were interpreted in terms of a dual control at the levels of mRNA transcription and protein turnover.

A biological activity of p53 in cell cycle regulation is strongly suggested by the effect of the microinjection of monoclonal antibodies against p53. This microinjection inhibits serum-stimulated DNA synthesis in quiescent Swiss 3T3 cells and prevents the transition from the G0 to the S phase of the cell cycle (Mercer *et al.*, 1982).

In SV40-transformed cells, the steady-state level of p53 is controlled at the posttranscriptional level. No change in the rate of transcription, as compared to nontransformed control cells, was observed but the half-life of the protein itself changed from 20 min in mouse 3T3 cells to 24 hr in SV40-transformed cells (Oren *et al.*, 1981; Oren and Levine, 1983). Thus, accumulation of p53 in SV40-transformed cells is the result of the protection of the protein through complex formation with the viral protein. The same is also true in advenovirus 5- or EBV-transformed cells (Oren and Levine, 1983).

The importance of p53-SV40 large T complex formation in cellular transformation is further suggested by recent observations on the role played by p53 in other systems. In the Abelson murine leukemia virus-transformed L12 cell line, it has been shown that the expression of the p53 cellular gene is prevented by insertional mutation of Moloney murine leukemia virus-like sequences (Wolf *et al.*, 1984). The L12 cell line, when injected into syngeneic mice, produces only local tumors which eventually regress. Reintroduction of a functional p53 gene in the L12 cells produces p53-expressing cell lines which cause lethal progressing tumors.

The highly conserved p53 gene (Chandrasekaran *et al.*, 1981; Fanning *et al.*, 1981; Jörnvall *et al.*, 1982) as a matter of fact can behave as an oncogene in *in vitro* cell transformation. Recombinant DNAs containing the p53 gene, placed under the transcriptional control of a viral promoter (Moloney murine sarcoma- and Moloney murine leukemia-LTR_S, or polyoma virus early promoter), can cooperate with the human T24-Harvey-*ras* oncogene upon transfection of primary rodent cells. The transformed cells thus obtained are tumorigenic in syngeneic animals (Eliyahu *et al.*, 1984; Parada *et al.*, 1984). Another expression vector, in which a p53 cDNA copy is linked to the Rous sarcoma virus LTR, is able to immortalize Wistar adult rat xiphisternum chondrocytes (Jenkins *et al.*, 1984).

A change in the normal regulation of p53 accumulation, caused by complex formation with viral proteins, might be important in cellular transformation; p53 could play the same role as the *myc* oncogene in the control of cellular division. In agreement with this hypothesis, the turnover rate of p53 returns to normal in tsA SV40-transformed cell lines incubated at the restrictive temperature (Oren *et al.*, 1981). Along the same line of argument, it was observed that precrisis mouse cell cultures from different strains of mice showed different levels of p53. Cells which display the higher levels were more easily transformed by SV40 (Chen *et al.*, 1983a). Nevertheless, the stability of p53 complexed to SV40 large T is probably not sufficient to maintain a transformed phenotype. In cells transformed by the tsA1499 SV40 mutant, the transformed phenotype is cold sensitive, but p53 stabilization is not (Bouck *et al.*, 1984).

The behavior of other types of SV40 large T mutants is also of interest. While cloned DNAs coding for the super-T forms of large T are able to induce both p53 stabilization and the transformation of secondary rat kidney cells (May *et al.*, 1983) or BALB/c 3T3 cells (Chaudry *et al.*, 1982b), deletion mutants, which produce truncated large T-related proteins, can transform Rat-1 cells (Clayton *et al.*, 1982) or mouse C3H 10 T 1/2 cells (Sompayrac *et al.*, 1983) without observable stabilization of p53. It could be concluded that if the activated p53 gene has the same function as *myc* in immortalization (Eliyahu *et al.*, 1984; Parada *et al.*, 1984; Jenkins *et al.*, 1984), stabilization of p53 is not required to transform already established cell lines.

This interpretation of the role of p53 stabilization by SV40 large T in cellular immortalization unfortunately does not apply to adenovirus 5-transformed cells. In the adenovirus case, the stabilizing viral protein has been identified as the product of the E1b gene, despite the fact that the immortalizing function of adenoviruses is exerted by the E1a gene product (Van der Eb *et al.*, 1977; Shiroki *et al.*, 1979).

In polyoma-transformed cells, large T, when it exists at all, does not associate with p53 (Dilworth *et al.*, 1984). On the other hand, polyoma middle T has recently been identified in a complex with the product of a cellular oncogene, c-*src*.

A protein kinase activity has been demonstrated in immunoprecipitates containing middle T (Smith *et al.*, 1979; Eckhart *et al.*, 1979; Schaffhausen and Benjamin, 1979). There is a correlation between the kinase activity measured *in vitro* in immunoprecipitates from cells infected with polyoma mutants and their transforming ability. In particular, no kinase activity is observed with hr-t mutants which are totally defective in transformation (Schaffhausen and Benjamin, 1981; Benjamin, 1982).

In the process of *in vitro* phosphorylation, middle T itself becomes phosphorylated on a major site, which was identified as the tyrosine residue at position 315 in the peptide chain, although the phosphorylation of Tyr^{315} is not observed *in vivo* (Schaffhausen and Benjamin, 1981). This tyrosine residue is preceded by a cluster of six glutamic acid residues, a sequence which is reminiscent of the amino acid sequences surrounding sites of tyrosine phosphorylation in the transforming proteins of several retroviruses (Patschinsky *et al.*, 1982; Bishop, 1983). This sequence has been either altered by mutations (Oostra *et al.*, 1983; Carmichael *et al.*, 1984) or suppressed totally or partially by deletions (Griffin and Maddock, 1979; Ding *et al.*, 1982; Nilsson *et al.*, 1983; Mes-Masson *et al.*, 1984) in order to determine its significance in transformation. The general conclusion, which can be drawn from contradictory results, is that the Glu-Glu-Glu-Glu-Glu-Glu-Tyr[315] sequence is not essential for transformation, when Rat-1 cells are used in transformation tests (Oostra *et al.*, 1983; Mes-Masson *et al.*, 1984). Nevertheless, a mutant in which Tyr[315] is replaced by Phe is transformation-defective when the rat F111 line is employed (Carmichael *et al.*, 1984). These differences could be explained by differences in the expression of the c-*src* protooncogene in different cell types (Mes-Masson *et al.*, 1984).

In spite of these observations, which suggest an important role in transformation for the middle T-associated tyrosyl-kinase activity, no increase in the amount of phosphotyrosine, comparable to that observed in Rous sarcoma virus-transformed cells, is detected in polyoma virus-transformed ones (Sefton *et al.*, 1980).

Although it was suggested initially that the kinase activity could be carried by the middle T molecule itself, it was soon demonstrated that, in the course of a purification procedure, based on immunoaffinity and applied to middle T, the kinase activity was not associated with the main protein fraction. Rather, it was associated with a minor fraction which sedimented on sucrose density gradients with a velocity suggestive of an apparent molecular weight in the vicinity of 200K (Walter *et al.*, 1982).

With the use of monoclonal antibodies directed against either middle T or synthetic peptides related to middle T or $pp60^{c\text{-}src}$, it was finally demonstrated that polyoma middle T associated with the product of the c-*src* oncogene (Courtneidge and Smith, 1983). A small fraction only of the cellular $pp60^{c\text{-}src}$ is engaged in the complex, but its kinase activity is stimulated, as compared to the free protein (Bolen *et al.*, 1984). This stimulation is probably important in transformation; the dl23 mutant, which is defective in transformation (Griffin and Maddock, 1979), produces a middle T, which can still form a complex with $pp60^{c\text{-}src}$ (Courtneidge and Smith, 1983). But the associated kinase activity is impaired relative to the wild-type case (Schaffhausen and Benjamin, 1981).

H. Stimulation of Cellular Gene Expression in Transformed Cells

It must be recalled that, upon infection of either permissive or nonpermissive resting cells by polyoma or SV40, cellular DNA replication followed by mitosis is quickly observed (Dulbecco *et al.*, 1965; Winocour *et al.*, 1965; Gershon *et al.*, 1965, 1966; Sheinin, 1966; Hatanaka and Dulbecco, 1966; Henry *et al.*, 1966). This cellular response to infection, which prepares the cell to efficiently replicate the viral DNA, is probably related to the influence exerted by viral functions expressed in transformed cells on their growth regulation (Weil, 1978).

Correlatively, infected cells display an increased activity of several enzymes involved in DNA replication (Hartwell *et al.*, 1965; Kit *et al.*,

1966). Since this increase in activity shows a requirement for protein synthesis (Frearson *et al.*, 1966), it means that viral functions stimulate the biosynthesis of cellular proteins.

This stimulation might occur through a control of the steady-state level of the corresponding mRNAs. As a matter of fact, several groups have isolated cDNA clones which correspond to mRNAs which are present at higher levels in SV40-transformed cells than in the normal parental untransformed cell lines (Schutzbank *et al.*, 1982; Scott *et al.*, 1983). Among these mRNAs, some may occur at elevated levels in many transformed cell lines, irrespective of the nature of the transforming agent. A case in point is provided by the accumulation, in the SV40-transformed SV3T3 C1 38, derived from the BALB/c 3T3 cell line, of a 1.6-kb mRNA. This particular mRNA, which probably encodes a Qa/T1a class I major histocompatibility complex antigen, is also accumulated in all transformed mouse fibroblasts examined (Brickell *et al.*, 1983). But other mRNAs are apparently more specific of SV40 transformation (Schutzbank *et al.*, 1982) as exemplified by the mRNA encoding a 58K mitochondrial protein (Zuckerman *et al.*, 1984).

Given the type of analyses which have been performed, it is not possible to decide whether a change in transcription or in turnover is responsible for the increased mRNA steady-state levels. But since it is known that SV40 large T can directly stimulate the transcription of the SV40 late mRNAs (Brady *et al.*, 1984; Keller and Alwine, 1984), it is conceivable that SV40 large T can influence the rate of transcription of some cellular genes.

Directly related to these possible regulatory effects of SV40 large T on cellular gene transcription, is the reactivation of silent rRNA genes in somatic cell hybrids after SV40 or polyoma infection (Soprano *et al.*, 1980). In the case of SV40, the reactivation of rRNA genes is controlled by large T, because it is abolished at the restrictive temperature when tsA mutants of the virus are used. Small t is not involved, since d154-59 mutants, which do not produce normal small t, are active (Soprano *et al.*, 1980). Using deletion mutants of SV40, which produce truncated large T antigens, Soprano *et al.* (1983) were able to demonstrate that, while a truncated large T encompassing the first 272 N-terminal amino acid residues is sufficient to induce cellular DNA synthesis in resting cells, the rRNA gene reactivation requires sequences extending between amino acids 72 and 509.

It is not clear whether the stimulation of rRNA synthesis plays any role in cell transformation. But transformation by SV40 and polyoma also induces the production of transforming growth factors (TGFs), which can be recognized by their effect on the anchorage-independent growth of NRK or Rat-1 cells (Kaplan *et al.*, 1981; Kaplan and Ozanne, 1982). The production of TGFs by SV40- or polyoma-transformed cells can lead to an autocrine system, in which autostimulation of cell growth occurs (Kaplan *et al.*, 1982). In polyoma transformants, the production of TGFs is not under the control of large T or small t, but under the control of middle T, as shown by transformation of cells (Kaplan and Ozanne, 1982) by recombinant plasmids encoding only one type of tumor antigen (Treisman *et al.*, 1981).

I. The State of the Viral Genome in Transformed Cells

The vast majority of transformed cell lines, established from SV40- or polyoma-induced tumors or after *in vitro* infection of cell cultures, do not produce virions in detectable amounts. The usual persistence of viral gene expression requires that all or part of the virus genome be maintained in the transformed cells.

The first evidence for the presence of viral DNA sequences in SV40- and polyoma-transformed cells was obtained by hybridization experiments, in which highly radioactive RNA, transcribed *in vitro* from the viral DNA by *E. coli* RNA polymerase, was matched to DNA extracted from the transformed cells (Westphal and Dulbecco, 1968). This important observation established for the first time the persisting presence of viral DNA sequences in transformed cells.

Because of technical pitfalls, which were recognized later, the direct hybridization technique proposed by Westphal and Dulbecco (1968) was unable to provide reliable values of the number of viral genome equivalents per transformed cell or per cellular diploid genome equivalent. A technique, based on the measurement of the rate of reannealing of small amounts of denatured radiolabeled viral DNA in the presence of increasing amounts of unlabeled cellular DNA (Gelb *et al.*, 1971), gave more reliable values. In SV40-transformed mouse cells, the number of viral genome equivalents was found to vary between 1.1 and 8–10 copies per cellular diploid genome equivalent. In several polyoma-transformed mouse cells, 0.6 to 2.9 viral genome equivalents were measured (Kamen *et al.*, 1974), while in transformed rat cells, the number of polyoma virus genome equivalents could be as high as 50 (Zouzias *et al.*, 1977). More detailed observations on the presence of various parts of the viral genome were made with the help of specific viral DNA radiolabeled fragments, instead of the total viral genome. It was then realized that in the SV40-transformed SVT2 cell line, if the use of the total viral genome in reannealing experiments led to an average value of 1.6–2.2 viral DNA copies per cellular diploid genome equivalent, early viral sequences were represented 6 times more frequently than the late sequences, which occurred at the level of only one copy per cellular diploid genome equivalent (Botchan *et al.*, 1974). The unequal representation of early and late viral information in the subline SVT2/S has since been confirmed by cloning the SV40 viral information (Sager *et al.*, 1981).

Evidence for a covalent linkage between viral sequences and cellular chromosomal DNA was first obtained for SV40-transformed cells by Sambrook *et al.* (1968) and later for polyoma-transformed hamster cells (Shani *et al.*, 1972; Folk, 1973; Manor *et al.*, 1973). These observations were instrumental in proving that the persistence of viral information in transformed cells could be the result of their insertion in the cellular genome, thus providing for their replication at each cellular division along with the cellular DNA.

Free viral DNA copies can also be observed in some transformed cells, particularly at early times after transformation of rat embryo cells by polyoma (Zouzias *et al.*, 1977; Birg *et al.*, 1979). Similarly, a large number (2000 copies/cell) of unintegrated supercoiled viral DNA molecules were observed in human fibroblasts transformed by a fragment of SV40 DNA, covering the early region of the genome (Zouzias *et al.*, 1980). Upon further passages in culture, the number of free viral DNA copies usually decreases in polyoma-transformed rat cells (Zouzias *et al.*, 1977; Birg *et al.*, 1979).

In situ hybridization was used to analyze the distribution of free viral DNA copies in such transformed cell populations. Both in polyoma-transformed rat cells and in SV40 fragment-transformed human cells, a small fraction of the total population was found to contain a large number of free copies (Neer *et al.*, 1977; Zouzias *et al.*, 1977, 1980). In polyoma-transformed rat cells, 0.04 to 0.25% of the total cell population were positive in *in situ* hybridization, each positive cell containing up to 25,000 copies (Neer *et al.*, 1977).

In cells transformed with the tsa polyoma mutant, the presence of free viral DNA copies is only observed at the permissive temperature. The persistence of free viral DNA is thus dependent on the activity of large T (Folk and Bancuk, 1976; Gattoni *et al.*, 1980), as well as on a functional viral replication origin (Pellegrini *et al.*, 1984). At the restrictive temperature, the viral information is exclusively maintained in cellular chromosome-integrated forms.

Free viral DNA copies are mainly observed in transformed cells which derive from semipermissive cell types. Their presence is interpreted as due to replication from the integrated viral copies in the fraction of the transformed cell population, which contains sufficient amounts of permissivity factors. The precise mechanism by which this replication occurs will be discussed in Section III.J.

A more precise analysis of the structure of the viral genome copies integrated in cellular chromosomes become possible upon adoption of Southern's tehnique (1975). Under appropriate conditions, as little as 0.02 copy of viral DNA per diploid equivalent of the cell genome could be detected in a 5-μg sample of transformed cell DNA (Ketner and Kelly, 1976; Botchan *et al.*, 1976). This approach has more recently been complemented by molecular cloning of viral insertions, enabling sequence determination of both viral information and flanking cellular sequences (e.g., Botchan *et al.*, 1980; Clayton and Rigby, 1981; Stringer, 1982b).

Although some transformed cells contain a unique integrated copy of the viral genome at a single chromosomal site, e.g., the SV40-transformed rat 14B cell line (Bochan *et al.*, 1976), many contain head-to-tail tandem repeats. The formation of tandem repeats appears to be favored by viral DNA replication occurring at the time of integration. They are actually found more frequently in semipermissive transformed cells: in polyoma-transformed rat cells, tandem repeats of up to 5 full-length copies have been observed (Birg *et al.*, 1979). Similar observations were made with SV40-transformed human cells (Campo *et al.*, 1978). In F2408 rat cells, which have been transformed by transfection with tsa polyoma DNA, tandem integration was favored when the transformation was performed at the permissive temperature. At the restrictive temperature, single-copy integrations were mostly observed (Della Valle *et al.*, 1981).

But tandem integrations have also been observed in fully nonpermissive cells, such as the SV40-transformed mouse SV3T3 cell clones examined by Clayton and Rigby (1981). Chia and Rigby (1981) have demonstrated the production, in SV40-infected mouse cells, of high-molecular-weight head-to-tail viral DNA concatemers, which form even at 39.5°C after infection with the tsA58 mutant. Double infection experiments, performed with polyoma wild-type and h-rt mutants (Friderici *et al.*, 1984) or tsa and hr-t mutants (Fluck *et al.*, (1983), have led to the conclusion that oligomer formation, leading to head-to-tail tandem integration, results from recombination.

In DNA-transfection experiments, integration is not absolutely dependent on an active large T protein, as shown for polyoma DNA by Israel *et al* (1979). Nevertheless, the efficiency of transformation of F2408 rat cells by tsa polyoma DNA is 20-fold higher at the permissive temperature than at the restrictive one (Della Valle *et al.*, 1981). In similar experiments, in which a recombinant DNA associating 1.5 copies of the polyoma DNA mutant tsP155, temperature-sensitive for viral DNA replication (Eckhart, 1974), and mouse cellular unique and repetitive sequences, integration occurred randomly with respect to the transfected recombinant DNA at 33°C; it preferentially involved the cellular sequences at 39°C (Wallenburg *et al.*, 1984). It appears therefore that an active large T protein can modify the substrate for integration.

Viral DNA integrated copies frequently are not colinear with the viral genome, even in the case of single-copy integrations, as shown by the SV40-transformed rat 14B cell line (Botchan *et al.*, 1980). In many SV40-transformed rat and mouse cell lines, tandem duplications are observed in the large T-coding sequences, with production of large T-like high-molecular-weight proteins (super-T; see Section III.D).

Although it has been concluded from the first detailed studies on integration, that integration within one established cell line was stable (Botchan *et al.*, 1976), much evidence to the contrary has now accumulated. Early after infection and selection of transformed clones or cell lines, the state of the viral genome evolves, usually from a complex to a simpler situation (Zouzias *et al.*, 1977; Birg *et al.*, 1979; Hiscott *et al.*,

1981; Bender and Brockman, 1981). Even in long-established cell lines, modifications in the integrated viral copies occur at relatively high frequencies (Clayton and Rigby, 1981; Sager *et al.*, 1981). When transformed cells are followed during *in vitro* subculturing, the observed alterations are compatible with the hypothesis that their evolution might occur through selection of those cells which contain optimal quantities of the domains of viral proteins which are responsible for rapid cell growth (Clayton and Rigby, 1981).

There is no specific site on the viral genome through which integration occurs and most experimenters have concluded that integration is essentially random with respect to viral DNA sequences, although some regions are preferred. Clayton and Rigby (1981) have noticed in SV40-transformed mouse BALB/c 3T3 cell lines a clustering of integration sites in the 5′-half of the late viral region and in the middle of the large-T-coding region. Maruyama and Oda (1984) in the independently established SV40-transformed mouse kidney W-2D-11 cell line demonstrated that the left-hand and right-hand junctions occur at positions on the viral DNA which are within a few nucleotides from previously observed junctions in the rat 14B cell line (Botchan *et al.*, 1980). Other exceptions to completely random integrations have been recognized in SV40-transformed rat fibroblast lines (Mougneau *et al.*, 1980) and in polyoma-transformed Syrian hamster cell lines (Rey-Bellet and Türler, 1984).

On the cellular DNA side, there is no specificity either with respect to particular chromosomes. Somatic cell hybrids between SV40-transformed human cells and mouse cells have been used to map SV40 integration sites to particular human chromosomes. In different transformed cell lines, sites were identified on chromosome 5 (Hwang and Kucherlapati, 1980), 7 (Croce *et al.*, 1973; Campo *et al.*, 1978; Rabin *et al.*, 1984), 8 (Kucherlapati *et al.*, 1978), 12 (Croce, 1981), and 17 (McDougall *et al.*, 1976; Croce, 1977). In two different cell lines, *in situ* hybridization localized SV40-integration to band 031 on chromosome 7 (Rabin *et al.*, 1984), but analysis of the integration by the more sensitive Southern blotting technique showed that the sites were actually different (Campo *et al.*, 1978). It is clear therefore that integration is also random with respect to cellular DNA.

Sequence determination at the junction sites did not detect strong homologies between cellular and viral DNAs in SV40-transformed cells (Botchan *et al.*, 1980; Stringer, 1982b; Maruyama and Oda, 1984; Hasson *et al.*, 1984), eliminating legitimate recombination as a possible mechanism of integration. Even in the rat 14B cell line, the integration of new genomes upon superinfection does not occur at the site of the first integration (Botchan *et al.*, 1979).

If long homologous sequences between the viral and cellular sequences located at the site of integration have never been identified, patchy homologies, over 3 to 5 base pairs, are usually found (Stringer, 1982b; Maruyama and Oda, 1984). Moreover, one junction occurs at the level of a TG repetition in the cellular DNA sequence in the Fischer rat embryo fibroblast SVRE9 line (Stringer, 1982b), while in the mouse kidney W-2D-11 cell line studied by Maruyama and Oda (1984) the left-hand and right-hand junctions occur in a CCC repeat and an AAT repeat, respectively. Similar information on the type of sequences most frequently found at the junctions between cellular DNA and SV40 DNA was also obtained from the study of evolutionary variants (Lavi and Winocour, 1972) as well as from that of circular molecules excised from the integrated state in such cell lines as the rat 14B line (Bullock *et al.*, 1984).

In the case of the SV40-transformed rat fibroblast cell line SVRE9, the chromosomal site of integration has been cloned from the parental untransformed cell line (Stringer, 1982b). Sequence determination of the cloned DNA fragment revealed that a deletion of a ca. 3 kb of cellular DNA at the site of integration occurred, indicating that the two junctions between viral and cellular DNA were created by two independent crossover events.

From the available data, it appears that the most likely mechanism for SV40 DNA integration is illegitimate recombination occurring near short direct repeats in the SV40 genome. The model suggested by Hasson *et al.*, (1984) assumes that single-stranded loops, resulting from slippage between the two DNA strands, induced by the direct repeats, formed on both viral and cellular DNAs engaged in the process of integration. The single-stranded loops may be substrates for single-strand-specific nucleases, creating free ends on both DNAs. These free ends could then match together through the limited available nucleotide homology. The resulting heteroduplex would be resolved by repair in order to produce a continuous heterologous junction.

The process of integration of viral DNA from SV40 and polyoma therefore bears no relationship either to the site-specific integration of temperate phages in bacterial host DNA or to the integration of the retroviral proviruses in eukaryotic cell DNA.

J. Virus Rescue and Viral Genome Excision from Transformed Cells

Cocultivation of SV40-transformed nonproducer cells with permissive monkey cells frequently leads to the appearance of SV40 virions. The rescue of infectious virus from transformed cells is thus possible and is actually more efficient when the tranformed cells and permissive cells are induced to fuse and to form heterokaryons by Sendai virus or polyethylene glycol (Tournier *et al.*, 1967; Koprowski *et al.*, 1967; Watkins and Dulbecco, 1968). It is assumed that the required viral genome replication is stimulated by permissivity factors which are provided by the permissive cells and which diffuse to the transformed cell nucleus (Croce and Koprowski, 1973; Poste *et al.*, 1974).

Virus rescue is more easily observed with SV40-transformed than with polyoma-transformed cells. This is probably related to the semi-permissivity of most cell types to polyoma. In consequence, many transformed cell lines actually contain defective viral genomes, which were selected for during the establishment of the transformed cell lines and which are in essence nonrescuable. Nevertheless, polyoma-transformed cells from which virions can be recovered do exist (Fogel and Sachs, 1969).

Fusion to permissive cells is not the only way by which virions can be rescued from transformed cells. Exposure to many DNA-damaging agents, such as UV light (Fogel and Sachs, 1970), mitomycin C (Burns and Black, 1968; Rothschild and Black, 1970; Zouzias *et al.*, 1980; Baran *et al.*, 1983), and chemical carcinogens (Lavi, 1981), can induce the production of infectious virions from nonproducer transformed cells. Both approaches have been combined by Lambert *et al.* (1983), who showed that polyoma rescue can be enhanced by fusing normal rat fibroblasts, pretreated with a DNA-damaging agent, to polyoma-transformed fibroblasts.

In all instances, the viral genome replication and excision from the integrated state are prerequisites to virus rescue. It has been demonstrated that replication and excision require a functional large T protein since no replication or excision is observed at the restrictive temperature in cells transformed by temperature-sensitive mutants in which the large T protein is thermosensitive (Basilico *et al.*, 1979; Lavi, 1981). Miller *et al.* (1984) have also used pBR322–SV40 chimeras to study the role of large T in viral DNA excision, by comparing two plasmids which carry direct repetitions of SV40 DNA of similar sizes. In one of them (pSVED), the early region and the early promoter are separated, in such a way that no large T protein can be produced from the unrearranged plasmid. From the other plasmid (pSVLD), the expression of large T is possible. From cells transformed with the pSVED plasmid and which were large T negative, no viral genome excision was observed upon fusion with permissive CV-1 cells. pSVLD-transformed cells on the contrary produced viral DNA under similar conditions. In both cases, fusion to the large-T-containing COS-7 cells led to the production of viral DNA and virions. The requirement for a functional replication origin has also been demonstrated in polyoma-transformed cells, which are large T positive. Some, which contain tandem arrays with defective replication origins, cannot excise the viral DNA sequences (Pellegrini *et al.*, 1984).

The requirements for both a functional large T and an active replication origin can be accounted for by the onion skin replication model proposed by Botchan *et al.* (1979). According to this model, when permissivity factors appear in a nonproducer transformed cell, multiple rounds of replication occur at the viral origin of replication in the integrated genome and can propagate in the adjoining cellular sequences (Baran *et al.*, 1983). Excision results from illegitimate recombination between the ends of the replicated copies and can result from crossovers between viral or cellular DNA sequences, producing a heterogeneous population of circular molecules of various sizes. The points of crossover are defined by redundancies of 2–3 base pairs in both parental molecules (Bullock *et al.*, 1984). Excision of the viral genome from the integrated state therefore occurs by recombination mechanisms, which are similar to those which permit integration.

The onion skin replication model is in agreement with most of the reported evidence. There is nevertheless one possible exception. Lania *et al.* (1982) have described a rat cell line transformed by polyoma, which contains a single insert with a tandem duplication. This cell line, which expresses no large T, nevertheless produces free viral DNA and virions when fused to permissive mouse cells. The authors propose that, in this particular cell line, free viral DNA is produced first, without previous replication, by homologous recombination and later amplified by mouse cell permissivity factors.

K. Abortive versus Permanent Transformation

The establishment of permanently transformed cell lines after infection with SV40 or polyoma or after transfection with viral DNA is a rare event, as already pointed out (see Section III.C). Nevertheless, early after infection, nonpermissive or semipermissive cells acquire some of the properties associated with transformation. As shown by Stoker and Dulbecco (1969), polyoma-infected BHK 21/13 cells can perform up to six consecutive divisions in semisolid medium. Later, most of the cells regain the characteristics of untransformed cells and only a few retain the transformation phenotype indefinitely. In a similar way, SV40-infected mouse BALB/c 3T3 cells acquire the capacity to form microfoci on plastic in growth-factor-free medium (Smith *et al.*, 1971). The transient acquisition by infected cells of at least some of the properties of transformed cells has been called by Stoker "abortive transformation" and interpreted as the result of the temporary expression of early viral functions.

The limiting step in the transition from abortive to permanently transformed cells has conventionally been identified as the integration of the viral sequences in the cellular DNA in a chromosomal location and in a state which permits efficient expression of the viral transforming functions. Retroviral vectors, which carry and express early functions of SV40 or polyoma, have recently been constructed (Kriegler *et al.*, 1984; Donoghue *et al.*, 1984). The MV40 construct (Kriegler *et al.*, 1984) uses Harvey murine sarcoma virus LTR, the splice donor-, tRNA primer-, and packaging-sites and is packaged in Moloney murine retrovirus envelope. When Rat-2 or NIH 3T3 cells are infected with MV40, up to 80% of the population form microcolonies in soft agar, after 1 week. Virtually all these microcolonies develop into macrocolonies after 3 weeks, as compared to a maximum of 0.1% of the infected cell population when SV40 itself is used. Similar results were obtained with a polyoma construct, PyMLV (Donoghue *et al.*, 1984), which expresses the polyoma middle T protein, under the control of Moloney murine leukemia virus expression elements.

In these experiments, the great increase in permanent transformation efficiency is mainly due to the efficiency of infection and integration, ensured by the retroviral vector. But the influence exerted by the retroviral LTR on the expression of the papovavirus early proteins should not be neglected. In MV40-transformed cell lines, the steady-state level of SV40 early proteins is 5- to 10-fold greater than in SV40-transformed cell lines (Kriegler *et al.*, 1984). The role played by the retroviral LTR enhancer elements is also indicated by experiments in which a recombinant SV40 virus, SVLTR1, containing the Harvey murine sarcoma virus LTR was used to transform NIH 3T3 cells, with an efficiency 10- to 20-fold higher than the transformation efficiency displayed by wild-type SV40 (Kriegler and Botchan, 1983). These observations are in keeping with the compared efficiencies in murine cells of murine retroviral enhancers versus the SV40 enhancer (Khoury and Gruss, 1983).

Abortive transformation experiments have been proposed as a way to study maintenance functions, the transient expression of which would be responsible for the abortive transformation phenotype (Fluck and Benjamin, 1979). Since the SV40 tsA58 mutant can still induce abortive transformation at the restrictive temperature, while the dl54-59 mutant, dl884, was found unable to do so, Fluck and Benjamin (1979) suggested that in SV40, the large T protein is responsible for establishment, while the small t protein is responsible for maintenance. Although Sompayrac and Danna (1983b) confirmed the observation that the tsA58 mutant abortively transforms at the restrictive temperature, they found that a d1884–tsA58 double mutant is thermosensitive for abortive transformation. They also observed the abortive transformation of rat F111 cells by the d1884 mutant by letting the infected cells perform two generations before plating in soft agar. Rubin *et al.* (1982) have also examined the respective roles of small t and large T in abortive transformation with the help of a defective SV40 virus recombinant, SV402, which can produce small t but no detectable large T by immunoprecipitation assay. The SV402 defective virus was found unable to induce abortive transformation in mouse BALB/c 3T3 A31 cells. These contradictory observations can probably be explained by the leakiness of the SV40 tsA58 mutant, favored by the high multiplicity of infection which is usually employed to ensure that a large fraction of the cell population is actually infected. There is therefore no strong argument to be derived from abortive transformation experiments in favor of distinct roles of SV40 large T and small t in establishment and maintenance.

L. Role of Polyoma Large T and Middle T Proteins

The availability of cloned cellular oncogenes, such as *myc* or Ha-*ras*, has enabled the study of the cooperation between oncogenes in *in vitro* cellular transformation (Land *et al.*, 1983; Ruley, 1983). A distinction has thus been suggested between two classes of oncogenes. The class I oncogenes, such as *myc*, are able to confer on primary rodent embryo fibroblast cultures the capacity to multiply indefinitely *in vitro* (immortalization). The class II oncogenes, such as Ha-*ras*, can morphologically transform already established rodent fibroblast cell lines. Class I and class II oncogenes can cooperate to directly convert primary rodent embryo fibroblasts into transformed tumorigenic permanent cell lines.

Appropriate recombinant DNAs which can individually express the polyoma early antigens, middle T (Treisman *et al.*, 1981), large T, and small t (Zhu *et al.*, 1984), have been constructed and used to study the role of each early antigen in cellular transformation (Treisman *et al.*, 1981; Rassoulzadegan *et al.*, 1982, 1983). The results clearly established that large T is a class I oncogene which can convert rat embryo fibroblasts into continuously growing cell lines. These cell lines display the same phenotype as continuous 3T3 lines, with respect to morphology, saturation density, and anchorage dependence. They are nevertheless much less dependent on serum growth factors and require a lower serum concentration for growth than 3T3 cell lines (Rassoulzadegan *et al.*, 1983). The normal full-size large T antigen is not necessary to ensure this immortalization and can be replaced by an N-terminal fragment, encompassing about 40% of large T (Rassoulzadegan *et al.*, 1983).

On the contrary, middle T belongs to the class II oncogenes. It can transform established cell lines, such as the rat FR3T3 cell line, into cell lines which are morphologically altered in the same way as FR3T3 cells transformed with wild-type polyoma. Middle T-transformed FR3T3 cells are no longer contact-inhibited or anchorage dependent, but still show the same requirement for serum growth factors as nontranformed FR3T3 cells (Rassoulzadegan *et al.*, 1982).

M. Role of SV40 Large T Protein in Cellular Transformation

Wild-type SV40 is able to convert primary rodent cells into continuously growing cell lines. Such cell lines can easily be selected for by replating the infected cells at low cell density. In such conditions, noninfected primary rat embryo fibroblasts produce very few colonies (0.1–0.9% plating efficiency), which cannot be maintained in culture, while SV40-infected cells plate with an efficiency between 2.5 and 10% and most of the colonies which are picked can be established in culture (Risser *et al.*, 1974). Similar results can be obtained with d154-59 SV40 mutants, which produce no small t (Petit *et al.*, 1983). Thus, SV40 large T is endowed with the capacity to immortalize primary rodent cells, as is polyoma large T. Similarly to polyoma large T, the immortilization function resides in the N-terminal half of the protein, as shown by Colby and Shenk (1982). They were able to immortalize rat embryo cells with a cloned fragment of SV40 DNA which extends from 0.17 to 0.37 map unit and which codes for a truncated 40–50K large T in the immortalized cell lines.

The phenotype of SV40-immortalized cell lines is highly variable. Among 12 rat cell clones examined by Risser *et al.* (1974), some were anchorage dependent, while others had a high plating efficiency in methylcellulose. Similarly, when continuous rat FR3T3 (Perbal and Rassoulzadegan, 1980) or mouse 3T3 cells (Risser and Pollack, 1974; Risser *et al.*, 1974) were infected with SV40, the large-T-positive cell lines which were obtained showed widely different phenotypes. Perbal and Rassoulzadegan (1980) have compared the ability of SV40 and polyoma to produce tranformed cells. They observed that, under similar conditions, SV40 produced many "minimal" transformants, which grow poorly in suspension. Polyoma-transformed cells were more uniform and usually fully transformed.

Although it appears that large T of SV40 is able both to immortalize and to transform rodent cells, it does so less efficiently than the combination of large T and middle T from polyoma. The end result of SV40 transformation appears to be more dependent on the type of cells used and on the level of large T accumulated in the transformed cells.

Sugano and Yamaguchi (1984) have contructed a mutant, in A941, by filling in the staggered ends produced in SV40 DNA by *Ava* II cleavage at map position 0.636. They assume that as a consequence an Asp residue has been inserted between amino acid residues 16 and 17 in large T. This mutant, which cannot replicate on CV-1 cells, can be propagated on COS-1 cells. Although it has lost the capacity to transform the established rat cell line 3Y1, it can still immortalize primary newborn rat brain cells. Immortalization and transformation have therefore been dissociated in this mutant and the transforming ability of large T apparently depends on the integrity of the N-terminus. Pipas *et al.* (1983) have also described a deletion mutant, dl1135, in which 11 amino acid residues, located between positions 16 and 28, are missing and which cannot transform the rat REF52 cell line.

As previously mentioned, the C-terminus of large T appears unimportant in immortalization (Colby and Shenk, 1982). The transforming activity of SV40 deletion mutants which are missing various parts of the C-terminal portion of large T has been examined (Clayton *et al.*, 1982; Pipas *et al.*, 1983; Sompayrac and Danna, 1983b; Sompayrac *et al.*, 1983; Chang *et al.*, 1984; Hirschhorn *et al.*, 1984). All these authors do not agree on the minimum size of truncated large T which is sufficient to promote transformation of established rodent cell lines. As usual, the nature of the cells to be transformed (Chang *et al.*, 1984) and the experimental conditions (Sompayrac and Danna, 1983b) are important factors, as well as the amount of truncated protein which does accumulate in infected cells (Hirschhorn *et al.*, 1984). In any case, the shortest large T fragment which has been reported as able to fully transform mouse C3H 10 T 1/2 cells extends up to amino acid 272, followed by 27 amino acid residues, unrelated to the large T sequence and coded for by monkey DNA, which has been inserted during the construction of the F8d1 mutant (Sompayrac *et al.*, 1983). This mutant, which can be grown in COS-1 cells, transforms the C3H 10 T 1/2 cells with an efficiency of 2%, as compared with wild-type SV40, but the cell lines which are derived from the observed dense foci grown in low serum, reach high saturation densities in 10% serum and form colonies in soft agar (Sompayrac *et al.*, 1983). It also abortively transforms Fischer rat F111 cells as efficiently as the wild-type virus (Sompayrac and Danna, 1983b).

The F8d1 mutant lacks the large T C-terminal sequences which carry the ATPase activity (Clark *et al.*, 1983) and the ATP-binding site (Clertant *et al.*, 1984). It also lacks the sequences which are required to induce the transcription of silent rRNA genes in human–mouse cell hybrids (Soprano *et al.*, 1983). On the contrary, it still contains the sequences which are required to stimulate cellular DNA replication in quiescent infected cells (Galanti *et al.*, 1981; Soprano *et al.*, 1983), as well as the sequences which are involved in the binding of SV40 large T to dsDNA (Chaudry *et al.*, 1982a).

On the other hand, the ability of SV40 large T to bind specifically to sequences near the origin of viral DNA replication is not required for efficient transformation of established or primary rodent cells. Replacement of Asn^{153} by Thr (Prives *et al.*, 1983) or of Lys^{214} by Gln (Stringer, 1982a) suppresses large T *ori* binding but does not impair tranformation of Rat-2 cells or of primary rat embryo fibroblasts. However, the $Asn^{153} \rightarrow Thr^{153}$ mutation does not prevent the unspecific binding of large T to DNA-cellulose (Prives *et al.*, 1983).

The effect of point mutations and small deletions, located in the region which encodes amino acid residues between positions 92 and 158, has been thoroughly explored by Kalderon and Smith (1984). The overall conclusion of these studies is that the ability to transform established Rat-1 cells is relatively insensitive to alterations of individual or groups of amino acid residues in this part of the large T molecule, while many single point mutants and all the deletion and multiple point mutants are unable to stimulate viral DNA replication. The mutants which are severely impaired in transformation of Rat-1 cells include mutants in which Ser^{106} is replaced by Phe or Ser^{189} by Asn and mutants with lesions in the sequence Glu^{107} to Ser^{112}. Mutants in which some of the main phosphorylation sites in the N-terminal portion of large T (Ser^{111}, Ser^{112}, Ser^{123}, Thr^{124}) are eliminated are not impaired in transformation, indicating that phosphorylation at these sites is not required. An interesting class of mutants is described by Kalderon and Smith (1984) as supertransforming.

They have alterations in the sequence from residues 144 to 156 and all are replication defective. The authors suggest that their increased ability to transform, as compared to wild-type SV40, is related to a defect in the regulation of expression of the early regions.

N. Role of Small t Protein in Cellular Transformation

The role of small t has been studied in greater detail in SV40 than in polyoma, because of the availability of the dl54-59 mutants of SV40 which affect the production of small t without affecting that of large T.

The construction of a modified polyoma genome, which only encodes small t, has permitted observation of the behavior of cells expressing large amounts of this protein. The cells become rounded and easily detach from the plastic culture dishes (Zhu *et al.*, 1984). In attempts to establish mouse or rat cell lines expressing only the polyoma small t protein, it was observed that the cells detach from the substrate and cannot be propagated on unmodified plastic surfaces (Rassoulzadegn *et al.*, 1983). These changes in cellular adhesiveness and morphology must be correlated with the effects of SV40 small t on the organization of actin cables. The microinjection into untransformed Rat-1 cells of either the pHR402 plasmid, which encodes small t only (Rubin *et al.*, 1982), or the purified small t, produced in *E. coli* (Bikel *et al.*, 1983), induces a disruption of the actin cables.

Nevertheless, expression of the SV40 small t gene in BALB/c 3T3 A31 cells is not sufficient to promote soft-agar growth, as judged by abortive transformation experiments performed with the SV40 mutant, SV402 (Rubin *et al.*, 1982). The SV402 mutant could be complemented in these abortive transformation experiments with the dl54-59 mutant, dl884, as if the induction of soft-agar growth required the cooperation of small t and large T.

Actually, the dl54-59 mutants of SV40, which produce either a modified small t or no small t at all, have been repeatedly tested for their transforming capacity. When mouse or rat embryo cells or continuous cell lines are used and anchorage independence is selected for, the small t mutants very often show a strongly reduced ability to produce large colonies (>2 mm in diameter) (Shenk *et al.*, 1976; Sleigh *et al.*, 1978; Feunteun *et al.*, 1978; Bouck *et al.*, 1978). But the requirement for small t is strongly influenced by the experimental conditions. The growth state of the cells in the course of the tranformation experiments is crucial (Martin *et al.*, 1979; Seif and Martin, 1979b; Sompayrac and Danna, 1983b), as well as the nature of the cells which are transformed (Christensen and Brockman, 1982; Sugano *et al.*, 1982). For example, small t mutants are able to induce the formation of fully transformed colonies from the Fischer rat 3Y1-K clone, but not from the 3Y1-1-6 subclone (Sugano *et al.*, 1982).

In summary, small t appears to provide a facilitator or promoter function which can also be provided by some actively growing cells (Seif and Martin, 1979b; Sompayrac and Danna, 1983b). This facilitator function can also be dispensed with, when the level of expression of large T is very high, as occurs in cells infected with a large-T-producing recombinant retrovirus (MV40) (Kreigler *et al.*, 1984). Therefore, the role of small t in *in vitro* cell transformation by either SV40 or polyoma genes is not essential.

O. Large T Protein and the Maintenance of Transformation in SV40- and Polyoma-Transformed Cells

The availability of thermosensitive mutants of SV40 and polyoma, which produce thermolabile large T proteins, enables determination of whether an active A gene product is continuously required to maintain part or all of the transformed phenotype. A number of early reports actually concluded that cells, which have been transformed at the permissive temperature with SV40 tsA mutants, lose their transformed phenotype at the restrictive temperature (Osborn and Weber, 1975; Kimura and Itagaki, 1975; Tegtmeyer, 1975; Brugge and Butel, 1975; Martin and Chou, 1975).

Actually, tsA mutants can produce two types of transformants. Some transformants, designated as N transformants by Seif and Cuzin (1977), display a thermosensitive transformation phenotype, whereas type A transformants keep their transformed phenotype at the restrictive temperature. Both types can be obtained with either polyoma tsa mutants (Seif and Cuzin, 1977) or SV40 tsA mutants (Rassoulzadegan *et al.*, 1978a).

The outcome of a transformation experiment in terms of N or A phenotype appears to depend on the experimental conditions. When the infected cells are maintained in active growth for a critical number of cell divisions before selection of the transformants, N transformants are more frequent than A transformants (Rassoulzadegan *et al.*, 1978b; Brockman, 1978; Seif and Martin, 1979a). Direct plating in soft agar, which inhibits the growth of untransformed cells, therefore favors the isolation of A transformants (Seif and Cuzin, 1977; Rassoulzadegan *et al.*, 1978a), although this conclusion has been challenged by Kelley *et al.* (1980). The multiplicity of infection, according to Rassoulzadegan and Cuzin (1980), also plays a role, as well as the type of cells which are infected. O'Neill *et al.* (1980) observed that primary mouse embryo cells preferentially gave rise to N-type transformed cells irrespective of the selection procedure, whereas mouse 3T3 cells behaved more erratically, especially when the transformed cells were selected by focus formation. In agreement with O'Neill *et al.* (1980), it must be recalled that cell immortalization is always thermosensitive when tsA mutants of SV40 or polyoma are used (Petit *et al.*, 1983; Rassoulzadegan *et al.*, 1983).

In several instances, the virus which could be rescued from nonthermosensitive type A transformants has been shown to be as thermosensitive as the mutant originally used to transform the cells (Tenen *et al.*, 1977; Brockman, 1978; Martin *et al.*, 1979).

Because all thermosensitive mutants are leaky, the difference between N and A transformants might be explained by different levels of expression and/or accumulation of the thermosensitive large T protein. Actually, Christensen and Brockman (1982) observed that N transformants derived from BALB/c 3T3 cells transformed with the tsA209 SV40 mutant do not completely stop to replicate their DNA and to divide at the restrictive temperature, but these residual activities are offset by cell detachment and death. They concluded that this behavior may be due to leakiness of the tsA mutation as well as to a permanent cellular alteration induced during viral transformation. Comparison of N and A transformants derived from different rodent cell lines (Chinese hamster lung cells: Tenen *et al.*, 1977; mouse BALB/c 3T3: Brockman, 1978; rat FR3T3: Imbert *et al.*, 1983) has in all cases established a correlation between type A transformation and a higher level of large T protein than observed in type N transformants.

The results are therefore in favor of a requirement for the continuous expression of SV40 large T in the maintenance of the transformed phenotype. This conclusion is also supported by the observation that nontransformed revertants selected from the SV40-transformed rat 14B cell line, no longer express large T (Steinberg *et al.*, 1978, 1979).

Contrary to the SV40 situation, in which a single protein carries both class I and class I oncogene functions, polyoma transformation implies the cooperation of the genes for large T and middle T. As previously mentioned, the immortalization of rat embryo fibroblasts by a tsa-derived plasmid encoding the large T protein only is thermosensitive (Rassoulzadegan *et al.*, 1983). But it is not clear why N transformants obtained from FR3T3 cells after polyoma tsa infection revert to a normal phenotype at the restrictive temperature in the presence of high serum concentrations.

IV. CONCLUSION

Contrary to acute transforming retroviruses, which owe their transforming capacity to the acquisition of cellular genes, SV40 and polyoma transforming genes do not show any extensive homology to known cellular protooncogenes. It remains to be seen whether homologies to unknown cellular genes will be found in the future. At present, it is reasonable to conclude that cellular transformation by polyoma viruses is due to the expression in cells which do not efficiently replicate the viral DNA, of viral functions which have evolved to ensure an efficient multiplication of virus in permissive cells.

The amount of genetic information encoded in these small viral genomes is such that the polyoma viruses are totally dependent on the cellular machinery for replicating their DNA, once replication has been initiated by large T, and for translating their structural proteins. In order to package their DNA inside viral capsids, they even make use of cellular histones.

It is not surprising therefore that SV40 or polyoma infection induces a cell division cycle in resting cells, thus ensuring that cellular functions required for successful virus multiplication are actually expressed. This induction is clearly linked to the expression of early viral functions. Transformation of nonpermissive cells probably is an accidental consequence of the acquisition by the virus of viral functions adapted to its own needs.

Although SV40 and polyoma are superficially similar, the genetic complexity of polyoma is greater and if the strategy of viral replication is similar in both cases, the detailed mechanisms used by both viruses to modify the cellular properties are different.

With regard to polyoma, full cellular transformation occurs through the cooperation of two genes. The product of the first gene, large T, is a nuclear protein, which may act to modify the expression of cellular genes at the chromatin or DNA level. It must be recalled in this respect that class I oncogenes, like *myc*, also code for nuclear proteins with affinity for dsDNA (Persson and Leder, 1984), which are able to immortalize cells and to regulate cellular gene expression (Kingston *et al.*, 1984).

The second polyoma gene codes for a protein, middle T, which is partly located in the plasma membrane, where it forms a complex with the product of a cellular protooncogene, pp60$^{c\text{-}src}$, which is a protein kinase, with specificity for tyrosine residues. It must be recalled in this context that some receptors for growth factors, located in the cell membrane, also display a tyrosine kinase activity (Cohen *et al.*, 1982; Ek *et al.*, 1982; Kasuga *et al.*, 1982; Downward *et al.*, 1984), while the product of the class II oncogenes, *ras*, is also located at the inner side of the plasma membrane (Papageorge *et al.*, 1982).

The only important transforming protein of SV40 is large T. But this protein certainly is multifunctional and located both in the cell nucleus and in the cell membrane. It could therefore be proposed that the fully transformed phenotype of SV40-transformed cells results from the cooperation between the nuclear form of large T, which binds to DNA and could regulate the expression of cellular genes, and the plasma membrane form of large T, which could influence the behavior of cellular membrane proteins. The dissociation between immortalizing and transforming functions, which has been observed in a mutant altered in the N-terminal part of large T (Sugano and Yamaguchi, 1984), and the loss of transforming activity in the dl135 mutant (Pipas *et al.*, 1983), would support such a model, since it is known that large T interacts with the plasma membrane through its N-terminus (Tevethia *et al.*, 1983, 1984). The preliminary observations of Lanford and Butel (1984) that a mutant altered in large T nuclear transport is still able to transform established rodent cells, but is impaired in primary cell transformation, is also in favor of this model, but it should be confirmed on other similar mutants before definitive conclusions are drawn.

REFERENCES

Abramczuk, J., Pan, S., Maul, G., and Knowles, B. B., 1984, Tumor induction by SV40 in mice is controlled by long term persistence of the viral genome and the immune response of the host, *J. Virol.* **49**:540.

Anderson, J. L., Martin, R. G., Chang, C., Mora, P. T., and Livingston, D. M., 1977, Nuclear preparations of SV40-transformed cells contain TSTA activity, *Virology* **76**:420.

Àsselin, C., Gelinas, C., Branton, P. E., and Bastin, M., 1984, Polyoma middle T antigen requires cooperation from another gene to express the malignant phenotype *in vivo*, *Mol. Cell. Biol.* **4**:755.

Aub, J. C., Tieslau, C., and Lankester, A., 1963, Reaction of normal and tumor cell surfaces to enzymes. I. Wheat germ lipase and associated mucopolysaccharides, *Proc. Natl. Acad. Sci. USA* **50**:613.

Baran, N., Neer, A., and Manor, H., 1983, "Onion skin" replication of integrated polyoma virus DNA and flanking sequences in polyoma-transformed rat cells: Termination within a specific cellular DNA segment, *Proc. Natl. Acad. Sci. USA* **80**:105.

Barrett, J. C., 1980, A preneoplastic stage in the spontaneous neoplastic transformation of Syrian hamster embryo cells in culture, *Cancer Res.* **40**:91.

Basilico, C., Gattoni, S., Zouzias, D., and Della Valle, G., 1979, Loss of integrated viral DNA sequences in polyoma-transformed cells is associated with an active viral A function, *Cell* **17**:645.

Bender, M. A., and Brockman, W. W., 1981, Rearrangement of integrated viral DNA sequences in mouse cell transformed by SV40, *J. Virol.* **38**:872.

Benjamin, T. L., 1966, Virus-specific RNA in cells productively infected or transformed by polyoma virus, *J. Mol. Biol.* **16**:359.

Benjamin, T. L., 1982, The hr-t gene of polyoma virus, *Biochim. Biophys. Acta* **695**:69.

Bikel, I., Roberts, T. M., Bladon, M. T., Green, R., Amann, E., and Livingston, D. M., 1983, Purification of biologically active simian virus 40 small tumor antigen, *Proc. Natl. Acad. Sci. USA* **80**:906.

Birg, F., Dulbecco, R., Fried, M., and Kamen, R., 1979, State and organization of polyoma virus DNA sequences in transformed rat cell lines, *J. Virol.* **29**:633.

Bishop, J. M., 1983, Cellular oncogenes and retroviruses, *Annu. Rev. Biochem.* **52**:301.

Black, P. H., Rowe, W. P., Turner, H. C., and Huebner, R. J., 1963, A specific complement fixing antigen present in SV40 tumor and transformed cells, *Proc. Natl. Acad. Sci. USA* **50**:1148.

Boast, S., La Mantia, G., Lania, L., and Blasi, F., 1983, High efficiency of replication and expression of foreign genes in SV40-transformed human fibroblasts, *EMBO J.* **2**:2327.

Bolen, J. B., Thiele, C. J., Israel, M. A., Yonemoto, W., Lipsich, L. A., and Brugge, J. S., 1984, Enhancement of cellular *src* gene product associated tyrosine kinase activity following polyoma virus infection and transformation, *Cell* **38**:767.

Botchan, M., Ozanne, B., Sugden, B., Sharp, P. A., and Sambrook, J., 1974, Viral DNA in transformed cells. III. The amounts of different regions of the SV40 genome present in a line of transformed mouse cells, *Proc. Natl. Acad. Sci. USA* **71**:4183.

Botchan, M., Topp, W., and Sambrook, J., 1976, The arrangement of SV40 sequences in the DNA of transformed cells, *Cell* **9**:269.

Botchan, M., Topp, W., and Sambrook, J., 1979, Studies on simian virus 40 excision from cellular chromosomes, *Cold Spring Harbor Symp. Quant. Biol.* **43**:709.

Botchan, M., Stringer, J., Mitchison, T., and Sambrook, J., 1980, Integration and excision of SV40 DNA from the chromosome of a transformed cell, *Cell* **20**:143.

Bouck, N., Beales, N., Shenk, T., Berg, P., and di Mayorca, G., 1978, New region of simian virus 40 genome required for efficient viral transformation, *Proc. Natl. Acad. Sci. USA* **75**:2473.

Bouck, N., Fikes, J., and Rundell, K., 1984, Large-T-antigen–p53 complex formation is not cold sensitive in a cold-sensitive transformant induced by SV40 mutant tsA 1499, *J. Virol.* **49**:997.

Bradley, M. K., Griffin, J. D., and Livingston, D. M., 1982, Relationship of oligomerization to enzymatic and DNA-binding properties of the SV40 large T antigen, *Cell* **28**:125.

Brady, J., Bolen, J. B., Radonovich, M., Salzman, N., and Khoury, G., 1984, Stimulation of SV40 late gene expression by SV40 tumor antigen, *Proc. Natl. Acad. Sci. USA* **81**:2040.

Brickell, P. M., Latcheman, D. S., Murphy, D., Willison, K., and Rigby, P. W. J., 1983, Activation of a Qa/Tla class I major histocompatibility antigen gene is a general feature of oncogenesis in the mouse, *Nature* **306**:756.

Brinster, R. L., Chen, H. Y., Messing, A., Van Dyke, T., Levine, A. J., and Palmiter, R. D., 1984, Transgenic mice harboring SV40-T-antigen gene develop characteristic brain tumors, *Cell* **37**:367.

Brockman, W. W., 1978, Transformation of BALB/c 3T3 cells by tsA mutants of SV40: Temperature sensitivity of the transformed phenotype and retransformation by WT virus, *J. Virol.* **25**:860.

Brugge, J. S., and Butel, J. S., 1975, Role of simian virus 40 gene A function in maintenance of transformation, *J. Virol.* **15**:619.

Bullock, P., Forrester, W., and Botchan, M., 1984, DNA sequence studies of SV40 chromosomal excision and integration in rat cells, *J. Mol. Biol.* **174:55.**

Burger, M. M., 1969, A difference in the architecture of the surface membrane of normal and virally transformed cells, *Proc. Natl. Acad. Sci. USA* **62**:994.

Burger, M. M., 1973, Surface changes in transformed cells detected by lectins, *Fed. Proc.* **32**:91.

Burns, W. H., and Black, P. H., 1968, Analysis of simian virus 40-induced transformation of hamster kidney tissue *in vitro*. V. Variability of virus recovery from cell clones inducible with mitomycin C and cell fusion, *J. Virol.* **2**:606.

Campo, M. S., Cameron, I. R., and Rogers, M. E., 1978, Tandem integration of complete and defective SV40 genomes in mouse–human somatic hybrids, *Cell* **15**:1411.

Carmichael, G. G., Schaffhausen, B. S., Dorsky, D. I., Oliver, D. B., and Benjamin, T. L., 1982, Carboxy terminus of polyoma middle-sized tumor antigen is required for attachment to membranes, associated proteinkinase activities and cell transformation, *Proc. Natl. Acad. Sci. USA* **79**:3579.

Carmichael, G., Schaffhausen, B. S., Mandel, G., Liang, T. J., and Benjamin, T. L., 1984, Transformation by polyoma virus is drastically reduced by substitution of phenylalanine for tyrosine at residue 315 of middle-sized tumor antigen, *Proc. Natl. Acad. Sci. USA* **81**:679.

Chandrasekaran, K., McFarland, V. W., Simmons, D. T., Dziaked, M., Gurney, E. G., and Mora, P. T., 1981, Quantitation and characterization of a species-specific and embryo stage-dependent 55-kd phosphoprotein also present in cells transformed by SV40, *Proc. Natl. Acad. Sci. USA* **78**:6953.

Chang, C., Marin, R. G., Livingston, D. M., Luborsky, S. W., Hu, C. -P., and Mora, P. T., 1979, Relationship between T-antigen and tumor-specific transplantation antigen in SV40-transformed cells, *J. Virol.* **29**:69.

Chang, L. -S., Pater, M. M., Hutchinson, N. I., and di Mayorca, G., 1984, Transformation by purified early genes of simian virus 40, *Virology* **133**:341.

Chaudry, F., Harvey, R., and Smith, A. E., 1982a, Structure and biochemical functions of four SV40 truncated large T-antigens, *J. Virol.* **44**:54.
Chaudry, F., Belsham, G. J., and Smith, A. E., 1982b, Biochemical properties of the 145,000-dalton super-T antigen from SV40-transformed BALB/c 3T3 clone 20 cells, *J. Virol.* **45**:1098.
Chen, S., Blanck, G., and Pollack, R. E., 1983a, Pre-crisis mouse cells show strain-specific covariation in the amount of 54-kd phosphoprotein and in susceptibility to transformation by SV40, *Proc. Natl. Acad. Sci. USA* **80**:5670.
Chen, S., Grass, D. S., Blanck, G., Hoganson, N., Manley, J. L., and Pollack, R. E., 1983b, A functional SV40 origin of replication is required for the generation of super T antigen with a molecular weight of 100,000 in transformed mouse cells, *J. Virol.* **48**:492.
Chen, S., Blanck, G., and Pollack, R., 1984, Reacquisition of a functional early region by a mouse transformant containing only defective SV40 DNA, *Mol. Cell. Biol.* **4**:666.
Chia, W., and Rigby, P. W. J., 1981, Fate of viral DNA in nonpermissive cells infected with SV40, *Proc. Natl. Acad, Sci. USA* **78**:6638.
Christensen, J. B., and Brockman, W. W., 1982, Effects of large and small T-antigens on DNA synthesis and cell division in SV40-transformed BALB/c 3T3 cells, *J. Virol.* **44**:574.
Clark, R., Peden, K., Pipas, J. M., Nathans, D., and Tjian, R., 1983, Biochemical activities of T-antigen proteins encoded by SV40 A gene deletion mutants, *Mol. Cell. Biol.* **3**:220.
Clayton, C. E., and Rigby, P. W., 1981, Cloning and characterization of integrated viral DNA from three lines of SV40-transformed mouse cells, *Cell* **25**:547.
Clayton, C. E., Murphy, D., Lovett, M., and Rigby, P. W. J., 1982, A fragment of the SV40 large T-antigen gene transforms, *Nature* **299**:59.
Clertant, P., Gaudray, P., May, E., and Cuzin, F., 1984, The nucleotide binding site detected by affinity labeling in the large T proteins of polyoma and SV40 viruses is distinct from their ATPase catalytic site, *J. Biol. Chem.* **259**:15196.
Coggin, J. H., Larson, V. M., and Hilleman, M. R., 1967, Immunologic responses in hamster to homologous tumor antigens measured *in vivo* and *in vitro*, *Proc. Soc. Exp. Biol. Med.* **124**:1295.
Cohen, S., Fava, R. A., and Sawyer, S. T., 1982, Purification and characterization of EGF receptor/protein kinase from normal mouse liver, *Proc. Natl. Acad. Sci. USA* **79**:6237.
Colby, W. W., and Shenk, T., 1982, Fragments of the simian virus 40 transforming gene facilitate transformation of rat embryo cells, *Proc. Natl. Acad. Sci. USA* **79**:5189.
Courtneidge, S. A., and Smith, A. E., 1983, Polyoma virus transforming protein associates with the product of the c-src cellular gene, *Nature* **303**:435.
Courtneidge, S. A., and Smith, A. E., 1984, The complex of polyoma virus middle-T antigen and $pp60^{c\text{-}src}$, *EMBO J.* **3**:585.
Crawford, L., Pim, D. C., Gurney, E. G., Goodfellow, P., and Taylor-Papadimitriou, J., 1981, Detection of a common feature in several human cell lines—a 53,000-dalton protein, *Proc. Natl. Acad. Sci. USA* **78**:41.
Croce, C. M., 1977, Assignment of the integration site for SV40 to chromosome 17 in GM54VA, a human cell line transformed by SV40, *Proc. Natl. Acad. Sci. USA* **74**:315.
Croce, C. M., 1981, Integration of oncogenic viruses in mammalian cells, *Int. Rev. Cytol.* **71**:1.
Croce, C. M., and Koprowski, H., 1973, Enucleation of cells made simple and rescue of SV40 by enucleated cells made even simpler, *Virology* **51**:227.
Croce, C. M., Girardi, A. J., and Koprowski, H., 1973, Assignment of the T-antigen gene of simian virus 40 to human chromosome 7, *Proc. Natl. Acad. Sci. USA* **70**:3617.
Cunningham, D. D., and Pardee, B., 1969, Transport changes rapidly initiated by serum addition to "contact inhibited" cells, *Proc. Natl. Acad. Sci. USA* **64**:1049.
Dawe, C. J., and Law, L. W., 1959, Morphologic changes in salivary gland tissue of the newborn mouse exposed to parotid-tumor agent *in vitro*, *J. Natl. Cancer Inst.* **23**:1157.
Defendi, V., 1960, Induction of tumors by polyoma in adult hamsters, *Nature* **189**:508.

Defendi, V., 1963, Effect of SV40 virus immunization on growth of transplantable SV40 and polyoma virus tumors in hamsters, *Proc. Soc. Exp. Biol. Med.* **113**:12.

DeLeo, A. B., Jay, G., Appella, G., Dubois, C., Law, L. W., and Old, J., 1979, Detection of a transformation-related antigen in chemically induced sarcomas and other transformed cells of the mouse, *Proc. Natl. Acad. Sci. USA* **76**:2420.

Della Valle, G., Fenton, R. G., and Basilico, C., 1981, Polyoma large T antigen regulates the integration of viral DNA sequences into the genome of transformed cells, *Cell* **23**:347.

Deppert, W., 1980, SV40 T-antigen related surface antigen: Correlated expression with nuclear T-antigen in cells transformed by an SV40 A-gene mutant, *Virology* **104**:497.

Deppert, W., and Henning, R., 1979, SV40-T-antigen related molecules on the surface of adenovirus-2–SV40 hybrid virus infected HeLa cells and on SV40-transformed cells, *Cold Spring Harbor Symp. Quant. Biol.* **44**:225.

Deppert, W., Hanke, K., and Henning, R., 1980, SV40 T-antigen-related cell surface antigen: Serological demonstration on SV40-transformed monolayer cells *in situ*, *J. Virol.* **35**:505.

Diamandopoulos, G. T., 1978, Incidence, latency and morphological types of neoplasms induced by SV40 inoculated intravenously into hamsters of three inbred strains and one outbred stock, *J. Natl. Cancer Inst.* **60**:445.

Dilworth, S. M., Cowie, A., Kamen, R. I., and Griffin, B. E., 1984, DNA binding activity of polyoma virus large tumor antigen, *Proc. Natl. Acad. Sci. USA* **81**:1941.

Ding, D., Dilworth, S. M., and Griffin, B. E., 1982, Mlt mutants of polyoma virus, *J. Virol.* **44**:1080.

Dippold, W. G., Jay, G., DeLeo, A. B., Khoury, G., and Old, L. J., 1981, p53 transformation-related protein: Detection by monoclonal antibody in mouse and human cells, *Proc. Natl. Acad. Sci. USA* **78**:1695.

Dixon, K., Ryder, B. -J., and Burch-Jaffe, E., 1982, Enhanced metastasis of tumours induced by a SV40 small t deletion mutant, *Nature* **296**:672.

Donoghue, D. J., Anderson, C.., Hunter, T., and Kaplan, P. L., 1984, Transmission of the polyoma virus middle T gene as the oncogene of a murine retrovirus, *Nature* **308**:748.

Downward, J., Parker, P., and Waterfield, M. D., 1984, Autophosphorylation sites on the epidermal growth factor receptor, *Nature* **311**:483.

Dulbecco, R., and Eckhart, W., 1970, Temperature dependent properties of cells transformed by a thermosensitive mutant of polyoma virus, *Proc. Natl. Acad. Sci. USA* **67**:1175.

Dulbecco, R., Hartwell, L. H., and Vogt, M., 1965, Induction of cellular DNA synthesis by polyoma virus, *Proc. Natl. Acad. Sci. USA* **53**:403.

Earle, W. R., Schilling, E. L., Stark, T. H., Straus, N., Brown, M. F., and Shelton, E., 1943a, Production of malignancy *in vitro*. IV. The mouse fibroblast cultures and changes seen in the living cell, *J. Natl. Cancer Inst.* **4**:165.

Earle, W. R., Nettleship, A., Schilling, E. L., Stark, T. H., Straus, N., Brown, M. F., and Shelton, E., 1943b, Production of malignancy *in vitro*. V. Results of injections of cultures into mice, *J. Natl. Cancer Inst.* **4**:213.

Eckhart, W., 1974, Properties of temperature-sensitive mutants of polyoma virus, *Cold Spring Harbor Symp. Quant. Biol.* **39**:37.

Eckhart, W., Hutchinson, M. A., and Hunter, T., 1979, An activity phosphorylating tyrosine in polyoma T antigen immunoprecipitates, *Cell* **18**:925.

Eddy, B. E., Stewart, S. E., and Berkeley, W., 1958, Cytopathogenicity in tissue cultures by a tumor virus from mice, *Proc. Soc. Exp. Biol. Med.* **98**:848.

Eddy, B. E., Borman, G. S., Grubbs, G. E., and Young, R. D., 1962, Identification of the oncogenic substance in rhesus monkey kidney cell cultures as simian virus 40, *Virology* **17**:65.

Ek, B., Westermark, B., Wasteson, A., and Heldin, C. -H., 1982, Stimulation of tyrosine-specific phosphorylation by PDGF, *Nature* **295**:419.

Eliyahu, D., Raz, A., Gruss, P., Givol, D., and Oren, M., 1984, Participation of p53 cellular tumor antigen in transformation of normal embryonic cells, *Nature* **312**:646.

Ellman, M., Bikel, I., Figge, J., Roberts, T., Schlossman, R., and Livingston, D. M., 1984, Localization of the SV40 small t antigen in the nucleus and cytoplasm of monkey and mouse cells, *J. Virol.* **50:**623.

Fanning, E., Burger, C., and Gurney, E. G., 1981, Comparison of T-antigen-association host phosphoproteins from SV40-infected and transformed cells of different species, *J. Gen. Virol.* **55:**367.

Feunteun, J., Kress, M., Gardes, M., and Monier, R., 1978, Viable dl mutants in the Simian virus 40 early region, *Proc. Natl. Acad. Sci. USA* **75:**4455.

Fluck, M. M., and Benjamin, T. L., 1979, Comparison of two early gene functions essential for transformation in polyoma virus and SV40, *Virology* **96:**205.

Fluck, M. M., Shaik, R., and Benjamin, T. L., 1983, An analysis of transformed clones obtained by coinfections with hr-t and Ts-a mutants of polyoma virus, *Virology* **130:**29.

Fogel, M., and Sachs, L., 1969, The activation of virus synthesis in polyoma transformed cells, *Virology* **37:**327.

Fogel, M., and Sachs, L., 1970, Induction of virus synthesis in polyoma transformed cells by ultraviolet light and mitomycin C, *Virology* **40:**174.

Folk, W. R., 1973, Induction of virus synthesis in polyoma-transformed BHK-21 cells, *J. Virol.* **11:**424.

Folk, W. R., and Bancuk, J. E., 1976, Polyoma genome in hamster BHK-21-C13 cells: Integration into cellular DNA and induction of the viral replicon, *J. Virol.* **20:**133.

Foster, D. O., and Pardee, A. B., 1969, Transport of aminoacids by confluent and nonconfluent 3T3 and polyoma virus-transformed 3T3 cells growing on glass, *J. Biol. Chem.* **244:**2675.

Frearson, P. M., Kit, S., and Dubbs, D. R., 1966, Induction of dihydrofolate reductase activity by SV40 and polyoma virus, *Cancer Res.* **26:**1653.

Freed, M. I., Lubin, I., and Simmons, D. T., 1983, Stoichiometry of large t and pp53 in complexes isolated from SV40-transformed rat cells, *J. Virol.* **46:**1061.

Friderici, K., Oh, S. Y., Ellis, R., Guacci, V., and Fluck, M. M., 1984, Recombination induces tandem repeats of integrated viral sequences in polyoma-transformed cells, *Virology* **137:**67.

Galanti, N., Jonak, G. J., Soprano, K. J., Floros, J., Kaczmarek, L., Weissman, S., Reddy, V. B., Tilghman, S. M., and Baserga, R., 1981, Characterization and biological activity of cloned simian virus 40 DNA fragments, *J. Biol. Chem.* **256:**6469.

Gattoni, S., Colantuoni, V., and Basilico, C., 1980, Relationship between integrated and nonintegrated viral DNA in rat cells transformed by polyoma virus, *J. Virol.* **34:**615.

Gelb, L. D., Kohne, D. E., and Martin, M. A., 1971, Quantitation of simian virus 40 sequences in African green monkey, mouse and virus-transformed cell genomes, *J. Mol. Biol.* **57:**129.

Gelinas, C., Bouchard, L., and Bastin, M., 1981, Tumorigenic activity of cloned polyoma virus DNA in newborn rats, *Experientia* **37:**1074.

Gershon, D., Hausen, P., Sachs, L., and Winocour, E., 1965, On the mechanism of polyoma virus-induced synthesis of cellular DNA, *Proc. Natl. Acad. Sci. USA* **54:**1584.

Gershon, D., Sachs, L., and Winocour, E., 1966, The induction of cellular DNA synthesis by SV40 in contact inhibited and in X-irradiated cells, *Proc. Natl. Acad. Sci. USA* **56:**918.

Gluzman, Y., 1981, SV40-transformed simian cells support the replication of early SV40 mutants, *Cell* **23:**175.

Gluzman, Y., and Ahrens, B., 1982, SV40 early mutants that are defective for viral DNA synthesis but competent for transformation of cultured rat and simian cells, *Virology* **123:**78.

Gluzman, Y., Davidson, J., Oren, M., and Winocour, E., 1977, Properties of permissive monkey cells transformed by UV-irradiated SV40, *J. Virol.* **22:**256.

Gluzman, Y., Fisque, R. J., and Sambrook, J., 1980, Origin-defective mutants of SV40, *Cold Spring Harbor Symp. Quant. Biol.* **44:**293.

Gooding, L. R., 1977, Specificities of killing by cytotoxic lymphocytes generated *in vivo* and *in vitro* to syngeneic SV40-transformed cells, *J. Immunol.* **118:**920.
Graham, F. L., and Van der Eb, A. J., 1973, A new technique for the assay of infectivity of human adenovirus 5 DNA, *Virology* **52:**456.
Greenspan, D. S., and Carroll, R. B., 1981, Complex of SV40 large tumor antigen and 48,000-dalton host tumor antigen, *Proc. Natl. Acad. Sci. USA* **78:**105.
Griffin, B. E., and Maddock, C., 1979, New classes of viable deletion mutants in the early region of polyoma virus, *J. Virol* **31:**645.
Grisoni, M., Meneguzzi, G., De Lapeyriere, O., Binetruy, B., Rassoulzadegan, M., and Cuzin, F., 1984, The transformed phenotype in culture and tumorigenicity of FR3T3 transformed with bovine papilloma virus type I, *Virology* **135:**406.
Gross, L., 1953, A filterable agent, recovered from Ak leukemic extracts, causing salivary gland carcinomas in C3H mice, *Proc. Soc. Exp. Biol. Med.* **83:**414.
Gruss, C., Aumann, E. B., and Knippers, R., 1984, DNA binding properties of a mutant T antigen from the SV40-transformed human cell line SV80, *J. Virol.* **50:**943.
Habel, K., 1961, Resistance of polyoma virus immune animals to transplanted polyoma tumors, *Proc. Soc. Exp. Biol. Med.* **106:**722.
Habel, K., 1965, Specific complement-fixing antigens in polyoma tumors and transformed cells, *Virology* **25:**55.
Habel, K., and Eddy, B. E., 1963, Specificity of resistance to tumor challenge of polyoma and SV40 virus-immune hamsters, *Proc. Soc. Exp. Biol. Med.* **113:**1.
Hargis, B. J., and Malkiel, S., 1979, Sarcomas induced by injection of simian virus 40 into neonatal CFW mice, *J. Natl. Cancer Inst.* **63:**965.
Hartwell, L. H., Vogt, M., and Dulbecco, R., 1965, Induction of cellular DNA synthesis by polyoma virus. II. Increase in the rate of enzyme synthesis after infection with polyoma virus in mouse kidney cells, *Virology* **27:**262.
Hasson, J. F., Mougneau, E., Cuzin, F., and Yaniv, M., 1984, Simian virus 40 illegitimate recombination occurs near short direct repeats, *J. Mol. Biol.* **177:**53.
Hatanaka, M., and Dulbecco, R., 1966, Induction of DNA synthesis by SV40, *Proc. Natl. Acad. Sci. USA* **56:**736.
Henning, R., and Lange-Mütschler, J., 1983, Tightly associated lipids may anchor SV40 large T antigen in plasma membrane, *Nature* **305:**736.
Henning, R., Lange-Mütschler, J., and Deppert, W., 1981, SV40-transformed cells express SV40 T antigen-related antigens on the cell surface, *Virology* **108:**325.
Henry, P. P., Black, P. H., Oxman, M. N., and Weisman, S., 1966, Stimulation of DNA synthesis in mouse cell line 3T3 by simian virus 40, *Proc. Natl. Acad. Sci. USA* **56:**1170.
Hirschhorn, R. R., Mercer, W. E., Liu, H. -T., and Baserga, R., 1984, Transforming potential of deletion mutants of the SV40 T antigen coding gene in Syrian hamster cells, *Virology* **134:**220.
Hiscott, J. B., Murphy, D., and Defendi, V., 1981, Instability of integrated viral DNA in mouse cells transformed by SV40, *Proc. Natl. Acad. Sci. USA* **78:**1736.
Hwang, S. P., and Kucherlapati, R., 1980, Localization and organization of integrated SV40 sequences in a human cell line, *Virology* **105:**196.
Hynes, R. O., 1973, Alteration of cell-surface proteins by viral transformation and by proteolysis, *Proc. Natl. Acad. Sci. USA* **70:**3170.
Imbert, J., Clertant, P., De Bovis, B., Planche, J., and Birg, E., 1983, Stabilization of the large T protein in temperature-independent (type A) FR3T3 rat cells transformed with the SV40 tsA 30 mutant, *J. Virol.* **47:**442.
Inbar, M., Rabinowitz, Z., and Sachs, L., 1969, The formation of variants with a reversion of properties of transformed cells. III. Reversion of the cell surface membrane, *Int. J. Cancer* **4:**690.
Inui, K. -I., Moller, D. E., Tillotson, L. G., and Isselbacher, K. J., 1979, Stereospecific hexose transport by membrane vesicles from mouse fibroblasts: Membrane vesicles retain

increased hexose transport associated with viral transformation, *Proc. Natl. Acad. Sci. USA* **76**:3972.
Israel, M. A., Simmons, D. T., Hourihan, S. L., Rowe, W. P., and Martin, M. A., 1979, Interrupting the early region of polyoma DNA enhances tumorigenicity, *Proc. Natl. Acad. Sci. USA* **76**:3713.
Isselbacher, K. J., 1972, Increased uptake of aminoacids and 2-deoxy-D-glucose by virus-transformed cells in culture, *Proc. Natl. Acad. Sci. USA* **69**:585.
Ito, Y., and Spurr, N., 1979, Polyoma virus T antigens expressed in transformed cells: Significance of middle T antigen in transformation, *Cold Spring Harbor Symp. Quant. Biol.* **44**:149.
Ito, Y., Spurr, N., and Dulbecco, R., 1977a, Characterization of polyoma virus T-antigen, *Proc. Natl. Acad. Sci. USA* **74**:1259.
Ito, Y., Brocklehurst, J. R., and Dulbecco, R., 1977b, Virus-specific proteins in the plasma membrane of cells lytically infected or transformed by polyoma virus, *Proc. Natl. Acad. Sci. USA* **74**:4666.
Jaenisch, R., and Mintz, B., 1974, Simian virus 40 DNA sequences in DNA of healthy adult mice derived from preimplantation blastocysts infected with viral DNA, *Proc. Natl. Acad. Sci. USA* **71**:1250.
Jay, G., Khoury, G., DeLeo, A. B., Dippold, W. G., and Old, J., 1981, p53 transformation-related protein: Detection of an associated phosphotransferase activity, *Proc. Natl. Acad. Sci. USA* **78**:2932.
Jenkins, J. R., Rudge, K., and Curie, G. A., 1984, Cellular immortalization by a cDNA clone encoding the transformation associated phosphoprotein p53, *Nature* **312**:651.
Jörnvall, H., Luka, J., Klein, G., and Appella, E., 1982, A 53-K protein common to chemically and virally transformed cells shows extensive sequence similarities between species, *Proc. Natl. Acad. Sci. USA* **79**:287.
Kalderon, D., and Smith, A. E., 1984, *In vitro* mutagenesis of a putative DNA binding domain of SV40 large-T, *Virology* **139**:109.
Kalderon, D., Richardson, W. D., Markham, A. F., and Smith, A. E., 1984, Sequence requirements for nuclear localization of SV40 large T antigen, *Nature* **311**:33.
Kamen, R. D., Lindstrom, D., Shore, H., and Old, R. W., 1974, Virus-specific RNA in cells productively infected with or transformed by polyoma virus, *Cold Spring Harbor Symp. Quant. Biol.* **38**:187.
Kamen, R., Favaloro, J., Parker J., Treisman, R., Lania, L., Fried, M., and Mellor, A., 1980, Comparison of polyoma virus transcription in productively infected mouse cells and transformed rodent cell lines, *Cold Spring Harbor Symp. Quant. Biol.* **44**:312.
Kaplan, P. L., and Ozanne, B., 1982, Polyoma-virus-transformed cells produce transforming growth factor(s) and grow in serum-free medium, *Virology* **123**:372.
Kaplan, P. L., Topp, W. C., and Ozanne, B., 1981, Simian virus 40 induces the production of a polypeptide transforming factor(s), *Virology* **108**:484.
Kaplan, P. L., Anderson, M., and Ozanne, B., 1982, Transforming growth factor(s) production enables cells to grow in the absence of serum: An autocrine system, *Proc. Natl. Acad. Sci. USA* **79**:485.
Kasuga, M., Zick, Y., Blithe, D. L., Crettaz, M., and Kahn, C. R., 1982, Insulin stimulates tyrosine phosphorylation of the insulin receptor in a cell-free system, *Nature* **298**:667.
Keller, J. M., and Alwine, J. C., 1984, Activation of the SV40 late promoter: Direct effects of T-antigen in the absence of viral DNA replication, *Cell* **36**:381.
Kelley, S., Bender, M. A. R., and Brockman, W. W., 1980, Transformation of BALB/c-3T3 cells by tsA mutants of SV40: Effect of transformation technique on the transformed phenotype, *J. Virol.* **33**:550.
Ketner, G., and Kelly, T. J., 1976, Integrated simian virus 40 sequences in transformed cell DNA: Analysis using restriction endonucleases, *Proc. Natl. Acad. Sci. USA* **73**:1102.
Khera, K. S., Ashkenazi, A., Rapp, F., and Melnick, J. L., 1963, Immunity in hamsters to cells transformed *in vitro* and *in vivo* by SV40: Tests for antigenic relationship among the papova viruses, *J. Immunol.* **91**:604.

Khoury, G., and Gruss, P., 1983, Enhancer elements, *Cell* **33**:313.

Khoury, G., Byrne, J. C., Takemoto, K. K., and Martin, M. A., 1973, Patterns of simian virus 40 deoxyribonucleic acid transcription. II. In transformed cells, *J. Virol.* **11**:54.

Kimura, G., and Itagaki, A., 1975, Initiation and maintenance of cell transformation by simian virus 40, *Proc. Natl. Acad. Sci. USA* **72**:673.

Kingston, R. E., Baldwin, A. S., Jr., and Sharp, P. A., 1984, Regulation of heat shock protein 70 gene expression by c-myc, *Nature* **312**:280.

Kit, S., Dubbs, D. R., Frearson, P. M., and Melnick, J. L., 1966, Enzyme induction in SV40-infected green monkey kidney cultures, *Virology* **29**:69.

Klockmann, U., and Deppert, W., 1983, Acylated SV40 large T-antigen: A new subclass associated with a detergent-resistant lamina of the plasma membrane, *EMBO J.* **2**:1151.

Knowles, B. B., Koncar, M., Pfizenmaier, K., Solter, D., Aden, D. P., and Trinchieri, G., 1979, Genetic control of the cytotoxic T cell response to SV40 tumor-associated specific antigen, *J. Immunol.* **122**:1798.

Koch, M. A., and Sabin, A. B., 1963, Specificity of virus-induced resistance to transplantation of polyoma and SV40 tumors in adult hamsters, *Proc. Soc. Exp. Biol. Med.* **113**:4.

Koprowski, H., Jensen, F. C., and Steplewski, Z., 1967, Activation of production of infectious tumor virus SV40 in heterokaryon cultures, *Proc. Natl. Acad. Sci. USA* **58**:127.

Kraemer, P. M., Travis, G. L., Ray, F. A., and Cram, L. S., 1983, Spontaneous neoplastic evolution of Chinese hamster cells in culture: Multistep progression of phenotype, *Cancer Res.* **43**:4822.

Krieg, P., Amtmann, E., Jonas, D., Fischer, H., Zang, K., and Sauer, G., 1981, Episomal simian virus 40 genomes in human brain tumors, *Proc. Natl. Acad. Sci. USA* **78**:6446.

Kriegler, M., and Botchan, M., 1983, Enhanced transformation by a SV40 recombinant virus containing a Harvey murine sarcoma virus long terminal repeat, *Mol. Cell. Biol.* **3**:325.

Kriegler, M., Perez, C. F., Hardy, C., and Botchan, M., 1984, Transformation mediated by the SV40 antigens: Separation of the overlapping SV40 early genes with a retroviral vector, *Cell* **38**:483.

Kucherlapati, R. S., Hwang, S. P., Shimuzu, N., McDougall, J. K., and Botchan, M. R., 1978, Another chromosomal assignment for a simian virus 40 integration site in human cells, *Proc. Natl. Acad. Sci. USA* **75**:4460.

Lambert, M. E., Sebastiano, G. -C., Kirschmeier, P., and Weinstein, I. B., 1983, Polyoma excision is enhanced by fusing normal rat fibroblasts pretreated with a DNA-damaging agent to unexposed polyoma-transformed fibroblasts, *Carcinogenesis* **4**:587.

Land, H., Parada, L. F., and Weinberg, R. A., 1983, Tumorigenic conversion of primary embryo fibroblasts requires at least two cooperating oncogenes, *Nature* **304**:596.

Lane, D. P., and Crawford, L. V., 1979, T-antigen is bound to a host protein in SV40-transformed cells, *Nature* **278**:261.

Lanford, R. E., and Butel, J. S., 1981, Effect on nuclear localization of large tumor antigen on growth potential of SV40-transformed cell, *Virology* **110**:147.

Lanford, R. E., and Butel, J. S., 1982, Intracellular transport of SV40 large tumor antigen: A mutation which abolishes migration to the nucleus does not prevent association with the cell surface, *Virology* **119**:169.

Lanford, R. E., and Butel, J. S., 1984, Construction and characterization of an SV40 mutant defective in nuclear transport of T antigen, *Cell* **37**:801.

Lange-Mütschler, J., Deppert, W., Hanke, K., and Henning, R., 1981, Detection of SV40 T-antigen-related antigen by an ^{125}I-protein A binding assay and by immunofluorescence microscopy on the surface of SV40-transformed monolayer cells, *J. Gen. Virol.* **52**:301.

Lange-Mütschler, J., and Henning, R., 1982, Cell surface binding affinity of simian virus 40 T-antigen, *Virology* **117**:173.

Lange-Mütschler, J., and Henning, R., 1984, Cell surface binding SV40 large T antigen becomes anchored and stably linked to lipid of the target cells, *Virology* **136**:404.

Lania, L., Hayday, A., and Fried, M., 1981, Loss of functional large T-antigen and free viral genomes from cells transformed *in vitro* by polyoma virus after passage *in vivo* as tumor cells, *J. Virol.* **39**:422.

Lania, L., Boast, S., and Fried, M., 1982, Excision of polyoma virus genomes from chromosomal DNA by homologous recombination, *Nature* **295**:349.

Lavi, S., 1981, Carcinogen-mediated amplification of viral DNA sequences in simian virus 40-transformed Chinese hamster embryo cells, *Proc. Natl. Acad. Sci. USA* **78**:6144.

Lavi, S., and Winocour, E., 1972, Acquisition of sequences homologous to host deoxyribonucleic acid by closed circular simian virus 40 deoxyribonucleic acid, *J. Virol.* **9**:309.

Lewis, A. M., Jr., and Martin, R. G., 1979, The oncogenicity of SV40 deletion mutants that induce altered 17 K t proteins, *Proc. Natl. Acad. Sci. USA* **76**:4299.

Linzer, D. I. H., and Levine, A. J., 1979, Characterization of a 54K dalton cellular SV40 tumor antigen present in SV40-transformed cells and uninfected embryonal carcinoma cells, *Cell* **17**:43.

Lovett, M., Clayton, C. E., Murphy, D., Rigby, P. W. J., Smith, A. E., and Chaudry, F., 1982, Structure and synthesis of a simian virus 40 super T-antigen, *J. Virol.* **44**:963.

McCormick, F., and Harlow, E., 1980, Association of a murine 53,000-dalton phosphoprotein with SV40 large T antigen in transformed cells, *J. Virol.* **34**:213.

McDougall, J. K., Gallimore, P. H., Dunn, A. R., Webb, T. P., Kucherlapati, R. S., Nichols, E. A., and Ruddle, F. H., 1976, Mapping viral integration sites in somatic cell hybrids, in: *Human Gene Mapping* 3 (D. Bergsma, ed.), pp. 206–210, Karger, Basel.

McNutt, N. S., Culp, L. A., and Black, P. H., 1973, Contact-inhibited revertant cell lines isolated from SV40-transformed cells. IV. Microfilament distribution and cell shape in untransformed, transformed and revertant BALB/c 3T3 cells, *J. Cell Biol.* **56**:412.

MacPherson, I., and Stoker, M., 1962, Polyoma transformation of hamster cell clones—An investigation of genetic factors affecting cell competence, *Virology* **16**:147.

Major, E. O., and Matsumura, P., 1984, Human embryonic kidney cells: Stable transformation with an origin-defective simian virus 40 DNA and use as hosts for human papovavirus replication, *Mol. Cell. Biol.* **4**:379.

Manor, H., Fogel, M., and Sachs, L., 1973, Integration of viral into chromosomal deoxyribonucleic acid in an inducible line of polyoma-transformed cells, *Virology* **63**:174.

Martin, R. G., and Chou, J. Y., 1975, Simian virus 40 functions required for the establishment and maintenance of malignant transformation, *J. Virol.* **15**:599.

Martin, R. G., Setlow, V. P., and Edwards, C. A. F., 1979, Roles of the simian virus 40 tumor antigens in transformation of Chinese hamster lung cells: Studies with simian virus 40 double mutants, *J. Virol.* **31**:596.

Maruyama, K., and Oda, K., 1984, Two types of deletion within integrated viral sequences mediate reversion of SV40-transformed mouse cells, *J. Virol.* **49**:479.

May, E., Jeltsch, J. M., and Gannon, F., 1982, Characterization of a gene encoding a 115 K super-T antigen expressed by a SV40-transformed rat cell line, *Nucleic Acids Res.* **9**:4111.

May, E., Lasne, C., Prives, C., Borde, J., and May, P., 1983, Study on the functional activities concomitantly retained by the 115,000 Mr super T antigen, an evolutionary variant of SV40 large T antigen expressed in transformed rat cells, *J. Virol.* **45**:901.

May, P., Resche-Rigon, M., Borde, J., Breugnot, C., and May, E., 1984, Conversion through homologous recombination of the gene encoding SV40 115,000-molecular-weight super T antigen to a gene encoding a normal-size large T antigen variant, *Mol. Cell. Biol.* **4**:1141.

Mercer, W. E., Nelson, D., DeLeo, A. B., Old, L. J., and Baserga, R., 1982, Microinjection of monoclonal antibody to protein p53 inhibits serum-induced DNA synthesis in 3T3 cells, *Proc. Natl. Acad. Sci. USA* **79**:6309.

Mes, A. -M., and Hassell, J. A., 1982, Polyoma viral middle-T antigen is required for transformation, *J. Virol.* **42**:621.

Mes-Masson, A. -M., Schaffhausen, B., and Hassel, J. A., 1984, The major site of tyrosine phosphorylation in polyomavirus middle T antigen is not required for transformation, *J. Virol.* **52**:457.

Miller, J., Bullock, P., and Botchan, M., 1984, Simian virus 40 T-antigen is required for viral excision from chromosomes, *Proc. Natl. Acad. Sci. USA* **81**:7534.

Montano, X., and Lane, D. P., 1984, Monoclonal antibody to SV40 small t, *J. Virol.* **51**:760.

Moore, J. L., Chowdury, K., Martin, M. A., and Israel, M. A., 1980, Polyoma large tumor antigen is not required for tumorigenesis mediated by viral DNA, *Proc. Natl. Acad. Sci. USA* **77**:1336.

Mougneau, E., Birg, F., Rassoulzadegan, M., and Cuzin, F., 1980, Integration sites and sequence arrangement of SV40 DNA in a homogeneous series of transformed rat fibroblast lines, *Cell* **22**:917.

Nagata, Y., Diamond, B., and Bloom, B. R., 1983, The generation of human monocyte/macrophage cell lines, *Nature* **306**:597.

Neer, A., Baran, N., and Manor, H., 1977, *In situ* hybridization analysis of polyoma DNA replication in an inducible line of polyoma-transformed cells, *Cell* **11**:65.

Nicholson, E. L., 1974, Interaction of lectin with animal cell surfaces, *Int. Rev. Cytol.* **38**:89.

Nilsson, S. V., Tyndall, C., and Magnusson, G., 1983, Deletion mapping of a short polyoma virus middle T antigen segment important in transformation, *J. Virol.* **46**:284.

Novack, U., and Griffin, B. E., 1981, Requirement for the C-terminal region of middle T antigen in cellular transformation by polyoma virus, *Nucleic Acids Res.* **9**:2055.

Novack, U., Dilworth, S. M., and Griffin, B. E., 1980, Coding capacity of a 35 % fragment of the polyoma virus genome is sufficient to initiate and maintain cellular transformation, *Proc. Natl. Acad. Sci. USA* **77**:3278.

O'Neill, F. J., Cohen, S., and Renzetti, L., 1980, Temperature dependency for maintenance of transformation in mouse cells transformed by SV40 tsA mutants, *J. Virol.* **35**:233.

Oostra, B., Harvey, R., Ely, B., Markham, A., and Smith, A., 1983, Transforming activity of polyoma virus middle-T antigen probed by site-directed mutagenesis, *Nature* **304**:456.

Oren, M., and Levine, A. J., 1983, Molecular cloning of a cDNA specific for the murine p53 cellular tumor antigen, *Proc. Natl. Acad. Sci. USA* **80**:56.

Oren, M., Maltzman, W., and Levine, A. J., 1981, Post-translational regulation of the 54K cellular tumor antigen in normal and transformed cells, *Mol. Cell. Biol.* **1**:101.

Osborn, M., and Weber, K., 1975, Simian virus 40 gene A function and maintenance of transformation, *J. Virol.* **15**:636.

Ossowski, L., Quigley, J. P., Kellerman, G. M., and Reich, E., 1973, Fibrinolysis associated with oncogenic transformation: Requirement of plasminogen for correlated changes in cellular morphology, colony formation in agar, and cell migration, *J. Exp. Med.* **138**:1056.

Pan, S., and Knowles, B., 1983, Monoclonal antibody to SV40 T-antigen blocks lysis of cloned cytotoxic T-cell line specific for SV40 TSTA, *Virology* **125**:1.

Papageorge, A., Lowy, D., and Scolnick, E. M., 1982, Comparative biochemical properties of p21ras molecules coded for by viral and cellular ras genes, *J. Virol.* **44**:509.

Parada, L. F., Land, H., Weinberg, R. A., Wolf, D., and Rotter, V., 1984, Cooperation between gene encoding p53 tumor antigen and ras in cellular transformation, *Nature* **312**:649.

Patschinsky, T., Hunter, T., Esch, F. S., Cooper, J. A., and Sefton, B., 1982, Analysis of the sequence of amino acids surrounding sites of tyrosine phosphorylation, *Proc. Natl. Acad. Sci. USA* **79**:973.

Pellegrini, S., Dailey, L., and Basilico, C., 1984, Amplification and excision of integrated polyoma DNA sequences require a functional origin of replication, *Cell* **36**:943.

Perbal, B., and Rassoulzadegan, M., 1980, Distinct transformation phenotypes induced by polyoma virus and SV40 in rat fibroblasts and their control by an early gene function, *J. Virol.* **33**:697.

Persson, H., and Leder, P., 1984, Nuclear localization and DNA binding properties of a protein expressed by human c-myc oncogene, *Science* **225**:718.

Petit, C. A., Gardes, M., and Feunteun, J., 1983, Immortalization of rodent embryo fibroblasts by SV40 is maintained by the A-gene, *Virology* **127**:74.

Pipas, J. M., Peden, K. W. C., and Nathans, D., 1983, Mutational analysis of SV40 T antigen: Isolation and characterization of mutants with deletions in the T-antigen gene, *Mol. Cell. Biol.* **3**:203.

Pollack, R., Risser, R., Conlon, S., and Rifkin, D., 1974, Plasminogen activator production accompanies loss of anchorage regulation in transformation of primary rat embryo cells by simian virus 40, *Proc. Natl. Acad. Sci. USA* **71**:4792.

Pollack, R., Osborn, M., and Weber, K., 1975, Patterns of organization of actin and myosin in normal and transformed cultured cells, *Proc. Natl. Acad. Sci. USA* **72:**994.

Poste, G., Schaeffer, B., Reeve, P., and Alexander, D. J., 1974, Rescue of simian virus 40 (SV40) from SV40-transformed cells by fusion with replicating or nonreplicating monkey cells, *Virology* **60:**85.

Prives, C. L., Covey, L., Scheller, A., and Gluzman, Y., 1983, DNA-binding properties of simian virus 40 T-antigen mutants defective in viral DNA replication, *Mol. Cell. Biol.* **3:**1958.

Quinlan, D. C., Parnes, J. R., Shalom, R., Garvey, T. Q., Isselbacher, K. J., and Hochstadt, J., 1976, Sodium-stimulated amino acid uptake into isolated membrane vesicles from BALB/c 3T3 cells transformed by simian virus 40, *Proc. Natl. Acad. Sci. USA* **73:**1631.

Rabin, M., Uhlenbeck, O. C., Steffensen, D. M., and Mangel, W. F., 1984, Chromosomal sites of integration of SV40 DNA sequences mapped by *in situ* hybridization in two transformed hybrid cell lines, *J. Virol.* **49:**445.

Rassoulzadegan, M., and Cuzin, F., 1980, Transformation of rat fibroblast cells with early mutants of polyoma (tsa) and SV40 (tsA30): Occurrence of either A or N transformants depends on the multiplicity of infection, *J. Virol.* **33:**909.

Rassoulzadegan, M., Perbal, B., and Cuzin, F., 1978a, Growth control in SV40-transformed rat cells: Temperature-independent expression of the transformed phenotype in ts-A transformants derived by agar selection, *J. Virol.* **28:**1.

Rassoulzadegan, M., Seif, R., and Cuzin, F., 1978b, Conditions leading to the establishment of the N (A gene dependent) and A (A gene independent) transformed states after polyoma virus infection of rat fibroblasts, *J. Virol.* **28:**421.

Rassoulzadegan, M., Cowie, A., Carr, A., Glaichenhaus, N., Kamen, R., and Cuzin, F., 1982, The roles of individual polyoma virus early proteins in oncogenic transformation, *Nature* **300:**713.

Rassoulzadegan, M., Naghashfar, Z., Cowie, A., Carr, A., Grisoni, M., Kamen, R., and Cuzin, F., 1983, Expression of the large T protein of polyoma virus promotes the establishment in culture of "normal" rodent fibroblast cell lines, *Proc. Natl. Acad. Sci. USA* **80:**4354.

Reich, N., and Levine, A. J., 1982, specific interaction of the SV40 T antigen–cellular p53 protein complex with SV40 DNA, *Virology* **117:**286.

Reich, N. C., and Levine, A. J., 1984, Growth regulation of a cellular tumor antigen, p53, in nontransformed cells, *Nature* **308:**199.

Rey-Bellet, V., and Türler, H., 1984, A 61,000-dalton truncated large T-antigen is uniformly expressed in hamster cells transformed by polyomavirus, *J. Virol.* **50:**587.

Risser, R., and Pollack, R., 1974, A nonselective analysis of SV40 transformation of mouse 3T3 cells, *Virology* **59:**477.

Risser, R., Rifkin, D., and Pollack, R., 1974, The stable classes of transformed cells induced by SV40 infection of established 3T3 cells and primary rat embryonic cells, *Cold Spring Harbor Symp. Quant. Biol.* **39:**317.

Rothschild, H., and Black, P. H., 1970, Analysis of SV40-induced transformation of hamster kidney tissue *in vitro.* VII. Induction of SV40 virus from transformed hamster cell clones by various agents, *Virology* **42:**251.

Rowe, W. P., 1961, The epidemiology of mouse polyoma virus infection, *Bacteriol. Rev.* **25:**18.

Rubin, H. J., Figge, M. T., Bladon, L. B., Chen, M., Ellman, I., Bikel, I., Farrell, M. P., and Livingston, D. M., 1982, role of small t antigen in the acute transforming activity of SV40, *Cell* **30:**469.

Ruley, H. E., 1983, Adenovirus early region E1A enables viral and cellular transforming genes to transform primary cells in culture, *Nature* **304:**602.

Sack, G. H., Jr., 1981, Human cell transformation by SV40, *In Vitro* **17:**1.

Sager, R., Anisowicz A., and Howell, N., 1981, Genomic rearrangements in a mouse cell line containing integrated SV40 DNA, *Cell* **23:**41.

Sambrook, J., Westphal, H., Srinivasan, P. R., and Dulbecco, R., 1968, The integrated state of viral DNA in SV40-transformed cells, *Proc. Natl. Acad. Sci. USA* **60:**1288.

Santos, M., and Butel, J. S., 1984a, Dynamic nature of the association of large tumor antigen and p53 cellular protein with the surfaces of simian virus 40-transformed cells, *J. Virol.* **49**:50.

Santos, M., and Butel, J. S., 1984b, Antigenic structure of SV40 large tumor antigen and association with cellular protein p53 on the surface of SV40-infected and transformed cells, *J. Virol.* **51**:376.

Sarnow, P., Ho, Y. S., Williams, J., and Levine, A. J., 1982, Adenovirus Elb-58 kd tumor antigen and SV40 large tumor antigen are physically associated with the same 54 kd cellular protein in transformed cells, *Cell* **28**:387.

Schaffhausen, B. S., and Benjamin, T. L., 1979, Phosphorylation of polyoma T antigens, *Cell* **18**:935.

Schaffhausen, B., and Benjamin, T. L., 1981, Comparison of phosphorylation of two polyoma virus middle T antigens *in vivo* and *in vitro*, *J. Virol.* **40**:184.

Scheidtmann, K. -H., Hardung, M., Echle, B., and Walter, G., 1984, DNA-binding activity of SV40 large T antigen correlates with a distinct phosphorylation state, *J. Virol.* **50**:1.

Scheller, A., Covey, L., Barnet, B., and Prives, C., 1982, A small subclass of SV40 T antigen binds to the viral origin of replication, *Cell* **29**:375.

Schmidt-Ullrich, R., Thompson, W. S., Kahn, S. J., Monroe, M. J., and Wallach, D. F. H., 1982, SV40-specific isoelectric point-4.7-94,000-Mr membrane glycoprotein: Major peptide homology exhibited with the nuclear and membrane associated 94,000-Mr SV40 T-antigen in hamsters, *J. Natl. Cancer Inst.* **69**:839.

Schutzbank, T., Robinson, R., Oren, M., and Levine, A. J., 1982, SV40 large T antigen can regulate some cellular transcripts in a positive fashion, *Cell* **30**:481.

Scott, M. R. D., Westphal, K. -H., and Rigby, P. W. J., 1983, Activation of mouse genes in transformed cells, *Cell* **34**:557.

Sefton, B. M., Hunter, T., Beemon, K., and Eckhart, W., 1980, Evidence that the phosphorylation of tyrosine is essential for cellular transformation by Rous Sarcoma virus, *Cell* **20**:807.

Segawa, K., and Ito, Y., 1982, Differential subcellular localization of *in vivo*-phosphorylated and non-phosphorylated MT of polyoma virus and its relationship to MT phosphorylating activity *in vitro*, *Proc. Natl. Acad. Sci. USA* **79**:6812.

Seif, R., and Cuzin, F., 1977, Temperature-sensitive growth regulation in one type of transformed rat cells induced by the tsA mutant of polyoma virus, *J. Virol.* **24**:721.

Seif, R., and Martin, R. G., 1979a, Growth state of the cell early after infection with SV40 determines whether the maintenance of transformation will be A gene dependent or independent, *J. Virol.* **31**:350.

Seif, R., and Martin, R. G., 1979b, Simian virus 40 small t antigen is not required for the maintenance of transformation but may act as a promoter (cocarcinogen) during establishment of transformation in resting rat cells, *J. Virol.* **32**:979.

Shani, M., Rabinowitz, Z., and Sachs, L., 1972, Virus deoxyribonucleic acid sequences in subdiploid and subtetraploid revertants of polyoma-transformed cells, *J. Virol.* **10**:456.

Sheinin, R., 1966, DNA synthesis in rat embryo cells infected with polyoma virus, *Virology* **29**:167.

Shenk, T. E., Carbon, J., and Berg, P., 1976, Construction and analysis of viable deletion mutants of simian virus 40, *J. Virol.* **18**:664.

Shin, S. -I., Freedman, V. H., Risser, R., and Pollack, R., 1975, Tumorigenicity of virus-transformed cells in nude mice is correlated specifically with anchorage independent growth *in vitro*, *Proc. Natl. Acad. Sci. USA* **72**:4435.

Shiroki, K., Shimojo, H., Sawada, Y., Uemizu, Y., and Fujinaga, K., 1979, Incomplete transformation of rat cells by a small fragment of adenovirus 12 DNA, *Virology* **95**:127.

Sjogren, H. O., 1965, Transplantation methods as a tool for detection of tumor-specific antigens, *Prog. Exp. Tumor Res.* **6**:289.

Sjogren, H. O., Helström I., and Klein, G., 1961, Transplantation of polyoma virus-induced tumors in mice, *Cancer Res.* **21**:329.

Sleigh, M. J., Topp, W. C., Hanich, R., and Sambrook, J. F., 1978, Mutants of SV40 with an altered small t protein are reduced in their ability to transform cells, *Cell* **14**:79.

Small, M. B., Gluzman, Y., and Ozer, H. L., 1982, Enhanced transformation of human fibroblasts by origin-defective simian virus 40, *Nature* **296**:671.

Smith, A. E., Smith, R., Griffin, B., and Fried, M., 1979, Protein kinase activity associated with polyoma virus middle T antigen *in vitro, Cell* **18**:915.

Smith, H. S., Scher, C. D., and Todaro, G., 1971, Induction of cell division in medium lacking serum growth factor by SV40, *Virology* **44**:359.

Sompayrac, L., and Danna, K. J., 1983a, A SV40 dl 884/ts A 58 double mutant is temperature sensitive for abortive transformation, *J. Virol.* **46**:620.

Sompayrac, L., and Danna, K. J., 1983b, Simian virus 40 sequences between 0.168 and 0.424 map units are not required for abortive transformation, *J. Virol.* **46**:475.

Sompayrac, L. M., Gurney, E. G., and Danna, K. J., 1983, Stabilization of the 53,000-dalton nonviral tumor antigen is not required for transformation by SV40, *Mol. Cell. Biol.* **3**:290.

Soprano, K. J., Rossini, M., Croce, C., and Baserga, R., 1980, The role of large T antigen in SV40-induced reactivation of silent rRNA genes in human–mouse hybrid cells, *Virology* **102**:317.

Soprano, K. J., Galanti, N., Jonak, G. J., McKercher, S., Pipas, J. M., Peden, K. W. C., and Baserga, R., 1983, Mutational analysis of SV40 T antigen: Stimulation of cellular DNA synthesis and activation of rRNA genes by mutants with deletions in the T-antigen gene, *Mol. Cell. Biol.* **3**:214.

Soule, H., and Butel, J., 1979, Subcellular localization of simian virus 40 large tumor antigen, *J. Virol.* **30**:523.

Soule, H. R., Lanford, R. E., and Butel, J. S., 1980, Antigenic and immunogenic characteristics of nuclear and membrane-associated SV40 tumor antigen, *J. Virol.* **33**:887.

Southern, E. M., 1975, Detection of specific sequences among DNA fragments separated by gel electrophoresis, *J. Mol. Biol.* **98**:503.

Steinberg, B. M., Pollack, R., Topp, W., Botchan, M., 1978, Isolation and characterization of T antigen-negative revertants from a line of transformed rat cells containing one copy of the SV40 genome, *Cell* **13**:19.

Steinberg, B. M., Rifkin, D., Shin, S. -I., Boone, C., and Pollack, R., 1979, Tumorigenicity of revertants from an SV40-transformed line, *J. Supramol. Struct.* **11**:539.

Stoker, M., and Dulbecco, R., 1969, Abortive transformation by the ts-a mutant of polyoma, *Nature* **223**:397.

Stringer, J. R., 1982a, Mutant of simian virus 40 large T-antigen that is defective for viral DNA synthesis, but competent for transformation of cultured rat cells, *J. Virol.* **42**:854.

Stringer, J. R., 1982b, DNA sequence homology and chromosomal deletion at a site of SV40 DNA integration, *Nature* **296**:363.

Sugano, S., and Yamaguchi, N., 1984, Two classes of transformation-deficient, immortalization positive simian virus 40 mutants constructed by making three-base insertions in the T antigen, *J. Virol.* **52**:884.

Sugano, S., Yamaguchi, N., and Shimojo, H., 1982, Small t protein of SV40 is required for dense focus formation in a rat cell line, *J. Virol.* **41**:1073.

Sweet, B. H., and Hilleman, M. R., 1960, The vacuolating virus, SV40, *Proc. Soc. Exp. Biol. Med.* **105**:420.

Tegtmeyer, P., 1975, Function of simian virus 40 gene A in transforming infection, *J. Virol.* **15**:613.

Tegtmeyer, P., Schwartz, M., Collins, J. K., and Rundell, K., 1975, Regulation of tumor antigen synthesis by simian virus 40 gene A, *J. Virol.* **16**:168.

Templeton, D., and Eckhart, W., 1982, Mutation causing premature termination of the polyoma virus medium T antigen blocks cell transformation, *J. Virol.* **41**:1014.

Templeton, D., Voronova, A., and Eckhart, W., 1984, Construction and expression of a recombinant DNA gene encoding a polyoma virus middle-size tumor antigen with the carboxy terminus of the vesicular stomatitis virus glycoprotein G, *Mol. Cell. Biol.* **4**:282.

Tenen, D. G., Martin, R. G., Anderson, J., and Livingston, D. M., 1977, Biological and biochemical studies of cells transformed by SV40 ts gene A mutants and A mutant revertants, *J. Virol.* **22:**210.

Tevethia, M. J., and Tevethia, S. S., 1977, Biology of SV40 transplantation antigen (Tr Ag). III. Involvement of SV40 gene A in the expression of Tr Ag in permissive cell, *Virology* **81:**212.

Tevethia, S. S., and Rapp, F., 1966, Prevention and interruption of SV40 induced transplantation immunity with tumor cell extracts. *Proc. Soc. Exp. Biol. Med.* **123:**612.

Tevethia, S. S., Blasecki, J. W., Waneck, G., and Goldstein, A. L., 1974, Requirement of thymus derived A-positive lymphocytes for rejection of DNA virus (SV40) tumors in mice, *J. Immunol.* **113:**1417.

Tevethia, S. S., Flyer, D. C., and Tjian, R., 1980, Biology of SV40 transplantation antigen (Tr Ag). VI. Mechanism of induction of SV40 transplantation immunity in mice by purified SV40 T antigen (D2 protein), *Virology* **107:**13.

Tevethia, S. S., Tevethia, M. J., Lewis, A. J., Reddy, V. B., and Weissman, B. M., 1983, Biology of simian virus 40 (SV40) transplantation antigen (Tr Ag). IX. Analysis of Tr Ag in mouse cells synthesizing truncated SV40 large T antigen, *Virology* **128:**319.

Tevethia, S. S., Lewis, A. J., Campbell, A. E., Tevethia, M. J., and Rigby, P. W. J., 1984, Simian virus 40 specific cytotoxic lymphocyte clones localize two distinct TSTA sites on cells synthesizing a 48 KD SV40 T antigen, *Virology* **133:**443.

Todaro, G., and Green, H., 1963, Quantitative studies on the growth of mouse embryo cells in culture and their development into established lines, *J. Cell Biol.* **17:**299.

Todaro, G. J., and Green, H., 1966, High frequency of SV40 transformation of mouse cell line 3T3, *Virology* **28:**756.

Tournier, P., Cassingena, R., Wicker, R., Coppey, J., and Suarez, H. G., 1967, Etude du mécanisme de l'induction chez des cellules de hamster syrien transformées par le virus SV40: Propriétés d'une lignée cellulaire clonale, *Int. J. Cancer* **2:**117.

Treisman, R., Novak, U., Favaloro, J., and Kamen, R., 1981, Transformation of rat cells by an altered polyoma virus genome expressing only the middle-T protein, *Nature* **292:**595.

Trinchieri, G., Adan, D. P., and Knowles, B. B., 1976, Cell mediated cytotoxicity to SV40-specific tumor associated antigens, *Nature* **261:**312.

Van der Eb, A. J., Mulder, C., Graham, F. L., and Houweling, A., 1977, Transformation with specific fragments of adenovirus DNAs. I. Isolation of specific fragments with transforming activity of adenovirus 2 and 5 DNA, *Gene* **2:**115.

Van der Eb, A. J., Van Ormondt, H., Schrier, P. I., Lupker, J. H., Jochemsen, H., Van der Elsen, P. J., Deleys, R. J., Maat, J., Van Deveren, C. P., Dijkema, R., and De Waard, A., 1979, Structure and function of the transforming genes of human adenovirus and SV40, *Cold Spring Harbor Symp. Quant. Biol.* **44:**383.

van Roy, F., Fransen, L., and Fiers, W., 1984, Protein kinase activities in immune complexes of simian virus 40 large T-antigen and transformation-associated cellular p53 protein, *Mol. Cell. Biol.* **4:**232.

Verderame, M. F., Kohtz, D. S., and Pollack, R. E., 1983, 94,000 and 100,000-molecular weight SV40 T antigens are associated with the nuclear matrix in transformed and revertant mouse cells, *J. Virol.* **46:**575.

Vogt, M., and Dulbecco, R., 1960, Virus–cell interactions with a tumor-producing virus, *Proc. Natl. Acad. Sci. USA* **46:**365.

Vogt, M., and Dulbecco, R., 1963, Steps in the neoplastic transformation of hamster embryo cells by polyoma virus, *Proc. Natl. Acad. Sci. USA* **49:**171.

Wallenburg, J. C., Nepveu, A., and Chartrand, P., 1984, Random and nonrandom integration of a polyomavirus DNA molecule containing highly repetitive cellular sequences, *J. Virol.* **50:**678.

Walter, G., Hutchinson, M. A., Hunter, T., and Eckhart, W., 1982, Purification of polyoma virus medium-size tumor antigen by immuno affinity chromatography, *Proc. Natl. Acad. Sci. USA* **79:**4025.

Watkins, J. F., and Dulbecco, R., 1968, Production of SV40 virus in heterokaryons of transformed and susceptible cells, *Proc. Natl. Acad. Sci. USA* **58:**1396.

Weil, R., 1978, Viral "tumor antigens": A novel type of mammalian regulatory protein, *Biochim. Biophys. Acta* **516**:301.

Westphal, H., and Dulbecco, R., 1968, Viral DNA in polyoma and SV40-transformed cell lines, *Proc. Natl. Acad. Sci. USA* **59**:1158.

Winocour, E., Kaye, A. M., and Stollar, V., 1965, Synthesis and transmethylation of DNA in polyoma-infected cultures, *Virology* **27**:156.

Wolf, D., Harkis, N., and Rotter, V., 1984, Reconstitution of p53 expression in a nonproducer Ab-MuLV-transformed cell line by transfection of a functional p53 gene, *Cell* **38**:119.

Yabe, Y., Neriishi, S., Sato, Y., Liebelt, A., Taylor, H. G., and Trentin, J. J., 1961, Distribution of hemagglutination-inhibiting antibodies against polyoma virus in laboratory mice, *J. Natl. Cancer Inst.* **26**:621.

Yang, Y. C., Hearing, P., and Rundell, K., 1979, Cellular proteins associated with SV40 early gene products in newly infected cells, *J. Virol.* **32**:147.

Zhu, Z., Veldman, G. M., Cowie, A., Carr, A., Schaffhausen, B., and Kamen, R., 1984, Construction and functional characterization of polyomavirus genomes that separately encode the 3 early proteins, *J. Virol.* **51**:170.

Zouzias, D., Prasad, I., and Basilico, C., 1977, State of the viral DNA in rat cells transformed by polyoma virus. II. Identification of the cell containing non-integrated viral DNA and the effect of viral mutations, *J. Virol.* **24**:142.

Zouzias, D., Jha, K. K., Mulder, C., Basilico, C., and Ozer, H. L., 1980, Human fibroblasts transformed by the early region of SV40 DNA: Analysis of "free" viral DNA sequences, *Virology* **104**:439.

Zuckerman, S. H., Linder, S., and Eisenstadt, J. M., 1984, Transformation-associated changes in nuclear-coded mitochondrial proteins in 3T3 cells and SV40-transformed 3T3 cells, *Biochim. Biophys. Acta* **804**:285.

CHAPTER 5

Studies with BK Virus and Monkey Lymphotropic Papovavirus

KUNITO YOSHIIKE AND KENNETH K. TAKEMOTO

I. BK VIRUS

BK virus (BKV) is a human polyomavirus isolated by Gardner *et al.* (1971) from the urine of a renal allograft recipient undergoing immunosuppressive therapy, and named after the initials of the patient. The discovery of BKV, together with the isolation of another human polyomavirus—JC virus (JCV)—by Padgett *et al.* (1971) from brain tissue of a patient with progressive multifocal leukoencephalopathy (PML) (see Chapter 6), led to the initiation of extensive studies on the natural history and molecular biology of these viruses, since they were the first human viruses found to resemble morphologically the potentially oncogenic polyomaviruses.

The relationship between BKV and humans is apparently a peaceful one. BKV infects humans in childhood and persists in a latent form probably in the kidney for years without causing any serious disease. Therefore, BKV is not pathogenic in humans, whereas JCV is believed to be associated with PML.

Biological and biochemical studies have shown that BKV resembles simian virus 40 (SV40), which has been extensively investigated over the

KUNITO YOSHIIKE • Department of Enteroviruses, National Institute of Health, Shinagawa-ku, Tokyo 141, Japan. KENNETH K. TAKEMOTO • Laboratory of Molecular Microbiology, National Institute of Allergy and Infectious Diseases, National Institutes of Health, Bethesda, Maryland 20892.

past two decades. Despite their resemblance, BKV has some unique biological characteristics. Comparative studies of these two closely related viruses with different host ranges have broadened the scope of the molecular biology of polyomaviruses.

In this review, we will deal with the principal findings of studies with BKV during the past decade, some of which have been reviewed (Padgett and Walker, 1976; Gardner, 1977a,b; Takemoto, 1978; Howley, 1980; Padgett, 1980; Goudsmit, 1982; Takemoto and Yoshiike, 1982), and will emphasize the recent work.

A. Natural History

1. Isolation and Various Strains of BKV

During cytological and virological investigation of post allograft patients, Gardner *et al.* (1971) found the urine of one patient to contain a larger number of virus particles morphologically resembling particles of the polyoma subgroup of the papovaviruses. The virus was successfully isolated in Vero cell cultures (a continuous cell line of African green monkey kidney) after a prolonged incubation period of 1–3 months. Interestingly, primary human embryonic lung fibroblast cultures were less efficient at the initial isolation. The isolated virus, designated BK virus, was found to be different from any of the previously described members of the polyomaviruses.

Subsequent isolations of this virus have been almost exclusively from the urine samples of patients undergoing immunosuppressive therapy after renal transplantation (Dougherty and DiStefano, 1974; Lecatsas and Prozesky, 1975; Wright *et al.*, 1976) or from patients with Wis-

TABLE I. BK Virus Isolates

Strain	Source	Reference
BK Gardner's	Urine, renal allograft recipient	Gardner *et al.* (1971)
MM	Urine and brain tumor, Wiskott–Aldrich syndrome	Takemoto *et al.* (1974)
Dun	Urine, Wiskott–Aldrich syndrome	Takemoto *et al.* (1974), Howley *et al.* (1975b)
JM	Urine, Wiskott–Aldrich syndrome	Takemoto *et al.* (1974), Howley *et al.* (1975b)
RF	Urine, renal allograft recipient	Dougherty and DiStefano (1974)
MG	Urine, renal allograft recipient	Lecatsas and Prozesky (1975)
GS	Urine, renal allograft recipient	Wright *et al.* (1976)
DW	Urine, renal allograft recipient	Wright *et al.* (1976)
JL	Urine, bone marrow allograft recipient	Pauw and Choufoer (1978)
Dik	Urine, child with tonsillitis	Goudsmit *et al.* (1981), Goudsmit (1982)

kott–Aldrich syndrome, a sex-linked immunodeficiency disease (Takemoto *et al.*, 1974). More recently, BKV has been isolated from the urine of pregnant women (Coleman *et al.*, 1980) and from the urine of a child with tonsillitis (Goudsmit *et al.*, 1981). Table I lists some of the strains being used in various laboratories. Upon detailed examination of their genomes, these strains were shown to be naturally occurring BKV variants (see Section I.B.8).

2. Serological Surveys

Serological surveys have indicated that infection with BKV is widespread. Gardner (1973) examined more than 500 human sera for antibody to BKV by complement fixation (CF) and hemagglutination inhibition (HI) tests. Results of both tests showed that the antibody developed rapidly after the age of 1 year and, by age 10, a substantial proportion of children in England had CF antibody (71%) and HI antibody (83%). Surveys done in the United States (Shah *et al.*, 1973), Finland (Mäntyjärvi *et al.*, 1973), and Italy (Portolani *et al.*, 1974) all showed that antibody to BKV was in common in the population and that a majority of the infections probably occurred in early childhood. All the surveys reported a small decline in both the incidence and the mean titer of BKV HI antibodies after middle age. Thus, the virus is worldwide in distribution and very common in prevalence.

Serological evidence suggests that BKV is readily transmitted to children, but the route of transmission and the syndrome of the primary infection are not as yet known. From the epidemiology of SV40 and the mouse polyoma virus and from the fact that BKV has been recovered from urine, it is reasonable to suspect that BKV is transmitted by the oral or respiratory route. Indeed, BKV may cause acute respiratory disease among children as a result of the primary infection (Goudsmit, 1982).

Isolations of BKV from renal transplant patients and from pregnant women provide evidence that BKV replicates in the kidneys and is excreted in the urine (Coleman *et al.*, 1977, 1980; Hogan *et al.*, 1980). From the serological evidence (Borgatti *et al.*, 1979; Hogan *et al.*, 1980; Shah *et al.*, 1980), excretion of BKV is believed to result from reactivation of latent infection. It seems likely, therefore, that BKV that infects humans at early childhood and persists somewhere until its replication in the kidney is induced by immunodeficient states or other factors without causing obvious diseases.

B. Molecular Biology

1. The Genome of BKV

BKV was first recognized by electron microscopic examination of urine samples (Gardner *et al.*, 1971) to have morphological characteristics common to SV 40 and the mouse polyoma virus, and was considered to be a new member of the polyomavirus genus. BKV virions, grown in human embryonic kidney cell culture, were found to contain DNA molecules similar to those of SV40, which are about 5000-bp-long circular supercoiled molecules. BKV DNA has been characterized extensively by construction of restriction endonulease cleavage maps (Osborn *et al.*, 1974, 1976; Howley *et al.*, 1975a,b; Yang and Wu, 1978a,b,c; Freund *et al.*, 1979) (see Appendix), and by determination of base-sequence homology with other polyomaviruses monitored by reassociation kinetics (Howley *et al.*, 1975b, 1976), filter blot-hybridization in controlled stringency (Howley *et al.*, 1979; Law *et al.*, 1979), and electron microscopy (Khoury *et al.*, 1975; Newell *et al.*, 1978). The complete DNA sequence of BKV (MM) DNA was determined to be 4963 bp long (Yang and Wu, 1979a,b,c,d; Yang *et al.*, 1980), while the DNA of BKV (Dun) was shown to be 5153 bp long (Dhar *et al.*, 1978, 1979; Seif *et al.*, 1979a,b) (see Appendix).

Characterization of BKV DNA revealed that the genomic organization of BKV is strikingly similar to that of SV40. From the nucleotide sequence, the six gene loci for large and small T-antigens, VPx, VP1, VP2, and VP3 can be deduced (Yang and Wu, 1979c; Seif *et al.*, 1979b). The sequences of the deduced proteins in BKV share 73% amino acid homology with those in SV40, whereas the DNA sequences of the two viruses share 70% homology, indicating a close evolutionary relationship. However, the two viruses have different structures in the control region for transcription, which is located near the origin of DNA replication in the late region of the genome (see Section I.B.8.b.).

Cleavage of BKV DNA with various restriction endonucleases has yielded characteristic patterns, distinct from those of SV40 and JCV (see Appendixes). The cleavage maps of BKV DNA may be used as a rapid method for identification of new isolates. However, the patterns may differ among various isolates of BKV. *Hin*dIII cleaves the prototype (Gardner's strain) BKV DNA at four sites (Howley *et al.*, 1975a,b), generating four fragments arranged in the order a, b, d, and c (clockwise) (Fig. 1). The *Hin*dIII-C fragment contains the BKV DNA replication origin and its flanking region; the putative control region for transcription; and a part of the T-antigen gene. This fragment is sometimes variable in size among BKV isolates. Strains MM and JL deviate from the prototype strain more drastically in *Hin*dIII cleavage patterns. Strains MG and RF consisting of two complementing defectives (Section I.B.8.a) show cleaveage patterns different from those of the prototype BKV.

DNA homology between BKV and other polyomaviruses, SV40 and JCV, was studied by DNA–DNA hybridization under various stringency conditions (Howley *et al.*, 1979; Law *et al.*, 1979). Since the total sequences of the viruses are known (see Appendix), these data can be used to correlate the degree of homology detected by hybridization to that from the actual nucleotide sequences. This will provide the basis to assess the degree of homology between known viruses and new isolates by hybridization.

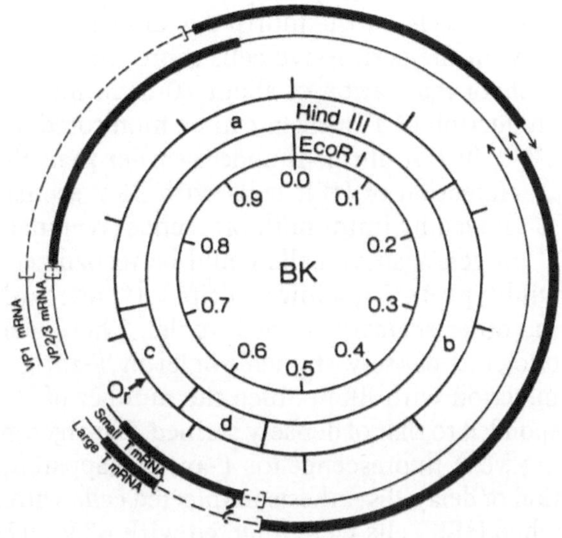

FIGURE 1. Physical and functional map of BKV (Gardner) genome. Map coordinates are oriented clockwise from the unique *Eco*RI cleavage site. *Hin*dIII cleavage sites are from Howley *et al.* (1975a,b). The origin of DNA replication is at 0.67 map unit. Data for mRNAs are from Manaker *et al.* (1979). The thick lines represent the structural genes for BKV proteins. The broken lines represent the spliced-out regions.

2. Productive Infection with BKV

BKV replicates most efficiently in human embryonic kidney (HEK) cells in culture (Takemoto and Mullarkey, 1973), although it can grow on embryonic lung cells (Coleman *et al.*, 1973), fetal brain cells (Takemoto *et al.*, 1979), fetal fibroblasts (Coleman, 1975), human foreskin cultures (Shah *et al.*, 1976), primary human urinary cells (Sack *et al.*, 1980), and on certain monkey cell lines (Seehafer *et al.*, 1978). BKV was initially isolated in Vero cells—an African green monkey kidney cell line (Gardner *et al.*, 1971)—and has been shown to grow better in monkey CV-1 cells at an elevated temperature (Miyamura and Takemoto, 1979). In cells other than HEK cells, cytopathic effects develop more slowly and the yields of virus are poorer.

Thus, the optimal cells for propagation and assay of BKV are primary or secondary HEK cells. Infection with BKV causes cytopathic effects: enlargement of nuclei with inclusions, cytoplasmic vacuolization, and cell-rounding followed by detachment from the plastic or glass surface. Virions are not readily released into the culture fluid and are mainly cell-associated. The yield of virus is about 10^5 virions per cell, which is similar to that of SV40 in primary African green monkey kidney cells. (The yield of SV40 is 3–10 times higher in CV-1 cells than in the primary cells.) Infectivity of BKV is assayed by plaque formation in HEK cells. BKV forms plaques which increase in number until day 20. The specific infectivity of purified Gardner's BKV (plaque isolate) is about 1 PFU/10^4 virions, which is lower than that of SV40 (1) PFU/10^2–10^3 virions).

The growth of BKV in HEK cells can be studied by immunofluorescence. In low-multiplicity infection, the antigen-positive cells reach maximum levels on the fourth day after infection. Therefore, BKV seems to grow in the permissive cells more slowly than SV40, which has a single cycle of replication of about 70 hr in monkey cells. Besides V-antigens, production of T-antigen can be monitored by immunofluorescence. Like SV40, BKV T-antigens appear earlier than V-antigen.

Infection of HEK cells with BKV seems to be poorly synchronized, as assayed by immunofluorescence (Watanabe *et al.*, 1979). Two types of T-antigen-positive cells could be recognized in HEK cells infected with wild-type BKV at a low multiplicity: one with densely stained nuclei and the other sparsely stained nuclei. There were as many sparsely stained nuclei as densely stained nuclei in T-antigen-positive cells 4 days after infection with BKV. Since the number of V-antigen-positive cells corresponded to that of densely stained T-antigen-positive cells, the cells showing weak fluorescence for T-antigen apparently are due to abortive infection or delayed synthesis in infected cells with nondefective BKV. However, when HEK cells were infected with BKV DNA, almost all the T-antigen-positive cells in the cultures were densely stained and comparable in number with V-antigen-positive cells (Watanabe *et al.*, 1981). These results suggest that the process of infection is more synchronized with DNA than with virions, and that the adsorption or uncoating process of virions may cause a delay in the productive cycle of BKV in HEK cells.

The molecular events in BKV-infected HEK cells have not been studied as extensively as has been done in the SV40–monkey cell system. However, what has been reported for SV40 is also believed to be applicable to BKV infection, because of the high degree of DNA homology and the similarity of the genomic organization between the two viruses. The center of BKV DNA replication origin containing four T-antigen recognition sequences (DeLucia *et al.*, 1983; Ryder *et al.*, 1983)—5′ *GAGGC* C *GAGGC* C *GCCTC* T *GCCTC* 3′ (L strand)—is almost identical to that of SV40—5′ *GAGGC* C *GAGGC* C *GCCTC* G *GCCTC* 3′. There is considerable nucleotide homology in the flanking regions of the two viruses (see Chapter 3 and Appendix). Like SV40 DNA, therefore, BKV DNA replication proceeds bidirectionally from this point (0.67 map unit from *Eco*RI site), terminating at a point 180° from the origin. The signals for splicing of mRNAs resemble structurally and topographically those of SV40 (Seif *et al.*, 1979a). The spliced structure of BKV mRNAs (two early RNAs and two late RNAs) in HEK cells (Manaker *et al.*, 1979) is consistent with the genomic organization deduced from the total nucleotide sequence.

The mRNAs of BKV from infected HEK cells were mapped by the nuclease S1 technique (Manaker *et al.*, 1979). Two mRNAs synthesized in the early stage, which were spliced out between 0.535 and 0.53 map unit and between 0.585 and 0.53, are for small T- and large T-antigen, respectively. The body sequences of late 16S RNA map between 0.93 and 0.16 and those of the late 19 S RNA, between 0.765 and 0.16. In analogy with SV40, the 16 S RNA is for VP1 and the 19 S RNA, for VP2 and VP3. These *in vivo* mRNAs were different from RNAs transcribed *in vitro* from BKV DNA or BKV minichromosome by *E. coli* RNA polymerase (Meneguzzi *et al.*, 1978a,b).

3. BKV Proteins

From the total sequence (Yang and Wu, 1979c; Seif *et al.*, 1979b), BKV DNA is believed to be capable of encoding six proteins: small and large T-antigens in the early genome and VP1, VP2, VP3, and VPx in the late side of the genome. The gene for VPx corresponds to the agnogene of SV40. The amino acid sequences deduced for these BKV proteins show striking similarity to those of SV40 and JCV (see Appendix).

The immunological and biochemical data on BKV proteins obtained in earlier studies are consistent with the deduced structures of BKV proteins. Relatedness and cross-reactivity of BKV T-antigens to those of SV40 and JCV have been shown by immunofluorescence and immunoprecipitation (Beth *et al.*, 1977; Rundell *et al.*, 1977; Simmons *et al.*, 1977, 1978; Simmons and Martin, 1978; Farrell *et al.*, 1978). Analyses of the methionine-labeled tryptic peptides of BKV T-antigens have shown that, as in SV40, a subset of amino acid sequences in BKV small T-antigen is present in large T-antigen (Simmons *et al.*, 1977, 1978; Simmons and Martin, 1978). These data are in agreement with the genomic organization deduced from the nucleotide sequence and the analyses of BKV mRNAs (Manaker *et al.*, 1979). Like SV40, six to seven species of polypeptides differing in mobility were found in disrupted BKV virions in gel electrophoresis (Mullarkey *et al.*, 1974; Barbanti-Brodano *et al.*, 1975; Seehafer *et al.*, 1975; Wright and di Mayorca, 1975); the three larger proteins VP1, VP2, and VP3 are coded by the BKV genome and others are cellular proteins. In electrophoresis, BKV VP1 (the major capsid protein) moved faster and, therefore, appeared smaller than SV40 VP1, but from the nucleotide sequence, the VP1 of the two viruses are believed to be composed of 362 amino acids. All members of the polyomaviruses share a common antigenic determinant located in their major capsid polypeptides. BKV virion antigens cross-reacted with antisera prepared against disrupted SV40 virions (Shah *et al.*, 1977a,b).

The functions of the BKV T-antigens, although not investigated thoroughly, are believed to parallel those of SV40 T-antigens. Therefore, BKV T-antigens are required for initiation of DNA replication and the switch from early to late lytic transcriptions. Indeed, BKV and SV40 T-antigens not only cross-react with each other immunologically, but share common biological properties. BKV T-antigen can substitute functionally for the defective SV40 T-antigen for SV40 replication (Mason and Takemoto, 1976; Lai *et al.*, 1979). Like SV40, BKV can provide factors required for adenovirus 2 to grow in monkey cells, where the replication of adenovirus alone is restricted (Miyamura and Takemoto, 1979). Purified SV40 A protein (T-antigen) has been shown to bind *in vitro* to the BKV origin of DNA replication (Ryder *et al.*, 1983). It was recently reported, however, that SV40 T-antigen allows replication of BKV DNA but may not be competent for switching to late gene functions (Major and Matsumura, 1984).

Two cellular proteins (56K and 32K) are specifically coimmunoprecipitated by SV40 and BKV T-antigens (Rundell *et al.*, 1981).

4. Oncogenicity and Transformation

The oncogenic potential of BKV has been well established in young hamsters by injection through various routes: subcutaneous, intravenous, intracerebral, and intraperitoneal. The tumor cells contain BKV-specific T-antigen and BKV DNA sequences, but the rescue of infectious virus from cultured tumor cells was infrequent (Uchida *et al.*, 1979). Initially, BKV seemed to be much less oncogenic than SV40 in hamsters, as the incidence of tumor induction was very low in early studies with Gardner's prototype strain (Shah *et al.*, 1975; Näse *et al.*, 1975; van der Noordaa, 1976; Greenlee *et al.*, 1977). However, Gardner's BKV was found later to induce a variety of malignant tumors at high frequencies, when highly purified and concentrated virus was injected into newborn hamsters intracerebrally (Uchida *et al.*, 1976, 1979; Corallini *et al.*, 1977; Watanabe *et al.*, 1979, 1982) or into 3-week-old hamsters intravenously (Corallini *et al.*, 1978). With strains MM (Costa *et al.*, 1976) and RF (Dougherty, 1976), the tumor incidence was intermediate. BKV induced a variety of tumors: fibrosarcoma, ependymoma, neuroblastoma, pineal gland tumor, insulinoma, peritoneal tumor, adenocarcinoma, angiosarcoma, osteosarcoma, lymphoma, seminoma, and so on. The frequency and type of tumors are probably dependent on the concentration of virus and the route of injection, but more importantly, on the genetic properties of the virus.

Although the oncogenic potential of BKV was tested mostly with the virus stocks originating from Gardner's prototype strain, genetic heterogeneity probably arose in the stocks during passages. Uchida *et al.* (1979) found that BKV (Gardner's strain) induced various types of tumors in hamsters, using a series of virus samples originating from a single stock, and that the spectrum of tumors varied with the virus sample. From the passage history and the relationship among the virus samples, the polyoncogenicity and especially the insulinoma-inducing capacity seemed to be best accounted for by the presence of BKV mutants differing in tumorigenic capacity and having different target cells. This hypothesis has been supported by the fact that wild-type plaque isolates and viable deletion mutants (pm-522 and its derivatives; see Section I.B.8.b) showed markedly different spectra of induced tumors (Watanabe *et al.*, 1979, 1982). Wild-type plaque isolates were less tumorigenic (about one-fifth of pm-522) and induced ventricular tumors but no insulinomas in hamsters, whereas viable deletion mutants induced ventricular tumors and insulinomas. These findings suggest that the types of tumors are determined by the genetic properties of BKV variants.

Besides hamsters, mice and rats are known to be susceptible to tumor induction by BKV (Corallini *et al.*, 1977; Noss *et al.*, 1981; Noss and Stauch, 1984). BKV induced fibrosarcoma, osteosarcoma, liposarcoma, choroid plexux papilloma, glioma, and nephroblastoma in outbred and inbred rats. In inbred rats, the tumor rate is negatively correlated with the level of anti-T antibody 3 months after inoculation (Noss and Stauch, 1984).

The capacity of BKV to transform nonpermissive or semipermissive cultured cells has been well demonstrated. Major and di Mayorca (1973) first reported neoplastic transformation of hamster BHK-21 cells by BKV, but they did not demonstrate the presence of BKV-specific T-antigen or BKV genomes in the transformed cells. Portolani *et al.* (1975) presented evidence for transformation of hamster cells by producing transformed and tumorigenic cells containing T-antigen and rescuable BKV. Since then, a number of studies have shown that BKV can transform cultured cells from hamsters, rats, mice, rabbits, and monkeys, using Gardner's strain (Tanaka *et al.*, 1976; van der Noordaa, 1976; Seehafer *et al.*, 1977, 1979a; Portolani *et al.*, 1978; Watanabe and Yoshiike, 1982) and strains MM and RF (Mason and Takemoto, 1977; Costa *et al.*, 1977; Bradley and Dougherty, 1978). Furthermore, BKV DNA (Takemoto and Martin, 1976; van der Noordaa, 1976; van der Noordaa *et al.*, 1979; Watanabe and Yoshiike, 1982) or its appropriate fragments containing the early genes (Grossi *et al.*, 1982b) could also transform rodent cells.

Apparently, transformation of rodent cells by BKV is inefficient. In the study by Portolani *et al.* (1975), eight cell passages were required before transformed hamster-cell colonies appeared. Takemoto and Martin (1976) reported that hamster, mouse, and rat cells failed to be transformed by BKV, although hamster cells were transformed by BKV DNA.

Transforming capacity of BKV varies with the strain (Watanabe and Yoshiike, 1982). One wild-type plaque isolate (wt-501) from a Gardner's stock, which was tumorigenic in hamsters (Watanabe *et al.*, 1982), did not readily transform hamster embryo cells or rat cell line 3Y1. However, a viable deletion mutant (pm-522) transformed hamster and rat cells much more efficiently than the wild type (see Section I.B.8.b). The data strongly suggest that variant viruses arising during passages can affect the transforming capacity of the virus stock in different laboratories. It should be noted that in many laboratories BKV had been passaged in monkey Vero cells, which are not the best cells for the growth of BKV. Possibly, propagation of BKV in less efficient host cells allows generation or selection of certain BKV mutants.

The transforming capacity of BKV is probably lower than that of SV40. Table II compares the transforming capacity of SV40 and BKV DNAs in the rat 3Y1 cell system (S. Watanabe, H. Uehara, and K. Yoshiike, unpublished data). Use of DNA allows direct comparison of transforming

TABLE II. Transforming Capacity of BKV and SV40 DNAs for Rat 3Y1 Cells[a]

Source of DNA (form I from virions)	No. of foci/μg DNA per 5-cm dish
BKV wt-501 (plaque isolate of Gardner's[b])	0
BKV pm-522 (viable deletion mutant[b])	7 ± 3
SV40 (plaque isolate of strain 777)	87 ± 21

[a] Standard calcium phosphate method was used. Each dish was infected with 0.25 to 2 μg of form I viral DNA. Stained foci were counted 35 days after infection. Number was the average of five dishes ± S.D.

[b] Watanabe *et al.* (1979, 1982, 1984a,b), Watanabe and Yoshiike (1982).

potential without the possible involvement of virion proteins for adsorption–penetration. Under these conditions, DNA from wild-type plaque isolates of BKV rarely yielded foci of transformed cells. As can be seen, even the BKV DNA with the highest transforming capacity was much less efficient than SV40 DNA for transformation of rat cells. Furthermore, foci of BKV-transformed cells were less dense than those of SV40-transformed cells. It is possible that this difference results from the qualitative difference of T-antigens between the two primate polyomaviruses.

Like SV40 and the mouse polyoma virus, BKV induces tumor-specific transplantation antigen (TSTA). Cross-protection tests in immunized hamsters with cells transformed by primate polyomaviruses have shown that they have their own TSTAs (Takemoto and Mullarkey, 1973; Karjalainen *et al.*, 1978; Law *et al.*, 1978). However, BKV-transformed mouse cells were found to have a TSTA immunologically related to the one induced by SV40 (Law *et al.*, 1978; Seehafer *et al.*, 1979b; Kato *et al.*, 1979).

5. BKV Genomes in Tumor or Transformed Cells

The DNA from BKV-transformed hamster cells (Howley and Martin, 1977) and from BKV-induced hamster tumors (Yogo *et al.*, 1980b) were examined by reassociation kinetics for viral sequences. In such studies the amount of viral DNA can be quantitated. The transformed cells and tumor cells were found to contain from a few to 150 copies of BKV genomes per diploid genome.

Studies using Southern blot analyses have revealed the state of BKV DNA in transformed and tumor cells. These cells were found to contain both integrated and free BKV genomes (Chenciner *et al.*, 1980a,b; Yogo *et al.*, 1980a; ter Schegget *et al.*, 1980; Meneguzzi *et al.*, 1981; Beth *et al.*, 1981; Grossi *et al.*, 1982a,b), which varied in number depending on the cell line. The presence of free viral genome is much more frequent in BKV-transformed hamster or rodent cells than in SV40-transformed hamster cells. This is probably because hamster cells are more permissive to BKV than to SV40.

Without the integrated viral genome, the nonintegrated episomal BKV genome alone seems to be capable of maintaining the transformed state of the cells. It was reported that one cell line derived from a BKV-induced hamster osteosarcoma contained nonintegrated defective BKV DNA, mostly in a monomeric form (Yogo *et al.*, 1980a; Furuno and Yogo, 1983). Although polymeric defective DNA was found in this cell line, the integrated form was undetectable (<0.1 BKV genome/diploid cell).

Another exceptional hamster tumor is the BKV-induced T-antigen-negative tumor (choroid plexus papilloma), which contains one copy of BKV DNA per cell as revealed by DNA–DNA reassociation kinetics (Yogo *et al.*, 1980b, 1981). The BKV DNA was found by blot hybridization to be integrated into cellular DNA at a site in the middle of the early region of the viral genome. Thus, the integrated BKV genome cannot direct the synthesis of a full-sized large T-antigen. The data suggest that the complete large T-antigen is not necessarily required for transformation of hamster cells.

6. Transformation of Human Cells

Since human cells are permissive for productive infection of BKV, their transformation by BKV is an exceptional event. Generally, foci of transformed cells emerge after the majority of cells are killed as a result of productive infection. Thus, the transformed human cells are often resistant to superinfection with BKV.

The properties of transformed human cells are markedly different from report to report. Purchio and Fareed (1979) obtained HEK cells transformed by BKV. The cells were T-antigen positive, contained the integrated BKV genome, and could grow in a medium with lowered serum concentration (2%), but could not form colonies in agar, nor induce tumors in nude mice. Therefore, they seem to be partially transformed. Takemoto *et al.* (1979) reported that morphologically transformed cells appeared from rare survivors after lytic infection of human fetal brain cells by BKV. These transformed cells were tumorigenic in nude mice, contained exclusively unintegrated episomal viral DNA, but expression of T-antigen, as examined by immunofluorescence, was negative in these cells. Grossi *et al.* (1982c) described human embryonic fibroblasts transformed by BKV DNA and BKV subgenomic fragments. The transformed cells produced large and small T-antigens, and contained considerable amount of free BKV DNA in monomeric and polymeric forms. Integrated BKV DNA was absent in most cell lines. Hara *et al.* (1983) found that a small plaque-forming viable deletion mutant (pm-522) of BKV transformed HEK cells more readily than the wild-type BKV. The viral DNA in the transformed HEK cells were mostly unintegrated.

The differences of properties among various human cells transformed by BKV are probably due to the fact that the transformants were derived from rare survivors. Under these circumstances, the pressure to select a certain exceptional variant clone of human cells is enormously high, as compared with the infection of rodent cells with BKV. Mutation of BKV may also affect transformation efficiency of permissive human cells.

The BKV transforming capacity for human cells is apparently much lower than that of SV40. This host cell specificity seems to be determined by the T-antigens (A. de Ronde, C. Sol, M. Macdonald, J. ter Schegget, A. van Strien, and J. van der Noordaa, personal communication). Among the constructed chimeric plasmids containing the transcriptional control region (see Section I.B.8.b) and the T-antigen gene from the two viruses, only those having the SV40 T-antigen gene could transform human foreskin cells.

7. Possible Involvement of BKV in Human Cancer

Since BKV is oncogenic in rodents, hamsters, mice, and rats, and is capable of transforming human cells, BKV may be potentially oncogenic for humans. The possible involvement of BKV and other polyomaviruses, SV40 and JCV, has been investigated rather extensively by searching for T antibody in sera from cancer patients, for T-antigen in human tumor cells, and for viral DNA sequences in human tumor cell DNA. Such studies have been systematically reviewed by Takemoto (1980). Although a few positive results have been reported with the three types of testing, there is no definitive evidence that any known polyomavirus is involved in human neoplasia.

From the known models of transformation by SV40 or the mouse polyoma virus, demonstration of the presence of viral genomes in cancer cells is the first step for further investigation on the possible role of polyomaviruses in carcinogenesis. We can now detect viral DNA contained in cancer cells at less than one viral genome per cell, owing to the efficient labeling of viral DNA *in vitro* (Rigby *et al.*, 1977), by either reassociation kinetics or blot hybridization. Fiori and di Mayorca (1976) and M. M. Pater *et al.* (1980) reported that the DNA from human tumors and tumor cell lines contain BKV DNA sequences, but their observations were not confirmed by others (Israel *et al.*, 1978; Wold *et al.*, 1979; Grossi *et al.*, 1981).

8. Naturally Occurring BKV Mutants with Rearranged DNA

a. Defective Viruses

Defective BKV occurs spontaneously during passages in cultured cells. Presence of abundant heterogeneous defective (shorter) DNA was first detected by electrophoresis of virion DNA from Gardner's strain grown at a low multiplicity of infection (MOI) in HEK cells (Howley *et al.*, 1975a). van der Noordaa *et al.* (1978) systematically attempted to generate defective BKV by undiluted (high MOI) passage starting with a virus stock prepared at a low MOI. After ten undiluted passages, two types of defectives became the dominant species. These defective viruses could not induce T-antigen synthesis, nor transform hamster cells, but interfered with the replication of nondefective DNA in HEK cells. Another class of defectives capable of directing T-antigen synthesis and viral DNA synthesis also occurred in the stocks of Gardner's strain (Watanabe *et al.*, 1981). All these defective viruses contained rearranged DNA. Molecular cloning BKV strain GS DNA also revealed the presence of defective DNA as a dominant species of DNA in the stock (Akrigg *et al.*, 1981).

The occurrence of defective viruses with rearranged DNA is not limited to BKV grown in cell cultures. Two isolates from the urine of renal allograft recipients, strains RF and MG, turned out to consist of two classes of defective viruses which complemented each other for productive infection in HEK cells (A. Pater *et al.*, 1980, 1981, 1983; M. M. Pater *et al.*, 1981). The RF strain consisted of two viruses containing rearranged DNA designated R1a and R2; R1a had an intact early region and three DNA replication origins and R2 had an intact late region and two origins. Both R1a and R2 were required for growth in HEK cells, but only R1a DNA was found in transformed hamster cells.

b. Nondefective Variants with Changes in the Control Region for Transcription

The noncoding region between the T-antigen gene and the agnogene in the SV40 genome contains the DNA replication origin and the control signals for early and late transcription. The region of the late side of the origin consists of unique repeating elements: three 21-bp repeats containing six 5′ CCGCCC 3′ elements and two 72-bp tandem repeats. The 21-bp repeats constitute the essential part of the promoter for early and late transcription (Byrne *et al.*, 1983; Brady *et al.*, 1984; Hartzell *et al.*, 1984) and the 72-bp element acts as a host discriminatory enhancer for transcription (Benoist and Chambon, 1981; Gruss *et al.*, 1981; Laimins *et al.*, 1982; Byrne *et al.*, 1983; Weiher *et al.*, 1983). The changes in the topographically corresponding region have been shown to affect various important biological parameters of other polyomarviruses—the host ranges of the mouse polyoma virus (Katinka *et al.*, 1980; Sekikawa and Levine, 1981; Fujimura *et al.*, 1981; Tanaka *et al.*, 1982), JCV (Yoshiike *et al.*, 1982), and BKV (Watanabe and Yoshiike, 1982; Watanabe *et al.*, 1984a,b).

The viable strains of BKV isolated from humans in various parts of the world probably are more or less different from one another at the nucleotide sequence level, especially in the control region for transcription. Gardner's strain and the two major variant strains Dun and MM differ in the noncoding region near the origin of DNA replication (Seif *et al.*, 1979b; Yang and Wu, 1979c). The putative control region of Gardner's strain contains three tandemly repeating 68-bp elements, the central element of which has a deletion of 18 bp. Compared with Gardner's BKV, strain Dun has a deletion of 43 bp and strain MM has deletion and insertion. Other strains (GS, RF, JL, and Dik) also have rearrangements in the control region (M. M. Pater *et al.*, 1979; A. Pater *et al.*, 1983; ter Schegget *et al.*, 1985). (It is known that variant strains have other changes. For instance, MM has a deletion in the small T-antigen gene, and JL has an extra *Eco*RI site.)

Comparison of the nucleotide sequences in the control regions of various BKV strains (Yang and Wu, 1979c; Seif *et al.*, 1979b; A. Pater *et al.*, 1983; Watanabe *et al.*, 1984b; ter Schegget *et al.*, 1985) shows that they are all related to one another and have diverged from a common ancestral proto-BKV. Figure 2 illustrates the nucleotide sequence of the control region of the hypothetical proto-BKV constructed by incorporating all the elements (oligonucleotides) found with BKV variants and Figure 3 shows the relationship of variants to the proto-BKV sequence. Among the nondefective strains whose control regions have been sequenced, strain Dik (ter Schegget *et al.*, 1985) has the richest variety of oligonucleotide sets and therefore most resembles the proto-BKV, except for some point mutations. Others seem to have been generated through deletions and repeated duplications from the proto-BKV. A series of viable deletion mutants presumably originating from Gardner's strain showed a wide variation of rearrangements in this region (Watanabe *et al.*, 1984b). Their rearrangements, which affected productive infection and transformation, were essentially deletions and duplications.

The probable mechanism of deletion and duplication is as follows. Perhaps recombination by breakage and rejoining occurs within a single molecule rather than between two free molecules, for the genetic recombination frequency between two polyomavirus mutants is very low. Recombination may occur between the two newly duplicated daughter segments of a replicating molecule (replicative intermediate) at nonhomologous points. The resulting molecule is a dimer containing one molecule with deletion and the other with duplication. Dimers may be conveted into monomers by recombination at any two homologous points within a dimer molecule. Then two daughter molecules, one with deletion and the other with tandem duplication, may be generated. (An alternate way to generate deletion and duplication molecules: recombination between the two daughter segments occurs at homologous points and subsequent recombination within a dimer, at nonhomologous points.) If the resulting mutant has any selective advantage over the others, it will become a dominant species of the population.

Rosenthal *et al.* (1983) studied the enhancer function of the BKV sequences in the noncoding region near the DNA replication origin, using strain Dun. The isolated 216-bp fragment containing all three repeats (triplication of 68-bp elements) without a TATA box or replication origin, was inserted into the expression vector consisting of the SV40 promoter (without the 72-bp enhancer elements) and the bacterial gene encoding chloramphenicol acetyltransferase (CAT). Upon transfection with the

```
                 260        250        240        230        220
      5' GGCCTTT GTCCAGTTTA ACTGGGGACA AGGCCAAGAT TCCTAGGCTC GCAAAACATG

210        200        190        180        170        160        150
TCTGTCTGGC TGCTTTCCAC TCCTTTGGCC AGTTTCCACT TCCTCTTGTG TTTATTTGAG AATTTCAGGG

140        130        120        110        100        90         80
GCGGGGTTTC ACTATTAACT GCCACTGGCT GGCTGCCCAG TCATGCACTT TCCTTCCTGA GGTCATGGTT

70         60         50         40         30         20         10
TGGCTGCATT CCATGGGTAA GCAGCTCCTC CCTGTGGCCT TTTTTTTTAT AATATATAAG AGGCCGAGGC

0          -10        -20
CGCCTCTGCC TCCACCCTTT CTCTCAAGTA 3'
```

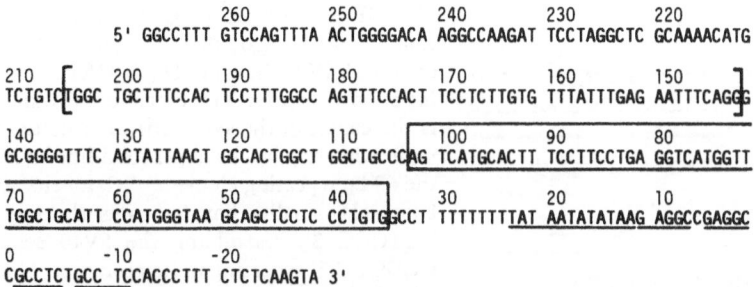

FIGURE 2. Nucleotide sequence of BKV transcriptional control region. The sequence represents the L strand of hypothetical proto-BKV between the VPx gene (starting at nucleotide 268 toward left) and the center of the symmetrical structure of DNA replication origin. The hypothetical sequence was constructed from the data of Yang and Wu (1979c), Seif *et al.* (1979b), A. Pater *et al.* (1983), Watanabe *et al.* (1984b), and ter Schegget *et al.* (1985). Numbering is as described in Watanabe *et al.* (1984): cytosine in the center of the replication origin was taken as zero, and the numbering increases toward the late genome. Brackets indicate the unique 63-bp segment of Dik and RF. The box indicates the 68-bp element repeating three times in Gardner's strain. Underlined are the TATA box and T-antigen recognition sequence 5′ GAGGC 3′ (DeLucia *et al.*, 1983).

constructed recombinant, CAT was produced in human HeLa cells, monkey CV-1 cells, and mouse L cells. Their study clearly demonstrated that the BKV sequence (triplication of 68-bp elements) can enhance CAT expression from the heterologous SV40 promoter.

The role of the above BKV segment containing triplication of 68-bp elements is not believed to be limited to the enhancer activity *in vivo*, although study of the functional anatomy of BKV enhancer/promoter has not yet been completed. It is conceivable that the region containing 68-bp elements and their flanking area functions both as the enhancer and as the promoter in BKV, since there is no such structure in BKV DNA as the 21-bp repeats in the SV40 genome. Indeed, the BKV fragment (from Gardner's strain) including the BKV replication origin, a TATA box, and three 68-bp elements can induce CAT, when ligated upstream of the CAT gene, in transfected monkey CV-1 and HEK cells (Watanabe and Yoshiike, 1986). However, the triplication of 68-bp elements in Gardner's strain is not always needed for BKV to grow in HEK cells. The control regions of strains JL, Dik, and pm-522 and its derivatives have only one complete 68-bp element (in some cases with the shorter repeats resulting from deletions and duplications).

The 68-bp element of BKV has signals resembling the sequence in the SV40 promoter (5′ CCTCCC 3′ of BKV, 5′ CCGCCC 3′ of SV40), the SV40 enhancer core [5′ G*TGGA(T)A(T)A(T)*G 3′ (Weiher *et al.*, 1983)], and the adenovirus type 5 EIA enhancer core [5′ A(C)*GGAAGTGA*A(C) 3′ (Hearing and Shenk, 1983)]. Therefore, a single 68-bp element seems to function as both a promoter and an enhancer. However, the removal of two 68-bp elements and their upstream sequences (thus, the BKV frag-

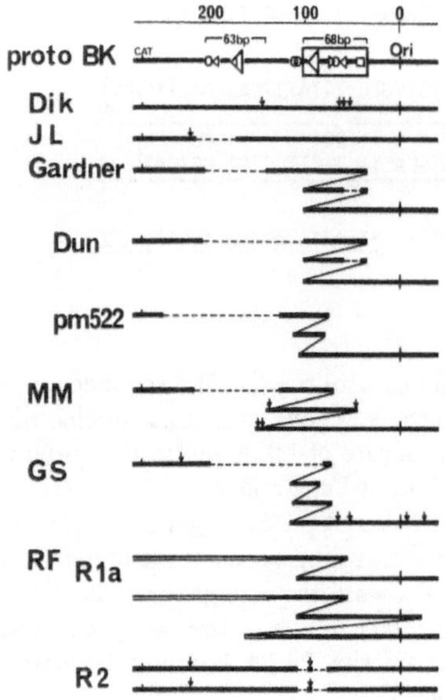

FIGURE 3. DNA rearrangements in the transcriptional control region of BKV strains. The sequence of proto-BKV is shown in Fig. 2. CAT represents the intiation codon for VPx extending toward left. Ori is the center of the symmetrical structure of DNA replication origin. The larger solid-line box indicates the 68-bp repeating element of Gardner's strain and Dun. The smaller box indicates the sequence 5′ CCTCCC 3′, resembling the SV40 sequence 5′ CCGCCC 3′ repeating in its promoter. The smaller triangles represent the sequence 5′ TGGA(T)A(T)A(T) 3′, found in the SV40 enhancer core. The larger triangles represent the sequences resembling the adenovirus E1A enhancer. Small circles indicate the sequence 5′ TGGCTG(C) 3′, frequently found in the control regions of JCV and BKV (Watanabe *et al.*, 1984b). The 63-bp segment is the unique region of RF and Dik. The data are from Yang and Wu (1979c) for Gardner and MM, Seif *et al.* (1979b) for Dun, A. Pater *et al.* (1983) for GS and RF, Watanabe *et al.* (1984b) for pm-522, and ter Schegget *et al.* (1985) for Dik and JL. Small arrows indicate point mutations (base substitutions) or single base deletions: for Dik, arrows indicate from left to right, T, A, A, and T at nucleotides 145, 65, 62, and 52, respectively; for JL, C at 222; for MM, AT, deletion (dl), dl, and T at 138–137, 49, 150, and 145, respectively; for GS, T, A, T, G, and G at 229, 65, 52, −6, and −24, respectively; for R2, C and dl at 222 and 95, respectively. Dashed lines indicate deletions. Ranges of deletion are, from left to right: for JL, from nucleotides 213 to 175; for Gardner, from 204 to 142 and from 59 to 42; for Dun, from 207 to 102 and from 59 to 42; for pm-522, from 247 to 128; for MM, from 216 to 128; for GS, from 198 to 82; and for R2, from 107 to 99 and from 92 to 79. The unfilled areas for R1a indicate rearranged DNA (VP1 gene). The linkages from breakage and rejoining are, from left to right: for Gardner and Dun, from nucleotides 35 to 102 and from 35 to 102; for pm-522, from 74 to 113 and from 79 to 109; for MM, from 70 to 141 and from 47 to 152; for GS, from 72 to 116, from 84 to 114, and from 72 to 116; for R1a, from 52 to 110, and from 52 to 110 and from −22 to 23 bp upstream of 141 in VP1 gene.

ment ligated in the CAT expression vector contains only one 68-bp element, a TATA box, and the DNA replication origin) markedly reduced expression of the CAT gene in human cells (Watanabe and Yoshiike, 1986). For BKV to be viable in human cells, the control region probably requires at least one complete 68-bp element plus its partially duplicated structure (or some upstream flanking sequence in the proto-BKV), as seen in viable deletion mutants of BKV.

The structure of the BKV control region affects the transforming capacity of BKV. The mutant designated pm-522, which was rescued from a hamster pineocytoma induced by Gardner's BKV, proved to have a number of characteristics distinct from those of the wild-type (plaque isolate of Gardner's strain) BKV (Watanabe *et al.*, 1979, 1982; Watanabe and Yoshiike, 1982). In HEK cell cultures, pm-522 formed turbid plaques smaller than clear plaques of wild-type BKV. This mutant was approximately five times as tumorigenic in hamsters as wild-type BKV and could induce hamster insulinomas, which so far have not been induced by the plaque isolates of wild-type BKV. Despite its somewhat inefficient growth in HEK cells, BKV pm-522 produced many more foci of permanently transformed cells in hamster or rat cell cultures than wild-type BKV. Characterization of the constructed recombinant viruses between pm-522 and wild type (wt-501) and determination of nucleotide sequence have shown that the DNA rearrangement (deletions and duplications) in a limited area of the control region was solely responsible for both the altered plaque morphology and the high transforming capacity (Watanabe and Yoshiike, 1982; Watanabe *et al.*, 1984b). The transcriptional control region of pm-522 had one set of 68-bp element and two sets of shorter 37-bp repeats (Fig. 3). Apparently, pm-522 is a host range mutant that can express its transforming gene in nonpermissive cells more efficiently than wild-type BKV.

The high transforming capacity of pm-522, compared with that of the wild-type plaque isolate of Gardner's BKV, can probably be attributed to the capacity of unintegrated pm-522 genomes to express continuously their early functions in nonpermissive cells and to transform them transiently before stable transformation is established (Watanabe *et al.*, 1984a). Rat cells infected with either pm-522 or wild-type BKV contained unintegrated BKV DNA in their nuclei during the first 2-week period, before foci of transformed cells become detectable. The free viral genomes continued to be transcribed to direct the synthesis of T-antigen, which rendered the cells transiently transformed, in the mutant infection. In the wild-type infection, however, transcription of the T-antigen gene occurring initially seemed to be repressed soon after infection, and T-antigen production ceased. Apparently, wild-type BKV is susceptible in rat cells to some repressive mechanism, from which pm-522 can escape owing to the rearrangement in the control region.

C. Summary and Conclusions

Since the first isolation in 1971, BKV has been repeatedly isolated from the urine of patients with immunodeficiency and from pregnant women. Serological surveys conducted in various countries have shown that infection with BKV occurs early in childhood and may cause respiratory disease. BKV persists in humans, probably in the kidneys, as a latent nonpathogenic virus, which may be reactivated during immunodeficiency. There is no convincing evidence that BKV is associated with human cancers.

BKV grows best in HEK cell cultures and poorly in monkey cells, whereas SV40 grows best in African green monkey kidney cell cultures and poorly in human cells. Despite their different host ranges, biological and biochemical events occurring in productively infected cells are probably very similar. BKV and SV40 virions are morphologically identical, and their proteins and T-antigens cross-react immunologically. BKV T-antigens can complement the growth of conditional lethal mutants of SV40. The two viruses can transform various rodent cells in culture, but the transforming ability of BKV is lower than that of SV40. Episomal viral genomes are more readily found in BKV-transformed cells than in SV40-transformed cells.

Comparison of the total nucleotide sequences of BKV and SV40 has indisputably demonstrated the evolutionary relatedness of the two viruses—they share 70% nucleotide sequence homology. The genomic organizations of the two viruses are identical and amino acid sequence homology can be found with all the virus-coded proteins. Despite their overall striking similarities, the area to the late side of the DNA replication origin (the transcriptional control region) is the region of extensive divergence between the two viruses.

The structure of the promoter/enhancer in the BKV genome is different from that in the SV40 genome, and has diverged even among BKV strains. The unique structure in this region has evolved in different ways to produce effective viral transcriptional control elements in different host cells after BKV and SV40 diverged from a common polyomavirus ancestor. The various arrangements in BKV strains suggest that if certain essential elements are retained, the enhancer/promoter structure may vary without loss of function in permissive cells. In some cases, however, such changes may affect transforming capacity of BKV in nonpermissive cells.

II. LYMPHOTROPIC PAPOVAVIRUS

Polyomaviruses have been isolated from a wide variety of species, including rodents, rabbits, monkeys, cattle, birds, and humans. All of these isolates have narrow host ranges for replication and grow well in cultures of fibroblastic or epithelial cells derived from their natural hosts. A unique, new polyomavirus has recently been isolated from a B-lymphoblastic cell line derived from a lymph node of an African green monkey by zur Hausen and Gissmann (1979). Unlike other members of the polyomavirus genus, this virus grew only in B-lymphoblastoid cell lines and was therefore referred to as lymphotropic papovavirus (LPV). Electron microscopic observations indicated that the particle morphology resembled papovaviruses; its reactivity with broad-reacting antiserum prepared in animals inoculated with SDS-disrupted SV40 virions clearly established that the virus belonged to the polyomavirus genus. Analysis of the viral DNA by restriction enzymes and hybridization experiments showed no similarities with SV40, providing evidence that this was a unique, new polyomavirus. Serological surveys have shown that antigenically related lymphotropic polyomaviruses are widely prevalent in virtually all primates, including humans.

A. Biology

1. Replication

LPV, like other members of the polyomavirus genus, exhibits an extremely narrow host range. Virus growth has thus far been shown only in certain B-lymphoblastic cell lines of monkey or human origin. No growth has been observed in cultures of primary monkey or human peripheral blood lymphocytes. The virus has a strict requirement for dividing (established) B lymphoblasts in order for viral replication to occur (zur Hausen *et al.*, 1980; Takemoto *et al.*, 1982). T cells or non-B, non-T (null) cells are also resistant to infection by LPV (zur Hausen *et al.*, 1980; Brade *et al.*, 1981; Takemoto *et al.*, 1982). Among the established B-lymphoblastic cell lines that have been examined, only two have supported the growth of LPV: BJA-B (Klein *et al.*, 1974) and TL-1 (Hayashi *et al.*, 1980). Nonlymphoid cells such as human or monkey fibroblasts or epithelial-like cells are completely resistant to infection.

2. Antigenic Relationship of LPV to Other Polyomaviruses

LPV is antigenically unrelated to any of the known primate polyomaviruses. Fluorescent-antibody tests of LPV cross-reactivity with SV40, BKV, or JCV were completely negative. However, tests with antiserum prepared against SDS-disrupted SV40 virions were positive, proving that LPV possessed the antigen shared by all members of the polyomavirus genus; this was definitive proof that LPV belonged to this genus (zur Hausen *et al.* 1980; Takemoto *et al.*, 1982).

3. Oncogenic and Transforming Properties of LPV

LPV has thus far not been shown to have any oncogenic properties *in vivo.* Eighty-five newborn hamsters were inoculated with LPV at birth and after an observation period of over 1 year, none of the animals developed tumors (Takemoto *et al.*, 1982), although they responded to infection by the production of high levels of antiviral as well as anti-T antibody. Newborn BALB/c mice inoculated with LPV also failed to develop tumors.

Although LPV has not been shown to be oncogenic in animals, its transforming ability was demonstrated in hamster embryo cells (Takemoto and Kanda, 1984). In this system, LPV appeared to be a relatively weak transforming virus, since transformants appeared in the culture after a long delay of about 2 months. The transformed cells contained intranuclear T-antigens demonstrable by reactivity with anti-LPV hamster serum which contained anti-T antibody. Inoculation of newborn and weanling hamsters with the transformed cells resulted in the development of rapidly growing tumors. The serum from tumor-bearing animals contained anti-T antibody which reacted against the T-antigens in LPV-transformed as well as in lytically infected cells.

Previous results (zur Hausen and Gissmann, 1979; Takemoto *et al.*, 1982) failed to show common T-antigens of LPV and the primate polyomaviruses BKV, JCV, and SV40. In these tests, negative results were obtained when T-antisera of these viruses were reacted with LPV-infected lymphoblastoid cells which presumably were synthesizing T-antigens. When these same sera were tested against LPV-transformed hamster embryo cells, the results were positive, although relatively weaker compared to tests in their respective homologous systems. The data indicated that the LPV T-antigen was immunologically not as closely related to the T-antigens of the other primate polyomaviruses. This correlates well with the observation that the early region of LPV DNA showed only weak homology to those of SV40 and BKV (Kanda *et al.*, 1983).

The large T-antigen in transformed cells as well as in cells undergoing lytic infection is an 84K protein (Segawa and Takemoto, 1983; Takemoto and Kanda, 1984). The small T-antigen was not detected in these experiments and its size is still to be determined.

4. Evidence for the Prevalence of LPV-like Viruses in Humans and Other Primates

Serological tests have shown that approximately 30% of normal human sera contain antibody to LPV (zur Hausen *et al.*, 1980; Brade *et al.*, 1980; Takemoto *et al.*, 1982). The antibody in human sera is specific since it neutralizes LPV and also immunoprecipitates the 41K major structural polypeptide (zur Hausen *et al.*, 1980; Takemoto *et al.*, 1982; Segawa and Takemoto, 1983).

Examination of sera from Old World and New World monkeys as well as apes also revealed the presence of antibody to LPV. The evidence is thus clear that lymphotropic polyomaviruses related to the African green monkey LPV are prevalent in most, if not all, primates including humans.

Attempts to correlate LPV antibody in human sera to disease have not revealed any definite association of human LPV infection to specific disease. zur Hausen *et al.* (1980) examined sera from patients with cancer and other diseases and found a higher incidence of LPV antibody in sera from patients with lymphadenitis. In another study (Brade *et al.*, 1980), a higher percentage of sera from patients who were being tested for hepatitis were positive. Since LPV is a member of a well-known group of oncogenic viruses, some interest has been focused on the possibility of the human LPV being associated with cancer. Limited studies have not shown that patients with various types of neoplasias have a higher incidence of antibody to LPV (zur Hausen *et al.*, 1980; K. K. Takemoto, unpublished data).

B. Biochemical Studies

Like other members of the polyomavirus genus, these capsid proteins of LPV consist of three polypeptides, VP1, VP2, and VP3. Their sizes have been determined to be 41K, 35K, and 26K, respectively (Segawa and Takemoto, 1983). As previously mentioned, the early genes of LPV code for an 84K large T-antigen which is present in infected as well as in transformed cells (Segawa and Takemoto, 1983; Takemoto and Kanda, 1984).

Circular LPV DNA (about 5000 bp long) replicates bidirectionally from the origin located near its unique *Bam*HI cleavage site (Kanda *et al.*, 1983). The structure of the LPV DNA replication origin resembled that of the mouse polyoma virus rather than that of SV40 (Furuno *et al.*, 1984). Polyadenylated transcripts of LPV RNA in BJA-B cells have been mapped (Abraham *et al.*, 1984).

Homology between the DNA LPV of and that of SV40 or BKV was not detected under stringent conditions of hybridization (40% formamide or at T_m-36°C) (Kanda *et al.*, 1983). This is in contrast to the extensive homology observed among the DNAs of the other primate polyomaviruses (SV40, BKV, JCV) under the same stringency conditions (Law *et al.*, 1979). Homology between the DNA of LPV and those of SV40 and BKV was seen only under low stringency conditions (20% formamide). This provides further evidence that LPV is only distantly related to the other primate polyomaviruses.

The DNA sequence of the entire LPV genome has recently been determined for one strain of LPV (K38) by Pawlita *et al.* (1984, 1985b). Comparison of the sequence of LPV with those of SV40 and the mouse polyoma virus revealed that LPV, SV40, and the mouse polyoma virus are about equally related to each other (41% to 44%). The LPV genomic organization is similar to that of SV40 (or of BKV in Fig. 1), and LPV DNA can code for five proteins: large and small T-antigens, VP1, VP2, and VP3. The LPV genome did not have genes similar to the SV40 agnogene or polyoma middle T-antigen gene. Using the DNA from a different strain of LPV (LO2), Furuno *et al.* (1984) have determined the nucleotide sequence of the region encompassing the origin of DNA replication, the

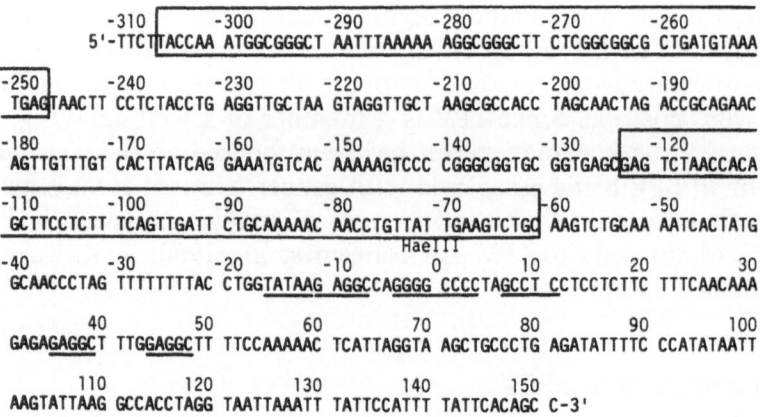

```
        -310      -300      -290      -280      -270      -260
    5'-TTCTTACCAA ATGGCGGGCT AATTTAAAAA AGGCGGGCTT CTCGGCGGCG CTGATGTAAA

-250      -240      -230      -220      -210      -200      -190
TGAGTAACTT CCTCTACCTG AGGTTGCTAA GTAGGTTGCT AAGCGCCACC TAGCAACTAG ACCGCAGAAC

-180      -170      -160      -150      -140      -130      -120
AGTTGTTTGT CACTTATCAG GAAATGTCAC AAAAAGTCCC CGGGCGGTGC GGTGAGCGAG TCTAACCACA

-110      -100      -90       -80       -70       -60       -50
GCTTCCTCTT TCAGTTGATT CTGCAAAAAC AACCTGTTAT TGAAGTCTGC AAGTCTGCAA AATCACTATG
                                              HaeIII
-40       -30       -20       -10       0         10        20        30
GCAACCCTAG TTTTTTTTAC CTGGTATAAG AGGCCAGGGG CCCCTAGCCT CCTCCTCTTC TTTCAACAAA

        40        50        60        70        80        90       100
GAGAGAGGCT TTGGAGGCTT TTCCAAAAAC TCATTAGGTA AGCTGCCCTG AGATATTTTC CCATATAATT

       110       120       130       140       150
AAGTATTAAG GCCACCTAGG TAATTAAATT TATTCCATTT TATTCACAGC C-3'
```

FIGURE 4. Nucleotide sequence around LPV DNA replication origin between VP2 and T-antigen genes. The sequence represents the L strand of the hypothetical prototype constructed from the sequences of LO2 (Furuno *et al.*, 1984), K38 (Pawlita *et al.*, 1985), and LO4 (A. Furuno and K. Yoshiike, unpublished data). LO4 lacks large repeating elements and therefore among the three closely resembles the one shown here. Numbering is as described in Furuno *et al.* (1984). One of the *Hae*III cleavage sites in the DNA replication origin was taken as zero. Nucleotide numbering increases to the early side of the genome and decreases to the late side of the genome. The boxed areas indicate the 60-bp element (nucleotides −306 to −247) in LO2 and the 63-bp element (−123 to −61) in K38. Underlined are the TATA box (−16 to −12) and T-antigen recognition sequences (5′ GAGGC 3′ or 5′ GGGGC 3′).

control region for early transcription, and the gene coding for the LPV small T-antigen. The two LPVs had different structures of the putative transcriptional control region. The 63-bp repeating element of the LPV K38 (Pawlita *et al.*, 1985) differed in sequence and location from the 60-bp tandem repeats of the LO2 (Furuno *et al.*, 1984). Perhaps duplications have occurred at different sites, evolving from a common ancestor (Figs. 4 and 5). In the CAT assay, Pawlita *et al.* (1984) reported enhanced activity of the fragment from the putative control region in T- and B-lymphoblastoid cells, but not in nonlymphoid cells such as HeLa or hamster embryo cells.

An interesting host-range mutant of LPV has recently been isolated by Kanda and Takemoto (1985) which, unlike the original LPV-02 strain (Takemoto *et al.*, 1982), is capable of replicating in T and null cells. The changes in the DNA of the host-range mutant have been determined to be in two regions, one in the enhancer/promoter region and the other in the VP1 coding sequences. Characterization of the constructed recombinant viruses has shown that the change in VP1 is probably responsible for the extended host range. The studies with LPV mutants may provide useful information concerning the cellular and viral factors which govern host range.

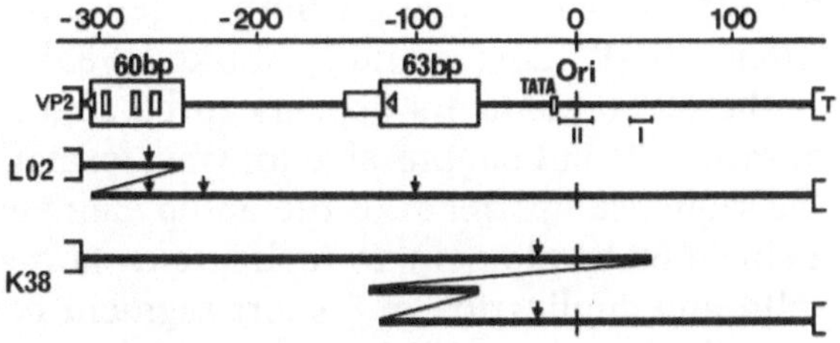

FIGURE 5. DNA duplication in the putative control region of LPV strains. The sequence for the molecule at the top is shown in Fig. 4. The larger solid-line boxes are 60-bp and 63-bp repeating elements in LO2 and K38, respectively. The smaller rectangles represent GC-rich oligonucleotides, which resemble those in the SV40 promoter (Furuno *et al.*, 1984). The triangles indicate the sequence 5′ TGGA(T)A(T)A(T) 3′ found in the SV40 enhancer core. I and II refer to T-antigen binding sites. Small arrows indicate the position of point mutations (single base substitutions); for LO2, T, T, and A at nucleotides −268, −233, and −101, respectively; for K38, C at nucleotide −24. The linkages resulting from breakage and rejoining are from nucleotides −247 to −306 for LO2; from 48 to −130 and from −61 to −123 for K38.

C. Summary and Conclusions

Although the African green monkey LPV was isolated only recently, much progress has been made in understanding the biological and biochemical properties of this interesting and unique polyomavirus. Serological studies have firmly established that humans are hosts to a lymphotropic polyomavirus closely related to the monkey LPV. Thus, humans are infected with three distinct polyomaviruses. It is important that future experiments be directed toward the isolation of the putative human LPV. The information obtained from studies on the monkey LPV should serve as a useful model for the isolation and investigation of the human LPV.

The unique host range of LPV provides an opportunity to study the molecular basis of virus–cell interactions which determine growth or restriction. Investigations on host-range mutants of LPV have already provided evidence that enhancer elements together with changes in viral structural proteins may be the critical factors which determine the host range of LPV.

Note Added in Proof

Since completion of this chapter some new information has been obtained as to the promoter/enhancer, and the host-range of BKV and LPV. Wild-type BKV (Gardner strain), which has three 68-bp promoter/enhancer elements tandemly repeated in its transcriptional control region (Figs. 2 and 3), forms clear, large plaques in human cell cultures but rarely transforms rat cells. From this BKV DNA, deletion mutants with fewer than three 68-bp elements have been constructed and characterized. The mutant DNA with one 68-bp element was found to form minute, turbid plaques in human cell cultures and transform rat cells efficiently (Watanabe and Yoshiike, 1985). Thus, the multiplicity effect of the 68-bp elements appears to be augmentative for virus growth in human cells but suppressive for transformation of rat cells. Furthermore, the sequence upstream of the 68-bp element was required for BKV with a single 68-bp element to replicate in human cells and to transform rat cells, and duplication of a short segment containing the SV40 enhancer core (Weiher *et al.*, 1983) in the middle of 68-bp element improved early transcription but suppressed transformation (S. Watanabe and K. Yoshiike, unpublished data; Watanabe and Yoshiike, 1986). The structure of the LPV transcriptional control region varied with the virus clone, and a nondefective LPV without duplications in the promoter/enhancer (Fig. 4) was found (A. Furuno, T. Kanda, and K. Yoshiike, unpublished data). The LPV enhancer has been shown to be host-discriminatory and to function well only in hematopoietic cells including both B and T cells (Mosthaf *et al.*, 1985). The tropism of LPV for B cells seems to be determined at the adsorption-penetration level, for substitution of three amino acids was found to be responsible for the altered host-range of an LPV mutant capable of replicating in T cells (Kanda *et al.*, 1986).

ACKNOWLEDGMENTS. We are very grateful to Drs. Harald zur Hausen and Jan van der Noordaa for supplying us their respective preprints and unpublished data. We are also indebted to Dr. Sumie Watanabe for compiling the sequence data.

REFERENCES

Abraham, G., Yarom, R., and Manor, H., 1984, Mapping of polyadenylated transcripts of a monkey lymphotropic papova virus, *Virology* **136**:442.

Akrigg, A., Gardner, S. D., and Greenaway, P. J., 1981, Molecular cloning of infectious DNA from human papovavirus BK in *Escherichia coli*, *J. Gen. Virol.* **55**:247.

Barbanti-Brodano, G., Minelli, G. P., Portolani, M., Lambertini, L., and Toppini, M., 1975, Structural proteins of a human papovavirus (BK virus): A comparison with the structural proteins of simian virus 40, *Virology* **64**:269.

Benoist, C., and Chambon, P., 1981, *In vivo* sequence requirements of the SV40 early promoter region, *Nature* **290**:304.

Beth, E., Cikes, M., Schloen, L., di Mayorca, G., and Giraldo, G., 1977, Interspecies-, species-, and type-specific T antigenic determinants of human papovaviruses (JC and BK) and of simian virus 40, *Int. J. Cancer* **20**:551.

Beth, E., Giraldo, G., Schmidt-Ullrich, R., Pater, M. M., Pater, A., and di Mayorca, G., 1981, BK virus-transformed inbred hamster brain cells. I. Status of the viral DNA and the association of BK virus early antigens with purified plasma membranes, *J. Virol.* **40**:276.

Borgatti, M., Costanzo, F., Portolani, M., Vullo, C., Osti, L., Masi, M., and Barbanti-Brodano, G., 1979, Evidence for reactivation of persistent infection during pregnancy and lack of congenital transmission of BK virus, a human papovavirus, *Microbiologica* **2**:173.

Brade, L., Müler-Lantzsch, N., and zur Hausen, H., 1980, B-Lymphotropic papovavirus and possibility of infections in humans, *J. Med. Virol.* **6**:301.

Brade, L., Vogl, W., Gissmann, L., and zur Hausen, H., 1981, Propagation of B-lymphotropic papovavirus (LPV) in human B-lymphoma cells and characterization of its DNA, *Virology* **114**:228.

Bradley, M. K., and Dougherty, R. M., 1978, Transformation of African green monkey kidney cells with the RF strain of human papovavirus BKV, *Virology* **85**:231.

Brady, J., Radonovich, M., Thoren, M., Das, G., and Salzman, N. P., 1984, Simian virus 40 major late promoter: An upstream DNA sequence required for efficient *in vitro* transcription, *Mol. Cell. Biol.* **4**:133.

Byrne, B. J., Davis, M. S., Yamaguchi, J., Bergsma, D. J., and Subramanian, K. N., 1983, Definition of the simian virus 40 early promoter region and demonstration of a host range bias in the enhancement effect of the simian virus 40 72-base-pair repeat, *Proc. Natl. Acad. Sci. USA* **80**:721.

Chenciner, N., Meneguzzi, G., Corallini, A., Grossi, M. P., Grassi, P., Barbanti-Brodano, G., and Milanesi, G., 1980a, Integrated and free viral DNA in hamster tumors induced by BK virus, *Proc. Natl. Acad. Sci. USA* **77**:975.

Chenciner, N., Grossi, M. P., Meneguzzi, G., Corallini, A., Manservigi, R., Barbanti-Brodano, G., and Milanesi, G., 1980b, State of viral DNA in BK virus-transformed rabbit cells, *Virology* **103**:138.

Coleman, D. V., 1975, The cytodiagnosis of human polyomavirus infection, *Acta Cytol.* **19**:93.

Coleman, D. V., Gardner, S. D., and Field, A. M., 1973, Human polyomavirus infection in renal allograft recipients, *Br. Med. J.* **3**:371.

Coleman, D. V., Daniel, R. A., Gardner, S. D., Field, A. M., and Gibson, P. E., 1977, Polyoma virus in urine during pregnancy, *Lancet* **2**:709.

Coleman, D. V., Wolfendale, M. R., Daniel, R. A., Dhanjal, N. K., Gardner, S. D., Gibson, P. E., and Field, A. M., 1980, A prospective study of human polyomavirus infection in pregnancy, *J. Infect. Dis.* **142**:1.

Corallini, A., Barbanti-Brodano, G., Bortolini, W., Nenci, I., Cassai, E., Tampieri, M., Portolani, M., and Borgatti, M., 1977, High incidence of ependymomas induced by BK virus, a human papovavirus, *J. Natl. Cancer Inst.* **59**:1561.

Corallini, A., Altavilla, G., Cecchetti, M. G., Fabris, G., Grossi, M. P., Balboni, P. G., Lanza, G., and Barbanti-Brodano, G., 1978, Ependymomas, malignant tumors of pancreatic islets, and osteosarcomas induced in hamsters by BK virus, a human papovavirus, *J. Natl. Cancer Inst.* **61**:875.

Costa, J., Yee, C., Tralka, T. S., and Rabson, A. S., 1976, Hamster ependymomas produced by intracerebral inoculation of a human papovavirus (MMV), *J. Natl. Cancer Inst.* **56**:863.

Costa, J., Howley, P. M., Legallais, F., Yee, C., Young, N., and Rabson, A. S., 1977, Oncogenicity of a nude mouse cell line transformed by a human papovavirus, *J. Natl. Cancer Inst.* **58**:1147.

DeLucia, A. L., Lewton, B. A., Tjian, R., and Tegtmeyer, P., 1983, Topography of simian virus 40 A protein–DNA complexes: Arrangement of pentanucleotide interaction sites at the origin of replication, *J. Virol.* **46**:143.

Dhar, R., Lai, C.-J., and Khoury, G., 1978, Nucleotide sequence of the DNA replication origin for human papovavirus BKV: Sequence and structural homology with SV40, *Cell* **13**:345.

Dhar, R., Seif, I., and Khoury, G., 1979, Nucleotide sequence of the BK virus DNA segment encoding small t antigen, *Proc. Natl. Acad. Sci. USA* **76**:565.

Dougherty, R. M., 1976, Inductin of tumors in Syrian hamsters by a human renal papovavirus, RF strain, *J. Natl. Cancer Inst.* **57**:395.

Dougherty, R. M., and DiStefano, H. S., 1974, Isolation and characterization of a papovavirus from human urine, *Proc. Soc. Exp. Biol. Med.* **146**:481.

Farrell, M. P., Mäntyjärvi, R. A., and Pagano, J. S., 1978, T antigen of BK papovavirus in infected and transformed cells, *J. Virol.* **25**:871.

Fiori, M., and di Mayorca, G., 1976, Occurrence of BK virus DNA in DNA obtained from certain human tumors, *Proc. Natl. Acad. Sci. USA* **73**:4662.

Freund, J., di Mayorca, G., and Subramanian, K. N., 1979, Mapping and ordering of fragments of BK virus DNA produced by restriction endonucleases, *J. Virol.* **29**:915.

Fijimura, F. K., Deininger, P. L., Friedmann, T., and Linney, E., 1981, Mutation near the polyoma DNA replication origin permits productive infection of F9 embryonal carcinoma cells, *Cell* **23**:809.

Furuno, A., and Yogo, Y., 1983, Free viral DNA present in BKV hamster osteosarcoma (Os-513) cell clones, *Jpn. J. Med. Sci. Biol.* **36**:105.

Furuno, A., Miyamura, T., and Yoshiike, K., 1984, Monkey B-lymphotropic papovavirus DNA: Nucleotide sequence of the region around the origin of replication, *J. Virol.* **50**:451.

Gardner, S. D., 1973, Prevalence in England of antibody to human polyomavirus (B.K.), *Br. Med. J.* **1**:77.

Gardner, S. D., 1977a, Implication of papovaviruses in human diseases, in: *Comparative Diagnosis of Viral Diseases,* Volume I (E. Kurstak and C. Kurstak, ed.), pp. 41–84, Academic Press, New York.

Gardner, S. D., 1977b, The new human papovaviruses: Their nature and significance, in: *Recent Advances in Clinical Viology* (A. P. Waterson, ed.), pp. 93–115, Churchill Livingstone, Edinburgh.

Gardner, S. D., Field, A. M., Coleman, D. V., and Hulme, B., 1971, New human papovavirus (B.K.) isolated from urine after renal transplantation, *Lancet* **1**:1253.

Goudsmit, J., 1982, The human papovavirus BK: Natural history and pathogenesis in man, Thesis, University of Amsterdam, Amsterdam.

Goudsmit, J., Baak, M. L., Slaterus, K. W., and van der Noordaa, J., 1981, Human papovavirus isolated from urine of a child with acute tonsillitis, *Br. Med. J.* **283**:1363.

Greenlee, J. E., Narayan, O., Johnson, R. T., and Herndon, R. M., 1977, Induction of brain tumors in hamsters with BK virus, a human papovavirus, *Lab. Invest.* **36**:636.

Grossi, M. P., Meneguzzi, G., Chenciner, N., Corallini, A., Poli, F., Altavilla, G., Alberti, S., Milanesi, G., and Barbanti-Brodano, G., 1981, Lack of association between BK virus and ependymomas, malignant tumors of pancreatic islets, osteosarcomas, and other human tumors, *Intervirology* **15**:10.

Grossi, M. P., Corallini, A., Meneguzzi, G., Chenciner, N., Barbanti-Brodano, G., and Milanesi, G., 1982a, Tandem integration of complete viral genomes can occur in nonpermissive hamster cells transformed by linear BKV DNA with cohesive ends, *Virology* **120**:500.

Grossi, M. P., Corallini, A., Valieri, A., Balboni, P. G., Poli, F., Caputo, A., Milanesi, G., and Barbanti-Brodano, G., 1982b, Transformation of hamster kidney cells by fragments of BK virus DNA, *J. Virol.* **41**:319.

Grossi, M. P., Caputo, A., Meneguzzi, G., Corallini, A., Carra, L., Portolani, M., Borgatti, M., Milanei, G., and Barbanti-Brodano, G., 1982c, Transformation of human embryonic fibroblasts by BK virus, BK virus DNA and a subgenomic BK virus DNA fragment, *J. Gen. Virol.* **63**:393.

Gruss, P., Dhar, R., and Khoury, G., 1981, Simian virus 40 tandem repeated sequences as an element of the early promoter, *Proc. Natl. Acad. Sci. USA* **78**:943.

Hara, K., Yogo, Y., and Uchida, S., 1983, Transformation of human embryonic kidney cells by a viable deletion mutant of BK virus, *Microbiol. Immunol.* **27**:1067.

Hartzell, S. W., Byrne, B. J., and Subramanian, K. N., 1984, Mapping of the late promoter of simian virus 40, *Proc. Natl. Acad. Sci. USA* **81**:23.

Hayashi, Y., Matsumura, Y., Nishihira, T., Watanabe, I., Ohi, R., Kasai, M., Kumagai, K., and Kamada, N., 1980, Burkitt's lymphoma cell line bearing surface IgA and negative for nuclear antigen of Epstein–Barr virus (EBNA), *Jpn. J. Exp. Med.* **50**:423.

Hearing, P., and Shenk, T., 1983, The adenovirus type 5 E1A transcriptional control region contains a duplicated enhancer element, *Cell* **33**:695.

Hogan, T. F., Borden, E. C., McBain, J. A., Padgett, B. L., and Walker, D. L., 1980, Human polyoma virus infections with JC virus and BK virus in renal transplant patients, *Ann. Intern. Med.* **92**:373.

Howley, P. M., 1980, Molecular biology of SV40 and the human polyomaviruses BK and JC, in: *Viral Oncology* (G. Klein, ed.), pp. 489–550, Raven Press, New York.

Howley, P. M., and Martin, M. A., 1977, Uniform representation of the human papovavirus BK genome in transformed hamster cells, *J. Virol.* **23**:205.

Howley, P. M., Mullarkey, M. F., Takemoto, K. K., and Martin, M. A., 1975a, Characterization of human papovavirus BK DNA, *J. Virol.* **15**:173.

Howley, P. M., Khoury, G., Byrne, J. C., Takemoto, K. K., and Martin, M. A., 1975b, Physical map of the BK virus genome, *J. Virol.* **16**:959.

Howley, P. M., Khoury, G., Takemoto, K. K., and Martin, M. A., 1976, Polynucleotide sequences common to the genomes of simian virus 40 and the human papovaviruses JC and BK, *Virology* **73**:303.

Howley, P. M., Israel, M. A., Law, M.-F., and Martin, M. A., 1979, A rapid method for detecting and mapping homology between heterologous DNAs: Evaluation of polyomavirus genomes, *J. Biol. Chem.* **254**:4876.

Israel, M. A., Martin, M. A., Takemoto, K. K., Howley, P. M., Aaronson, S. A., Solomon, D., and Khoury, G., 1978, Evaluation of normal and neoplastic human tissue for BK virus, *Virology* **90**:187.

Kanda, T., and Takemoto, K. K., 1985, Monkey B-lymphotropic papovavirus mutant capable of replicating in T-lymphoblastoid cells. *J. Virol.* **55**:960.

Kanda, T., Furuno, A., and Yoshiike, K., 1986, Mutation in the VP-1 gene is responsible for the extended host range of a monkey B-lymphotropic papovavirus mutant capable of growing in T-lymphoblastoid cells, *J. Virol.*, in press.

Kanda, T., Yoshiike, K., and Takemoto, K. K., 1983, Alignment of the genome of monkey B-lymphotropic papovavirus to the genomes of simian virus 40 and BK virus, *J. Virol.* **46**:333.

Karjalainen, H. E., Laaksonen, A. M., and Mäntyjärvi, R. A., 1978, Tumour specific transplantation antigen in hamster tumour cells induced with BK virus, *J. Gen. Virol.* **41**:171.

Katinka, M., Yaniv, M., Vasseur, M., and Blangy, D., 1980, Expression of polyoma early functions in mouse embryonal carcinoma cells depends on sequence rearrangements in the beginning of the late region, *Cell* **20**:393.

Kato, K., Iwamura, Y., and Kurimura, T., 1979, Induction of SV40-related transplantation immunity in mice by BK virus-transformed cells, *Jpn. J. Med. Sci. Biol.* **32**:311.

Khoury, G., Howley, P. M., Garon, C., Mullarkey, M. F., Takemoto, K. K., and Martin, M. A., 1975, Homology and relationship between the genomes of papovaviruses, BK virus and simian virus 40, *Proc. Natl. Acad. Sci. USA* **72**:2563.

Klein, G., Lindahl, T., Jondal, M., Leibold, W., Menezes, J., Nilsson, K., and Sundstrom, C., 1974, Continuous lymphoid cell lines with characteristics of B-cell (bone-marrow derived), lacking the Epstein–Barr virus genome and derived from three human lymphomas, *Proc. Natl. Acad. Sci. USA* **71**:3283.

Lai, C.-J., Goldman, N. D., and Khoury, G., 1979, Functional similarity between the early antigens of simian virus 40 and human papovavirus BK, *J. Virol.* **30**:141.

Lamins, L. A., Khoury, G., Gorman, C., Howard, B., and Gruss, P., 1982, Host-specific activation of transcription by tandem repeats from simian virus 40 and Moloney murine sarcoma virus, *Proc. Natl. Acad. Sci. USA* **79**:6453.

Law, L. W., Takemoto, K. K., Rogers, M. J., Henriksen, O., and Ting, C. C., 1978, Differences in the capacity of simian virus 40 (SV40) tumor antigens on cells, membranes and soluble form to induce transplantation immunity in hamsters and mice, *Int. J. Cancer* **22**:315.

Law, M.-F., Martin, J. D., Takemoto, K. K., and Howley, P. M., 1979, The co-linear alignment of the genomes of papovaviruses JC, BK, and SV40, *Virology* **96**:576.

Lecatsas, G., and Prozesky, O. W., 1975, Excretion of morphological variants of human polyoma virus, *Arch. Virol.* **47**:393.

Major, E. O., and di Mayorca, G., 1973, Malignant transformation of BHK_{21} clone 13 cells by BK virus—a human papovavirus, *Proc. Natl. Acad. Sci. USA* **70**:3210.

Major, E. O., and Matsumura, P., 1984, Human embryonic kidney cells: Stable transformation with an origin-defective simian virus 40 DNA and use as hosts for human papovavirus replication, *Mol. Cell. Biol.* **4**:379.

Manaker, R. A., Khoury, G., and Lai, C.-J., 1979, The spliced structure of BK virus mRNAs in lytically infected and transformed cells, *Virology* **97**:112.

Mäntyjärvi, R. A., Meurman, O. H., Vihma, L., and Bergland, B., 1973, A human papovavirus (B.K.), biological properties and seroepidemiology, *Ann. Clin. Res.* **5**:283.

Mason, D. H., Jr., and Takemoto, K. K., 1976, Complementation between BK human papovavirus and a simian virus 40 tsA mutant, *J. Virol.* **17**:1060.

Mason, D. H., Jr., and Takemoto, K. K., 1977, Transformation of rabbit kidney cells by BKV(MM) human papovavirus, *Int. J. Cancer* **19**:391.

Meneguzzi, G., Barbanti-Brodano, G., and Milanesi, G., 1978a, Transcription of BK virus DNA by *Escherichia coli* RNA polymerase: Size and sequence analysis of RNA, *J. Virol.* **25**:940.

Meneguzzi, G., Pignatti, P. F., Barbanti-Brodano, G., and Milanesi, G., 1978b, Minichromosome from BK virus as a template for transcription *in vitro*, *Proc. Natl. Acad Sci. USA* **75**:1126.

Meneguzzi, G., Chenciner, N., Corallini, A., Grossi, M. P., Barbanti-Brodano, G., and Milanesi, G., 1981, The arrangement of integrated viral DNA is different in BK virus-transformed mouse and hamster cells, *Virology* **111**:139.

Mosthaf, L., Pawlita, M., and Gruss, P., 1985, A viral enhancer element specifically active in human haematopoietic cells, *Nature*, **315**:597.

Miyamura, T., and Takemoto, K. K., 1979, Helper function for adenovirus replication in monkey cells by BK human papovavirus, *Virology* **98**:279.

Mullarkey, M. F., Hruska, J. F., and Takemoto, K. K., 1974, Comparison of two human papovaviruses with simian virus 40 by structural protein and antigenic analysis, *J. Virol.* **13**:1014.

Näse, L. M., Kärkkäinen, M., and Mäntyjärvi, R. A., 1975, Transplantable hamster tumors induced with the BK virus, *Acta Pathol. Microbiol. Scand. Sect. B* **83**:347.

Newell, N., Lai, C.-J., Khoury, G., and Kelly, T. J., Jr., 1978, Electron microscope study of the base sequence homology between simian virus 40 and human papovavirus BK, *J. Virol.* **25**:193.

Noss, G., and Stauch, G., 1984, Oncogenic activity of the BK type of human papova virus in inbred rat strains, *Arch. Virol.* **81**:41.

Noss, G., Stauch, G., Mehraein, P., and Georgii, A., 1981, Oncogenic activity of the BK type of human papova virus in newbor Wistar rats, *Arch. Virol.* **69**:239.

Osborn, J. E., Robertson, S. M., Padgett, B. L., ZuRhein, G. M., Walker, D. L., and Weisblum, B., 1974, Comparison of JC and BK human papovaviruses with simian virus 40: Restriction endonuclease digestion and gel electrophoresis of resultant fragments, *J. Virol.* **13**:614.

Osborn, J. E., Robertson, S. M., Padgett, B. L., Walker, D. L., and Weisblum, B., 1976, Comparison of JC and BK human papovaviruses with simian virus 40: DNA homology studies, *J. Virol.* **19**:675.

Padgett, B., 1980, Human papovaviruses, in: *DNA Tumor Viruses* (J. Tooze, ed.), pp. 339–370, Cold Spring Harbor Laboratory, Cold Spring Harbor, N.Y.

Padgett, B. L., and Walker, D. L., 1976, New human papovaviruses, *Prog. Med. Virol.* **21**:1.

Padgett, B. L., Walker, D. L., ZuRhein, G. M., Eckroade, R. J., and Dessel, B. H., 1971, Cultivation of papova-like virus from human brain with progressive multifocal leukoencephalopathy, *Lancet* **1**:1257.

Pater, A., Pater, M. M., and di Mayorca, G., 1980, Arrangement of the genome of the human papovavirus RF virus, *J. Virol.* **36**:480.

Pater, A., Pater, M. M., Dougherty, R. M., and di Mayorca, G., 1981, Transformation of rodent cells by RFV, the human papovavirus with dual genome, *Virology* **113**:86.
Pater, A., Pater, M. M., Chang, L.-S., Slawin, K., and di Mayorca, G., 1983, Multiple origins of the complementary defective genomes of RF and origin proximal sequences of GS, two human papovavirus isolates, *Virology* **131**:426.
Pater, M. M., Pater, A., and di Mayorca, G., 1979, Comparative analysis of GS and BK virus genomes, *J. Virol.* **32**:220.
Pater, M. M., Pater, A., Fiori, M., Slota, J., and di Mayorca, G., 1980, BK virus DNA sequences in human tumors and normal tissues and cell lines, in: *Viruses in Naturally Occurring Cancers* (M. Essex, G. Todaro, and H. zur Hausen, eds.) pp. 329–341, Cold Spring Harbor Laboratory, Cold Spring Harbor, N.Y.
Pater, M. M., Pater, A., and di Mayorca, G., 1981, Genome analysis of MG virus, a human papovavirus, *J. Virol.* **39**:968.
Pauw, W., and Choufoer, J., 1978, Isolation of a variant of BK virus with altered restriction endonuclease pattern, *Arch. Virol.* **57**:35.
Pawlita, M., Mosthaf, L., Clad, A., and Gruss, P., 1984, Genome structure and host range restriction of the lymphotropic papovavirus (LPV): Identification of a viral lymphocyte specific enhancer element, *Curr. Top. Microbiol. Immunol.* **114**:26.
Pawlita, M., Clad, A., and zur Hausen, H., 1985, Complete DNA sequence of lymphotropic papovavirus (LPV): Prototype of a new species of the polyomavirus genus, *Virology* **143**:196.
Portolani, M., Marzocchi, A., Barbanti-Brodano, G., and La Placa, M., 1974, Prevalence in Italy of antibodies to a new human papovavirus (BK virus), *J. Med. Microbiol.* **7**:543.
Portolani, M., Barbanti-Brodano, G., and La Placa, M., 1975, Malignant transformation of hamster kidney cells by BK virus, *J. Virol.* **15**:420.
Portolani, M., Borgatti, M., Corallini, A., Cassai, E., Grossi, M. P., Barbanti-Brodano, G., and Possati, L., 1978, Stable transformation of mouse, rabbit and monkey cells and abortive transformation of human cells by BK virus, a human papovavirus, *J. Gen. Virol.* **38**:369.
Purchio, A. F., and Fareed, G. C., 1979, Transformation of human embryonic kidney cells by human papovavirus BK, *J. Virol.* **29**:763.
Rigby, P. W. J., Dieckmann, M., Rhodes, C., and Berg, P., 1977, Labeling deoxyribonucleic acid to high specific activity in vitro by nick translation with DNA polymerase I, *J. Mol. Biol.* **113**:237.
Rosenthal, N., Kress, M., Gruss, P., and Khoury, G., 1983, BK viral enhancer element and a human cellular homolog, *Science* **222**:749.
Rundell, K., Tegtmeyer, P., Wright, P. J., and di Mayorca, G., 1977, Identification of the human papovavirus T antigen and comparison with the simian virus 40 protein A, *Virology* **82**:206.
Rundell, K., Major, E. O., and Lampert, M., 1981, Association of cellular 56,000- and 32,000-molecular-weight proteins with BK virus and polyoma virus t-antigens, *J. Virol.* **37**:1090.
Ryder, K., DeLucia, A. L., and Tegtmeyer, P., 1983, Binding of SV40 A protein to the BK virus origin of DNA replication, *Virology* **129**:239.
Sack, G. H., Jr., Felix, J. S., and Lanahan, A. A., 1980, Plaque formation and purification of BK virus in cultured human urinary cells, *J. Gen Virol.* **50**:185.
Seehafer, J., Salmi, A., Scraba, D. G., and Colter, J. S., 1975, A comparative study of BK and polyoma viruses, *Virology* **66**:192.
Seehafer, J., Salmi, A., and Colter, J. S., 1977, Isolation and characterization of BK virus-transformed hamster cells, *Virology* **77**:356.
Seehafer, J., Carpenter, P., Downer, D. N., and Colter, J. S., 1978, Observations on the growth and plaque assay of BK virus in cultured human and monkey cells, *J. Gen. Virol.* **38**:383.
Seehafer, J., Downer, D. N., Salmi, A., and Colter, J. S., 1979a, Isolation and characterization of BK virus-transformed rat and mouse cells, *J. Gen. Virol.* **42**:567.

Seehafer, J., Downer, D. N., Gibney, D. J., and Colter, J. S., 1979b, Evidence for the expression of TSTA in BKV-transformed cells: Cross-reaction with SV40 TSTA, *Virology* **95**:241.

Segawa, K., and Takemoto, K. K., 1983, Identification of B-lymphotropic papovavirus-coded proteins, *J. Virol.* **45**:872.

Seif, I., Khoury, G., and Dhar, R., 1979a, BKV splice sequences based on analysis of preferred donor and acceptor sites, *Nucleic Acids Res.* **6**:3387.

Seif, I., Khoury, G., and Dhar, R., 1979b, The genome of human papovavirus BKV, *Cell* **18**:963.

Sekikawa, K., and Levine, A. J., 1981, Isolation and characterization of polyoma host range mutants that replicate in nullipotential embryonal carcinoma cells, *Proc. Natl. Acad. Sci. USA* **78**:1100.

Shah, K. V., Daniel, R. W., and Warszawski, R., 1973, High prevalence of antibodies to BK virus, an SV40 related papovavirus, in residents of Maryland, *J. Infect. Dis.* **128**:784.

Shah, K. V., Daniel, R. W., and Strandberg, J. D., 1975, Sarcoma in a hamster inoculated with BK virus, a human papovavirus, *J. Natl. Cancer Inst.* **54**:945.

Shah, K. V., Hudson, C., Valis, J., and Strandberg, J. D., 1976, Experimental infection of human foreskin cultures with BK virus, a human papovavirus, *Proc. Soc. Exp. Biol. Med.* **153**:180.

Shah, K. V., Daniel, R. W., and Kelly, T. J., Jr., 1977a, Immunological relatedness of papovaviruses of the simian virus 40–polyoma subgroup, *Infect. Immun.* **18**:558.

Shah, K. V., Ozer, H. L., Ghazey, H. N., and Kelly, T. J., Jr., 1977b, Common structural antigen of papovaviruses of the simian virus 40–polyoma subgroup, *J. Virol.* **21**:179.

Shah, K. V., Daniel, R. W., Madden, D., and Stagno, S., 1980, Serological investigation of BK papovavirus infection in pregnant women and their offspring, *Infect. Immun.* **30**:29.

Simmons, D. T., and Martin, M. A., 1978, Common methionine-tryptic peptides near the amino-terminal end of primate papovavirus tumor antigens, *Proc. Natl. Acad. Sci. USA* **75**:1131.

Simmons, D. T., Takemoto, K. K., and Martin, M. A., 1977, Relationship between the methionine tryptic peptides of SV40 and BK virus tumor antigens, *J. Virol.* **24**:319.

Simmons, D. T., Takemoto, K. K., and Martin, M. A., 1978, Properties of simian virus 40 and BK virus tumor antigens from productively infected and transformed cells, *Virology* **85**:137.

Takemoto, K. K., 1978, Human papovaviruses, *Int. Rev. Exp. Pathol.* **18**:281.

Takemoto, K. K., 1980, Human polyoma viruses: Evaluation of their possible involvement in human cancer, in: *Viruses in Naturally Occurring Cancers* (M. Essex, G. Todaro, and H. zur Hausen, eds.), pp. 311–318, Cold Spring Harbor Laboratory, Cold Spring Harbor, N.Y.

Takemoto, K. K., and Kanda, T., 1984, Lymphotropic papovavirus transformation of hamster embryo cells, *J. Virol.* **50**:100.

Takemoto, K. K., and Martin, M. A., 1976, Transformation of hamster kidney cells by BK papovavirus DNA, *J. Virol.* **17**:247.

Takemoto, K. K., and Mullarkey, M. F., 1973, Human papovavirus, BK strain: Biological studies including antigenic relationship to simian virus 40, *J. Virol.* **12**:625.

Takemoto, K. K., and Yoshiike, K., 1982, Human polyomaviruses, in: *Medical Virology* (L. M. de la Maza and E. M. Peterson, eds.), pp. 239–255, Elsevier, Amsterdam.

Takemoto, K. K., Rabson, A. S., Mullarkey, M. F., Blaese, R. M., Garon, C. F., and Nelson, D., 1974, Isolation of papovavirus from brain tumor and urine of a patient with Wiskott–Aldrich syndrome, *J. Natl. Cancer Inst.* **53**:1205.

Takemoto, K. K., Linke, H., Miyamura, T., and Fareed, G. C., 1979, Persistent BK papovavirus infection of transformed human fetal brain cells. I. Episomal viral DNA in cloned lines deficient in T-antigen expression, *J. Virol.* **29**:1177.

Takemoto, K. K., Furuno, A., Kato, K., and Yoshiike, K., 1982, Biological and biochemical studies of African green monkey lymphotropic papovavirus, *J. Virol.* **42**:502.

Tanaka, K., Chowdhury, K., Chang, K. S. S., Israel, M., and Ito, Y., 1982, Isolation and characterization of polyoma virus mutants which grow in murine embryonal carcinoma and trophoblast cells, *EMBO J.* **1**:1521.

Tanaka, R., Koprowski, H., and Iwasaki, Y., 1976, Malignant transformation of hamster brain cells in vitro by human papovavirus BK, *J. Natl. Cancer Inst.* **56**:671.

ter Schegget, J., Voves, J., van Strien, A., and van der Noordaa, J., 1980, Free viral DNA in BK virus-induced hamster tumor cells, *J. Virol.* **35**:331.

ter Schegget, J., Sol, C. J. A., Baan, E. W., van der Noordaa, J., and van Ormondt, H., 1985, Naturally occurring BK virus variants (JL and Dik) with deletions in the putative early enhancer-promoter sequences, *J. Virol.* **53**:302.

Uchida, S., Watanabe, S., Aizawa, T., Kato, K., Furuno, A., and Muto, T., 1976, Induction of papillary ependymomas and insulinomas in the Syrian golden hamster by BK virus, a human papovavirus, *Gann* **67**:857.

Uchida, S., Watanabe, S., Aizawa, T., Furuno, A., and Muto, T., 1979, Polyoncogenicity and insulinoma-inducing ability of BK virus, a human papovavirus, in Syrian golden hamsters, *J. Natl. Cancer Inst.* **63**:119.

van der Noordaa, J., 1976, Infectivity, oncogenicity and transforming ability of BK virus and BK virus DNA, *J. Gen. Virol.* **30**:371.

van der Noordaa, J., Sol, C. J. A., van Strien, A., and Walig, C., 1978, Interference by defectives of BK virus, in: *Antiviral Mechanisms in the Control of Neoplasia* (P. Chandra, ed.), pp. 301–313, Plenum Press, New York.

van der Noordaa, J., De Jong, W., Pauw, W., Sol, C. J. A., and van Strien, A., 1979, Transformation and T antigen induction by linearized BK virus DNA, *J. Gen. Virol.* **44**:843.

Watanabe, S., and Yoshiike, K., 1982, Change of DNA near the origin of replication enhances the transforming capacity of human papovavirus BK, *J. Virol.* **42**:978.

Watanabe, S., and Yoshiike, K., 1985, Decreasing the number of 68-base-pair tandem repeats in the BK virus transcriptional control region reduces plaque size and enhances transforming capacity *J. Virol.* **55**:823.

Watanabe, S., and Yoshiike, K., 1986, Evolutionary changes of the transcriptional control region in a minute-plaque viable deletion mutant of BK virus, *J. Virol.*, in press.

Watanabe, S., Yoshiike, K., Nozawa, A., Yuasa, Y., and Uchida, S., 1979, Viable deletion mutant of human papovavirus BK that induces insulinomas in hamsters, *J. Virol.* **32**:934.

Watanabe, S., Yoshiike, K., Yuasa, Y., and Uchida, S., 1981, Natural occurrence of deletion mutant of human papovavirus BK capable of inducing T antigen, *J. Gen. Virol.* **54**:431.

Watanabe, S., Kotake, S., Nozawa, A., Muto, T., and Uchida, S., 1982, Tumorigenicity of human BK papovavirus plaque isolates, wild-type and plaque morphology mutant, in hamsters, *Int. J. Cancer* **29**:583.

Watanabe, S., Yogo, Y., and Yoshiike, K., 1984a, Expression of viral early functions in rat 3Y1 cells infected with human papovavirus BK, *J. Virol.* **49**:78.

Watanabe, S., Soeda, E., Uchida, S., and Yoshiike, K., 1984b, DNA rearrangement affecting expression of the BK virus transforming gene, *J. Virol.* **51**:1.

Weiher, H., König, M., and Gruss, P., 1983, Multiple point mutations affecting the simian virus 40 enhancer, *Science* **219**:626.

Wold, W. S. M., Mackey, J. K., Brackmann, K. H., Takemori, N., Rigden, P., and Green, M., 1978, Analysis of human tumors and human malignment cell lines for BK virus-specific DNA sequences, *Proc. Natl. Acad. Sci. USA* **75**:454.

Wright, P. J., and di Mayorca, G., 1975, Virion polypeptide composition of the human papovavirus BK: Comparison with simian virus 40 and polyoma virus, *J. Virol.* **15**:828.

Wright, P. J., Bernhardt, G., Major, E. O., and di Mayorca, G., 1976, Comparison of the serology, transforming ability, and polypeptide composition of human papovaviruses isolated from urine, *J. Virol.* **17**:762.

Yang, R. C. A., and Wu, R., 1978a, BK virus DNA: Cleavage map and sequence analysis, *Proc. Natl. Acad. Sci. USA* **75**:2150.

Yang, R. C. A., and Wu, R., 1978b, Cleavage map of BK Virus DNA with restriction endonucleases *Mbo*I and *Hae*III, *J. Virol.* **27**:700.

Yang, R. C. A., and Wu, R., 1978c, Physical mapping of BK virus DNA with *Sac*I, *Mbo*II, and *Alu*I restriction endonucleases, *J. Virol.* **28**:851.

Yang, R. C. A., and Wu, R., 1979a, BK virus DNA sequence coding for the amino-terminus of the T-antigen, *Virology* **92**:340.

Yang, R. C. A., and Wu, R., 1979b, BK virus DNA sequence: Extent of homology with simian virus 40 DNA, *Proc. Natl. Acad. Sci. USA* **76**:1179.

Yang, R. C. A., and Wu, R., 1979c, BK virus DNA: Complete nucleotide sequence of a human tumor virus, *Science* **206**:456.

Yang, R. C. A., and Wu, R., 1979d, Comparative study of papovavirus DNA: BKV(MM), BKV(WT) and SV40, *Nucleic Acids Res.* **7**:651.

Yang, R. C. A., Young, A., and Wu, R., 1980, BK virus DNA sequence coding for the t and T antigens and evaluation of methods for determining sequence homology, *J. Virol.* **34**:416.

Yogo, Y., Furuno, A., Watanabe, S., and Yoshiike, K., 1980a, Occurrence of free, defective viral DNA in a hamster tumor induced by human papovavirus BK, *Virology* **103**:241.

Yogo, Y., Hondo, R., Uchida, S., Watanabe, S., Furuno, A., and Yoshiike, K., 1980b, Presence of viral DNA sequences in hamster tumors induced by BK virus, a human papovavirus, *Microbiol. Immunol.* **24**:861.

Yogo, Y., Furuno, A., Nozawa, A., and Uchida, S., 1981, Organization of viral genome in a T antigen-negative hamster tumor induced by human papovavirus BK, *J. Virol.* **38**:556.

Yoshiike, K., Miyamura, T., Chan, H. W., and Takemoto, K. K., 1982, Two defective DNAs of human polyomavirus JC adapted to growth in human embryonic kidney cells, *J. Virol.* **42**:395.

zur Hausen, H., and Gissmann, L., 1979, Lymphotropic papovaviruses isolated from African green monkey and human cells, *Med. Microbiol. Immunol.* **167**:137.

zur Hausen, H., Gissmann, L., Mincheva, A., and Böcker, J. F., 1980, Characterization of a lymphotropic papovavirus, in: *Viruses in Naturally Occurring Cancers* (M. Essex, G. Todaro, and H. zur Hausen, eds.), pp. 365–372, Cold Spring Harbor Laboratory, Cold Spring Harbor, N.Y.

CHAPTER 6

The Biology and Molecular Biology of JC Virus

DUARD L. WALKER AND RICHARD J. FRISQUE

I. INTRODUCTION

JC virus (JCV) was first isolated in 1971 (Padgett *et al.*, 1971) from diseased brain tissue obtained at autopsy from a patient with progressive multifocal leukoencephalopathy (PML). Early studies (Padgett *et al.*, 1971; Walker *et al.*, 1973a) demonstrated that the isolate was a papovavirus but not one of the previously known members of that family. It soon became clear (Padgett and Walker, 1973; Walker *et al.*, 1973a) that although PML is relatively uncommon, JCV is ubiquitous in the human population and most persons have been infected with the virus. Other interesting features emerged, such as evidence that most primary infections occur in childhood, that JCV is highly oncogenic in hamsters and some primates, that it has a very restricted host range, and that it shares some antigens and other characteristics with SV40 and another human polyomavirus, BK virus (BKV).

Subsequent investigations have enlarged our knowledge, particularly about the association of JCV with disease, its epidemiology, the structure of its genome, the relationship of JCV to other polyomaviruses, and its oncogenicity. All of this now adds up to the substantial body of knowledge that will be detailed in this chapter.

DUARD L. WALKER • Department of Medical Microbiology, University of Wisconsin Medical School, Madison, Wisconsin 53706. RICHARD J. FRISQUE • Department of Molecular and Cell Biology, Pennsylvania State University, University Park, Pennsylvania 16802.

II. BIOLOGICAL CHARACTERISTICS

A. Distribution and Natural History

1. Epidemiology

Because JCV has been relatively difficult to isolate and cultivate, most information about its distribution and natural history has come from seroepidemiology, i.e., using the presence of serum antibodies against the virus as evidence of past or present infection. Hemagglutination-inhibiting antibodies have been the ones most commonly measured because they can be measured rapidly and easily and they have been shown (Padgett and Walker, 1973) to correlate reasonably well with neutralizing antibodies.

All evidence points to the human population as the natural reservoir of JCV. There is no evidence that it is transmitted to people from animals or that it circulates in lower animals or nonhuman primates. Padgett *et al.* (1977b) did not find antibodies against the virus in a wide variety of animals, including species closely associated with people. Five species of nonhuman primates were devoid of serum antibodies.

Several serum surveys have indicated that JCV is widespread throughout the world and in most populations infection is common at an early age. Padgett and Walker (1973) tested for antibodies in persons of all ages in the state of Wisconsin, and found the highest rate of conversion to seropositivity to be during the first 10–14 years of life. By the time children reached that age, 65% had acquired antibodies against JCV. Seroconversions occurred at a slower rate thereafter, but by 50–59 years about 75% of persons of both sexes had antibody evidence of infection by JCV.

Serum antibodies against JCV are prevalent in many parts of the world. Antibodies were found in 66% of 142 adults from New York City and in 75% of 180 adults from Los Angeles, California (B. L. Padgett and D. L. Walker, unpublished observations). Antibodies against JCV were present in 71% of adult cancer patients in northern India (K. V. Shah, personal communication) and in 92% of 48 adults from urban areas of Brazil (B. L. Padgett and D. L. Walker, unpublished observations). Gardner (1977) found a prevalance of antibodies in England similar to that reported by Padgett and Walker for Wisconsin. Taguchi *et al.* (1982) tested sera from 384 children and 96 adults in Tokyo, Japan, and found that by 6 years of age about 50% of children had antibodies against JCV. The level reached 90% by 31–40 years of age. These reports indicated a uniformly high prevalence in urban areas, but when tests were made for serum antibodies in persons in small and isolated populations, much more variation was found. Brown *et al.* (1975) found no antibodies in several small isolated native populations in Brazil, Paraguay, and western New Guinea, a very low prevalence in the Solomon Islands (5%), but a prevalence of

50% in the New Hebrides Islands and 75% in Malaysian aborigines. Candeias *et al.* (1977) tested for antibodies in three very small, isolated Indian populations in Brazil. Antibody against JCV was absent in two of the tribes and found in only three persons (4.4%) in the third. This low prevalence was in sharp contrast to urban Brazilian populations where 85% of adults had antibody.

These serologic surveys provided evidence regarding prevalence of JCV in the world population and, in addition, those that included children provided a strong indication that most infections occur in childhood (Walker and Padgett, 1983b). Furthermore, the differences between small populations and large urban populations suggest that population density and environmental conditions affect the transmission of JCV. This point has been strongly reinforced by studies of Mashiko *et al.* (1982) on several population groups in Tokyo, Japan. They found that 39% of 3- to 5-year-old children in the densely populated central area of Tokyo had antibodies against JCV, whereas antibody was found in only 11% of children of that age living in less densely populated Tokyo suburbs. Among 18- to 20-year-old students at the Teikyo School for Medical Technicians, 67% of those coming from large cities had antibody but only 37% of students from small cities and towns carried antibody.

Padgett *et al.* (1977b) found antibodies against JCV in 92% of 60 sera collected from young adults in Wisconsin in 1950, thereby demonstrating that human infections with JCV were prevalent prior to the introduction of poliovirus vaccines in 1954 and removing any suspicion that JCV in the human population could be related to the widespread use of vaccines contaminated with SV40.

2. Transmission

In spite of the evidence that JCV is a virus limited to the human population and transmitted freely among persons, particularly among children, we know almost nothing about its routes of transmission. As will be discussed in a later section, the virus can be found in urine, particularly during periods of immunosuppression, and this may be the major source of infection for other persons. However, other sources of virus have not been ruled out. Because pregnant women sometimes release JCV in their urine and show serologic evidence of reactivation of JCV infection during pregnancy (Coleman *et al.*, 1980; Daniel *et al.*, 1981; Gibson *et al.*, 1981), the transplacental route of infection seemed a possibility. However, IgM antibodies against JCV have not been demonstrable in the cord blood of babies born to mothers showing evidence of active infection (Coleman *et al.*, 1980; Daniel *et al.*, 1981). This, together with failure to find evidence of virus in fetal tissues (Padgett and Walker, 1980), make the transplacental route unlikely as a common means of transmission.

B. Host Range

1. Experimental Animals

The lack of antibodies in the serum of animals other than humans has already been pointed out. Further evidence that JCV is a strictly human virus has come from the results of animal inoculations. With the exception of tumors induced by JCV in hamsters and some species of monkeys (to be discussed later), no evidence of progressive infections has been found in any of several species inoculated with JCV by multiple routes (Walker, Padgett, and ZuRhein, unpublished observations). Animals tested included adult rabbits, guinea pigs, and hamsters inoculated by multiple routes and observed for 6 to 12 months; adult mink and adult ferrets inoculated intracerebrally and observed for 1 year; newborn Swiss mice inoculated by multiple routes and observed for 1 year; newborn rhesus monkeys inoculated by several routes and observed for 6 years. The newborn rhesus monkeys developed high levels of antibody against both capsid and T antigens. This suggested some degree of viral multiplication, but no recognizable disease or tumors developed.

2. Cells in Culture

The host cell range of JCV is very restricted. The virus multiplies only in human cells, and productive infection in cell cultures is limited to very few cell types. In fact, work with JCV has been severely limited by the lack of convenient cell cultures, but this deficiency is slowly being overcome.

JCV was originally isolated in cultures of primary human fetal glial (PHFG) cells and most of the subsequent work with the virus has used these cells. JCV can be grown in serial passage in PHFG cells providing the cultures contain a large proportion of spongioblasts (Padgett *et al.*, 1977b). However, the cytopathic effect (CPE) in unstained cells is quite subtle, even after the virus has been passaged multiple times. About 14–21 days after inoculation, spongioblasts slowly enlarge and lose their characteristic spindle shape. Most spongioblasts ultimately are destroyed, but this is a slow and gradual process that takes 2–3 weeks after CPE first appears. If the cells are fixed and stained, intranuclear inclusion bodies are visible in the swollen spongioblasts (Fig. 1). Astrocytes usually show little change, but some enlarge and develop very large, abnormal nuclei (Fig. 1). Examination of cells by electron microscopy reveals myriads of virions in the nuclei of many of the spongioblasts, but rarely in the nuclei of astrocytes (Padgett *et al.*, 1971; Walker *et al.*, 1973a). In spongioblast nuclei, virions occur both singly and in densely packed crystalline arrays and have a mean diameter of 38–43 nm. Filamentous forms are common.

JCV is strongly cell associated. Mechanical disruption of cells and use of neuraminidase to free the virus from cell debris (Osborn *et al.*,

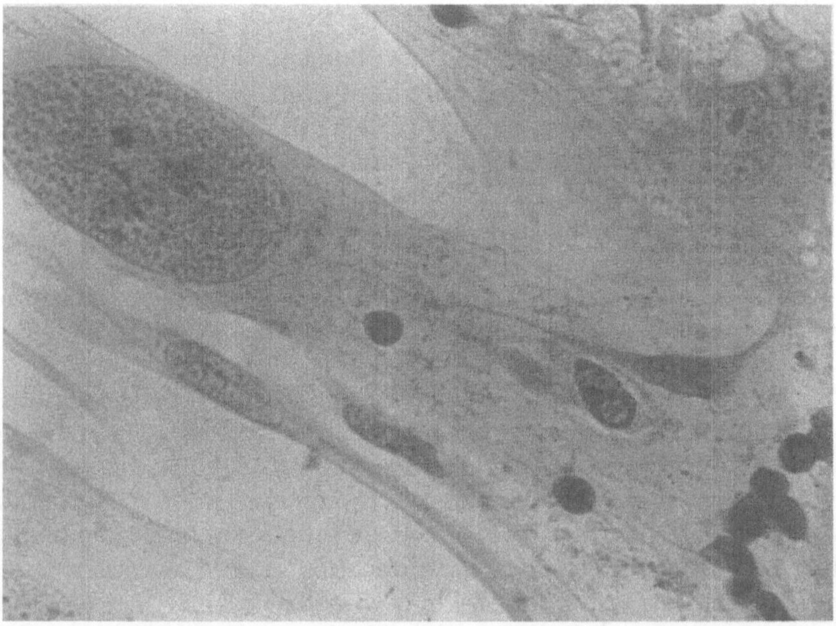

FIGURE 1. A culture of human fetal glial cells 14 days after inoculation with JCV. The culture is a mixture of small spongioblasts (lower right corner) and larger astrocytes (lightly stained nuclei in upper right corner). One infected spongioblast (right center) has a swollen nucleus containing two inclusion bodies. A very large, abnormal astrocyte nucleus occupies the upper left corner.

1974; Padgett *et al.*, 1976) are required to obtain good yields of virus from cell cultures.

The infectivity of JCV is similar to mouse polyoma virus and SV40 in being relatively stabile. It is unaffected by 50°C for 1 hr in water, but is destroyed by 50°C for 1 hr in 1 M $MgCl_2$ (Padgett and Walker, 1976).

JCV has a tendency to produce empty capsids and defective genomes in PHFG cell cultures, particularly if passaged at high multiplicity of infection (Osborn *et al.*, 1974). This will be discussed in detail in a later section.

PHFG cultures have several shortcomings as a culture system for JCV. First, they are of limited availability and are relatively difficult to cultivate. Second, the CPE is subtle and not distinct enough to produce plaques. Third, they are mixed cell cultures in which the cells differ in degree of permissiveness. This latter defect has prevented accurate measurement of the multiplication cycle of JCV. Such efforts as have been made to study the cycle indicate that it is a very long one that is measured in days rather than hours. In a study of JCV growth at 37°C in PHFG cells by Padgett *et al.* (1977b), only 10% of cell nuclei contained T antigen 2 days after inoculation with 4 infectious units of virus per cell. At 5 days, 57% contained T antigen but only 2.3% contained capsid antigen.

JCV actually grows better and produces better yields of virus from PHFG cells at 39°C than at 37°C (Grinnell *et al.*, 1982). Isolation of virus from human brain extracts is also more efficient if incubation is at 39°C.

Many human and animal cells have been tested in a search for a better culture system (Padgett and Walker, 1976; Padgett *et al.*, 1977b). Table I lists many of the cells tested and found not to be permissive for JCV. Many more cells than those listed in Table I have been tested by various investigators, but many negative results have not been reported. It should be noted that no serially passaged line of fetal brain cells, such as Flow 3000, has been found to be permissive for JCV. The transformed SVG cells developed by Major *et al.* (1985) seem to be the first exception to this.

There has been absolutely no success in persuading JCV to multiply in animal cells, but a number of human cells have been found to support JCV multiplication in varying degrees. Those cell types that have been useful are listed in Table II along with some of their characteristics. Takemoto *et al.* (1979a) succeeded where others had failed and obtained a limited but useful degree of multiplication of JCV in primary human amnion cells. They found the yield of virus to be somewhat low and to contain a substantial portion of defective virus, but the virus could be serially passed in amnion cells and was suitable for use as antigen and for a variety of studies. Wroblewska *et al.* (1980) demonstrated that adult

TABLE I. Cells Nonpermissive for JCV

Primary human embryonic cells	Human cell lines
Fibroblasts	Chang conjunctiva
Foreskin	I-407 (embryonic intestine)
Muscle	L-132 (embryonic lung)
Skin	A-549 (embryonic lung)
Kidney	J-111 (monocytic leukemia)
Lung	IMR-32 (neuroblastoma)
Liver	T98G (astrocytoma)
Adrenal	RT4 (embryonic bladder)
Intestine	FL (amnion)
Human diploid cells	Flow 3000 (fetal brain)
WI-38 (lung)	Primate cells
L-809 (lung)	Primary African green monkey kidney
AG534 (muscle)	Primary rhesus monkey kidney
Nonprimate cells	Primary rhesus monkey brain (adult)
Hamster fetal brain	Vero
Adult mink brain	BSC-1
Mouse embryo	CV-1
3T3	
Bovine fetal kidney	
Adult sheep brain	

TABLE II. Human Cells Permissive for JCV

Cell type	Cell passage	Cytopathic effect	Time to maximum yield (weeks)	Maximum yield	Reference
Primary fetal glial	Limited	Subtle	3–4	10^5–10^7 $TCID_{50}$/ml,[a] 200-fold increase in hemagglutination units	Padgett *et al.* (1977b)
Primary amnion[b]	Limited	Subtle	3	Low	Takemoto *et al.* (1979a)
Adult brain	15–25 subcultures	Cell shrinkage	4	10- to 40-fold increase in hemagglutination units	Wroblewska *et al.* (1980)
Urine-derived epithelium	Limited	Distinct degeneration	3–4	100- to 1000-fold increase in hemagglutination units	Beckman *et al.* (1982)
Embryonic kidney[b]	Limited	Distinct degeneration, plaques	3	2×10^3–4.5×10^4 PFU/ml[c]	Miyamura *et al.* (1980)
SVG cells	Unlimited	Information not available	3	500-fold increase in hemagglumation units	Major *et al.* (1985)

[a] Tissue culture infective doses per milliliter.
[b] JCV requires "adaptation" to these cells.
[c] Plaque-forming units per milliliter.

human brain cell cultures will yield a 10- to 40-fold increase in JCV during 4 weeks of infection after being inoculated with 10^3 hemagglutinating units of virus. They found that human brain tissue obtained from autopsies could be carried through 15–25 subcultures and was permissive for JCV through at least the 17th subculture. The cell type in which JCV multiplied was not determined. Miyamura *et al.* (1980) adapted JCV to human embryonic kidney cells. After 8 passages, JCV produced good virus yields in kidney cells and formed plaques in monolayers. Although JCV could be plaque-purified from such cultures, the DNA found in virus clones consisted of two or more species of DNA. Transitional epithelial cells obtained from the urine of newborn human infants were found to be quite permissive for JCV (Beckman *et al.*, 1982; Beckman and Shah, 1983). As with PHFG cells, urine-derived epithelial cells are not available to most laboratories, but when available they provide an alternative to PHFG cells. For maximum virus yields, a few serial passages are needed to adapt JCV to the cells. Although urine-derived cells were found not to be as sensitive as PHFG cells for primary isolation of virus, they were shown to be sufficiently sensitive to allow isolation of JCV from some clinical specimens.

All of the culture systems just described employ primary cell cultures or cells with limited passage potential. However, Major *et al.* (1985) have developed a continuously culturable line of cells (SVG cells) that support the multiplication of JCV. The cells are human embryonic astroglial cells transfected with DNA from an origin-defective mutant of SV40. All of the cells in a culture produce SV40 T protein, but do not yield SV40 virus. These cultures provide a uniform cell population that is permissive for JCV and one that produces a JCV yield comparable to that of PHFG cells at a rate very similar to that in PHFG cells. These cells should provide a very useful alternative to PHFG cells for *in vitro* cultivation of JCV.

C. Hemagglutination

JCV agglutinates chicken, guinea pig, and human type O erythrocytes, but not those of hamsters, sheep, African green monkeys, or rhesus monkeys (Padgett *et al.*, 1977b). Hemagglutination occurs best at 4°C and the activity has been shown to be a property of the virion (Osborn *et al.*, 1974). The hemagglutinating activity is quite stable. It is not inactivated at 56°C for 7 hr, but is destroyed at 80°C for 15 min (Padgett *et al.*, 1977b). In cell or tissue extracts it may be masked by inhibitors that can be dissociated by heating or by treatment with receptor-destroying enzyme (Padgett *et al.*, 1977b).

There have been no direct studies of the composition of JCV receptors on erythrocytes or tissue cells and no determination of the number of virions required for one hemagglutination unit. However, the susceptibility of tissue and serum inhibitors of hemagglutinating activity to periodate and receptor-destroying enzyme suggests that the cellular receptors are similar to those of mouse polyoma virus and BKV (Hartley *et al.*, 1959; Mäntyjärvi *et al.*, 1972).

D. Measurement of Virus and Antibodies

The hemagglutinating activity of JCV provides an easy, inexpensive, and rapid method for detecting and measuring JCV (Padgett and Walker, 1973). It suffers the lack of precision and sensitivity of all hemagglutination assays of viruses, but its simplicity and convenience are so much greater than other methods that it has been the technique most widely used.

JCV adapted to growth in human embryonic kidney cells causes sufficient CPE to form plaques under an overlay (Miyamura *et al.*, 1980), but this culture system has not been widely used. Measurement of infectivity in PHFG cells is slow and difficult. It frequently has been measured by production of CPE in cell cultures in tubes or microtiter plates (Padgett and Walker, 1973), but this requires an experienced person to recognize CPE. However, JCV can be assayed by counting individual antigen-containing cells using fluorescent antibody against capsid antigen (Osborn *et al.*, 1974; Padgett *et al.*, 1977b). This is done at about 10 days after inoculation at a time when capsid antigen has accumulated in infected cell nuclei and before a second cycle of infection has developed. It is a tedious task to count fluorescent cells microscopically, but the method approximates the precision of a plaque assay.

The hemagglutination-inhibition (HI) assay has been the most widely used method for measuring antibody against JCV. It is quick, easy, and inexpensive and titers have been shown (Padgett and Walker, 1973) to have a linear relationship with neutralizing antibodies over a wide antibody range. However, nonspecific inhibitors of hemagglutination are present in the serum of many species and must be removed to obtain an accurate assay of antibodies. Potassium periodate treatment (Padgett and Walker, 1973) will usually reduce the inhibitors in human serum sufficiently to allow accurate HI assays, but many animal sera contain nonspecific inhibitors that resist potassium periodate treatment. Acetone extraction (Hammon and Sather, 1969) will usually remove such inhibitors and allow measurement of specific antibody.

Measurement of neutralizing antibodies has been accomplished by testing the capacity of serial dilutions of serum to prevent CPE in PHFG cells read at 21 days after inoculation (Padgett and Walker, 1973) or the appearance of capsid antigen, detectable by immunofluorescence, in human amnion cells at 7 days after inoculation (Daniel *et al.*, 1981).

The enzyme-linked immunosorbent assay (ELISA) has been used to detect JCV in human urine (Arthur *et al.*, 1983) and it has been adapted to the measurement of anti-BKV antibodies in human serum. It is to be expected that it could be applied to measurements of antibodies against JCV, but it has not been used in published work up to this time.

E. Antigenic Characteristics

1. Structural Antigens

a. Genus-Specific Antigen

As a member of the polyomavirus genus, JCV shares with other members a genus-specific, cross-reacting antigen (Shah *et al.*, 1977a,b). This is an antigenic determinant in the major capsid polypeptide (VP1) of each member of the genus. It is best demonstrated by the immunofluorescent technique in the nuclei of infected cells in culture. Demonstrating it requires use of an antiserum prepared against purified VP1 or against virions disrupted by detergent or alkali. The determinant appears to be buried or not accessible in the intact virion since use of intact virions to immunize rabbits does not result in production of antibody against this determinant. The cross-reactions seen using immunofluorescence on virus-infected cells are strong, indicating that the portion of VP1 that it represents is very similar in all of the viruses of the genus. This antigen is particularly useful because it is resistant to formalin and it can be detected in the nuclei of infected cells in formalin-fixed, paraffin-embedded tissue by means of peroxidase–antiperoxidase immunostaining (Gerber *et al.*, 1980).

b. Cross-Reacting Surface Antigen

JCV also shares one or more minor surface antigens with BKV and SV40. This antigenic relationship has not been demonstrated in other members of the genus, thus setting JCV, BKV, and SV40 somewhat apart as a subgroup. Minor relatedness was first recognized by Gardner *et al.* (1971) as weak cross-reactions between BKV and SV40 in immunoelectron microscopy and HI tests. It was then shown that all three viruses share a minor, cross-reacting antigenic determinant that is exposed on the surface of virions. Use of intact virions in immunizing rabbits will produce antibodies against the antigen, but hyperimmunization is required. The cross-reactions can be shown by immunoelectron microscopy (Penney and Narayan, 1973), HI tests (Padgett and Walker, 1973; Walker *et al.*, 1973a), infectivity neutralization (Takemoto and Mullarkey, 1973; Dougherty and DiStefano, 1974), and immunofluorescence (Takemoto and Mullarkey, 1973; Padgett and Walker, 1976).

c. *Species-Specific Surface Antigen*

There is at least one antigenic determinant on the surface of JCV, BKV, and SV40 that is specific for each virus and allows them to be distinguished from each other and from other viruses by serologic methods. The specific reactions have been demonstrated by immunofluorescent tests on infected cells (Padgett *et al.*, 1971), neutralization of infectivity (Padgett *et al.*, 1977a), HI tests (Padgett *et al.*, 1977a; Gibson and Gardner, 1983), and electron microscope agglutination (Penney and Narayan, 1973; Field *et al.*, 1974).

Antiserum against the specific antigen can be produced in rabbits by one or two intravenous injections of purified, intact virions. Multiple injections of virus or use of adjuvants lead to the appearance of cross-reacting antibodies.

2. Nonvirion Antigens

JCV acts like other papovaviruses in inducing nonstructural T antigens in the nuclei of lytically infected cells as well as in transformed cells and tumor cells induced in animals. The T antigens of JCV, BKV, and SV40 are antigenically very similar (Takemoto and Mullarkey, 1973; Walker *et al.*, 1973a). Antiserum developed in tumor-bearing hamsters against the T antigen of one virus reacts very strongly with the T antigens of all three viruses. Beth *et al.* (1977) used a quantitative complement-fixation technique to show that although the T antigens are strongly related, they do possess subspecificities that can be distinguished. No cross-reactivity has been found between the T antigen of mouse polyoma virus and any of the three primate viruses. The T antigens will be discussed further in a later section.

The tumor-specific transplantation antigen (TSTA) on the surface of JCV-induced tumor cells has received relatively little study. However, Padgett *et al.* (1977a) investigated the specificity of the TSTA by cross-protection tests in weanling hamsters. Hamsters immunized with JCV, BKV, or SV40 were challenged 5 weeks later with known numbers of JCV- or SV40-induced hamster tumor cells. Both JCV-immune and SV40-immune hamsters showed resistance to challenge with homologous but not heterologous tumor cells, and the BKV-immune hamsters were not resistant to either heterologous tumor cell.

3. Antigenic Variability

More than 50 isolations of JCV have been made in widely separated parts of the world from human brain tissue and human urine (Narayan *et al.*, 1973; Nagashima *et al.*, 1981; Padgett and Walker, 1983; Gibson and Gardner, 1983; Padgett and Walker, unpublished observations). These isolates have been identified by using the name, or an abbreviation of the name, of the location of the laboratory making the isolation, plus a number. Thus, Mad-1 refers to the first isolation made in Madison, Wisconsin, Col-1 to the first made in Colindale in London, England, and Tokyo-1 to the first made in Tokyo, Japan.

Although many of the isolates have not been compared in detail, at this time it appears that the major antigenic determinants of JCV show relatively little variability and that there is only one major serotype circulating in the world. Each isolate, whether isolated in the United States, Europe, or Japan, has reacted well against prototype antiserum with but one exception. One strain, isolated from the brain tissue of a patient with PML (Padgett and Walker, 1983), was serologically distinguishable from prototype strain, Mad-1, and from other isolates. This strain, designated Mad-ll, clearly is JCV, but it reacts with prototype JCV antiserum to a lesser degree than other isolates in HI and immunofluorescence tests.

Coleman *et al.* (1980) reported the isolation of a virus that at first appeared to be a variant of JCV. These investigators have since studied the virus and its DNA in considerable detail and have concluded that it is neither JCV nor BKV and that it is a separate and distinct virus that they have termed AS virus (Gibson and Gardner, 1983).

III. PHYSICAL PROPERTIES

The physical properties of JCV virions and DNA clearly indicate that this recently discovered human virus belongs to the genus polyomavirus of the family Papovaviridae (Table III). The initial isolation and cultivation of JCV from the brain tissue of a PML patient revealed virions with icosahedral symmetry and diameters ranging from 38–43 nm (Padgett *et al.*, 1971). When propagated in glial cell cultures, the virus particles were often seen in crystalline arrays and as filamentous and tubular forms within the nuclei of infected cells.

Sedimentation analysis of virus particles isolated from tissue culture has indicated that both full and empty capsids are present; however, the defective forms are not detected when particles are isolated directly from diseased brain tissue (Grinnell *et al.*, 1982, 1983a,b). These findings are in agreement with studies demonstrating that defective viral genomes are generated during passage of the virus in tissue culture (see below).

The nucleic acid extracted from intact virions or from infected cells bands in CsCl–ethidium bromide gradients at a density of 1.60 g/cm^3 and has the characteristics of covalently closed, circular DNA (Osborn *et al.*, 1974). Original size measurements using electron microscopic (Osborn *et al.*, 1974) and sedimentation (Takemoto *et al.*, 1979a) techniques yielded molecular weight estimates that were significantly lower than those for other polyomaviruses (e.g., 2.9×10^6 versus 3.6×10^6 for SV40). However, because of the presence of shorter, presumably defective species in

TABLE III. Physical Properties of JCV Virions and DNA[a]

Size	
Icosahedral particles	38–43 nm (diam)
Capsomers	8–9 nm (diam)
Filamentous forms	29 nm (diam)
DNA (circular, double-stranded)	5130 np
Density	
In CsCl gradient	
Virions	1.345–1.35 g/cm^3
DNA	1.60 g/cm^3
In sucrose–D_2O gradient	
Virions	1.20 g/cm^3
Stability in:	
Ether	Stable
50°C, 1 hr	Stable
50°C, 1 hr, 1 M $MgCl_2$	Unstable
Sedimentation rate values	
In neutral sucrose gradient	
Virions	250 S
Empty capsids	150 S
DNA	19–20 S
In alkaline sucrose gradient	
DNA	51 S

[a] Compiled from Padgett *et al.* (1971, 1976, 1977b), Osborn *et al.* (1974), Padgett and Walker (1976), Dörries *et al.* (1979), Takemoto *et al.* (1979a), Grinnell *et al.* (1982, 1983a), Frisque *et al.* (1984), Tenser, Sommerville, Mummaw, and Frisque (unpublished).

these JCV DNA preparations, it was suggested that the size estimates would be revised upwards if measurements were made on full-length, infectious molecules only. Recent sequence analysis of such molecules yielded a value of 5130 nucleotide pairs (np) corresponding to a molecular weight of approximately 3.4 × 10^6 (Frisque *et al.*, 1984).

A. JCV Genome

1. Organization

Biological studies of JCV have been seriously hampered by the lack of a readily available permissive cell system, and therefore our understanding of SV40, BKV, and mouse polyoma virus, which are relatively easy to propagate in tissue culture, is much more complete. Because of these limitations *in vitro*, advances in JCV research have been most apparent in two areas: (1) in studies investigating the oncogenic potential of the virus in experimental animals and (2) in analyses of the viral regulatory sequences utilizing recombinant DNA techniques.

Determination of the genetic organization of JCV is critical to our understanding of the unique biology of this human pathogen. Initial immunological studies indicated that JCV, BKV, and SV40 are closely related viruses. Antiserum obtained from tumor-bearing hamsters cross-reacts strongly with the T proteins produced by all three viruses (Takemoto and Mullarkey, 1973; Walker *et al.*, 1973a), although these proteins can be distinguished by adsorption tests (Dougherty, 1976; Beth *et al.*, 1977) or by certain monoclonal antibodies (Major *et al.*, 1984). Only minor cross-reactivity is observed when the antigenic determinants on the virion surface are compared; however, a genus-specified capsid antigen that is found on the major viral peptide (VPl) and is probably located internally, can be demonstrated (Shah *et al.*, 1977a,b). While these studies emphasized the similarities between JCV, BKV, and SV40, early comparisons of the viral genomes pointed to their differences. Restriction enzyme analyses revealed obvious differences in their cleavage patterns (Osborn *et al.*, 1974; Martin *et al.*, 1979, 1982) while reassociation acceleration (Howley *et al.*, 1976) and competition hybridization (Osborn *et al.*, 1976) experiments detected only low levels of sequence homology (JCV × BKV: 20–25%; JCV × SV40: 11–40%). Later experiments by Law *et al.* (1979) using less stringent conditions that allowed detection of homologies involving greater mismatch (≤ 26%) in the DNA sequence, demonstrated extensive homology throughout the three genomes. The strongest homology was localized to the region corresponding to the N-terminus of the late capsid protein, VP2, of SV40 (0.76–0.85 m.u.) while the weakest homology was found in the region about 0.67–0.76 m.u., sequences that specify several viral regulatory signals. These experiments allowed a colinear alignment of the three genomes (the unique *Eco*RI site of each viral DNA is taken as map position 0) and suggested that the genetic organizations of the viruses were similar.

Completion of the nucleotide sequence of prototype Mad-1 DNA has enabled a direct comparison to be made of the genetic organizations of JCV, BKV, and SV40 (Frisque *et al.*, 1984; see Appendix). The JCV genome contains 5130 np and has a GC content of 40.4% (the values for SV40 are 5243 np and 40.8%). The overall homology with the BKV and SV40 sequences is 75 and 69%, respectively. As with these latter two viruses, a total of six JCV proteins can be deduced from the sequence data: two early, nonstructural proteins, T and t antigens; three capsid proteins, VP1, VP2, and VP3; and an agnoprotein. The genetic map of JCV is shown in Fig. 2.

a. T Antigen Binding Sites

Recent efforts have focused on identifying the similarities and differences in the polyomavirus genomes at the nucleotide level. Attention was first directed to the region near 0.67 m.u. of the JCV genome, an area expected to contain a number of replication and transcription signals. This region in SV40 includes three sites to which the regulatory protein

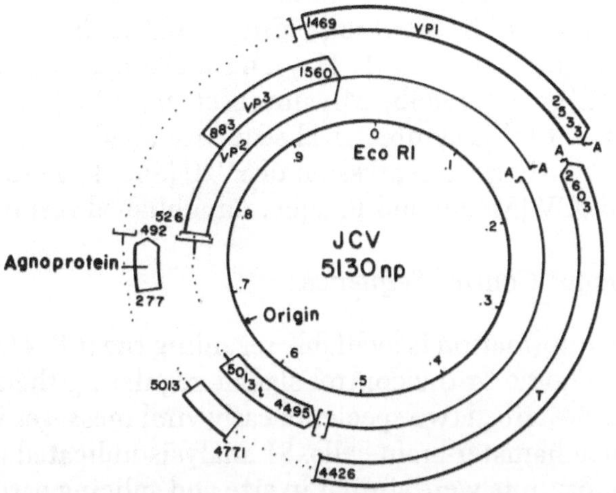

FIGURE 2. Circular map of the JCV genome (Mad-1 strain). The single *Eco*RI site is taken as map position 0.0 on the JCV genome. The map is divided into two nearly equal parts, depending on whether gene expression occurs primarily before (early) or after (late) viral DNA replication. Broad bars depict the coding regions for the six proposed JCV proteins. The dots at the beginning of bars indicate uncertainty as to the exact 5′ end of the mRNAs. Brackets enclosing dots represent intervening sequences, and single lines indicate untranslated 5′ and 3′ portions of the early and late messages. Reprinted from Frisque *et al.* (1984).

T antigen and the closely related protein D2T bind. Methylation protection and DNase footprinting experiments have identified multiple copies of a pentanucleotide sequence, 5′-(G > T)(A > G)GGC-3′, involved in this binding (Tjian, 1978; DeLucia *et al.*, 1983). These experiments have been repeated with BKV (Ryder *et al.*, 1983) and the same consensus sequence is found at locations within the BKV DNA that are analogous to the SV40 binding sites I and II. These studies identified an additional contact sequence, 5′-AAGGC-3′, which is part of the third BKV site. D2T protein also binds to the JCV origin region (*Hin*dIII-C fragment, 0.63–0.67 m.u.; Frisque, 1983a), thereby establishing the presence of a T antigen binding site(s) in the JCV genome. Subsequent sequencing results confirmed that the *Hin*dIII-C fragment contains a sequence that closely resembles the first binding sites of SV40 and BKV DNA. A second site was detected at the boundary of this fragment (0.67 m.u.) and it is nearly identical to site II of BKV and SV40. A third site was not readily apparent in the JCV sequence; neither of the two pentanucleotide sequences was found at the predicted position in the JCV DNA.

b. Origin of DNA Replication

The sequences near 0.67 m.u. in JCV, BKV, and SV40 DNA are highly conserved and contain a number of dyad symmetries and true palindromes (Fig. 3). One of these symmetries (25, 23, and 27 np in JCV, BKV, and

SV40, respectively) contains the second T antigen binding site and appears to function as the site from which bidirectional replication of the viral DNAs is initiated (Tooze, 1981; Hay and DePamphilis, 1982; Frisque, 1983b; Grinnell *et al.*, 1983b; Miyamura *et al.*, 1983). Introduction of small deletions into this symmetrical sequence abolishes viral DNA replication but not early gene expression of SV40 (Shenk, 1978; Gluzman *et al.*, 1980a) and JCV (Mandl and Frisque, unpublished results).

c. Transcriptional Control Sequences

Very little information is available regarding the mRNAs transcribed from the JCV genome or the control signals regulating their production. Frisque (1983a) described two species of early viral messages isolated from JCV-transformed hamster brain cells. S1 analysis indicated that the large and small T transcripts were similar in size and splicing patterns to SV40 messages obtained from similar kinds of transformants. Lengths of the intact mature mRNAs on neutral gels were 2200 and 2500 np for JCV and 2300 and 2600 np for SV40. (The larger mRNA in each pair represents the t transcript which contains a smaller intron.) Splice junctions could be approximated by running the S1 reaction products on alkaline gels. The 5′ portions of the T and t messages were 370 and 670 np long for JCV and 350 and 650 np long for SV40; the lengths of the shared 3′ ends of the early mRNAs were 1850 and 1950 np for JCV and SV40, respectively.

Recent advances in a number of eukaryotic systems have allowed us to make several predictions (from sequence data) concerning the structure of JCV mRNAs and the sequences governing their production. On the basis of their primary structure, location with respect to associated genes, and overall organization, a group of consensus sequences have been identified which promote transcriptional activity by RNA polymerase II. These include the Goldberg–Hogness sequence (also called the TATA box or AT-rich region; 5′-TATAAATA-3′), the CAAT box (5′-GGPyCAATCT-3′), the PyPyCCXCCC sequence, and the enhancer element. Because of their importance to efficient gene expression and because of JCV's restricted activity *in vitro,* these regulatory elements in the JCV genome are receiving considerable attention.

The Goldberg–Hogness sequence is required for the accurate initiation of early transcription (Gluzman *et al.*, 1980b; Benoist and Chambon, 1981; Ghosh *et al.*, 1981) and for the efficient replication of SV40 DNA (Myers and Tjian, 1980; Bergsma *et al.*, 1982). It is usually located 20–30 nucleotides upstream from the site at which transcription of eukaryotic mRNAs begins. The TATA box of prototype JCV (Mad-1) is unusual in that it is located within a large tandem repeat (see below) and therefore is duplicated. It is not known whether the duplication interferes with the positioning function of this sequence. Using the TATA box found closest to the early region (nucleotides No. *15–29*) as a landmark and relying on the start sites already mapped for SV40 early mRNAs (Reddy *et al.*, 1979; Ghosh *et al.*, 1981), two potential initiation sites can be identified in the JCV sequence (within nucleotides No. *5115* and *5121*). Locating the 5′ termini of the T and t mRNAs at these positions would agree with available S1 nuclease data (Frisque, 1983a; Mandl and Frisque, unpublished results). Recently, Kenney and co-workers have mapped the 5′ start sites of JCV late mRNAs isolated from lytically infected PHFG cells (S. Kenney, V. Natarajan, and N. P. Salzman, personal communication). Using the S1 nuclease technique, multiple 5′ termini were identified within a region spanning nearly 300 nucleotides (from near the origin of DNA replication to nucleotide No. 240). As with SV40 and BKV, there is not a readily apparent Goldberg–Hogness sequence at the proper position relative to the proposed late JCV mRNA start sites.

Located between the SV40 TATA and enhancer sequences are three copies of a 21-np repeat which contribute to efficient DNA replication and early and late transcription (Bergsma *et al.*, 1982; Byrne *et al.*, 1983; Everett *et al.*, 1983; Hartzell *et al.*, 1983, 1984). Each repeat contains two copies of the sequence 5′-PyPyCCXCCC-3′, a sequence present in the regulatory regions of a number of viruses (Seif *et al.*, 1979; Tolun *et al.*, 1979; Yang and Wu, 1979; Soeda *et al.*, 1980; McKnight and Kingsbury, 1982). Dynan and Tjian (1983b) have recently isolated a promoter-specific transcription factor from whole cell extracts which binds to the 21-np repeat region and stimulates early and late transcription. The Spl factor also stimulates BKV early transcription (Dynan and Tjian, 1983a) although the effect is reduced tenfold over SV40 (BKV has two copies of PyPyCCXCCC, SV40 has six). The JCV regulatory region does not contain the PyPyCCXCCC sequence although PyPyCCXXCCC is present in the tandem repeat; the first copy is in a position similar to that of the SV40 sequence. Perhaps this sequence contributes to the lytic activity of prototype JCV in PHFG cells.

Another signal required for efficient promoter function in some eukaryotic systems is the CAAT box, a sequence located approximately 80 nucleotides from the mRNA initiation site (Benoist *et al.*, 1980; Dierks *et al.*, 1981; Mellon *et al.*, 1981; Grosveld *et al.*, 1982). While several good candidates exist in the 21-np repeats of SV40 (Benoist *et al.*, 1980; Byrne *et al.*, 1983), potential CAAT boxes within the tandem repeats of JCV (5′-AGCCATCCCT-3′, No. 34–43; and 5′-GCTCATGCT-3′, No. 52–60) and BKV (5′-GGTCATGGT-3′; Seif *et al.*, 1979; Yang and Wu, 1979) demonstrate only a partial homology with the consensus sequence.

The large tandem repeats found to the late side of the replication origins of SV40 and BKV have been identified as enhancer or activator elements because of their ability to stimulate transcription of associated genes (Banerji *et al.*, 1981; Benoist and Chambon, 1981; Rosenthal *et al.*, 1983). The 98-np tandem repeat found immediately adjacent to the JCV origin of DNA replication (Fig. 3) has also been identified as an enhancer element (Kenney *et al.*, 1984). Although its structure (a tandem repeat)

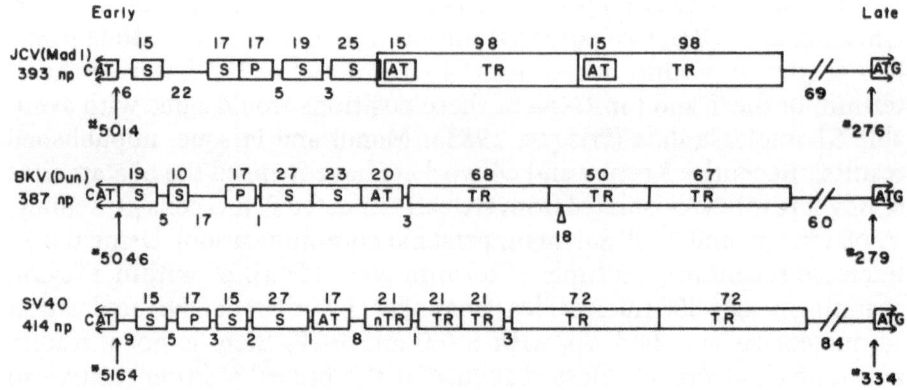

FIGURE 3. Comparison of the JCV, BKV, and SV40 regulatory regions. The noncoding regions of the three polyomaviruses are shown. The letters CAT within the open box to the left represent the initiation codon (opposite strand and polarity) for the early proteins, T and t antigens. To the far right is the ATG initiation codon for the agnoprotein located within the late leader sequence. Comparisons among the three viral DNAs include dyad symmetries (S), true palindromes (P), TATA boxes (AT), tandem repeats (TR), and nontandem repeats (nTR) (repeats which are not immediately adjacent to each other). Numbers above the linear arrangements refer to the sizes (in np) of the indicated structures. Numbers below refer to the distances between the structures. The triangle below the middle tandem repeat of BKV indicates a deletion of 18 nucleotides. This set of nucleotides is present in the adjacent repeats. Reprinted from Frisque *et al.* (1984).

and its location (near the replication origin) are similar to the enhancers of SV40 and BKV, little homology is evident between the three viral repeats. However, a core sequence (5′-GTGG$\genfrac{}{}{0pt}{}{\text{TTT}}{\text{AAA}}$G-3′; Weiher *et al.*, 1983) identified in a number of viral and cellular enhancers can be recognized in each virus (5′-GTGCTTTG-3′ for JCV; Frisque *et al.*, 1984). The enhancer sequences are particularly relevant to a discussion of JCV's restricted activity *in vitro* because of their involvement in the host range, tissue tropism, and oncogenic properties of the polyomaviruses (Watanabe *et al.*, 1979; Katinka and Yaniv, 1980; Katinka *et al.*, 1981; Fujimura *et al.*, 1981; de Villiers *et al.*, 1982; Laimins *et al.*, 1982; Watanabe and Yoshiike, 1982; Byrne *et al.*, 1983; Rosenthal *et al.*, 1983). Recent experiments by Kenney *et al.* (1984) have shown that, unlike the SV40 and BKV enhancer elements which function in a variety of cells, the JCV enhancer functions well only in PHFG cells. It is possible that alterations in the enhancer region may be partially responsible for the apparent adaptation of JCV to growth in cells of neural origin (in the general population, JCV probably multiplies in kidney or lung tissue and may have a different enhancer sequence). Significantly, JCV variants isolated from PML brain tissue do vary from one another in their enhancer sequences (Martin *et al.*, 1985) and it will be important to isolate the strain(s) of JCV detected in normal, non-PML tissue (McCance, 1983) for comparison.

Most eukaryotic mRNAs have a sequence of polyadenylic acid (~ 200 residues) at their 3′ ends. The consensus sequence 5′-AAUAAA-3′ has been shown to be part of the signal for cleavage and polyadenylation at appropriate sites (11–30 nucleotides downstream from the hexanucleotide) on the mRNA precursors (Proudfoot and Brownlee, 1974; Fitzgerald and Shenk, 1981). This polyadenylation signal is found on each strand of JCV DNA just beyond the termination codons for the T and VP1 proteins. Their location implies that the early and late transcripts overlap at their 3′ ends.

Splicing, another form of mRNA processing, is a process in which the nucleotide sequences of mature eukaryotic mRNAs are derived from noncontiguous regions of the template DNA. Sequences called introns are excised from precursor RNA molecules and the remaining sequences are spliced back together to form the mature transcript. Again, certain consensus sequences can be recognized near the donor (5′-AG ↓ GTAAGT-3′) and acceptor (5′-6PyXCAG ↑ -3′) splice sites (Lewin, 1983). Although considerable flexibility is noted in these sequences, the junction dinucleotides, GT and AG, are nearly invariant at the splice sites that have been analyzed. Using both the splicing information obtained for other polyomaviruses and the JCV sequence data, splicing patterns for this virus can be predicted (Frisque *et al.*, 1984). The donor splice sites for the T and t messages can be localized at nucleotides 4771 and 4494, respectively, while the acceptor site could be at nucleotide 4426. Although these two proteins share N-terminal sequences, their C-termini vary due to differential splicing of their messages. These predictions agree with S1 nuclease results (Frisque, 1983a). A third early RNA, representing a viral middle T transcript (as seen in mouse polyoma virus), has not been detected and the sequence data do not suggest a middle T protein.

The splicing patterns of the late JCV transcripts are predicted solely on the basis of the nucleotide sequence. A candidate for the shared donor splice site of the VP1, VP2, and VP3 messages occurs at nucleotide 492. The potential acceptor sites are at nucleotides 522 for the VP2/3 message(s) and 1427 for the VP1 transcript.

B. Viral Proteins

A genetic map specifying the probable locations of the six JCV proteins is shown in Fig. 2. The map is divided into two nearly equal parts depending on whether the encoded proteins are produced before DNA replication (early) or after the onset of replication (late). A comparison of these proteins with those of SV40 and BKV emphasizes the relatedness of these three viruses (Table IV).

1. Early Proteins

The first indications that the polyomavirus T antigens were similar came from immunofluorescence studies. Antibodies obtained from hamsters bearing JCV-, BKV-, and SV40-induced tumors cross-reacted strongly

TABLE IV. JCV, BKV, and SV40 Viral Proteins

	JCV		BKV(Dun)			SV40		
	MW[a]	No. of amino acids	MW	No. of amino acids		MW	No. of amino acids	
VP1[b]	39,606	354	40,106	362	(78)[c]	39,903	362	(75)
VP2	37,366	344	38,345	351	(79)	38,525	352	(72)
VP3	25,743	225	26,718	232	(75)	26,961	234	(66)
T	79,305	688	80,499	695	(83)	81,617	708	(72)
t	20,236	172	20,469	172	(78)	20,447	174	(67)
Agnoprotein[d]	8,081	71	7,396	66	(59)	7,335	62	(46)

[a] Molecular weights are calculated from sequence data using the IBI/Pustell sequence analysis system.
[b] In each virus there are two potential initiation codons for VP1 (ATG AAG ATG . . .). Calculations were based on the second (downstream) ATG.
[c] Numbers in parentheses indicate the percentage of amino acids shared with the corresponding JCV protein.
[d] Encoded within the leader sequences of late viral mRNAs.

with the heterologous nuclear tumor antigens present in the established tumor cell lines (Takemoto and Mullarkey, 1973; Walker *et al.*, 1973a; Takemoto and Martin, 1976; Beth *et al.*, 1977, 1978). Simmons and Martin (1978) showed that the JCV T antigen, isolated by immunoprecipitation from tumor cells, was similar in size to the SV40 and BKV T proteins (two species, 94K and 97K). Analysis of the methionine-labeled tryptic peptides of the T proteins revealed two peptides that were shared by all three viruses. Although a small t protein (17K) could be detected in the SV40 and BKV tumor lines, it was not found in the JCV tumor cells. Later studies with primary hamster brain (PHB) cells transformed by JCV and SV40 *in vitro*, indicated that the JCV T protein was slightly smaller than its SV40 counterpart (Frisque *et al.*, 1980). Immunoprecipitation of transformed cell extracts with tumor antiserum identified two T proteins of 92K and 96K in all JCV transformants. In addition, some of the cell lines contained larger immunoprecipitable proteins. Several proteins were also isolated from the SV40 transformants and were equal to or larger in size (94K, 99K, 108K) than the T antigen extracted from lytically infected CV-1 cells (94K). These studies also failed to detect a JCV t antigen using antisera that revealed the SV40 t protein. Similar findings have been reported for cell lines derived from JCV-induced owl monkey tumors (Major, 1983b; Major *et al.*, 1984).

There has been one preliminary report of the immunoprecipitation of t antigen from JCV-transformed hamster cells (Frisque, 1983a). In this experiment, only 1 of 20 animals responded to produce anti-t antibodies. Furthermore, resolution of the protein band on the gradient acrylamide gel was poor due to the presence of a cellular band migrating at nearly the same position. These results suggest that the difficulty in detecting a JCV t protein may be due to a combination of factors including poor expression, lability, weak antigenicity, and inadequate gel resolution.

A third early polyomavirus protein, middle T, has been reported for mouse polyoma virus. A similar protein does not appear to be encoded by the JCV, BKV, or SV40 genomes. However, the large T protein of the latter three viruses does appear to associate with a cellular protein (cellular middle T or tau antigen) of 50–60K (Lane and Crawford, 1979; Frisque *et al.*, 1980; Simmons, 1980). Two-dimensional peptide mapping indicates that all three viruses induce the same cellular middle T protein (Simmons, 1980).

The probable primary structures for the JCV early proteins have been deduced from the nucleotide sequence (Miyamura *et al.*, 1983; Frisque *et al.*, 1984; Figs. 4, 5). This analysis agrees with the immunoprecipitation and S1 nuclease data and emphasizes the relatedness of JCV, BKV, and SV40 (Table IV).

The multifunctional T protein of SV40 has been implicated in a number of activities including initiation of viral DNA replication, stimulation of host DNA synthesis, modulation of early and late transcription, and establishment and maintenance of cellular transformation (Tooze, 1981). It is thought that some of these functions are mediated through the specific binding of the protein to the origin region of the genome (Tjian, 1978; Shortle *et al.*, 1979; Myers and Tjian, 1980; Rio *et al.*, 1980). Involved in this binding might be a stretch of basic amino acids (reviewed by Prives *et al.*, 1982; Morrison *et al.*, 1983) that lies within a highly conserved region of the JCV, BKV, and SV40 T proteins (19 or 20 amino acids are identical; amino acids 120–139 for JCV).

Although considerable mismatch can be detected in the coding sequences of the JCV, BKV, and SV40 T antigens, computer-generated hydropathy plots of these proteins show almost complete identity over their entire lengths [i.e., the plots can be nearly superimposed upon one another at each numbered amino acid (Frisque, unpublished)]. There are three exceptions: (1) at amino acids No. 296–306, JCV and BKV have charged amino acids, SV40 has uncharged polar and hydrophobic amino acids; (2) at No. 619–627, JCV has uncharged polar amino acids, BKV and SV40 have hydrophobic amino acids; (3) near the C-terminus of each protein is a long stretch of charged amino acids, but because of an 18-amino-acid deletion in the JCV and BKV sequences this stretch is shorter in these two viral proteins. The importance of these differences (if any) remains to be determined. Interestingly, recent studies with SV40 indicate that deletion of sequences at the C-terminus of large T protein limits the host range of this virus (Manos and Gluzman, 1985; Tornow *et al.*, 1985).

The first 81 amino acids of the JCV small t protein overlap with the large T sequence; the remaining 91 amino acids are unique to small t due to differential splicing of the two early mRNAs (Fig. 2). The large degree of homology observed among the N-termini of the two early proteins for all three viruses (89% for JCV × BKV; 82% for JCV × SV40) is

JCV	Met	Asp	Lys	Val	Leu	Asn	Arg	Glu	Glu	Ser	Met	Glu	Leu	Met	Asp	Leu	Leu	Gly	Leu	Asp	Arg	Ser	Ala	Trp	Gly	Asn	Ile	Pro	Val	Met
BKV											Met	Glu								Glu		Ala					Leu		Leu	
SV40											Leu	Gln								Glu		Ser					Ile		Leu	
JCV	Arg	Lys	Ala	Tyr	Leu	Lys	Lys	Cys	Lys	Glu	Leu	His	Pro	Asp	Lys	Gly	Gly	Asp	Glu	Asp	Lys	Met	Lys	Arg	Met	Asn	Phe	Leu	Tyr	Lys
BKV						Arg					Phe									Asp				Arg			Thr			
SV40						Lys					Phe									Glu				Lys			Thr			
JCV	Lys	Met	Glu	Gln	Gly	Val	Lys	Val	Ala	His	Gln	Pro	Asp	Phe	Gly	Thr	N	Trp	Asn	Ser	Ser	Glu	Val	Pro	Thr	Tyr	Gly	Thr	Asp	Glu
BKV				Gln	Asp			Val								Thr	N		Ser	Ser	Ser		Val						Glu	
SV40				Asp	Gly			Tyr								Gly	Phe		Asp	Ala	Thr		Ile						Asp	
JCV	Trp	Glu	Ser	Trp	Trp	Asn	Thr	Phe	Asn	Glu	Lys	Trp	Asp	Glu	Asp	Leu	Phe	Cys	His	Glu	Glu	Met	Phe	Ala	Ser	Asp	Asp	Glu	Asn	Thr
BKV			Ser			Ser	Ser				Lys	Trp	Asp		Asp				His		Asp		Phe	Ala			Glu		Ala	
SV40			Gln			Asn	Ala				N	N	N		Asn				Ser		Glu		Pro	Ser			Asp		Ala	
JCV	Gly	N	Ser	Gln	His	Ser	Thr	Pro	Pro	Lys	Lys	Lys	Lys	Lys	Val	Glu	Asp	Pro	Lys	Asp	Phe	Pro	Val	Asp	Leu	His	Ala	Phe	Leu	Ser
BKV	Ala	Asp											Arg										Ser	Asp		His	Gln			
SV40	Ala	Asp											Arg										Ser	Glu		Leu	Ser			
JCV	Gln	Ala	Val	Phe	Ser	Asn	Arg	Thr	Val	Ala	Ser	Phe	Ala	Val	Tyr	Thr	Thr	Lys	Glu	Lys	Ala	Gln	Ile	Leu	Tyr	Lys	Lys	Leu	Met	Glu
BKV	Gln								Leu		Cys			Val								Gln	Ile					Leu		
SV40	His								Leu		Cys			Ile								Ala	Leu					Ile		
JCV	Lys	Tyr	Ser	Val	Thr	Phe	Ile	Ser	Arg	His	Gly	Phe	Gly	Gly	His	Asn	Ile	Leu	Phe	Phe	Leu	Thr	Pro	His	Arg	His	Arg	Val	Ser	Ala
BKV											Met	Cys	Ala	Gly				Ile												
SV40											Asn	Ser	Tyr	Asn				Leu												
JCV	Ile	Asn	Asn	Tyr	Cys	Gln	Lys	Leu	Cys	Thr	Phe	Ser	Phe	Leu	Ile	Cys	Lys	Gly	Val	Asn	Lys	Glu	Tyr	Leu	Phe	Tyr	Ser	Ala	Leu	Cys
BKV				Phe	Cys																				Leu					Thr
SV40				Tyr	Ala																				Met					Thr
JCV	Arg	Gln	Pro	Tyr	Ala	Val	Val	Glu	Glu	Ser	Ile	Gln	Gly	Gly	Leu	Lys	Glu	His	Asp	Phe	Asn	Pro	Glu	Glu	Pro	Glu	Glu	Thr	Lys	Gln
BKV		Asp		Tyr	His	Thr	Ile				Ile	Gln									Ser				Pro					
SV40		Asp		Phe	Ser	Val	Ile				Leu	Pro									Asn				Ala					
JCV	Val	Ser	Trp	Lys	Leu	Val	Thr	Gln	Tyr	Ala	Leu	Glu	Thr	Lys	Cys	Glu	Asp	Val	Phe	Leu	Leu	Met	Gly	Met	Tyr	Leu	Asp	Phe	Gln	Glu
BKV						Ile		Glu			Val					Glu			Phe			Leu					Glu			Tyr
SV40						Val		Glu			Met					Asp			Leu			Leu					Glu			Tyr
JCV	Asn	Pro	Gln	Gln	Cys	Lys	Lys	Cys	Glu	Lys	Lys	Asp	Gln	Pro	Asn	His	Phe	Asn	His	His	Glu	Lys	His	Tyr	Tyr	Asn	Ala	Gln	Ile	Phe
BKV	Asn	Val	Glu	Glu		Lys			Gln			Asp			Tyr		Phe	Lys	Tyr					Phe	Ala			Ile		
SV40	Ser	Phe	Glu	Met		Leu			Ile			Glu			Ser		Tyr	Lys	Tyr					Tyr	Ala			Ala		
JCV	Ala	Asp	Ser	Lys	Asn	Gln	Lys	Ser	Ile	Cys	Gln	Gln	Ala	Val	Asp	Thr	Val	Ala	Ala	Lys	Gln	Arg	Val	Asp	Ser	Ile	His	Met	Thr	Arg
BKV		Glu						Ser										Leu			Lys				Thr	Leu	His	Met		
SV40		Asp						Thr										Leu			Lys				Ser	Leu	Gln	Leu		
JCV	Glu	Glu	Met	Leu	Val	Glu	Arg	Phe	Asn	Phe	Leu	Leu	Asp	Lys	Met	Asp	Leu	Ile	Phe	Gly	Ala	His	Gly	Asn	Ala	Val	Leu	Glu	Gln	Tyr
BKV		Glu			Thr	Glu				His	Ile			Lys			Leu	Ile			Ala	His		Asn		Val	Leu		Gln	Tyr
SV40		Gln			Thr	Asn				Asp	Leu			Arg			Ile	Met			Ser	Thr		Ser		Asp	Ile		Glu	Trp

JCV	Met	Ala	Gly	Val	Ala	Trp	Ile	His	Cys	Leu	Leu	Pro	Gln	Met	Asp	Thr	Val	Ile	Tyr	Asp	Phe	Leu	Lys	Cys	Ile	Val	Leu	Asn	Ile	Pro
BKV							Leu						Lys			Ser		Ile	Phe				His		Ile		Phe		Val	
SV40							Leu						Lys			Ser		Val	Tyr				Lys		Met		Tyr		Ile	
JCV	Lys	Lys	Arg	Tyr	Trp	Leu	Phe	Lys	Gly	Pro	Ile	Asp	Ser	Gly	Lys	Thr	Thr	Leu	Ala	Ala	Ala	Leu	Leu	Asp	Leu	Cys	Gly	Gly	Lys	Ser
BKV		Arg																			Gly			Asp						Ala
SV40		Lys																			Ala			Glu						Ala
JCV	Leu	Asn	Val	Asn	Met	Pro	Leu	Glu	Arg	Leu	Asn	Phe	Glu	Leu	Gly	Val	Gly	Ile	Asp	Gln	Phe	Met	Val	Val	Phe	Glu	Asp	Val	Lys	Gly
BKV					Leu		Met	Glu			Thr						Ala				Tyr	Met								
SV40					Leu		Leu	Asp			Asn						Ala				Phe	Leu								
JCV	Thr	Gly	Ala	Glu	Ser	Arg	Asp	Leu	Pro	Ser	Gly	His	Gly	Ile	Ser	Asn	Leu	Asp	Cys	Leu	Arg	Asp	Tyr	Leu	Asp	Gly	Ser	Val	Lys	Val
BKV			Ala			Lys						His			Asn				Ser											
SV40			Gly			Arg						Gln			Asn				Asn											
JCV	Asn	Leu	Glu	Arg	Lys	His	Gln	Asn	Lys	Arg	Thr	Gln	Val	Phe	Pro	Pro	Gly	Ile	Val	Thr	Met	Asn	Glu	Tyr	Ser	Val	Pro	Arg	Thr	Leu
BKV				Lys			Leu						Ile					Leu							Pro			Lys		
SV40				Lys			Leu						Ile					Ile							Ser			Lys		
JCV	Gln	Ala	Arg	Phe	Val	Arg	Gln	Ile	Asp	Phe	Arg	Pro	Lys	Ala	Tyr	Leu	Arg	Lys	Ser	Leu	Ser	Cys	Ser	Glu	Tyr	Leu	Leu	Glu	Lys	Arg
BKV						Arg								Ile			Arg	Lys	Ser		Gln	Asn			Phe					
SV40						Lys								Asp			Lys	His	Cys		Glu	Arg			Phe					
JCV	Ile	Leu	Gln	Ser	Gly	Met	Thr	Leu	Leu	Leu	Leu	Leu	Ile	Trp	Phe	Arg	Pro	Val	Ala	Asp	Phe	Ala	Ala	Ala	Ile	His	Glu	Arg	Ile	Val
BKV		Leu				Met	Thr				Leu				Phe					Asp			Thr	Asp		Gln	Ser			
SV40		Ile				Ile	Ala				Met				Tyr					Glu			Gln	Ser		Gln	Ser			
JCV	Gln	Trp	Lys	Glu	Arg	Leu	Asp	Leu	Glu	Ile	Ser	Met	Tyr	Thr	Phe	Ser	Thr	Met	Lys	Ala	Asn	Val	Gly	Met	Gly	Arg	Pro	Ile	Leu	Asp
BKV	Glu							Ser		Ile		Met	Tyr	Thr	Phe	Ser	Arg			Tyr		Ile	Cys			Lys	Cys	Ile		
SV40	Glu							Lys		Phe		Leu	Ser	Val	Tyr	Gln	Lys			Phe		Val	Ala			Ile	Gly	Val		
JCV	Phe	Pro	Arg	Glu	Glu	Asp	Ser	Glu	Ala	Glu	Asp	Ser	N	N	N	N	N	N	N	N	N	N	N	N	N	N	N	N	N	N
BKV	Ile	Thr		Glu	Glu		Ser	Glu	Thr				N	N	N	N	N	N	N	N	N	N	N	N	N	N	N	N	N	N
SV40	Trp	Leu		Asn	Ser		Asp	Asp	Asp				Gln	Gln	Asn	Ala	Asp	Lys	Asn	Glu	Asp	Gly	Gly	Glu	Lys	Asn	Met	Glu	Asp	Ser
JCV	Gly	His	Gly	Ser	Ser	Thr	Glu	Ser	Gln	Ser	Gln	Cys	Phe	Ser	Gln	Val	Ser	Glu	Ala	Ser	Gly	N	Ala	N	Asp	Thr	Gln	Glu	N	N
BKV			Gly	Ser	Ser	Thr	Glu					Gys	Ser	Ser		Val	Ser	Asp	Thr	Ser	Ala	Pro	Ala	Glu	Asp	Ser		Arg	Ser	Asp
SV40			Glu	Thr	Gly	Ile	Asp					Gly	Ser	Phe		N	N	N	N	N	Ala	Pro	Gln	N	Ser	Ser		Ser	Val	His
JCV	N	N	Asn	Cys	Thr	Phe	His	Ile	Cys	Lys	Gly	Phe	Gln	Cys	Phe	Lys	Lys	Pro	Lys	Thr	Pro	Pro	Pro	Lys	N	N	N			
BKV	Pro	His	Ser	Gln	Glu	Leu		Leu		Lys			Gln				Arg		Lys					Lys	N	N	N			
SV40	Asp	His	Asn	Gln	Pro	Tyr		Ile		Arg			Thr				Lys		Pro					Glu	Pro	Glu	Thr			

FIGURE 4. Comparison of the large T proteins of JCV (Mad-1), BKV (Dun), and SV40. The proposed sequences for the large T proteins of the three polyomarviruses are aligned for maximum homology. In those instances when the amino acid is the same for all three T proteins, only the JCV sequence is shown. Reprinted from Frisque *et al.* (1984).

JCV	Met	Asp	Lys	Val	Leu	Asn	Arg	Glu	Glu	Ser	Met	Glu	Leu	Met	Asp	Leu	Leu	Gly	Leu	Asp	Arg	Ser	Ala	Trp	Gly	Asn	Ile	Pro	Val	Met
BKV											Met	Glu								Glu		Ala					Leu		Leu	
SV40											Leu	Gln								Glu		Ser					Ile		Leu	

JCV	Arg	Lys	Ala	Tyr	Leu	Lys	Lys	Cys	Lys	Glu	Leu	His	Pro	Asp	Lys	Gly	Gly	Asp	Glu	Asp	Lys	Met	Lys	Arg	Met	Asn	Phe	Leu	Tyr	Lys
BKV						Arg					Phe									Asp				Arg			Thr			
SV40						Lys					Phe									Glu				Lys			Thr			

BKV	Lys	Met	Glu	Gln	Gly	Val	Lys	Val	Ala	His	Gln	Pro	Asp	Phe	Gly	Thr	N	Trp	Asn	Ser	Ser	Glu	Val	Gly	Cys	N	Asp	N	Phe	Pro
JCV				Gln	Asp			Val								Thr	N		Ser	Ser	Ser			N	Cys	Ala	Asp	N	Phe	Pro
SV40				Asp	Gly			Tyr								Gly	Phe		Asp	Ala	Thr			N	Phe	Ala	Ser	Ser	Leu	Asn

JCV	Pro	Asn	Ser	Asp	Thr	Leu	Tyr	Cys	Lys	Glu	Trp	Pro	Asn	Cys	Ala	Thr	Asn	Pro	Ser	Val	His	Cys	Pro	Cys	Leu	Met	Cys	Met	Leu	Lys
BKV	Leu	Cys	Pro		Thr	Leu				Glu			Ile		Ser	Lys	Lys	Pro		Val	His		Pro		Met	Leu		Gln		Arg
SV40	Pro	Gly	Val		Ala	Met				Gln			Glu		Ala	Lys	Lys	Met		Ala	Asn		Ile		Leu	Leu		Leu		Arg

JCV	Leu	Arg	His	Arg	Asn	Arg	Lys	Phe	Leu	Arg	Ser	Ser	Pro	Leu	Val	Trp	Ile	Asp	Cys	Tyr	Cys	Phe	Asp	Cys	Phe	Arg	Gln	Trp	Phe	Gly
BKV	Leu	Arg		Leu				Phe	Leu		Lys	Glu					Ile					Ile				Thr	Gln			
SV40	Met	Lys		Glu				Leu	Tyr		Lys	Asp					Val					Phe				Arg	Met			

JCV	Cys	Asp	Leu	Thr	Gln	Glu	Ala	Leu	His	Cys	Trp	Glu	Lys	Val	Leu	Gly	Asp	Thr	Pro	Tyr	Arg	Asp	Leu	Lys	Leu
BKV	Leu			Thr	Glu	Glu	Thr		Gln	Trp		Val	Gln	Ile	Ile		Glu		Pro	Phe					
SV40	Leu			Cys	Glu	Gly	Thr		Leu	Leu		Cys	Asp	Ile	Ile		Gln		Thr	Tyr					

FIGURE 5. Comparison of the small t proteins of JCV (Mad-1), BKV (Dun), and SV40. The proposed sequences for the small t proteins of the three polyomaviruses are aligned for maximum homology as described in the legend to Fig. 4. Reprinted from Frisque *et al.* (1984).

significantly reduced beyond the large T donor splice junction (69% for JCV × BKV; 53% for JCV × SV40). One might speculate that since small t is dispensable for the lytic growth of SV40 and BKV *in vitro* (Takemoto *et al.*, 1974; Shenk *et al.*, 1976), alterations in its unique coding sequences might be better tolerated than changes in sequences overlapping the multifunctional large T protein. Alternatively, small t might contribute to the host range of the polyomaviruses, and differences in its coding sequence might reflect a functional requirement in the various cells permissive for each virus.

2. Late Proteins

Little information has been reported for the JCV capsid proteins, in part due to the difficulty in propagating the virus in culture. Unlike the strong immunological cross-reactivity observed between the polyomavirus T antigens, antigenic similarities between intact capsid proteins are difficult to demonstrate (Penney and Narayan, 1973; Takemoto and Mullarkey, 1973; Padgett and Walker, 1976; Gardner, 1977). However, strong cross-reactions between the VP1 proteins of JCV, BKV, and SV40 can be detected following disruption of the virions (Shah *et al.*, 1977a,b).

Three capsid proteins, VP1, VP2, and VP3, are produced late in polyomavirus infections. A fourth protein, the agnoprotein, appears to interact with VP1 during the late stages of SV40 development (Margolskee and Nathans, 1983). It is apparent from Table IV that polypeptide VP1 is the most highly conserved protein of the three polyomaviruses. This protein, presumably, is encoded within the large open reading frame at the 3′ end of the late region; the other two capsid proteins are likely encoded within the second open reading frame found at the 5′ end of this region. By analogy with BKV and SV40, the JCV VP3 sequences would be a subset of the VP2 sequences. The C-termini of VP2 and VP3 overlap with the N-terminus of VP1; within the overlap is a stretch of 8 amino acids in BKV and SV40 that is missing in JCV (Frisque *et al.*, 1984; Figs. 6, 7).

Although considerable homology is detected within the first two-thirds of the polyomavirus agnoprotein sequences, the carboxy third is completely different in JCV, BKV, and SV40 (Fig. 8). It is interesting to speculate that since this protein may be involved in capsid assembly and since the three viral capsids are antigenically distinguishable, the differences observed in the C-termini may be significant.

C. Heterogeneity of the Genome

1. Defective Genomes

With one exception, JCV DNA isolated directly from the PML brain tissue is homogeneous in length (Rentier-Delrue *et al.*, 1981; Grinnell *et*

JCV	Met	Ala	Pro	Thr	Lys	Arg	Lys	Gly	Glu	Arg	N	N	N	N	N	N	N	N	Lys	Asp	Pro	Val	Gln	Val	Pro	Lys	Leu	Leu	Ile	Arg
BKV									Glu	Cys	Pro	Gly	Ala	Ala	Pro	Lys	Lys	Pro		Glu								Leu		Lys
SV40									Ser	Cys	Pro	Gly	Ala	Ala	Pro	Lys	Lys	Pro		Glu								Val		Lys
JCV	Gly	Gly	Val	Glu	Val	Leu	Glu	Val	Lys	Thr	Gly	Val	Asp	Ser	Ile	Thr	Glu	Val	Glu	Cys	Phe	Leu	Thr	Pro	Glu	Met	Gly	Asp	Pro	Asp
BKV			Val				Glu							Ala	Ile								Asn		Glu			Asp		
SV40			Ile				Gly							Ser	Phe								Asn		Gln			Asn		
JCV	Glu	His	Leu	Arg	Gly	Phe	Ser	Lys	Ser	Ile	Ser	Ile	Ser	Asp	Thr	Phe	Glu	Ser	Asp	Ser	Pro	Asn	Arg	Asp	Met	Leu	Pro	Cys	Tyr	Ser
BKV		Asn	Leu	Arg		Phe		Leu	Lys	Leu	Ser	Ala	Glu	Asn	Asp		Ser	Ser				Glu	Arg	Lys	Met					
SV40		His	Gln	Lys		Leu		Lys	Ser	Leu	Ala	Ala	Glu	Lys	Gln		Thr	Asp				Asp	Lys	Glu	Gln					
JCV	Val	Ala	Arg	Ile	Pro	Leu	Pro	Asn	Leu	Asn	Glu	Asp	Leu	Thr	Cys	Gly	Asn	Ile	Leu	Met	Trp	Glu	Ala	Val	Thr	Leu	Lys	Thr	Glu	Val
BKV	Thr																	Leu								Val	Gln			
SV40	Val																	Ile								Val	Lys			
JCV	Ile	Gly	Val	Thr	Ser	Leu	Met	Asn	Val	His	Ser	Asn	Gly	Gln	Ala	Thr	His	Asp	Asn	Gly	Ala	Gly	Lys	Pro	Val	Gln	Gly	Thr	Ser	Phe
BKV			Ile		Ser	Met	Leu		Leu		Ala	Gly	Ser		Lys	Val		Glu	His		Gly				Ile			Ser	Asn	
SV40			Val		Ala	Met	Leu		Leu		Ser	Gly	Thr		Lys	Thr		Glu	Asn		Ala				Ile			Ser	Asn	
JCV	His	Phe	Phe	Ser	Val	Gly	Gly	Glu	Ala	Leu	Glu	Leu	Gln	Gly	Val	Leu	Phe	Asn	Tyr	Arg	Thr	Lys	Tyr	Pro	Asp	Gly	Thr	Ile	Phe	Pro
BKV				Ala				Asp	Pro			Met					Met								Asp	Gly		Ile	Thr	
SV40				Ala				Glu	Pro			Leu					Ala								Ala	Gln		Val	Thr	
JCV	Lys	Asn	Ala	Thr	Val	Gln	Ser	Gln	Val	Met	Asn	Thr	Glu	His	Lys	Ala	Tyr	Leu	Asp	Lys	Asn	Lys	Ala	Tyr	Pro	Val	Glu	Cys	Trp	Val
BKV			Pro		Ala	Gln			Val				Asp				Tyr				Asn	Asn								
SV40			Ala		Val	Asp			Gln				Asp				Val				Asp	Asn								
JCV	Pro	Asp	Pro	Thr	Arg	Asn	Glu	Asn	Thr	Arg	Tyr	Phe	Gly	Thr	Leu	Thr	Gly	Gly	Glu	Asn	Val	Pro	Pro	Val	Leu	His	Ile	Thr	Asn	Thr
BKV				Ser	Arg				Ala						Phe												Val			
SV40				Ser	Lys				Thr						Tyr												Ile			
JCV	Ala	Thr	Thr	Val	Leu	Leu	Asp	Glu	Phe	Gly	Val	Gly	Pro	Leu	Cys	Lys	Gly	Asp	Asn	Leu	Tyr	Leu	Ser	Ala	Val	Asp	Val	Cys	Gly	Met
BKV									Gln								Ala		Ser			Val			Ala		Ile			Leu
SV40									Gln								Ala		Ser			Val			Val		Ile			Leu
JCV	Phe	Thr	Asn	Arg	Ser	Gly	Ser	Gln	Gln	Trp	Arg	Gly	Leu	Ser	Arg	Tyr	Phe	Lys	Val	Gln	Leu	Arg	Lys	Arg	Arg	Val	Lys	Asn	Pro	Tyr
BKV				Ser			Thr				Arg			Ala					Ile	Arg					Ser					
SV40				Thr			Thr				Lys			Pro					Ile	Thr					Ser					
JCV	Pro	Ile	Ser	Phe	Leu	Leu	Thr	Asp	Leu	Ile	Asn	Arg	Arg	Thr	Pro	Arg	Val	Asp	Gly	Gln	Pro	Met	Tyr	Gly	Met	Asp	Ala	Gln	Val	Glu
BKV							Ser								Gln								Tyr			Glu	Ser			
SV40							Ser								Gln								Ile			Ser	Ser			
JCV	Glu	Val	Arg	Val	Phe	Glu	Gly	Thr	Glu	Glu	Leu	Pro	Gly	Asp	Pro	Asp	Met	Met	Arg	Tyr	Val	Asp	Lys	Tyr	Gly	Gln	Leu	Gln	Thr	Lys
BKV					Phe	Asp	Gly			Arg								Ile			Ile		Lys	Gln			Leu	Gln		Lys
SV40					Tyr	Glu	Asp			Glu								Ile			Ile		Glu	Phe			Thr	Thr		Arg
JCV	Met	Leu																												
BKV		Leu																												
SV40		Gln																												

FIGURE 6. Comparison of the VP1 proteins of JCV (Mad-1), BKV (Dun), and SV40. The amino acid sequences of the VP1 capsid proteins of the three polyomaviruses are aligned for maximum homology as described in the legend to Fig. 4. In each viral DNA, there are two potential initiation codons for the VP1 protein which occur in the same reading frame. We have used the second methionine residue as the first amino acid in the protein sequence. Reprinted from Frisque *et al.* (1984).

JCV	Met	Gly	Ala	Ala	Leu	Ala	Leu	Leu	Gly	Asp	Leu	Val	Ala	Thr	Val	Ser	Glu	Ala	Ala	Ala	Ala	Thr	Gly	Phe	Ser	Val	Ala	Glu	Ile	Ala
BKV						Ala						Val		Ser																
SV40						Thr						Ile		Thr																
JCV	Ala	Gly	Glu	Ala	Ala	Ala	Thr	Ile	Glu	Val	Glu	Ile	Ala	Ser	Leu	Ala	Thr	Val	Glu	Gly	Ile	Thr	Ser	Thr	Ser	Glu	Ala	Ile	Ala	Ala
BKV							Ala				Gln	Ile			Leu						Ile		Ser							
SV40							Ala				Gln	Leu			Val						Leu		N							
JCV	Ile	Gly	Leu	Thr	Pro	Glu	Thr	Tyr	Ala	Val	Ile	Thr	Gly	Ala	Pro	Gly	Ala	Val	Ala	Gly	Phe	Ala	Ala	Leu	Val	Gln	Thr	Val	Thr	Gly
BKV						Gln	Thr					Ala				Gly		Ile							Ile				Ser	
SV40						Gln	Ala					Ser				Ala		Ile							Leu				Thr	
JCV	Gly	Ser	Ala	Ile	Ala	Gln	Leu	Gly	Tyr	Arg	Phe	Phe	Ala	Asp	Trp	Asp	His	Lys	Val	Ser	Thr	Val	Gly	Leu	Phe	Gln	Gln	Pro	Ala	<u>Met</u>
BKV	Ile		Ser	Leu			Val						Ser												Tyr			Ser	Gly	
SV40	Val		Ala	Val			Val						Ser												Tyr			Pro	Gly	
JCV	Ala	Leu	Gln	Leu	Phe	Asn	Pro	Glu	Asp	Tyr	Tyr	Asp	Ile	Leu	Phe	Pro	Gly	Val	Asn	Ala	Phe	Val	Asn	Asn	Ile	His	Tyr	Leu	Asp	Pro
BKV		Leu	Glu		Phe	Asn		Asp	Glu										Asn	Thr			Asn	Asn	Ile	Gln				
SV40		Val	Asp		Tyr	Arg		Asp	Asp										Gln	Thr			His	Ser	Val	Gln				
JCV	Arg	His	Trp	Gly	Pro	Ser	Leu	Phe	Ser	Thr	Ile	Ser	Gln	Ala	Phe	Trp	Asn	Leu	Val	Arg	Asp	Asp	Leu	Pro	Ala	Leu	Thr	Ser	Gln	Glu
BKV						Ser			Ala	Thr					Leu		His	Val	Ile	Arg	Asp		Ile		Ser	Ile				
SV40						Thr			Asn	Ala					Phe		Arg	Val	Ile	Gln	Asn		Ile		Arg	Leu				
JCV	Ile	Gln	Arg	Arg	Thr	Gln	Lys	Leu	Phe	Val	Glu	Ser	Leu	Ala	Arg	Phe	Leu	Glu	Glu	Thr	Thr	Trp	Ala	Ile	Val	Asn	Ser	Pro	Ala	Asn
BKV	Leu	Gln				Glu	Arg	Phe	Phe	Arg	Asp												Thr	Ile	Val		Ala		Ile	
SV40	Leu	Glu				Gln	Arg	Tyr	Leu	Arg	Asp												Thr	Val	Ile		Ala		Val	
JCV	Leu	Tyr	Asn	Tyr	Ile	Ser	Asp	Tyr	Tyr	Ser	Arg	Leu	Ser	Pro	Val	Arg	Pro	Ser	Met	Val	Arg	Gln	Val	Ala	Gln	Arg	Glu	Gly	Thr	Tyr
BKV	Phe			Tyr	Ile	Gln	Gln				Asp				Ile			Ser							Glu				Thr	Arg
SV40	Trp			Ser	Leu	Gln	Asp				Thr				Ile			Thr							Asn				Leu	Gln
JCV	Ile	Ser	Phe	Gly	His	Ser	Tyr	Thr	Gln	Ser	Ile	Asp	Asp	Ala	Asp	Ser	Ile	Gln	Glu	Val	Thr	Gln	Arg	Leu	Asp	Leu	Lys	Thr	N	N
BKV	Val	His				Thr		N	N	Ser			Asp					Glu	Glu			Gln		Met	Asp	Leu	Arg	Asn	Gln	Gln
SV40	Ile	Ser				Thr		Asp	N	Asn			Glu					Gln	Gln			Glu		Trp	Glu	Ala	Gln	Ser	Gln	Ser
JCV	Pro	Asn	Val	Gln	Ser	Gly	Glu	Phe	Ile	Glu	Arg	Ser	Ile	Ala	Pro	Gly	Gly	Ala	Asn	Gln	Arg	Ser	Ala	Pro	Gln	Trp	Met	Leu	Pro	Leu
BKV	N	Ser		His							Lys	Thr	Ile									Thr								
SV40	Pro	Asn		Gln							Lys	Phe	Glu									Thr								
JCV	Leu	Leu	Gly	Leu	Tyr	Gly	Thr	Val	Thr	Pro	Ala	Leu	Glu	Ala	Tyr	Glu	Asp	Gly	Pro	Asn	Lys	Lys	Lys	Arg	Arg	N	N	N	N	N
BKV							Thr			Pro			Glu								Gln				Arg	Val	Ser	Arg	Gly	Ser
SV40							Ser			Ser			Lys								Lys				Lys	Leu	Ser	Arg	Gly	Ser
JCV	N	N	Lys	Glu	N	Gly	Pro	Arg	Ala	Ser	Ser	Lys	Thr	Ser	Tyr	Lys	Arg	Arg	Ser	Arg	Ser	Ser	Arg	Ser						
BKV	Ser	Gln		Ala	Lys		Thr	Arg			Ala		Thr	Thr	Asn				Ser											
SV40	Ser	Gln		Thr	Lys		Thr	Ser			Ala		Ala	Arg	His				Asn											

FIGURE 7. Comparison of the late structural proteins VP2 and VP3 of JCV (Mad-1), BKV (Dun), and SV40. The proposed sequences for the VP2 and VP3 proteins of the three polyomaviruses are aligned for maximum homology as described in the legend to Fig. 4. In each virus, VP3 is encoded by the C-terminal sequences of VP2. The first methionine residue in the VP3 protein is underlined. Reprinted from Frisque *et al.* (1984).

A

JCV	Met	Val	Leu	Arg	Gln	Leu	Ser	Arg	Lys	Ala	Ser	Val	Lys	Val	Ser	Lys	Thr	Trp	Ser	Gly	Thr	Lys	Lys	Arg	Ala	Gln	Arg	Ile	Leu	Ile
BKV					Gln				Gln						Gly	Lys	Thr		Thr	Gly	Thr			Arg				Ile	Phe	Ile
SV40					Arg				Gln						Arg	Arg	Ser		Thr	Glu	Ser			Thr				Leu	Phe	Val

JCV	Phe	Leu	Leu	Glu	Phe	Leu	Leu	Asp	Phe	Cys	Thr	Gly	Glu	Asp	Ser	Val	Asp	Gly	Lys	N	Lys	Arg	Gln	Arg	His	Ser	Gly	Leu	Thr	Glu
BKV		Ile			Leu			Glu			Arg				Ser					Asn		Ser	Thr	Thr	Ala	Leu	Pro	Ala	Val	Lys
SV40		Val			Leu			Gln			Glu				Thr					Arg		Lys	Pro	Glu	Arg	Leu	Thr	Glu	Lys	Pro

JCV	Gln	Thr	Tyr	Ser	Ala	Leu	Pro	Glu	Pro	Lys	Ala	Thr
BKV	Asp	Ser	Val	Lys	Asp	Ser	N	N	N	N	N	N
SV40	Glu	Ser	N	N	N	N	N	N	N	N	N	N

B

JCV	Lys	Leu	Met	Leu	Val	Trp	Gly	Asp	Pro	Phe	Leu	Thr	Phe	Leu	Glu	Arg	Lys	Ile	Leu	Lys	Gln	Lys	Thr	Leu	Asp	Met	Asp	Gln	Ala	Leu
BKV	Asn	Ile	Ile	Tyr	Ala			Asn	Val			Ile	Leu	Gln		Lys	Arg		Gln		Leu									

JCV	Asn	His	Asn	His	Asn	Ala	Phe	Pro	Arg	Ser	Gln	Lys	N	N	Pro	Leu	Val	N	N	N	N	N	Gln	Thr	His	Arg	Lys	Thr	Ala	Leu
BKV		Pro					Leu		Lys			Ile	Leu	Gln			Leu	Lys	Ile	Pro	Lys	Gly			Pro	Ile	Val	Lys	Ser	Cys

JCV	Phe	Thr	Ser	Val	Lys	Ala	Phe	Asn	Val	Ser	Lys	Asn	Gln	Arg	Pro	Leu	Pro	Gln	Asn	Asn	N	N	N	N	N	N	N	N	N	N
BKV	Ile	Cys	N					Ser		Leu		Gly	Leu	Lys	His	His					Thr	Ser	Leu	Lys	Val	Ala	Tyr	Thr	Lys	Ala

JCV	N	N	N	N	Cys	Asn	Cys	N	N	N	N	N	N
BKV	Ala	Phe	Ile	Lys		Ile		Thr	Ile	Lys	Ala	Pro	Val

FIGURE 8. Comparisons of two potential proteins that might be encoded within the late leader sequences (A) or the 3′ ends of the early regions (B) of JCV (Mad-1), BKV (Dun), and SV40. (A) Comparison of the agnoproteins thought to be encoded by the late leader sequences of the three polyomaviruses. Agnoprotein has been identified in SV40. Amino acid sequences are aligned to show maximum homology as described in the legend to Fig. 4. (B) Comparison of polypeptide sequences which may be translated from the 3′ ends of the JCV and BKV early regions. The putative SV40 protein sequence (98 amino acids) is not included since little homology is evident with the two sequences of the human viruses. Reprinted from Frisque *et al.* (1984).

al., 1982, 1983a); however, several investigators have reported that DNA isolated from JCV propagated in culture is variable in size (Osborn *et al.*, 1974; Howley *et al.*, 1976; Martin *et al.*, 1979; Grinnell *et al.*, 1982). The most heterogeneous populations of DNA contain at least four discrete size classes of molecules, the smallest being 88% of full-length JCV DNA (Frisque *et al.*, 1979; Martin *et al.*, 1983). These shorter molecules are encapsidated (Martin *et al.*, 1983) and account for the early underestimates of the length of JCV DNA (Osborn *et al.*, 1974; Takemoto *et al.*, 1979a) and of the buoyant density of JC virions (Grinnell *et al.*, 1982, 1983a).

As with the other polyomaviruses, the multiplicity of infection (m.o.i.) and the cell population used for propagating the virus, influence the degree of heterogeneity observed in JCV DNA (Frisque *et al.*, 1979; Takemoto *et al.*, 1979a; Miyamura *et al.*, 1980; Martin *et al.*, 1983). Many strains of JCV, when passaged in spongioblast-rich cultures of PHFG cells at low m.o.i. (< 0.1 infectious unit/cell), yield DNAs of uniform length. However, the use of astrocyte-rich cultures and/or higher m.o.i. result in more heterogeneous preparations of the viral genomes. This latter finding is also observed when JCV is grown in human embryonic kidney (HEK) and human amnion cells. JCV adapted to grow in HEK cells has acquired the ability to form plaques on these cells although prototype JCV Mad-1 does not form plaques on any cells tested. Analysis of the DNA extracted from plaque-purified virus failed to reveal the presence of full-length infectious genomes; instead, two or more species of defective molecules were found which had deleted sequences from the early or late regions (Miyamura *et al.*, 1980). Presumably, these molecules complement each other for growth. Similar findings have been reported for the RF strain of BKV (Pater *et al.*, 1980). Recent sequence analysis of one of these defective JCV genomes (JC-HEK-A) has revealed that complex rearrangements have occurred within the DNA; new transcriptional signals have been added, portions of the T and VP1 proteins have been translocated to the regulatory region, and two additional replication origins have been acquired. Remarkably, JC-HEK-A has retained the ability to produce a functional T protein (T. Miyamura, personal communication).

It is interesting to note that the HEK-adapted virus, unlike prototype JCV, has the ability to inhibit the growth of BKV in HEK cells (T. Miyamura, personal communication). The interference appears to be specific for BKV (adenovirus infection is not affected) and to involve early steps in BKV replication (BKV T antigen and replication are inhibited by the coinfection).

In addition to the influences of m.o.i. and cell type, a third factor appears to contribute to the heterogeneity observed in some JCV DNA preparations. Martin *et al.* (1983) found that when a number of JCV isolates were grown under identical conditions, only about half of them yielded DNAs which had sustained deletions. The deletions were not random but usually included the region encompassing the *Bam*HI site

(0.505 m.u.; T protein coding sequences). These investigators proposed that the differential susceptibility to deletion among JCV isolates was a consequence of natural genetic variation in the virus.

Originally, there was some question whether the shorter molecules in heterogeneous DNA preparations represented defective viruses or viable mutants. Early comparisons of the infectivity of heterogeneous and homogeneous DNA preparations failed to detect any significant differences; however, this may have been due to the insensitivity of the assay system (fluorescent cell assay) or to the ability of defective molecules to complement one another for growth (Frisque *et al.*, 1979). More definitive experiments using cloned DNA indicated that at least some of the smaller molecules are defective (Yoshiike *et al.*, 1982; Frisque, unpublished results).

2. Variant Genomes

Most isolations of JCV have been from the brain tissue of PML patients; a few have been from the urine of PML patients, renal transplant recipients, or normal individuals. Viral antigens and DNA have also been detected in the kidneys, lungs, lymph nodes, and spleens of some of these individuals (Chesters *et al.*, 1983; Grinnell *et al.*, 1983c; McCance, 1983; Dörries, 1984). McCance (1983) found that JCV DNA was focally distributed in the cortex and medulla of nearly 10% of normal kidneys. Different kidneys contained from 0.3 to 40 genome equivalents per cell with most of the DNA appearing as full-length, unintegrated molecules (Chesters *et al.*, 1983; McCance, 1983). Some differences in restriction enzyme patterns were noted when the kidney DNAs were compared to prototype Mad-1 DNA. Analysis of brain material from these normal individuals indicated that JCV sequences were not present in this tissue.

A number of studies have examined the JCV DNA present in different PML patients. Several DNAs have been molecularly cloned directly from brain tissue and compared by restriction enzyme analysis (Rentier-Delrue *et al.*, 1981; Grinnell *et al.*, 1983a,b; Martin and Frisque, unpublished results). Using the endonuclease *Pvu*II, these clones were shown to differ within the regulatory (noncoding) sequences (0.67 to 0.73 m.u.). Some of these DNAs had also acquired a second *Hpa*I site (0.895 m.u.) and lost about 75 np in the region spanning 0.14 to 0.235 m.u. Despite these alterations, the DNAs were equally infectious in PHFG cells and therefore represent variants of the prototype strain of JCV.

Dörries (1984) has conducted a similar analysis of JCV DNA recovered from the brain and kidney tissue of a single PML patient. While most restriction enzyme cleavage sites remained the same as those in Mad-1 DNA, some differences were observed, including changes in the 0.67 to 0.73 m.u. region. Minor heterogeneity in cleavage patterns was apparent in both brain and kidney DNAs; however, within each organ the sizes of the DNAs remained constant (~ 5150 np for brain DNA, ~ 5030 np for kidney DNA). It was suggested that the initial infection with one JCV subtype was followed by the development of heterogeneous DNA molecules.

Because of the importance of the regulatory region in the biology of the polyomaviruses and because of the apparent hypervariability of these sequences in the JCV genome, a number of cloned JCV DNAs, isolated either from virus passaged in culture or directly from brain, have been sequenced in the region of 0.67 to 0.73 m.u. (nucleotides 5113 to 279 in Mad-1 DNA; Martin *et al.*, 1985). The strains of JCV represented included Mad-4, an oncogenic variant (Padgett *et al.*, 1977c); Mad-7 and 8, viruses isolated from the urine and brain of a single PML patient; and Mad-11, an antigenic variant (Padgett and Walker, 1983). A total of 13 cloned DNAs from 6 different people were compared. While the results confirmed that the JCV regulatory region is indeed hypervariable, two general types of JCV genomes could be distinguished (Table V). The four type I regulatory regions were identical and were represented by Mad-1 DNA. These DNAs have a duplicated TATA box, lack the PyPyCCXCCC pro-

TABLE V. Hypervariability in the Regulatory Region of JCV DNA

JCV DNA[a]	Type[b]	No. of TATA sequences	No. of PyPyCCXCCC	No. of CAAT sequences	Length of tandem repeat (np)	No. of core sequences[c]
Mad-1-Tc	I	2	0	4	98	2
Mad-1-Br	I	2	0	4	98	2
Her-1-Br	II	1	1	2	73	1
Mad-4-Tc	?[d]	1	0	4	79	2
Mad-7B	I	2	0	4	98	2
Mad-7D	II	1	1	2	28	0
Mad-8-Br	II	1	1	2	83	2
Mad-8AA-TC	II	1	1	2	84	2
Mad-8FF-TC	I	2	0	4	98	2
Mad-9-Br	II	1	1	2	74	1
Mad-9.7-TC	II	1	1	2	91	1
Mad-11-Br	II	1	1	1	40	1
Mad-II.3-TC	II	1	2	2	51 (nTR)[e]	1

[a] Full-length variant JCV DNAs were cloned into pBR322 through their unique *Eco*RI site (*Bam*HI in the case of Mad-9.7-TC). DNAs were isolated from virus passaged in tissue culture (TC) or directly from brain tissue (Br).
[b] Type I DNA is arbitrarily defined as that of prototype (Mad-1). Type II DNA contains extra *Pvu*II sites and/or short insertions or deletions in the regulatory region.
[c] The core is a conserved sequence found in a number of enhancer elements.
[d] Mad-4-TC has one TATA sequence like the type II molecules but has not acquired the PyPyCCXCCC sequence found in these DNAs.
[e] The repeats of Mad-11.3-TC are nontandem; they are separated by 55 np.

moter element, and contain a 98-np tandem repeat. Type II JCV regulatory regions (eight DNAs) all differ from one another but each has diverged from type I sequences at the same nucleotide (No. 36 in Mad-1). Surprisingly, all have lost one copy of the TATA sequences (the one most distal to the early mRNA initiation sites) and have acquired the PyPyCCXCCC sequence. Presumably, these rearrangements occur via recombination with host sequences. It will be important to determine whether these alterations contribute to the apparent adaptation of JCV to function in brain cells.

The genome of the oncogenic variant, Mad-4, is an exception to the type I–II classification. Although the upstream Goldberg–Hogness sequence has been lost, the PyPyCCXCCC sequence has not been added. It is not known whether this unique arrangement contributes to the ability of Mad-4 to induce tumors in the pineal gland of the hamster, although alterations in a similar region of BKV are known to affect the oncogenic potential *in vitro* and *in vivo* (Watanabe *et al.*, 1979; Watanabe and Yoshiike, 1982).

The DNA, Mad-11-Br, representing the antigenic variant, Mad-11, has two unusual features. First, to the late side of its tandem repeats is a unique set of 49 nucleotides not found in the DNA of other JCV strains. Within this stretch of DNA are 27 nucleotides that are shared with a 43-np insertion in the corresponding region of BKV(WT) [Seif *et al.*, 1979; Yang and Wu, 1979; the BKV sequence is considered an insertion since it is absent in BKV(Dun)]. Second, missing in the Mad-11-Br DNA are potential initiation sites for late mRNAs that were found in most other JCV DNAs sequenced. Although it seems likely that antigenic variations in this isolate result from differences in the late coding sequences, it is possible that alterations in the regulatory region might be partially responsible. For example, changes in the 5′ termini or splice sites of the late mRNAs might alter the levels of expression of structural proteins and/or the agnoprotein. This in turn might affect the assembly of the capsid, thereby exposing different antigenic determinants.

A 17-np sequence was identified in this study which was common to all 13 JCV DNAs analyzed. It included the sequence 5′-AACCA-3′ which is conserved in five different, infectious isolates of BKV (Seif *et al.*, 1979; Yang and Wu, 1979; Pater *et al.*, 1983; Frisque, unpublished results) and a second pentanucleotide, 5′-AGGGA-3′, which is repeated seven times in the regulatory region of prototype Mad-1.

There were no discernible features which differentiated DNAs derived from diseased brain tissue from those obtained from virus isolated and passaged in culture. In none of the DNAs were there alterations, such as repetitions of the replication origin or major substitutions of host DNA sequences, which occur in SV40 or BKV DNA upon undiluted passage in permissive cells (Tooze, 1981) or in JCV DNA adapted to grow in HEK cells (T. Miyamura, personal communication).

D. DNA Transfection

A great deal of effort has been expended in trying to understand the restricted host range of JCV. Since most cells do not even produce T antigen following JCV infection, it appears that restriction occurs very early in the lytic cycle. One way to bypass early blocks to viral infection has been by DNA transfection techniques, which avoid the stages of adsorption, penetration, and uncoating. The calcium phosphate and DEAE-dextran procedures have demonstrated the infectivity of JCV DNA for PHFG cells (Frisque *et al.*, 1979; Miller *et al.*, 1983b); however, DNA transfection has not expanded the host range of JCV beyond that exhibited by intact virus (Frisque *et al.*, 1979). These results along with recent sequence data point to the possibility that JCV's restricted host range involves the inability to express the early viral proteins because of a weak or defective transcriptional control signal(s). If this were true, then supplying a functional T protein in *trans* in human cells might allow a JCV infection to proceed through an entire lytic cycle. This approach has been used successfully with SV40. By transforming normally permissive CV-1 cells with replication-defective SV40 DNAs (Gluzman *et al.*, 1980a; Gluzman, 1981), the COS cell line, containing a functional T protein, was produced. These cells can be used to propagate SV40 mutants unable to express their own early proteins. A similar approach has been tried with JCV (Major, 1983a; Major and Matsumura, 1984; Major *et al.*, 1985). HEK and PHFG cells were transformed by the origin-defective SV40 mutants and tested for the ability to support JCV growth. (Since PHFG cells are already permissive for JCV, it was not important, initially, to obtain cells expressing a complementing T protein; it was important, however, to obtain a permanent cell line that remained permissive for JCV.) Both types of transformants produced a replication-proficient SV40 T protein and, following transfection with JCV DNA, the transformed brain cells (SVG, a continuous astroglial cell line) supported a lytic cycle. Although a greater number of SVG than PHFG cells appeared to support JCV DNA replication, the amount of virus produced as judged by hemagglutination titers was about the same in both types of cells. Following the same approach as outlined above, origin-defective JCV DNAs have been constructed and transfected into PHFG cells (Mandl and Frisque, unpublished results). Cells transformed by two of these mutants were established as permanent cell lines (POJ-1 and POJ-2), and were shown to support JCV and SV40 DNA replication. Preliminary results suggest that JCV undergoes a lytic cycle in these cells. The development of continuous permissive cell lines will greatly facilitate JCV studies.

While several studies have suggested that certain features of the JCV regulatory region might contribute to the restricted activity of the virus *in vitro*, recent experiments indicate that the antigen coding sequences may have an even greater influence on this behavior. To pursue this possibility, hybrid viral genomes have been constructed by replacing the regulatory region of one polyomavirus with that of another. Surprisingly, those hybrids containing a JCV regulatory region joined to SV40 or BKV protein coding sequences show significant activity (in contrast to the reverse hybrids—BKV or SV40 regulatory regions ligated to JCV protein coding sequences) in transformation and lytic assays (Chuke, Bollag, Walker, and Frisque, unpublished results). The generation of viable virus following transfection of PHFG cells with the hybrid DNAs indicates that the JCV regulatory signals and the SV40 or BKV proteins are capable of undergoing a productive interaction, contrary to the findings of Major and Matsumura (1984). Furthermore, the data suggest that the restricted behavior of JCV in culture may be attributed primarily to deficiencies in its T antigen coding sequences rather than in its regulatory sequences. Further support for this idea comes from sequence analysis: although most of the JCV and SV40 large T sequences are highly conserved, they differ at their C-termini and this is the portion of T antigen shown to affect SV40 host range properties (Manos and Gluzman, 1985; Tornow *et al.*, 1985).

IV. ONCOGENICITY

A. Experimental Animals

1. Hamsters

As soon as JCV was identified as a polyomavirus, it could be predicted that it would be oncogenic in experimental animals. This expectation was borne out in the first experimental test. Walker *et al.* (1973b) found that 10^6 $TCID_{50}$ of JCV inoculated intracerebrally and subcutaneously into newborn Syrian hamsters induced central nervous system (CNS) tumors in 83% of 63 animals but no subcutaneous tumors. The tumors became evident 4–6 months after inoculation. They were malignant and rapidly progressed to kill the animal. There were no distant metastases. The tumors occurred most frequently in the cerebellum, but tumors were found in all areas of the brain. Most brains contained multiple tumors that frequently differed in histological type. The tumors were transplantable subcutaneously in series in hamsters and some could be established in cell cultures. Small amounts of JCV could sometimes be isolated from tumor tissue, but cells in culture did not contain or release infectious virus and did not contain capsid antigen. Cultured cells did contain a nuclear T antigen demonstrable by immunofluorescence using serum from the tumor-bearing hamsters. JCV could be rescued from tumor cells in culture by fusing the tumor cells with human fetal glial cells.

Clonal cell lines were isolated from some of the established tumor cell cultures. One tumor cell line, HJC-15, was derived from a malignant glioma of mixed astrocytic and ependymal cells. Two cell clones from the HJC-15 cell line were selected for study of the physical state of the JCV genome in tumor cells (Wold *et al.*, 1980). Both clonal cell lines were found to contain multiple copies of JCV genomes that were full length or nearly full length. The data indicated that the viral DNA in both cell lines was integrated into cellular DNA and each clonal line contained multiple independent sites for integration that were different in the two clonal cell lines. The viral DNA was integrated in a tandem head-to-tail orientation and no free viral DNA was detectable.

Subsequent experiments in hamsters have been reviewed and summarized by ZuRhein (1983). Table VI summarizes work in hamsters and primates. Multiple experiments (ZuRhein and Varakis, 1975, 1979; Padgett *et al.*, 1977c) have confirmed that JCV is highly oncogenic in the hamster CNS. Medulloblastomas of the cerebellum were the most common tumors induced by intracerebral inoculation within 24 hr of birth. Up to 95% of animals in individual experiments developed these lesions. Medulloblastomas were also common in hamsters inoculated *in utero* and at 3 days after birth, but were not among tumors produced in animals inoculated at an older age (ZuRhein, 1983). The medulloblastomas were characterized by extreme cellularity, numerous mitoses, and scattered giant cells (ZuRhein and Varakis, 1975). Thalamic gliomas were the next most common tumor. Many of these resembled glioblastomas, malignant astrocytomas, and primitive neuroectodermal tumors seen in humans. Tumors of the olfactory–frontal region, many of which were primitive neuroectodermal tumors, were found in about 8% of animals (ZuRhein, 1983).

TABLE VI. Neurooncogenicity of JCV in Experimental Animals

Experimental animal	Route of inoculation	Major types of tumors[a]
Newborn hamster[b]	Intracerebral or intracerebral and subcutaneous	Medulloblastoma Glioblastoma Malignant astrocytoma Primitive neuroectodermal Central neuroblastoma Pineocytoma Plexus papilloma Papillary ependymoma
	Intraocular	Retinoblastoma Peripheral neuroblastoma
Adult owl monkey[c]	Intracerebral, intravenous, and subcutaneous	Malignant astrocytoma Malignant tumor—both astrocytic and neuroblastic cell types
Adult squirrel monkey[d]	Intracerebral	Malignant astrocytoma

[a] See the text for references.
[b] Syrian golden hamster, *Mesocricetus auratus.*
[c] *Aotus trivirgatus.*
[d] *Saimiri sciureus.*

An experiment using intraocular inoculation of JCV into newborn hamsters (Ohashi *et al.*, 1978) produced retinoblastomas in 20% of inoculated animals, but an additional result was the induction of neuroblastomas. Neuroblastomas had been seen sporadically after intracerebral and after combined intraperitoneal and subcutaneous inoculation, but after intraocular injection 10 of 31 hamsters developed neuroblastomas (Varakis *et al.*, 1976, 1978). The tumors occurred paravertebrally in the neck, thorax, and abdomen and in the mesentery and adrenal glands. Metastases were found in the liver, bone marrow, and lymph nodes. The neuroblastomas could be established in cell cultures and transplanted subcutaneously to weanling hamsters where they grew rapidly and gave extensive remote metastases.

Most experiments studying the oncogenic effect of JCV in hamsters have employed the prototype strain, Mad-1. After intracerebral inoculation into newborn hamsters, this strain sometimes induced tumors of the pineal gland (Varakis and ZuRhein, 1976), but this was a rare event. However, in an experiment comparing three JCV strains, 10 of 22 animals inoculated with strain Mad-4 developed pineocytomas (Padgett *et al.*, 1977c). These tumors became quite large and the enzyme hydroxyindole-*O*-methyltransferase, which is localized in the mammalian pineal gland, was demonstrated in each of seven tumors tested, and its activity correlated with the degree of differentiation of the tumor (Quay *et al.*, 1977). Subsequent experiments using later culture passages of strain Mad-4 have not produced such a high incidence of pineocytomas, but this experiment suggests that the oncogenic potential and the target cells of different isolates of JCV may vary somewhat. However, Nagashima *et al.* (1984) studied the neurooncogenicity of a strain of JCV isolated in Japan (Tokyo-1 strain), and found that intracerebral inoculation of the Tokyo-1 strain into newborn hamsters resulted in tumors quite similar in number and histological type to the prototype strain, Mad-1.

The oncogenicity of JCV in hamsters is remarkable in its efficiency of tumor production in the nervous system and in the variety of tumors induced. The neurooncogenicity of mouse polyoma virus in hamsters is limited to meningeal sarcomas, and the tumors induced by SV40 have been limited to ependymomas or plexus papillomas (Ikuta and Kumanishi, 1973). The meningeal sarcomas, ependymomas, and plexus papillomas that are characteristic of mouse polyoma virus and SV40 do occur in JCV-inoculated hamsters, but they are relatively uncommon. Early experiments with BKV yielded only papillary ependymomas and plexus adenomas in the hamster CNS, but Uchida *et al.* (1979) found that in addition to ependymomas and plexus adenomas, some central neuroblastomas and pineocytomas were induced. Thus, BKV is closer to JCV in its neurooncogenicity than are mouse polyoma virus and SV40. On the other hand, JCV has appeared to be less oncogenic than the other polyomaviruses by the subcutaneous or intraperitoneal route (Padgett and

Walker, 1976), but JCV has not been extensively tested by these routes and no direct comparisons have been made with the other viruses.

2. Primates

Since JCV is a primate virus, it was important to test its oncogenicity in primates. In one experiment, adult owl monkeys with low or undetectable antibody levels against SV40 were inoculated with JCV, BKV, or SV40 by intracerebral, subcutaneous, and intravenous routes. Two of four monkeys inoculated with JCV developed brain tumors after incubation periods of 16 and 25 months (London *et al.*, 1978). One tumor was an astrocytoma, grade 3 to 4, and the other was a malignant tumor containing both astrocytic and neuroblastic cell types. No extraneural tumors were found, and no tumor of any kind developed in three monkeys inoculated with BKV or in four inoculated with SV40 during 3 years of observation.

Additional experiments have shown (Miller *et al.*, 1984) that JCV induces brain tumors in about 25% of adult owl monkeys. Incubation periods range from 16 to 36 months. The tumors are highly malignant and rapidly progressive and are usually classifiable as astrocytomas. In impression smears made from tumor tissue, many cells contain nuclear T antigen (London *et al.*, 1978) and tumor cells established in culture contain nuclear T antigen (Major, 1983b; Major *et al.*, 1984). Miller *et al.* (1984) found that the JCV genome was integrated into the cellular DNA of each of six tumors examined. Integration occurred at a limited number of sites, indicating a clonal origin for tumors, but the tumors did not have integration sites in common. In all but one of the tumors, there was tandem, head-to-tail integration of two or more copies of the JCV genome. In the one tumor so analyzed, there appeared to be a complete copy of the JCV genome.

Squirrel monkeys were found to be the second nonhuman primate in which JCV induces CNS tumors (London *et al.*, 1983). Cerebral tumors developed in four of six adult squirrel monkeys after bilateral intracerebral inoculation. The incubation period ranged from 14 to 19 months. There were single tumors in each animal and each of the tumors was a malignant astrocytoma.

B. *In Vitro* Transformation

Although JCV readily induces a variety of tumors in animals, it has been difficult to demonstrate its transforming activity *in vitro*. Early studies indicated that JCV would transform PHFG and human vascular endothelial cells (Fareed *et al.*, 1978; Walker and Padgett, 1978). While these cells were able to support JCV replication, a certain proportion of the cells began to exhibit a transformed phenotype following viral infection. These cells expressed viral T antigen, showed an altered morphology, tended to form dense foci, and had increased life spans (up to 20 subcultures). It is not known whether nonpermissive cells within these populations were transformed or whether permissive cells were transformed by defective viruses in the JCV pool. Later transformation studies were able to establish permanent cell lines using JCV virions or JCV DNA. Frisque *et al.* (1980) transformed primary hamster brain (PHB) cells with four strains of JCV virions and with JCV DNA from prototype Mad-1. These cells demonstrated a number of properties characteristic of cells transformed by the polyomaviruses; compared to untransformed PHB cells, the JCV transformants had increased growth rates in high and low serum concentrations, higher saturation densities and cloning efficiencies on plastic and in methylcellulose, and elevated levels of plasminogen activator secretion. Cellular tumorigenicity appeared to correlate with the ability to grow in semisolid media. The only significant difference between JCV- and SV40-transformed PHB cells was a flatter morphology and greater retention of actin cables in the JCV transformants. Immunoprecipitation experiments revealed the presence of several species of JCV T antigen in these cells; the two major species had molecular weights of about 92K and 94K or nearly 2K less than the corresponding SV40 proteins. Both sets of viral transformants contained a cellular middle T protein as detected by immunoprecipitation with anti-SV40 T serum or monoclonal antibody to cellular middle T. Only in the SV40-transformed cells could a small t protein be detected; testing with a number of SV40 and JCV tumor antisera failed to detect the analogous protein in JCV transformants. In these and other experiments, it has been noted that the intensity of nuclear fluorescence and the amount of immunoprecipitable T protein in SV40 tumor and transformed cells were generally greater than in the corresponding JCV or BKV cells. These observations are in agreement with quantitative measurements of the different T proteins in tumor cells (Beth *et al.*, 1977) and with the determination that JCV T antigen is more labile than SV40 T antigen following extraction and purification from cells (Cikes *et al.*, 1977).

As noted earlier, permissive PHFG cells have been transformed by origin-defective JCV DNAs. The resulting cell lines, POJ-1 and POJ-2, like their COS-1 counterparts, only contain the smallest T antigen species observed in hamster transformants (Frisque *et al.*, 1980; Mandl and Frisque, unpublished results). These results support the idea that replication-proficient viral DNAs are required for the DNA rearrangements that lead to the formation of genes encoding the larger T antigen species (May *et al.*, 1981, 1984; Manos and Gluzman, 1985).

Howley *et al.* (1980) have reported the transformation of human amnion cells using Mad-1 DNA cloned into pBR322. Approximately 5 foci/μg DNA were obtained following transfection of 1×10^6 cells by the calcium phosphate method. The JCV DNA apparently existed as episomes within the transformed cells. Since the DNA used to construct the clone was shorter than full-length Mad-1 DNA and because it failed to induce V antigen in the semipermissive amnion cells, it is possible that transformation was caused by a defective JCV genome.

Several groups of investigators have analyzed cell lines derived from JCV-induced tumors. Simmons and Martin (1978) found that hamster tumor cells obtained from JCV-, SV40-, and BKV-inoculated animals all contained T antigens of molecular weights 94K and 97K. These experiments also failed to detect a JCV small t protein. In later experiments, Simmons (1980) found that the tau or middle T proteins in these cells were indistinguishable when compared by two-dimensional peptide mapping, and that they were related to the tau antigens in monkey, mouse, and rat cells.

Major (1983b) has studied the early proteins in JCV-induced owl monkey tumor cell lines. Of four different lines, only one was shown to contain T antigen (94K) by immunoprecipitation or immunofluorescence experiments; small t was not detected. A phenotypic analysis, similar to that conducted for JCV-transformed hamster cells (Frisque *et al.*, 1980), was performed on the owl monkey tumor cells (Major *et al.*, 1984). Two properties associated with a functional small t protein (in SV40), plasminogen activator activity and actin cable disorganization, were examined. The finding of increased plasminogen activator activity and disruption of actin cables indicated that either a JCV small t antigen was present (though undetected) in these cells or that JCV T antigen contributed to these properties. Because of the proposed role for the SV40 t antigen in the transformation process (under certain conditions), it will be important to resolve the question of whether a small t is present in JCV-transformed cells.

The p53 protein (cellular middle T or tau antigen) has also been detected in owl monkey tumor lines (Major *et al.*, 1984). Using a monoclonal antibody directed against this protein, it appeared that a complex of p53 and JCV T protein did not form in these cells. In contrast, the same antibody recognized the analogous complex in SV40-transformed cells. The significance of this finding remains to be determined.

1. Integration

Cells transformed *in vitro* or *in vivo* by the polyomaviruses almost invariably contain copies of the viral genome integrated into the cellular chromosomes (Tooze, 1981). Analysis of hamster and monkey tumor cell lines induced by JCV indicates that integration patterns resembled those observed for the other polyomaviruses. In two clones of the hamster tumor line HJC-15, JCV DNA was detected at multiple, independent sites in the cellular genome (Wold *et al.*, 1980). The DNA was arranged in a tandem head-to-tail fashion and free DNA was not found. Both cloned cell lines expressed T antigen; however, the clone containing fewer copies of the viral genome (4–5 copies/cell versus 9–10 copies/cell) failed to induce tumors upon reintroduction into hamsters.

Similar findings have been made in cell lines derived from owl monkey tumors (Miller *et al.*, 1983a, 1984). Again a tandem head-to-tail arrangement of the JCV DNA was observed and free DNA was not detected. Integration was at a limited number of sites in all but one tumor, suggesting a clonal origin for the tumors. Although most tumors contained at least two copies of JCV DNA per cell, cells from one tumor appeared to have just a single complete copy of the genome.

V. ASSOCIATION WITH DISEASE

JCV infects millions of people throughout the world, but the full range of its disease-producing potential is not yet well defined. The early accumulation of antibody against JCV in childhood clearly indicates that most primary infections occur in children, but little is known about the nature of the primary infection in either children or adults. It seems likely that in otherwise healthy persons it is subclinical or so mild that it is not distinguishable from other minor illnesses. Yet, since primary infections in otherwise healthy persons have not been identified and studied, there is no certainty that they are entirely innocuous, and this remains an area in need of study (Walker and Padgett, 1983b).

JCV is well established as an opportunistic pathogen causing the destructive and usually fatal brain disease, PML, that is found in persons who have impaired immunity.

The association of a papovavirus with PML was first demonstrated by ZuRhein and Chou (1965) and by Silverman and Rubinstein (1965). Both groups used electron microscopy to study PML and found particles characteristic of papovavirus virions in the nuclei of the abnormal oligodendrocytes that are the outstanding feature of PML. Additional morphological evidence that PML lesions contain a papovavirus was provided by Howatson *et al.* (1965) who used negative staining and electron microscopy to demonstrate that crude extracts of formalin-fixed, diseased brain tissue from three PML patients contained particles very similar in size and structure to the virions of mouse polyoma virus. Schwerdt *et al.* (1966) added to this evidence by extracting, concentrating, and partially purifying particles from unfixed brain tissue of a case of PML. Their concentrated particles were similar in size, morphology, and density to mouse polyoma virus and to SV40.

These studies provided the stimulus and basis for numerous attempts to isolate and cultivate a virus from diseased brain tissue. Such efforts used many species of experimental animals and a wide variety of cell cultures. All efforts were unsuccessful, however, until Padgett and her colleagues (Padgett *et al.*, 1971) inoculated primary cultures of human fetal glial cells and succeeded in cultivating the virus. Since that time, JCV has been identified as the virus in the lesions of more than 70 cases of PML and has been firmly established as the major etiological agent in PML.

The early lesions of PML are multiple small areas of demyelination usually at the border between the white matter and the cortex of the cerebrum. The pathognomonic feature of PML is the altered oligodendrocytes that are prominent in and around the lesions. These cells have nuclei that are enlarged two or three times normal size. The normal chromatin pattern is effaced and the nuclei usually stain deeply basophilic. They often contain irregular basophilic inclusion bodies. Multiple techniques such as electron microscopy, immunofluorescence, and immunoperoxidase demonstrate that the abnormal oligodendrocyte nuclei are filled with virions and viral antigen. As the infection progresses, the oligodendrocytes in the lesions are destroyed (Itoyama *et al.*, 1982). Since it is well established that the myelin sheaths of the CNS are formed and maintained by the oligodendrocytes (Ross *et al.*, 1962) and loss of oligodendrocytes leads to demyelination of axons, this virus-induced destruction of oligodendrocytes provides good explanation for the demyelinative character of the disease.

The abnormal oligodendrocytes are found in small foci of demyelination and in the advancing edge of larger lesions. They are usually absent from the older, central portions of large lesions where demyelination is complete, but in these areas macrophages and reactive astrocytes are common and frequently giant, bizarre astrocytes can be found. The giant astrocytes have pleomorphic, hyperchromatic nuclei and resemble malignant astrocytes of pleomorphic glioblastomas. Occasionally they contain mitotic figures. Virions have not been found in these giant astrocytes, but a few have been shown by the sensitive peroxidase–antiperoxidase technique to contain some viral antigen (Itoyama *et al.*, 1982; Budka and Shah, 1983), and Dörries *et al.* (1979) found evidence by *in situ* hybridization of JCV DNA in some of the bizarre astrocytes. Thus, it seems likely that these cells are infected, but the infection is abortive and results in a form of virus-induced transformation. These cells have been of interest because of the possibility that JCV may be oncogenic in the human brain, but the giant astrocytes have not been observed to form clusters or colonies as would be expected of cells initiating a tumor. However, Castaigne *et al.* (1974) found numerous small gliomas within areas of PML lesions in a patient's brain and Sima *et al.* (1983) reported a case with multiple malignant astrocytomas that appeared to be originating from PML lesions, thus lending some support to the possibility that the bizarre astrocytes may have the potential to form tumors. It is noteworthy that these two patients survived much longer after onset of CNS disease than the 3- to 6-month period that is usual for PML patients. It is conceivable that tumors that could arise from transformed astrocytes are not usually found associated with PML simply because the demyelination and tissue destruction caused by the permissive infection in oligodendrocytes results in death of the patient before tumors can become recognizable.

It is characteristic that the lesions of PML are variable in size, giving the picture of expanding older lesions and the continued development of new, small lesions in multiple areas of the brain. Large areas of white matter appear to be involved by coalescence of multiple expanding lesions. The clinical features of PML fit with this concept. The disease usually begins insidiously with weakness of a limb, visual impairment, or a change in personality. Once clinical signs appear, progression is usually quite steady and relentless. Each functional impairment progressively becomes more severe and new signs and symptoms of disease in other sites keep appearing and then increasing in severity. Survival is usually about 3–6 months (Richardson, 1965; Padgett and Walker, 1983).

As compared to many other viral infections, the progression of PML is relatively slow, but Grinnell *et al.*, (1982) provided evidence that this is not because of the selection of temperature-sensitive mutants or the production of defective interfering particles in diseased tissue. They found that JCV isolated directly from diseased brain tissue of seven cases of PML grew better at 39°C than at lower temperatures. The opposite would be expected if the virus in lesions was a temperature-sensitive mutant. The JCV DNA extracted directly from diseased brain tissue of 13 cases of PML was found to be homogeneous. Therefore, although JCV tends to produce many defective genomes during multiplication in cell cultures, this does not appear to occur in human brain infections.

An important feature of PML is the usual presence of an underlying disease or therapeutic regimen that has an immunosuppressive effect. More than half of PML cases are associated with lymphoproliferative diseases. However, in recent years the disease has appeared with increasing frequency in persons under long-term treatment with corticosteroids, patients under immunosuppression after organ allografts, and quite recently in persons with the acquired immune deficiency syndrome (AIDS). It is quite clear that a deficiency in cellular immunity is of particular importance in susceptibility to PML (Walker and Padgett, 1983a).

It would be difficult to account for the multifocal distribution of PML lesions in the brain, and sometimes in the spinal cord, by any other than the bloodborne route of infection. If the virus reaches the CNS through the blood, it must also be distributed to other tissues. Although pathologists have not found morphological evidence of infection or disease in other tissues, there are many indications that JCV is deposited in other organs. JCV is frequently released in the urine of healthy women during pregnancy (Coleman *et al.*, 1980) and during the immunosuppression of allograft transplantation (Hogan *et al.*, 1980; O'Reilly *et al.*, 1981). In addition, Chesters *et al.* (1983) extracted JCV DNA from 10% of kidneys collected from random autopsies, and Dörries *et al.* (1980) and Dörries and ter Meulen (1983) used *in situ* hybridization to demonstrate JCV DNA in renal tubules of a PML patient. Grinnell *et al.* (1983c) searched for evidence of extraneural infection in ten cases of PML. Although JCV antigen was not found by immunofluorescence, JCV DNA was demonstrable by extraction and hybridization techniques. The DNA was not integrated and was present in tissues at concentrations that ranged between 0.1 and 25 copies per cell genome equivalent. JCV DNA was found in the kidneys of seven of the ten patients, in the lung tissue of three patients, and in the liver, spleen, and lymph nodes of two patients. Two of the patients, both children, had JCV DNA in kidney, liver, lung, spleen, and lymph nodes as well as in the brain, but evidence of active, destructive disease was found only in the brain.

ACKNOWLEDGMENTS. Studies by the authors have been supported by U.S. Public Health Service Grant AI-11217, American Cancer Society Grant MV-168, and National Science Foundation Grant PCM-8122058.

REFERENCES

Arthur, R. R., Shah, K. V., Yolken, R. H., and Charache, P., 1983, Detection of human papovavirus BKV and JCV in urines by ELISA, in: *Polyomaviruses and Human Neurological Disease* (J. L. Sever and D. L. Madden, eds.), pp. 169–176, Liss, New York.

Banerji, J., Rusconi, S., and Schaffner, W., 1981, Expression of a β-globin gene is enhanced by remote SV40 DNA sequences, *Cell* **27**:299.

Beckman, A. M., and Shah, K. V., 1983, Propagation and primary isolation of JCV and BKV in urinary epithelial cell cultures, in: *Polyomaviruses and Human Neurological Disease* (J. L. Sever and D. L. Madden, eds.), pp. 3–14, Liss, New York.

Beckman, A. M., Shah, K. V., and Padgett, B. L., 1982, Propagation and primary isolation of papovavirus JC in epithelial cells derived from human urine, *Infect. Immun.* **38**:774.

Benoist, C., and Chambon, P., 1981, *In vivo* sequence requirements of the SV40 early promoter region, *Nature* **290**:304.

Benoist, C., O'Hare, K., Breathnach, R., and Chambon, P., 1980, The ovalbumin gene-sequence of putative control regions, *Nucleic Acids Res.* **8**:127.

Bergsma, D. J., Olive, D. M., Hartzell, S. W., and Subramanian, K. N., 1982, Territorial limits and functional anatomy of the simian virus 40 replication origin, *Proc. Natl. Acad. Sci. USA* **79**:381.

Beth, E., Cikes, M., Schloen, L., di Mayorca, G., and Giraldo, G., 1977, Interspecies-, species-, and type-specific T antigenic determinants of human papovaviruses (JC and BK) and of simian virus 40, *Int. J. Cancer* **20**:551.

Beth, E., Cikes, M., and Giraldo, G., 1978, Microfluorometric analysis of anti-complement and indirect immunofluorescence tests for human papovavirus (JCV and BKV) T antigens, *Int. J. Cancer* **21**:1.

Brown, P., Tasai, T., and Gajdusek, D. C., 1975, Seroepidemiology of human papovaviruses, *Am. J. Epidemiol.* **102**:331.

Budka, H., and Shah, K. V., 1983, Papovavirus antigens in paraffin sections of PML brains, in: *Polyomaviruses and Human Neurological Disease* (J. L. Sever and D. L. Madden, eds.), pp. 299–309, Liss, New York.

Byrne, B. J., Davis, M. S., Yamaguchi, J., Bergsma, D. J., and Subramanian, K. N., 1983, Definition of the simian virus 40 early promoter region and demonstration of a host range bias in the enhancement effect of the simian virus 40 72-base-pair repeat, *Proc. Natl. Acad. Sci. USA* **80**:721.

Candeias, J. A. N., Baruzzi, R. G., Pripas, S., and Iunes, M., 1977, Prevalence of antibodies to the BK and JC papovaviruses in isolated populations, *Rev. Saude Publica* **11**:510.

Castaigne, P., Rondot, P., Escourolle, R., Ribadeau, J., Dumas, F., Cathala, F., and Hauw, J.-J., 1974, Leucoencéphalopathie multifocale progressive et 'gliomes' multiples, *Rev. Neurol.* **130**:379.

Chesters, P. M., Heritage, J., and McCance, D. J., 1983, Persistence of DNA sequences of BK virus and JC virus in normal human tissues and in diseased tissues, *J. Infect. Dis.* **147**:676.

Cikes, M., Beth, E., Guignard, N., Walker, D. L., Padgett, B. L., and Giraldo, G., 1977, Purification of simian virus 40 and JC T-antigens from transformed cells, *J. Natl. Cancer Inst.* **59**:889.

Coleman, D. V., Wolfendale, M. R., Daniel, R. A., Dhanjal, N. K., Gardner, S. D., Gibson, P. E., and Field, A. M., 1980, A prospective study of human polyomavirus infection in pregnancy, *J. Infect. Dis.* **142**:1.

Daniel, R., Shah, K., Madden, D., and Stagno, S., 1981, Serological investigation of the possibility of congenital transmission of papovavirus JC, *Infect. Immun.* **33**:319.

DeLucia, A. L., Lewton, B. A., Tjian, R., and Tegtmeyer, P., 1983, Topography of simian virus 40 A protein–DNA complexes: Arrangement of pentanucleotide interaction sites at the origin of replication, *J. Virol.* **46**:143.

de Villiers, J., Olson, J., Tyndall, C., and Schaffner, P., 1982, Transcriptional enhancers from SV40 and polyomavirus show a cell type preference, *Nucleic Acids Res.* **10**:7965.

Dierks, P., van Ooyen, A., Maneti, N., and Weissmann, C., 1981, DNA sequences preceding the rabbit β-globin gene are required for formation in mouse L cells of β-globin RNA with the correct 5′ terminus, *Proc. Natl. Acad. Sci. USA* **78**:1411.

Dörries, K., 1984, Progressive multifocal leucoencephalopathy: Analysis of JC virus DNA from brain and kidney tissue, *Virus Res.* **1**:25.

Dörries, K., and ter Meulen, V., 1983, Progressive multifocal leucoencephalopathy: Detection of papovavirus JC in kidney tissue, *J. Med. Virol.* **11**:307.

Dörries, K., Johnson, R. T., and ter Meulen, V., 1979, Detection of polyoma virus DNA in PML brain tissue by (*in situ*) hybridization, *J. Gen. Virol.* **42**:49.

Dörries, K., Reinhardt, V., and ter Meulen, V., 1980, Evidence for JC virus genomes in organ material from a PML patient, in: *Search for the Cause of Multiple Sclerosis and Other Chronic Diseases of the Nervous System* (A. Boese, ed.), pp. 284–294, Verlag Chemie, Weinheim.

Dougherty, R. M., 1976, A comparison of human papovavirus T antigens, *J. Gen. Virol.* **33**:61.

Dougherty, R. M., and DiStefano, H. S., 1974, Isolation and characterization of a papovavirus from human urine, *Proc. Soc. Exp. Biol. Med.* **146**:481.

Dynan, W. S., and Tjian, R., 1983a, A promoter-specific transcription factor allows recognition of upstream sequences in the SV40 early promoter, in: *Enhancers and Eukaryotic Gene Expression* (Y. Gluzman and T. Shenk, eds.), pp. 209–214, Cold Spring Harbor Laboratory, Cold Spring Harbor, N.Y.

Dynan, W. S., and Tjian, R., 1983b, The promoter-specific transcription factor SP1 binds to upstream sequences in the SV40 early promoter, *Cell* **35**:79.

Everett, R. D., Baty, D., and Chambon, P., 1983, The repeated GC-rich motifs upstream from the TATA box are important elements of the SV40 early promoter, *Nucleic Acids Res.* **11**:2447.

Fareed, G. C., Takemoto, K. K., and Gimbrone, M. A., Jr., 1978, Interaction of simian virus 40 and human papovaviruses, BK and JC, with human vascular endothelial cells, in: *Microbiology—1978* (D. Schlessinger, ed.), pp. 427–431, American Society for Microbiology, Washington, D.C.

Field, A. M., Gardner, S. D., Goodbody, R. A., and Woodhouse, M. A., 1974, Identity of a newly isolated human polyomavirus from a patient with progressive multifocal leucoencephalopathy, *J. Clin. Pathol.* **27**:341.

Fitzgerald, M., and Shenk, T., 1981, The sequence 5′-AAUAAA-3′ forms part of the recognition site for polyadenylation of late SV40 mRNAs, *Cell* **24**:251.
Frisque, R. J., 1983a, Regulatory sequences and virus–cell interactions of JC virus, in: *Polyomaviruses and Human Neurological Disease* (J. L. Sever and D. L. Madden, eds.), pp. 41–59, Liss, New York.
Frisque, R. J., 1983b, Nucleotide sequences of the region encompassing the JC virus origin of DNA replication, *J. Virol.* **46**:170.
Frisque, R. J., Martin, J. D., Padgett, B. L., and Walker, D. L., 1979, Infectivity of the DNA from four isolates of JC virus, *J. Virol.* **32**:476.
Frisque, R. J., Rifkin, D. B., and Walker, D. L., 1980, Transformation of primary hamster brain cells with JC virus and its DNA, *J. Virol.* **35**:265.
Frisque, R. J., Bream, G. L., and Cannella, M. T., 1984, Human polyomavirus JC virus genome, *J. Virol.* **51**:458.
Fujimura, F. K., Deininger, P. L., Friedmann, T., and Linney, E., 1981, Mutation near the polyoma DNA replication origin permits productive infection of F9 embryonal carcinoma cells, *Cell* **23**:809.
Gardner, S. D., 1977, The new human papovaviruses: Their nature and significance, in: *Recent Advances in Clinical Virology* (A. P. Waterson, ed.), pp. 93–115, Churchill Livingstone, Edinburgh.
Gardner, S. D., Field, A. M., Coleman, D. V., and Hulme, B., 1971, New human papovavirus (B.K.) isolated from urine after renal transplantation, *Lancet* **1**:1253.
Gerber, M. A., Shah, K. V., Thung, S. N., and ZuRhein, G. M., 1980, Immunohistochemical demonstration of common antigen of polyomaviruses in routine histologic tissue sections of animals and man, *Am. J. Clin. Pathol.* **73**:794.
Ghosh, P. K., Lebowitz, P., Frisque, R. J., and Gluzman, Y., 1981, Identification of a promoter component involved in positioning the 5′ termini of simian virus 40 early mRNAs, *Proc. Natl. Acad. Sci. USA* **78**:100.
Gibson, P. E., and Gardner, S. D., 1983, Strain differences and some serological observations on several isolates of human polyomaviruses, in: *Polyomaviruses and Human Neurological Disease* (J. L. Sever and D. L. Madden, eds.), pp. 119–132, Liss, New York.
Gibson, P. E., Field, A. M., Gardner, S. D., and Coleman, D. V., 1981, Occurrence of IgM antibodies against BK and JC polyomaviruses during pregnancy, *J. Clin. Pathol.* **34**:674.
Gluzman, Y., 1981, SV40-transformed simian cells support the replication of early SV40 mutants, *Cell* **23**:175.
Gluzman, Y., Frisque, R. J., and Sambrook, J., 1980a, Origin-defective mutants of SV40, *Cold Spring Harbor Symp. Quant. Biol.* **44**:293.
Gluzman, Y., Sambrook, J. F., and Frisque, R. J., 1980b, Expression of early genes of origin-defective mutants of simian virus 40, *Proc. Natl. Acad. Sci. USA* **77**:3898.
Grinnell, B. W., Martin, J. D., Padgett, B. L., and Walker, D. L., 1982, Is progressive multifocal leukoencephalopathy a chronic disease because of defective interfering particles or temperature-sensitive mutants of JC virus?, *J. Virol.* **43**:1143.
Grinnell, B. W., Martin, J. D., Padgett, B. L., and Walker, D. L., 1983a, Naturally occurring and passage-induced variation in the genome of JC virus, in: *Polyomaviruses and Human Neurological Disease* (J. L. Sever and D. L. Madden, eds.), pp. 61–77, Liss, New York.
Grinnell, B. W., Padgett, B. L., and Walker, D. L., 1983b, Comparison of infectious JC virus DNAs cloned from human brain, *J. Virol.* **45**:299.
Grinnell, B. W., Padgett, B. L., and Walker, D. L., 1983c, Distribution of nonintegrated DNA from JC papovavirus in organs of patients with progressive multifocal leukoencephalopathy, *J. Infect. Dis.* **147**:669.
Grosveld, G., deBoer, E., Shewmaker, C., and Flavell, R., 1982, DNA sequences necessary for transcription of the rabbit β-globin gene *in vivo*, *Nature* **295**:120.
Hammon, W. M., and Sather, G. E., 1969, Arboviruses, in: *Diagnostic Procedures for Viral and Rickettsial Infections*, 4th ed. (E. H. Lennette and N. J. Schmidt, eds.), pp. 227–280, American Public Health Association, New York.

Hartley, J. W., Rowe, W. P., Chanock, R. M., and Andrews, B. E., 1959, Studies of mouse polyoma virus infection. IV. Evidence for mucoprotein erythrocyte receptors in polyoma virus hemagglutination, *J. Exp. Med.* **110**:81.

Hartzell, S. W., Yamaguchi, J., and Subramanian, K. N., 1983, SV40 deletion mutants lacking the 21-bp repeated sequences are viable, but have non-complementable deficiencies, *Nucleic Acids Res.* **11**:1601.

Hartzell, S. W., Byrne, B. J., and Subramanian, K. N., 1984, Mapping of the late promoter of simian virus 40, *Proc. Natl. Acad. Sci. USA* **81**:23.

Hay, R. T., and DePamphilis, M. L., 1982, Initiation of SV40 DNA replication *in vivo:* Location and structure of 5′-ends of DNA synthesized in the *ori* region, *Cell* **28**:767.

Hogan, T. F., Borden, E. C., McBain, J. A., Padgett, B. L., and Walker, D. L., 1980, Human polyoma virus infections with JC virus and BK virus in renal transplant patients, *Ann. Intern. Med.* **92**:373.

Howatson, A. F., Nagai, M., and ZuRhein, G. M., 1965, Polyoma-like virions in human demyelinating brain disease, *Can. Med. Assoc. J.* **93**:379.

Howley, P. M., Khoury, G., Takemoto, K. K., and Martin, M. A., 1976, Polynucleotide sequences common to the genomes of simian virus 40 and the human papovaviruses JC and BK, *Virology* **73**:303.

Howley, P. M., Rentier-Delrue, F., Heilman, C. A., Law, M.-F., Chowdhury, K., Israel, M. A., and Takemoto, K. K., 1980, Cloned human polyomavirus JC DNA can transform human amnion cells, *J. Virol.* **36**:878.

Ikuta, F., and Kumanishi, T., 1973, Experimental virus-induced brain tumors, *Prog. Neuropathol.* **2**:253.

Itoyama, Y., Webster, H., Sternberger, N., Richardson, E. P., Jr., Walker, D. L., and Padgett, B. L., 1982, Distribution of papovavirus, myelin-associated glycoprotein and myelin basic protein in progressive multifocal leukoencephalopathy lesions, *Ann. Neurol.* **11**:396.

Katinka, M., and Yaniv, M., 1980, Expression of polyoma early functions in mouse embryonal carcinoma cells depends on sequence rearrangements in the beginning of the late region, *Cell* **20**:393.

Katinka, M., Vasseur, M., Montreau, N., Yaniv, M., and Blangy, D., 1981, Polyoma DNA sequences involved in control of viral gene expression in murine embryonal carcinoma cells, *Nature* **290**:720.

Kenney, S., Natarajan, V., Strike, D., Khoury, G., and Salzman, N. P., 1984, JC virus enhancer–promoter active in human brain cells, *Science* **226**:1337.

Laimins, L. A., Khoury, G., Gorman, C., Howard, B., and Gruss, P., 1982, Host-specific activation of transcription by tandem repeats from simian virus 40 and Moloney murine sarcoma virus, *Proc. Natl. Acad. Sci. USA* **79**:6453.

Lane, D. P., and Crawford, L. V., 1979, T-antigen is bound to a host protein in SV40-transformed cells, *Nature* **278**:261.

Law, M.-F., Martin, J. D., Takemoto, K. K., and Howley, P. M., 1979, The co-linear alignment of the genomes of papovaviruses JC, BK, and SV40, *Virology* **96**:576.

Lewin, B., 1983, *Genes,* p. 415, Wiley, New York.

London, W., Houff, S., Madden, D., Fuccillo, D., Gravell, M., Wallen, W., Palmer, A., Sever, J., Padgett, B., Walker, D., ZuRhein, G., and Ohashi, T., 1978, Brain tumors in owl monkeys inoculated with a human polyomavirus (JC virus), *Science* **201**:1246.

London, W. T., Houff, S. A., McKeever, P. E., Wallen, W. C., Sever, J. L., Padgett, B. L., and Walker, D. L., 1983, Viral-induced astrocytomas in squirrel monkeys, in: *Polyomaviruses and Human Neurological Disease* (J. L. Sever and D. L. Madden, eds.), pp. 227–237, Liss, New York.

McCance, D. J., 1983, Persistence of animal and human papovaviruses in renal and nervous tissues, in: *Polyomaviruses and Human Neurological Disease* (J. L. Sever and D. L. Madden, eds.), pp. 343–357, Liss, New York.

McKnight, S., and Kingsbury, R., 1982, Transcriptional control signals of a eukaryotic protein-coding gene, *Science* **217**:316.

Major, E. O., 1983a, Human papovavirus infection of an SV40-transformed human embryonic kidney cell line, in: *Polyomaviruses and Human Neurological Disease* (J. L. Sever and D. L. Madden, eds.), pp. 15–27, Liss, New York.
Major, E. O., 1983b, JC virus T protein expression in owl monkey tumor cell lines, in: *Polyomaviruses and Human Neurological Disease* (J. L. Sever and D. L. Madden, eds.), pp. 289–298, Liss, New York.
Major, E. O., and Matsumura, P., 1984, Human embryonic kidney cells: Stable transformation with an origin-defective simian virus 40 DNA and use as hosts for human papovavirus replications, *Mol. Cell. Biol.* **4**:379.
Major, E. O., Mourrain, P., and Cummins, C., 1984, JC virus-induced owl monkey glioblastoma cells in culture: Biological properties associated with the viral early gene product, *Virology* **136**:359.
Major, E. O., Miller, A. E., Mourrain, P., Traub, R. G., deWidt, E., and Sever, J., 1985, Establishment of a line of human fetal glial cells that supports JC virus multiplication, *Proc. Natl. Acad. Sci. USA* **82**:1257.
Manos, M. M., and Gluzman, Y., 1985, Genetic and biochemical analysis of transformation-competent, replication-defective simian virus 40 large T antigen mutants, *J. Virol.* **53**:120.
Mäntyjärvi, R. A., Arstila, P. P., and Meurman, O. H., 1972, Hemagglutination by BK virus, a tentative new member of the papovavirus group, *Infect. Immun.* **6**:824.
Margolskee, R. F., and Nathans, D., 1983, Suppression of a VP1 mutant of simian virus 40 by missense mutations in serine codons of the viral agnogene, *J. Virol.* **48**:405.
Martin, J. D., Frisque, R. J., Padgett, B. L., and Walker, D. L., 1979, Restriction endonuclease cleavage map of the DNA of JC virus, *J. Virol.* **29**:846.
Martin, J. D., Brackmann, K. H., Grinnell, B. W., Frisque, R. J., Walker, D. L., and Green, M., 1982, Recombinant JC viral DNA: Verification and physical map of prototype, *Biochem. Biophys. Res. Commun.* **109**:70.
Martin, J. D., Padgett, B. L., and Walker, D. L., 1983, Characterization of tissue culture-induced heterogeneity in DNAs of independent isolates of JC virus, *J. Gen. Virol.* **64**:2271.
Martin, J. D., King, D., Slauch, J. M., and Frisque, R. J., 1985, Differences in regulatory sequences of naturally occurring JC virus variants, *J. Virol.* **53**:306.
Mashiko, J., Nakamura, K., Shinozaki, T., Araki, K., Fujii, R., Yasui, K., and Ogiwara, H., 1982, A serological study of JC virus. Part I. Prevalence of antibody with respect to age and native place, *Teikyo Med. J.* **5**:299 (in Japanese).
May, E., Kress, M., Daya-Grosjean, L., Monier, R., and May, P., 1981, Mapping of the viral mRNA encoding a super-T antigen of 115,000 daltons expressed in simian virus 40-transformed rat cell lines, *J. Virol.* **37**:24.
May, P., Resche-Rigon, M., Borde, J., Breugnot, C., and May, E., 1984, Conversion through homologous recombination of the gene encoding simian virus 40 115,000-molecular-weight super T antigen to a gene encoding a normal-size large T antigen variant, *Mol. Cell. Biol.* **4**:1141.
Mellon, P., Parker, V., Gluzman, Y., and Maniatis, T., 1981, Identification of DNA sequences required for transcription of the human α1-globin gene in a new SV40 host-vector system, *Cell* **27**:279.
Miller, N. R., London, W., Padgett, B. L., Walker, D. L., and Wallen, W. C., 1983a, The detection of JC viral genome in owl monkey tumors, in: *Polyomaviruses and Human Neurological Disease* (J. L. Sever and D. L. Madden, eds.), pp. 271–288, Liss, New York.
Miller, N. R., Major, E. O., and Wallen, W. C., 1983b, Transfection of human fetal glial cells with molecularly cloned JCV DNA, in: *Polyomaviruses and Human Neurological Disease* (J. L. Sever and D. L. Madden, eds.), pp. 29–40, Liss, New York.
Miller, N. R., McKeever, P. E., London, W., Padgett, B. L., Walker, D. L., and Wallen, W. C., 1984, Brain tumors of owl monkeys inoculated with JC virus contain the JC virus genome, *J. Virol.* **49**:848.

Miyamura, T. Yoshiike, K., and Takemoto, K. K., 1980, Characterization of JC papovavirus adapted to growth in human embryonic kidney cells, *J. Virol.* **35**:498.

Miyamura, T., Jikuya, H., Soeda, E., and Yoshiike, K., 1983, Genomic structure of human polyoma virus JC: Nucleotide sequence of the region containing replication origin and small-T-antigen gene, *J. Virol.* **45**:73.

Morrison, B., Kress, M., Khoury, G., and Jay, G., 1983, Simian virus 40 tumor antigen: Isolation of the origin-specific DNA-binding domain, *J. Virol.* **47**:106.

Myers, R. M., and Tjian, R., 1980, Construction and analysis of simian virus 40 origins defective in tumor antigen binding and DNA replication, *Proc. Natl. Acad. Sci. USA* **77**:6491.

Nagashima, K., Yamaguchi, K., Yasui, K., and Ogiwara, H., 1981, Progressive multifocal leukoencephalopathy: Neuropathology and virus isolation, *Acta Pathol. Jpn.* **31**:953.

Nagashima, K., Yasui, K., Kimura, J., Washizu, M., Yamaguchi, K., and Mori, W., 1984, Induction of brain tumors by a newly isolated JC virus (Tokyo-1 strain), *Am. J. Pathol.* **116**:455.

Narayan, O., Penney, J. B., Johnson, R. T., Herndon, R. M., and Weiner, L. P., 1973, Etiology of progressive multifocal leukoencephalopathy, *N. Engl. J. Med.* **289**:1278.

Ohashi, T., ZuRhein, G. M., Varakis, J. N., Padgett, B. L., and Walker, D. L., 1978, Experimental (JC virus-induced) intraocular and extraorbital tumors in the Syrian hamster, *J. Neuropathol. Exp. Neurol.* **37**:667.

O'Reilly, R. J., Lee, F. K., Grossbard, E., Kapoor, N., Kirkpatrick, D., Dinsmore, R., Stutzer, C., Shah, K. V., and Nahmias, A. J., 1981, Papovavirus excretion following marrow transplantation: Incidence and association with hepatic dysfunction, *Transplant. Proc.* **13**:262.

Osborn, J. E., Robertson, S. M., Padgett, B. L., ZuRhein, G. M., Walker, D. L., and Weisblum, B., 1974, Comparison of JC and BK human papovaviruses with simian virus 40: Restriction endonuclease digestion and gel electrophoresis of resultant fragments, *J. Virol.* **13**:614.

Osborn, J. E., Robertson, S. M., Padgett, B. L., Walker, D. L., and Weisblum, B., 1976, Comparison of JC and BK human papovaviruses with simian virus 40: DNA homology studies, *J. Virol.* **19**:675.

Padgett, B. L., and Walker, D. L., 1973, Prevalence of antibodies in human sera against JC virus, an isolate from a case of progressive multifocal leukoencephalopathy, *J. Infect. Dis.* **127**:467.

Padgett, B. L., and Walker, D. L., 1976, New human papovaviruses, *Prog. Med. Virol.* **21**:1.

Padgett, B. L., and Walker, D. L., 1980, Human papovavirus JCV: Natural history, tumorigenicity, and interaction with human cells in culture, *Cold Spring Harbor Conferences on Cell Proliferation* **7**:319.

Padgett, B. L., and Walker, D. L., 1983, Virologic and serologic studies of progressive multifocal leukoencephalopathy, in: *Polyomaviruses and Human Neurological Disease* (J. L. Sever and D. L. Madden, eds.), pp. 107–118, Liss, New York.

Padgett, B. L., Walker, D. L., ZuRhein, G. M., Eckroade, R. J., and Dessel, B. H., 1971, Cultivation of papova-like virus from human brain with progressive multifocal leukoencephalopathy, *Lancet* **1**:1257.

Padgett, B. L., Walker, D. L., ZuRhein, G. M., Hodach, A. E., and Chou, S. M., 1976, JC papovavirus in progressive multifocal leukoencephalopathy, *J. Infect. Dis.* **133**:686.

Padgett, B. L., Hunt, J. M., and Walker, D. L., 1977a, Specificity of the tumor-specific transplantation antigen induced by JC virus, a human polyomavirus, *Intervirology* **8**:182.

Padgett, B. L., Rogers, C. M., and Walker, D. L., 1977b, JC virus, a human polyomavirus associated with progressive multifocal leukoencephalopathy: Additional biological characteristics and antigenic relationships, *Infect. Immun.* **15**:656.

Padgett, B. L., Walker, D. L., ZuRhein, G. M., and Varakis, J. N., 1977c, Differential neuroncogenicity of strains of JC virus, a human polyoma virus, in newborn Syrian hamsters, *Cancer Res.* **37**:718.

Pater, A., Pater, M. M., and di Mayorca, G. D., 1980, Arrangement of the genome of the human papovavirus RF virus, *J. Virol.* **36**:480.

Pater, A., Pater, M. M., Chang, L., Slawin, K., and di Mayorca, G. D., 1983, Multiple origins of the complementary defective genomes of RF and origin proximal sequences of GS, two human papovavirus isolates, *Virology* **131**:426.

Penney, J. B., Jr., and Narayan, O., 1973, Studies of the antigenic relationships of the new human papovaviruses by electron microscopy agglutination, *Infect. Immun.* **8**:299.

Prives, C., Barnet, B., Scheller, A., Khoury, G., and Jay, G., 1982, Discrete regions of simian virus 40 large T antigen are required for nonspecific and viral origin-specific DNA binding, *J. Virol.* **43**:73.

Proudfoot, N. J., and Brownlee, G. G., 1974, Sequence at the 3′ end of globin mRNA shows homology with immunoglobulin light chain mRNA, *Nature* **252**:359.

Quay, W. B., Ma, Y. H., Varakis, J. N., ZuRhein, G. M., Padgett, B. L., and Walker, D. L., 1977, Modification of hydroxyindole-O-methyltransferase activity in experimental pineocytomas induced in hamsters by a human papovavirus (JC), *J. Natl. Cancer Inst.* **58**:123.

Reddy, V. B., Ghosh, P. K., Lebowitz, P., Piatak, M., and Weissman, S. M., 1979, Simian virus 40 early mRNAs. I. Genome localization of 3′ and 5′ termini and two major splices in mRNA from transformed and lytically infected cells, *J. Virol.* **30**:279.

Rentier-Delrue, F., Lubiniecki, A., and Howley, P. M., 1981, Analysis of JC virus DNA purified directly from human progressive multifocal leukoencephalopathy brains, *J. Virol.* **38**:761.

Richardson, E. P., Jr., 1965, Progressive multifocal leukoencephalopathy, in: *The Remote Effects of Cancer in the Nervous System* (L. Brain and F. H. Norris, Jr., eds.), pp. 6–16, Grune & Stratton, New York.

Rio, D., Robbins, A., Myers, R., and Tjian, R., 1980, Regulation of simian virus 40 early transcription *in vitro* by a purified tumor antigen, *Proc. Natl. Acad. Sci. USA* **77**:5706.

Rosenthal, N., Kress, M., Gruss, P., and Khoury, G., 1983, BK viral enhancer element and a human cellular homology, *Science* **222**:749.

Ross, L. L., Bornstein, M. B., and Lehrer, G. M., 1962, Electron microscopic observations on myelin formation in tissue cultures of developing rat cerebellum, *J. Cell Biol.* **14**:19.

Ryder, K., DeLucia, A. L., and Tegtmeyer, P., 1983, Binding of SV40 A protein to the BK virus origin of DNA replication, *Virology* **129**:239.

Schwerdt, P. R., Schwerdt, C. E., Silverman, L., and Rubinstein, L. J., 1966, Virions associated with progressive multifocal leukoencephalopathy, *Virology* **29**:511.

Seif, I., Khoury, G., and Dhar, R., 1979, The genome of human papovavirus BKV, *Cell* **18**:963.

Shah, K. V., Daniel, R. W., and Kelly, T. J., Jr., 1977a, Immunological relatedness of papovaviruses of the simian virus 40–polyoma subgroup, *Infect. Immun.* **18**:558.

Shah, K. V., Ozer, H. L., Ghazey, H. N., and Kelly, T. J., Jr., 1977b, Common structural antigen of papovaviruses of the simian virus 40–polyoma subgroup, *J. Virol.* **21**:179.

Shenk, T., 1978, Construction of a viable SV40 variant containing two functional origins of DNA replication, *Cell* **13**:791.

Shenk, T. E., Carbon, J., and Berg, P., 1976, Construction and analysis of viable deletion mutants of simian virus 40, *J. Virol.* **18**:664.

Shortle, D. R., Margolskee, R. F., and Nathans, D., 1979, Mutational analysis of the simian virus 40 replicon: Pseudorevertants of mutants with a defective replication origin, *Proc. Natl. Acad. Sci. USA* **76**:6128.

Silverman, L., and Rubinstein, L. J., 1965, Electron microscopic observations on a case of progressive multifocal leukoencephalopathy, *Acta Neuropathol.* **5**:215.

Sima, A. A. F., Finkelstein, S. D., and McLachlin, D. R., 1983, Multiple malignant astrocytomas in a patient with spontaneous progressive multifocal leukoencephalopathy, *Ann. Neurol.* **14**:183.

Simmons, D. T., 1980, Characterization of tau antigens isolated from uninfected and simian virus 40-infected monkey cells and papovavirus-transformed cells, *J. Virol.* **36**:519.

Simmons, D. T., and Martin, M. A., 1978, Common methionine-tryptic peptides near the amino-terminal end of primate papovavirus tumor antigens, *Proc. Natl. Acad. Sci. USA* **75**:1131.

Soeda, E., Arrand, J. R., Smolar, N., Walsh, J. E., and Griffin, B. E., 1980, Coding potential and regulatory signals of the polyoma virus genome, *Nature* **283**:445.

Taguchi, F., Kajioka, J., and Miyamura, T., 1982, Prevalence rate and age of acquisition of antibodies against JC virus and BK virus in human sera, *Microbiol. Immunol.* **26**:1057.

Takemoto, K. K., and Martin, M. A., 1976, Transformation of hamster kidney cells by BK papovavirus DNA, *J. Virol.* **17**:247.

Takemoto, K. K., and Mullarkey, M. F., 1973, Human papovavirus, BK strain: Biological studies including antigenic relationship to simian virus 40, *J. Virol.* **12**:625.

Takemoto, K. K., Rabson, A. S., Mullarkey, M. F., Blaese, R. M., Garon, C. F., and Nelson, D., 1974, Isolation of papovavirus from brain tumor and urine of a patient with Wiskott–Aldrich syndrome, *J. Natl. Cancer Inst.* **53**:1205.

Takemoto, K. K., Howley, P. M., and Miyamura, T., 1979a, JC human papovavirus replication in human amnion cells, *J. Virol.* **30**:384.

Takemoto, K. K., Linke, H., Miyamura, T., and Fareed, G. C., 1979b, Persistent BK papovavirus infection of transformed human fetal brain cells, *J. Virol.* **29**:1177.

Tjian, R., 1978, Protein–DNA interactions at the origin of simian virus 40 DNA replication, *Cold Spring Harbor Symp. Quant. Biol.* **43**:655.

Tolun, A., Alestrom, P., and Pettersson, V., 1979, Sequence of inverted terminal repetitions from different adenoviruses: Demonstration of conserved sequences and homology between SA7 termini and SV40 DNA, *Cell* **17**:705.

Tooze, J. (ed.), 1981, *DNA Tumor Viruses*, Cold Spring Harbor Laboratory, Cold Spring Harbor, N.Y.

Tornow, J., Polvino-Bodnar, M., Santangelo, G., and Cole, C. N., 1985, Two separable functional domains of simian virus 40 large T-antigen: Carboxyl-terminal region of simian virus 40 large T-antigen is required for efficient capsid protein synthesis, *J. Virol.* **53**:415.

Uchida, S., Watanabe, S., Aizawa, T., Furuno, A., and Muto, T., 1979, Polyoncogenicity and insulinoma-inducing ability of BK virus, a human papovavirus, in Syrian golden hamsters, *J. Natl. Cancer Inst.* **63**:119.

Varakis, J. N., and ZuRhein, G. M., 1976, Experimental pineocytoma of the Syrian hamster induced by a human papovavirus (JC), *Acta Neuropathol.* **35**:243.

Varakis, J. N., ZuRhein, G. M., Padgett, B. L., and Walker, D. L., 1976, Experimental (JC virus-induced) neuroblastomas in the Syrian hamster, *J. Neuropathol. Exp. Neurol.* **35**:314.

Varakis, J., ZuRhein, G. M., Padgett, B. L., and Walker, D. L., 1978, Induction of peripheral neuroblastomas in Syrian hamsters after injection as neonates with JC virus, a human polyoma virus, *Cancer Res.* **38**:1718.

Walker, D. L., and Padgett, B. L., 1978, Biology of JC virus, a human papovavirus, in: *Microbiology—1978* (D. Schlessinger, ed.), pp. 432–434, American Society for Microbiology, Washington, D.C.

Walker, D. L., and Padgett, B. L., 1983a, Progressive multifocal leukoencephalopathy, in: *Comprehensive Virology*, Vol. 18 (H. Fraenkel-Conrat and R. R. Wagner, eds.), pp. 161–193, Plenum Press, New York.

Walker, D. L., and Padgett, B. L., 1983b, The epidemiology of human polyomaviruses, in: *Polyomaviruses and Human Neurological Disease* (J. L. Sever and D. L. Madden, eds.), pp. 99–106, Liss, New York.

Walker, D. L., Padgett, B. L., ZuRhein, G. M., Albert, A. E., and Marsh, R. F., 1973a, Current study of an opportunistic papovavirus, in: *Slow Virus Diseases* (W. Zeman and E. H. Lennette, eds.), pp. 49–58, Williams & Wilkins, Baltimore.

Walker, D. L., Padgett, B. L., ZuRhein, G. M., Albert, A. E., and Marsh, R. F., 1973b, Human papova-virus (JC): Induction of brain tumors in hamsters, *Science* **181**:674.

Watanabe, S., and Yoshiike, K., 1982, Change of DNA near the origin of replication enhances the transforming capacity of human papovavirus BK, *J. Virol.* **42**:978.
Watanabe, S., Yoshiike, K., Nozawa, A., Yuasa, Y., and Uchida, S., 1979, Viable deletion mutant of human papovavirus BK that induces insulinomas in hamsters, *J. Virol.* **32**:934.
Weiher, H., König, M., and Gruss, P., 1983, Multiple point mutations affecting the simian virus 40 enhancer, *Science* **219**:626.
Wold, W. S. M., Green, M., Mackey, J. K., Martin, J. D., Padgett, B. L., and Walker, D. L., 1980, Integration pattern of human JC virus sequences in two clones of a cell line established from a JC virus-induced hamster brain tumor, *J. Virol.* **33**:1225.
Wroblewska, Z., Wellish, M., and Gilden, G., 1980, Growth of JC virus in adult human brain cell cultures, *Arch. Virol.* **65**:141.
Yang, R. C. A., and Wu, R., 1979, BK virus DNA: Complete nucleotide sequence of a human tumor virus, *Science* **206**:456.
Yoshiike, K., Miyamura, T., Chan, H. W., and Takemoto, K. K., 1982, Two defective DNAs of human polyomavirus JC adapted to growth in human embryonic kidney cells, *J. Virol.* **42**:395.
ZuRhein, G. M., 1983, Studies of JC virus-induced nervous system tumors in the Syrian hamster: A review, in: *Polyomaviruses and Human Neurological Disease* (J. L. Sever and D. L. Madden, eds.), pp. 205–221, Liss, New York.
ZuRhein, G. M., and Chou, S.-M., 1965, Particles resembling papova viruses in human cerebral demyelination disease, *Science* **148**:1477.
ZuRhein, G. M., and Varakis, J., 1975, Morphology of brain tumors induced in Syrian hamsters after inoculation with JC virus, a new human papovavirus, in: *Proceedings, VIIth International Congress of Neuropathology*, Budapest, 1974, Vol. I (S. Kornzey, S. Tariska, and G. Gosztonyi, eds.), pp. 479–481, Academic Kiado, Budapest and Excerpta Medica, Amsterdam.
ZuRhein, G. M., and Varakis, J. N., 1979, Perinatal induction of medulloblastomas in Syrian golden hamsters by a human polyoma virus (JC), *Natl. Cancer Inst. Monogr.* **51**:205.

APPENDIX

Annotated Nucleotide Sequence and Restriction Site Lists for Selected Papovavirus Strains

CHARLES E. BUCKLER AND NORMAN P. SALZMAN

The nucleotide sequences for SV40, polyoma, and BK viruses are those present in GenBank(R), The Genetic Sequence Data Bank, Release 34.0, August 1985 (The Los Alamos Laboratory). The sequence of JC virus is that of the Mad1 strain (Frisque *et al.*, 1984). The annotated nucleotide sequences contain landmarks for viral protein coding sequence start and stop sites as well as other important structural elements (Table I). The position of these landmarks were taken from the GenBank Features lists for SV40, polyoma, and BK virus, and from Frisque *et al.* (1984) for JC virus. Additional annotations are the locations of all sites of commercially available restriction enzymes that cut a given sequence only one time (shown by a dashed line above the recognition sequence). In an attempt to keep the annotations in a simple form none of the mRNA initiation or termination sites have been shown.

Each of the papovaviruses presented has a different system for positioning the first nucleotide (sequence numbering origin) and of sequence

CHARLES E. BUCKLER AND NORMAN P. SALZMAN • Laboratory of Molecular Microbiology and Laboratory of Biology of Viruses, National Institute of Allergy and Infectious Diseases, National Institutes of Health, Bethesda, Maryland 20892.

TABLE I. Annotations Used for the DNA Sequences

ORGRPL	Origin of replication.
RPT	One of a series of tandem repeated sequences found near the origin of replication.
T-ag binding	Potential or actual site of T antigen binding near the replication origin.
CDS start	The initiator (ATG) codon for one of the early or late viral proteins.
CDS end	Viral protein translation, carboxy terminal amino acid codon end.
Dashed line (----) above a base sequence	Designates a recognition site for a restriction enzyme that cuts the viral DNA only one time.

polarity. SV40 virus DNA contains a total of 5243 bp. Its sequence numbering origin is the central G base of a 27 nucleotide palindrome at or near the origin of replication. The first three nucleotides are the last three bases of the unique Bgl I restriction site for SV40 virus DNA. The sequence shown is that of strain 776 (Fiers *et al.*, 1978; Reddy *et al.*, 1978) with the numbering system presented by Buchman *et al.* (Buchman *et al.*, 1981) with corrections for the additional 17 bases starting at nucleotide 165 (van Heuverswyn and Fiers, 1979). The polarity of the sequence numbering is that of the late mRNAs.

Polyoma virus (A2 strain) DNA contains a total of 5297 bp (Griffin *et al.*, 1981; Ito and Griffen, 1984; Katinka and Yaniv, 1982; Soeda *et al.*, 1980; Tyndall *et al.*, 1981). Its sequence numbering origin is the center of the restriction enzyme recognition site at the junction *Hpa* II 3/5 fragments. This is at or very near the origin of replication. The polarity of the sequence numbering is that of the early mRNAs.

BK virus (Dunlop strain) DNA contains a total of 5153 bp (Seif *et al.*, 1979, 1980a,b). Its sequence numbering origin is the first base after the initiator codon for the T antigens. This is the numbering originally presented for this strain (Seif *et al.*, 1979, 1980a,b). This differs from the numbering used by the same authors in another review (Seif *et al.*, 1980b) where the numbering origin was adjusted to be like that of SV40 with base number one located at or near the origin of replication. The polarity of the sequence numbering is that of the late mRNAs.

BK virus (MM strain) DNA contains a total of 4963 bp (Yang and Wu, 1979). Its sequence numbering origin is at the unique *Eco*R I restriction site in the genome. The polarity of the sequence numbering is that of the late mRNAs.

JC virus (Mad1 strain) DNA contains a total of 5130 bp (Frisque *et al.*, 1984). Its sequence numbering origin is at or near the presumed origin of replication. The polarity of the sequence numbering is that of the late mRNAs.

REFERENCES

Buchman, A. R., Burnett, L., and Berg, P., 1981, Appendix A: The SV40 nucleotide sequence, in: *DNA Tumor Viruses* (2nd ed. revised) (J. Tooze, ed.), pp. 799–841, Cold Spring Harbor Laboratory, Cold Spring Harbor, N.Y.

Fiers, W., Contreras, R., Haegeman, G., Rogiers, R., van de Voorde, A., van Heuverswyn, H., van Herreweghe, J., Volckaert, G., and Ysebaert, M., 1978, Complete nucleotide sequence of SV40 DNA, *Nature* **273:**113–120.

Frisque, R. J., Bream, G. L., and Cannella, M. T., 1984, Human polyoma JC virus genome, *J. Virol.* **51:**459–469.

Griffin, B. E., Soeda, E., Barrell, B. G., and Staden, R., 1981, Appendix B: Sequence and analysis of polyoma virus DNA, in *DNA Tumor Viruses* (2nd ed. revised) (J. Tooze, ed.), pp. 843–910, Cold Spring Harbor Laboratory, Cold Spring Harbor, N.Y.

van Heuverswyn, H., and Fiers, W., 1979, Nucleotide sequence of the Hind-c fragment of simian virus 40 DNA: Comparison of the 5′-untranslated region of wild-type virus and of some deletion mutants, *Eur. J. Biochem.* **100:**51–60.

Ito, Y., and Griffin, B. E., 1984, Genetic map of polyoma virus, in: *Genetic Maps*, Volume 3 (S. J. O'Brien, ed.), pp. 66–76, Cold Spring Harbor Laboratory, Cold Spring Harbor, N.Y.

Katinka, M., and Yaniv, M., 1982, Deletions of n-terminal sequences of polyoma virus T-antigens reduce but do not abolish transformation of rat fibroblasts, *Mol. Cell Biol.* **2:**1238–1246.

Reddy, V. B., Thimmappaya, B., Dhar, R., Subramanian, K. N., Zain, S., Pan, J., Ghosh, P. K., Celma, M. L., and Weissman, S. M., 1978, The genome of simian virus 40, *Science* **200:**494–502.

Seif, I., Khoury, G., and Dhar, R., 1979, The genome of human papovavirus BKV, *Cell* **18:**963–977.

Seif, I., Khoury, G., and Dhar, R., 1980a, Errata, *Cell* **19:**567.

Seif, I., Khoury, G., and Dhar, R., 1980b, Sequence and analysis of the genome of human papovavirus BKV, in: *DNA Tumor Viruses* (2nd ed. revised) (J. Tooze, ed.), pp. 843–910, Cold Spring Harbor Laboratory, Cold Spring Harbor, N.Y.

Soeda, E., Arrand, R. J., Smolar, N., Walsh, J. E., and Griffin, B. E., 1980, Coding potential and regulatory signals of the polyoma virus genome, *Nature* **283:**445–453.

Tyndall, C., la Mantia, G., Thacker, C. M., Favaloro, J., and Kamen, R., 1981, A region of the polyoma virus genome between the replication origin and late protein coding sequences is required in cis for both early gene expression and viral DNA replication, *Nucleic Acids Res.* **9:**6231–6250.

Yang, R. C. A., and Wu, R., 1979, BK virus DNA: Complete nucleotide sequence of human tumor virus, *Science* **206:**456–462.

```
        10         20         30         40         50         60         70         80         90        100
GCCTCGGCCT CTGCATAAAT AAAAAAAATT AGTCAGCCAT GGGGCGGAGA ATGGGCGGAA CTGGGCGGAG TTAGGGGCGG GATGGGCGGA GTTAGGGGCG
CGGAGCCGGA GACGTATTTA TTTTTTTTAA TCAGTCGGTA CCCCGCCTCT TACCCGCCTT GACCCGCCTC AATCCCCGCC CTACCCGCCT CAATCCCCGC
........ORGRPL CORE............¦¦--->                        ORGRPL AUX                <---¦
                                          ¦---> 21 bp RPT  <---¦  ¦---> 21 bp RPT  <---¦¦---> 21 bp RPT  <-
T-ag II  <---¦                ¦--->     T-ag Binding site III   <---¦

       110        120        130        140        150        160        170        180        190        200
GGACTATGGT TGCTGACTAA TTGAGATGCA TGCTTTGCAT ACTTCTGCCT GCTGGGGAGC CTGGGGACTT TCCACACCTG GTTGCTGACT AATTGAGATG
CCTGATACCA ACGACTGATT AACTCTACGT ACGAAACGTA TGAAGACGGA CGACCCCTCG GACCCCTGAA AGGTGTGGAC CAACGACTGA TTAACTCTAC
--¦   ¦--->                          72 bp RPT                                    <---¦¦--->        72 bp RPT

       210        220        230        240        250        260        270        280        290     ------
CATGCTTTGC ATACTTCTGC CTGCTGGGGA GCCTGGGGAC TTTCCACACC CTAACTGACA CACATTCCAC AGCTGGTTCT TTCCGCCTCA GAAGGTACCT
GTACGAAACG TATGAAGACG GACGACCCCT CGGACCCCTG AAAGGTGTGG GATTGACTGT GTGTAAGGTG TCGACCAAGA AAGGCGGAGT CTTCCATGGA
                                                <---¦                                                Kpn I

                                     ¦---> Agnoprotein CDS Start
       310        320        330     ¦  340    ----         360        370        380        390        400
AACCAAGTTC CTCTTTCAGA GGTTATTTCA GGCCATGGTG CTGCGCCGGC TGTCACGCCA GGCCTCCGTT AAGGTTCGTA GGTCATGGAC TGAAAGTAAA
TTGGTTCAAG GAGAAAGTCT CCAATAAAGT CCGGTACCAC GACGCGGCCG ACAGTGCGGT CCGGAGGCAA TTCCAAGCAT CCAGTACCTG ACTTTCATTT
                                               Hpa II

       410        420        430        440        450        460        470        480        490        500
AAAACAGCTC AACGCCTTTT TGTGTTTGTT TTAGAGCTTT TGCTGCAATT TTGTGAAGGG GAAGATACTG TTGACGGGAA ACGCAAAAAA CCAGAAAGGT
TTTTGTCGAG TTGCGGAAAA ACACAAACAA AATCTCGAAA ACGACGTTAA AACACTTCCC CTTCTATGAC AACTGCCCTT TGCGTTTTTT GGTCTTTCCA

     Agnoprotein
        CDS End <---¦                                          ¦---> VP2 CDS Start
       510        520        530        540        550        560 ¦      570        580        590        600
TAACTGAAAA ACCAGAAAGT TAACTGGTAA GTTTAGTCTT TTTGTCTTTT ATTTCAGGTC CATGGGTGCT GCTTTAACAC TGTTGGGGGA CCTAATTGCT
ATTGACTTTT TGGTCTTTCA ATTGACCATT CAAATCAGAA AAACAGAAAA TAAAGTCCAG GTACCCACGA CGAAATTGTG ACAACCCCCT GGATTAACGA

       610        620        630        640        650        660        670        680        690        700
ACTGTGTCTG AAGCTGCTGC TGCTACTGGA TTTTCAGTAG CTGAAATTGC TGCTGGAGAG GCCGCTGCTG CAATTGAAGT GCAACTTGCA TCTGTTGCTA
TGACACAGAC TTCGACGACG ACGATGACCT AAAAGTCATC GACTTTAACG ACGACCTCTC CGGCGACGAC GTTAACTTCA CGTTGAACGT AGACAACGAT

       710        720        730        740        750        760     ------        780        790        800
CTGTTGAAGG CCTAACAACC TCTGAGGCAA TTGCTGCTAT AGGCCTCACT CCACAGGCCT ATGCTGTGAT ATCTGGGGCT CCTGCTGCTA TAGCTGGATT
GACAACTTCC GGATTGTTGG AGACTCCGTT AACGACGATA TCCGGAGTGA GGTGTCCGGA TACGACACTA TAGACCCCGA GGACGACGAT ATCGACCTAA
                                                                    EcoR V
```

```
       810        820        830  ------840        850        860        870        880        890        900
TGCAGCTTTA CTGCAAACTG TGACTGGTGT GAGCGCTGTT GCTCAAGTGG GGTATAGATT TTTTAGTGAC TGGGATCACA AAGTTTCTAC TGTTGGGTTTA
ACGTCGAAAT GACGTTTGAC ACTGACCACA CTCGCGACAA CGAGTTCACC CCATATCTAA AAAATCACTG ACCCTAGTGT TTCAAAGATG ACAACCCAAAT
                                  Hae II

                 ¦---> VP3 CDS Start
       910       ¦ 920        930        940        950        960        970        980        990       1000
TATCAACAAC CAGGAATGGC TGTAGATTTG TATAGGCCAG ATGATTACTA TGATATTTTA TTTCCTGGAG TACAAACCTT TGTTCACAGT GTTCAAGTATC
ATAGTTGTTG GTCCTTACCG ACATCTAAAC ATATCCGGTC TACTAATGAT ACTATAAAAT AAAGGACCTC ATGTTTGGAA ACAAGTGTCA CAAGTTCATAG

      1010       1020       1030       1040       1050       1060       1070       1080       1090       1100
TTGACCCCAG ACATTGGGGT CCAACACTTT TTAATGCCAT TTCTCAAGCT TTTTGGCGTG TAATACAAAA TGACATTCCT AGGCTCACCT CACAGGGAGCT
AACTGGGGTC TGTAACCCCA GGTTGTGAAA AATTACGGTA AAGAGTTCGA AAAACCGCAC ATTATGTTTT ACTGTAAGGA TCCGAGTGGA GTGTCCCTCGA

      1110       1120       1130       1140       1150       1160       1170       1180       1190       1200
TGAAAGAAGA ACCCAAAGAT ATTTAAGGGA CAGTTTGGCA AGGTTTTTAG AGGAAACTAC TTGGACAGTA ATTAATGCTC CTGTTAATTG GTATAAACTCT
ACTTTCTTCT TGGGTTTCTA TAAATTCCCT GTCAAACCGT TCCAAAAATC TCCTTTGATG AACCTGTCAT TAATTACGAG GACAATTAAC CATATTTGAGA

      1210       1220       1230       1240       1250       1260       1270       1280       1290       1300
TTACAAGATT ACTACTCTAC TTTGTCTCCC ATTAGGCCTA CAATGGTGAG ACAAGTAGCC AACAGGGAAG GGTTGCAAAT ATCATTTGGG CACACCCTATG
AATGTTCTAA TGATGAGATG AAACAGAGGG TAATCCGGAT GTTACCACTC TGTTCATCGG TTGTCCCTTC CCAACGTTTA TAGTAAACCC GTGTGGGATAC

      1310       1320       1330       1340       1350       1360       1370       1380       1390       1400
ATAATATTGA TGAAGCAGAC AGTATTCAGC AAGTAACTGA GAGGTGGGAA GCTCAAAGCC AAAGTCCTAA TGTGCAGTCA GGTGAATTTA TTGAAAAAATT
TATTATAACT ACTTCGTCTG TCATAAGTCG TTCATTGACT CTCCACCCTT CGAGTTTCGG TTTCAGGATT ACACGTCAGT CCACTTAAAT AACTTTTTTAA
                                                                                                          ¦-
      1410       1420       1430       1440       1450       1460       1470       1480       1490       1500
TGAGGCTCCT GGTGGTGCAA ATCAAAGAAC TGCTCCTCAG TGGATGTTGC CTTTACTTCT AGGCCTGTAC GGAAGTGTTA CTTCTGCTCT AAAAGGCTTAT
ACTCCGAGGA CCACCACGTT TAGTTTCTTG ACGAGGAGTC ACCTACAACG GAAATGAAGA TCCGGACATG CCTTCACAAT GAAGACGAGA TTTTCCGAATA

--> VP1 CDS Start
      1510       1520       1530       1540       1550       1560       1570       1580       1590       1600
GAAGATGGCC CCAACAAAAA GAAAAGGAAG TTGTCCAGGG GCAGCTCCCA AAAAACCAAA GGAACCAGTG CAAGTGCCAA AGCTCGTCAT AAAAGGGAGGA
CTTCTACCGG GGTTGTTTTT CTTTTCCTTC AACAGGTCCC CGTCGAGGGT TTTTTGGTTT CCTTGGTCAC GTTCACGGTT TCGAGCAGTA TTTTCCCTCCT

  VP2, VP3
   CDS End <---¦
      1610       1620      --- ---       1640       1650       1660       1670       1680       1690       1700
ATAGAAGTTC TAGGAGTTAA AACTGGAGTA GACAGCTTCA CTGAGGTGGA GTGCTTTTTA AATCCTCAAA TGGGCAATCC TGATGAACAT CAAAAAAGGCT
TATCTTCAAG ATCCTCAATT TTGACCTCAT CTGTCGAAGT GACTCCACCT CACGAAAAAT TTAGGAGTTT ACCCGTTAGG ACTACTTGTA GTTTTTTCCGA
                              Acc I
```


(Continued)

```
      1710       1720       1730       1740       1750       1760       1770       1780  ------         1800
TAAGTAAAAG CTTAGCAGCT GAAAAACAGT TTACAGATGA CTCTCCAGAC AAAGAACAAC TGCCTTGCTA CAGTGTGGCT AGAATTCCTT TGCCTAATTT
ATTCATTTTC GAATCGTCGA CTTTTTGTCA AATGTCTACT GAGAGGTCTG TTTCTTGTTG ACGGAACGAT GTCACACCGA TCTTAAGGAA ACGGATTAAA
                                                                                        EcoR I

      1810       1820       1830       1840       1850       1860       1870       1880       1890       1900
AAATGAGGAC TTAACCTGTG GAAATATTTT GATGTGGGAA GCTGTTACTG TTAAAACTGA GGTTATTGGG GTAACTGCTA TGTTAAACTT GCATTCAGGG
TTTACTCCTG AATTGGACAC CTTTATAAAA CTACACCCTT CGACAATGAC AATTTTGACT CCAATAACCC CATTGACGAT ACAATTTGAA CGTAAGTCCC

      1910       1920       1930       1940       1950       1960       1970       1980       1990       2000
ACACAAAAAA CTCATGAAAA TGGTGCTGGA AAACCCATTC AAGGGTCAAA TTTTCATTTT TTTGCTGTTG GTGGGGAACC TTTGGAGCTG CAGGGTGTGT
TGTGTTTTTT GAGTACTTTT ACCACGACCT TTTGGGTAAG TTCCCAGTTT AAAAGTAAAA AAACGACAAC CACCCCTTGG AAACCTCGAC GTCCCACACA

      2010       2020       2030       2040       2050       2060       2070       2080       2090       2100
TAGCAAACTA CAGGACCAAA TATCCTGCTC AAACTGTAAC CCCAAAAAAT GCTACAGTTG ACAGTCAGCA GATGAACACT GACCACAAGG CTGTTTTGGA
ATCGTTTGAT GTCCTGGTTT ATAGGACGAG TTTGACATTG GGGTTTTTTA CGATGTCAAC TGTCAGTCGT CTACTTGTGA CTGGTGTTCC GACAAAACCT

      2110       2120       2130       2140       2150       2160       2170       2180       2190       2200
TAAGGATAAT GCTTATCCAG TGGAGTGCTG GGTTCCTGAT CCAAGTAAAA ATGAAAACAC TAGATATTTT GGAACCTACA CAGGTGGGGA AAATGTGCCT
ATTCCTATTA CGAATAGGTC ACCTCACGAC CCAAGGACTA GGTTCATTTT TACTTTTGTG ATCTATAAAA CCTTGGATGT GTCCACCCCT TTTACACGGA

      2210       2220       2230       2240       2250        --- ---   2270       2280       2290       2300
CCTGTTTTGC ACATTACTAA CACAGCAACC ACAGTGCTTC TTGATGAGCA GGGTGTTGGG CCCTTGTGCA AAGCTGACAG CTTGTATGTT TCTGCTGTTG
GGACAAAACG TGTAATGATT GTGTCGTTGG TGTCACGAAG AACTACTCGT CCCACAACCC GGGAACACGT TTCGACTGTC GAACATACAA AGACGACAAC
                                                             Apa I

      2310       2320       2330       2340       2350       2360       2370       2380       2390       2400
ACATTTGTGG GCTGTTTACC AACACTTCTG GAACACAGCA GTGGAAGGGA CTTCCCAGAT ATTTTAAAAT TACCCTTAGA AAGCGGTCTG TGAAAAACCC
TGTAAACACC CGACAAATGG TTGTGAAGAC CTTGTGTCGT CACCTTCCCT GAAGGGTCTA TAAAATTTTA ATGGGAATCT TTCGCCAGAC ACTTTTTGGG

      2410       2420       2430       2440       2450       2460       2470       2480       2490       2500
CTACCCAATT TCCTTTTTGT TAAGTGACCT AATTAACAGG AGGACACAGA GGGTGGATGG GCAGCCTATG ATTGGAATGT CCTCTCAAGT AGAGGAGGTT
GATGGGTTAA AGGAAAAACA ATTCACTGGA TTAATTGTCC TCCTGTGTCT CCCACCTACC CGTCGGATAC TAACCTTACA GGAGAGTTCA TCTCCTCCAA

                                                                               VP1 CDS End <---:
      2510       2520       2530  ------         2550       2560       2570       2580       2590       2600
AGGGTTTATG AGGACACAGA GGAGCTTCCT GGGGATCCAG ACATGATAAG ATACATTGAT GAGTTTGGAC AAACCACAAC TAGAATGCAG TGAAAAAAAT
TCCCAAATAC TCCTGTGTCT CCTCGAAGGA CCCCTAGGTC TGTACTATTC TATGTAACTA CTCAAACCTG TTTGGTGTTG ATCTTACGTC ACTTTTTTTA
                                   BamH I
```

```
      2610       2620       2630       2640       2650       2660       2670       2680       2690       2700
GCTTTATTTG TGAAATTTGT GATGCTATTG CTTTATTTGT AACCATTATA AGCTGCAATA AACAAGTTAA CAACAACAAT TGCATTCATT TTATGTTTCA
CGAAATAAAC ACTTTAAACA CTACGATAAC GAAATAAACA TTGGTAATAT TCGACGTTAT TTGTTCAATT GTTGTTGTTA ACGTAAGTAA AATACAAAGT
                                                                                                    :--->

      2710       2720       2730       2740       2750       2760          - ----- 2780       2790       2800
GGTTCAGGGG GAGGTGTGGG AGGTTTTTTA AAGCAAGTAA AACCTCTACA AATGTGGTAT GGCTGATTAT GATCATGAAC AGACTGTGAG GACTGAGGGG
CCAAGTCCCC CTCCACACCC TCCAAAAAAT TTCGTTCATT TTGGAGATGT TTACACCATA CCGACTAATA CTAGTACTTG TCTGACACTC CTGACTCCCC
Large T EXON 2 CDS End                                                           Bcl I

      2810       2820       2830       2840       2850       2860       2870       2880       2890       2900
CCTGAAATGA GCCTTGGGAC TGTGAATCAA TGCCTGTTTC ATGCCCTGAG TCTTCCATGT TCTTCTCCCC ACCATCTTCA TTTTTATCAG CATTTTCCTG
GGACTTTACT CGGAACCCTG ACACTTAGTT ACGGACAAAG TACGGGACTC AGAAGGTACA AGAAGAGGGG TGGTAGAAGT AAAAATAGTC GTAAAAGGAC

      2910       2920       2930       2940       2950       2960       2970       2980       2990       3000
GCTGTCTTCA TCATCATCAT CACTGTTTCT TAGCCAATCT AAAACTCCAA TTCCCATAGC CACATTAAAC TTCATTTTTT GATACACTGA CAAACTAAAC
CGACAGAAGT AGTAGTAGTA GTGACAAAGA ATCGGTTAGA TTTTGAGGTT AAGGGTATCG GTGTAATTTG AAGTAAAAAA CTATGTGACT GTTTGATTTG

      3010       3020       3030       3040       3050       3060       3070       3080       3090       3100
TCTTTGTCCA ATCTCTCTTT CCACTCCACA ATTCTGCTCT GAATACTTTG AGCAAACTCA GCCACAGGTC TGTACCAAAT TAACATAAGA AGCAAAGCAA
AGAAACAGGT TAGAGAGAAA GGTGAGGTGT TAAGACGAGA CTTATGAAAC TCGTTTGAGT CGGTGTCCAG ACATGGTTTA ATTGTATTCT TCGTTTCGTT

      3110       3120       3130       3140       3150       3160       3170       3180       3190       3200
TGCCACTTTG AATTATTCTC TTTTCTAACA AAAACTCACT GCGTTCCAGG CAATGCTTTA AATAATCTTT GGGCCTAAAA TCTATTTGTT TTACAAATCT
ACGGTGAAAC TTAATAAGAG AAAAGATTGT TTTTGAGTGA CGCAAGGTCC GTTACGAAAT TTATTAGAAA CCCGGATTTT AGATAAACAA AATGTTTAGA

      3210       3220       3230       3240       3250       3260       3270       3280       3290       3300
GGCCTGCAGT GTTTTAGGCA CACTGTACTC ATTCATGGTG ACTATTCCAG GGGGAAATAT TTGAGTTCTT TTATTTAGGT GTTTCTTTTC TAAGTTTACC
CCGGACGTCA CAAAATCCGT GTGACATGAG TAAGTACCAC TGATAAGGTC CCCCTTTATA AACTCAAGAA AATAAATCCA CAAAGAAAAG ATTCAAATGG

      3310       3320       3330       3340       3350       3360       3370       3380       3390       3400
TTAACACTGC CATCCAAATA ATCCCTTAAA TTGTCCAGGT TATTAATTCC CTGACCTGAA GGCAAATCTC TGGACTCCCC TCCAGTGCCC TTTACATCCT
AATTGTGACG GTAGGTTTAT TAGGGAATTT AACAGGTCCA ATAATTAAGG GACTGGACTT CCGTTTAGAG ACCTGAGGGG AGGTCACGGG AAATGTAGGA

      3410       3420       3430       3440       3450       3460       3470       3480       3490       3500
CAAAAACTAC TAAAAACTGG TCAATAGCTA CTCCTAGCTC AAAGTTCAGC CTGTCCAAGG GCAAATTAAC ATTTAAAGCT TTCCCCCCAC ATAATTCAAG
GTTTTTGATG ATTTTTGACC AGTTATCGAT GAGGATCGAG TTTCAAGTCG GACAGGTTCC CGTTTAATTG TAAATTTCGA AAGGGGGGTG TATTAAGTTC
```

(Continued)

3510	3520	3530	3540	3550	3560	3570	3580	3590	3600
CAAAGCAGCT	GCTAATGTAG	TTTTACCACT	ATCAATTGGT	CCTTTAAACA	GCCAGTATCT	TTTTTTAGGA	ATGTTGTACA	CCATGCATTT	TAAAAAGTCA
GTTTCGTCGA	CGATTACATC	AAAATGGTGA	TAGTTAACCA	GGAAATTTGT	CGGTCATAGA	AAAAAATCCT	TACAACATGT	GGTACGTAAA	ATTTTTCAGT
3610	3620	3630	3640	3650	3660	3670	3680	3690	3700
TACACCACTG	AATCCATTTT	GGGCAACAAA	CAGTGTAGCC	AAGCAACTCC	AGCCATCCAT	TCTTCTATGT	CAGCAGAGCC	TGTAGAACCA	AACATTATAT
ATGTGGTGAC	TTAGGTAAAA	CCCGTTGTTT	GTCACATCGG	TTCGTTGAGG	TCGGTAGGTA	AGAAGATACA	GTCGTCTCGG	ACATCTTGGT	TTGTAATATA
3710	3720	3730	3740	3750	3760	3770	3780	3790	3800
CCATCCTATC	CAAAAGATCA	TTAAATCTGT	TTGTTAACAT	TTGTTCTCTA	GTTAATTGTA	GGCTATCAAC	CCGCTTTTTA	GCTAAAACAG	TATCAACAGC
GGTAGGATAG	GTTTTCTAGT	AATTTAGACA	AACAATTGTA	AACAAGAGAT	CAATTAACAT	CCGATAGTTG	GGCGAAAAAT	CGATTTTGTC	ATAGTTGTCG
3810	3820	3830	3840	3850	3860	3870	3880	3890	3900
CTGTTGGCAT	ATGGTTTTTT	GGTTTTTGCT	GTCAGCAAAT	ATAGCAGCAT	TTGCATAATG	CTTTTCATGG	TACTTATAGT	GGCTGGGCTG	TTCTTTTTTA
GACAACCGTA	TACCAAAAAA	CCAAAAACGA	CAGTCGTTTA	TATCGTCGTA	AACGTATTAC	GAAAAGTACC	ATGAATATCA	CCGACCCGAC	AAGAAAAAAT
3910	3920	3930	3940	3950	3960	3970	3980	3990	4000
ATACATTTTA	AACACATTTC	AAAACTGTAC	TGAAATTCCA	AGTACATCCC	AAGCAATAAC	AACACATCAT	CACATTTTGT	TTCCATTGCA	TACTCTGTTA
TATGTAAAAT	TTGTGTAAAG	TTTTGACATG	ACTTTAAGGT	TCATGTAGGG	TTCGTTATTG	TTGTGTAGTA	GTGTAAAACA	AAGGTAACGT	ATGAGACAAT
4010	4020	4030	4040	4050	4060	4070	4080	4090	4100
CAAGCTTCCA	GGACACTTGT	TTAGTTTCCT	CTGCTTCTTC	TGGATTAAAA	TCATGCTCCT	TTAACCCACC	TGGCAAACTT	TCCTCAATAA	CAGAAAATGG
GTTCGAAGGT	CCTGTGAACA	AATCAAAGGA	GACGAAGAAG	ACCTAATTTT	AGTACGAGGA	AATTGGGTGG	ACCGTTTGAA	AGGAGTTATT	GTCTTTTACC
4110	4120	4130	4140	4150	4160	4170	4180	4190	4200
ATCTCTAGTC	AAGGCACTAT	ACATCAAATA	TTCCTTATTA	ACCCCTTTAC	AAATTAAAAA	GCTAAAGGTA	CACAATTTTT	GAGCATAGTT	ATTAATAGCA
TAGAGATCAG	TTCCGTGATA	TGTAGTTTAT	AAGGAATAAT	TGGGGAAATG	TTTAATTTTT	CGATTTCCAT	GTGTTAAAAA	CTCGTATCAA	TAATTATCGT
4210	4220	4230	4240	4250	4260	4270	4280	4290	4300
GACACTCTAT	GCCTGTGTGG	AGTAAGAAAA	AACAGTATGT	TATGATTATA	ACTGTTATGC	CTACTTATAA	AGGTTACAGA	ATATTTTTCC	ATAATTTTCT
CTGTGAGATA	CGGACACACC	TCATTCTTTT	TTGTCATACA	ATACTAATAT	TGACAATACG	GATGAATATT	TCCAATGTCT	TATAAAAAGG	TATTAAAAGA
4310	4320	4330	4340	4350	4360	4370	4380	4390	4400
TGTATAGCAG	TGCAGCTTTT	TCCTTTGTGG	TGTAAATAGC	AAAGCAAGCA	AGAGTTCTAT	TACTAAACAC	AGCATGACTC	AAAAAACTTA	GCAATTCTGA
ACATATCGTC	ACGTCGAAAA	AGGAAACACC	ACATTTATCG	TTTCGTTCGT	TCTCAAGATA	ATGATTTGTG	TCGTACTGAG	TTTTTTGAAT	CGTTAAGACT
4410	4420	4430	4440	4450	4460	4470	4480	4490	4500
AGGAAAGTCC	TTGGGGTCTT	CTACCTTTCT	CTTCTTTTTT	GGAGGAGTAG	AATGTTGAGA	GTCAGCAGTA	GCCTCATCAT	CACTAGATGG	CATTTCTTCT
TCCTTTCAGG	AACCCCAGAA	GATGGAAAGA	GAAGAAAAAA	CCTCCTCATC	TTACAACTCT	CAGTCGTCAT	CGGAGTAGTA	GTGATCTACC	GTAAAGAAGA

```
      4510       4520       4530       4540       4550       4560       4570       4580       4590       4600
GAGCAAAACA GGTTTTCCTC ATTAAAGGCA TTCCACCACT GCTCCCATTC ATCAGTTCCA TAGGTTGGAA TCTAAAATAC ACAAACAATT AGAATCAGTA
CTCGTTTTGT CCAAAAGGAG TAATTTCCGT AAGGTGGTGA CGAGGGTAAG TAGTCAAGGT ATCCAACCTT AGATTTTATG TGTTTGTTAA TCTTAGTCAT
                                                 Large T EXON 2 CDS Start <---¦¦--->       Large T INTRON

      4610       4620       4630       4640       4650       4660       4670       4680       4690       4700
GTTTAACACA TTATACACTT AAAAATTTTA TATTTACCTT AGAGCTTTAA ATCTCTGTAG GTAGTTTGTC CAATTATGTC ACACCACAGA AGTAAGGTTC
CAAATTGTGT AATATGTGAA TTTTTAAAAT ATAAATGGAA TCTCGAAATT TAGAGACATC CATCAAACAG GTTAATACAG TGTGGTGTCT TCATTCCAAG
                                            ¦--->  Small t CDS End

      4710       4720       4730       -- --      4750         -- ----------      4780       4790       4800
CTTCACAAAG ATCAAGTCCA AACCACATTC TAAAGCAATC GAAGCAGTAG CAATCAACCC ACACAAGTGG ATCTTTCCTG TATAATTTTC TATTTTCATG
GAAGTGTTTC TAGTTCAGGT TTGGTGTAAG ATTTCGTTAG CTTCGTCATC GTTAGTTGGG TGTGTTCACC TAGAAAGGAC ATATTAAAAG ATAAAAGTAC
                                         Taq I                     BstX I

      4810       4820       4830       4840       4850       4860       4870       4880       4890       4900
CTTCATCCTC AGTAAGCACA GCAAGCATAT GCAGTTAGCA GACATTTTCT TTGCACACTC AGGCCATTGT TTGCAGTACA TTGCATCAAC ACCAGGATTT
GAAGTAGGAG TCATTCGTGT CGTTCGTATA CGTCAATCGT CTGTAAAAGA AACGTGTGAG TCCGGTAACA AACGTCATGT AACGTAGTTG TGGTCCTAAA

      4910       4920       4930       4940       4950       4960       4970       4980       4990       5000
AAGGAAGAAG CAAATACCTC AGTTGCATCC CAGAAGCCTC CAAAGTCAGG TTGATGAGCA TATTTTACTC CATCTTCCAT TTTCTTGTAC AGAGTATTCA
TTCCTTCTTC GTTTATGGAG TCAACGTAGG GTCTTCGGAG GTTTCAGTCC AACTACTCGT ATAAAATGAG GTAGAAGGTA AAAGAACATG TCTCATAAGT
  Lg-T IVS  <---¦¦---> Large T EXON 1 CDS End

      5010       5020       5030       5040       5050       5060       5070       5080       5090       5100
TTTTCTTCAT TTTTTCTTCA TCTCCTCCTT TATCAGGATG AAACTCCTTG CATTTTTTTA AATATGCCTT TCTCATCAGA GGAATATTCC CCCAGGCACT
AAAAGAAGTA AAAAAGAAGT AGAGGAGGAA ATAGTCCTAC TTTGAGGAAC GTAAAAAAAT TTATACGGAA AGAGTAGTCT CCTTATAAGG GGGTCCGTGA

      5110       5120       5130       5140       5150       5160       5170       5180       5190       5200
CCTTTCAAGA CCTAGAAGGT CCATTAGCTG CAAAGATTCC TCTCTGTTTA AAACTTTATC CATCTTTGCA AAGCTTTTTG CAAAAGCCTA GGCCTCCAAA
GGAAAGTTCT GGATCTTCCA GGTAATCGAC GTTTCTAAGG AGAGACAAAT TTTGAAATAG GTAGAAACGT TTCGAAAAAC GTTTTCGGAT CCGGAGGTTT
                              Large T EXON 1, Small t CDS Start <---¦                  ¦          ¦........
                                                                                       ¦ T-ag Binding Site

      5210       5220       5230       5240
AAAGCCTCCT CACTACTTCT GGAATAGCTC AGAGGCCGAG GCG
TTTCGGAGGA GTGATGAAGA CCTTATCGAG TCTCCGGCTC CGC
.............ORGRPL CORE.......................
I   <---¦                       ¦-> T-ag Binding Site II
```

Restriction Enzyme Sites in SV40 Virus DNA

Restriction enzyme[1,2]	Recognition pattern[3]	Number of sites	Base number matched[4]
Acc I	GT/(AC)(GT)AC	1	1628
[Afl II]	C/TTAAG	1	1699
Aha III	TTT/AAA	12	1657, 1798, 2363, 2727, 3157, 3472, 3543, 3589, 3907, 4646, 5057, 5147
Alu I	AG/CT	34	271, 406, 435, 612, 639, 792, 804, 1047, 1097, 1350, 1494, 1543, 1581, 1634, 1709, 1717, 1840, 1986, 2272, 2279, 2523, 2651, 3426, 3436, 3477, 3507, 3780, 4003, 4160, 4314, 4643, 5126, 5172, 5226
Apa I	GGGCC/C	1	2258
Ava II	G/G(AT)CC	6	557, 588, 1018, 2013, 3538, 5118
[Avr II]	C/CTAGG	2	1078, 5187
BamH I	G/GATCC	1	2533
Ban I	G/G(CT)(AG)CC	1	294
Ban II	G(AG)GC(TC)/C	2	776, 2258
Bbv I->	GCAGC(N)$_8$/	7	802, 1541, 1715, 2461, 3505, 3844, 4312
<-Bbv I	/(N)$_{12}$GCTGC	15	340, 442, 568, 613, 616, 619, 649, 664, 667, 733, 784, 1987, 2652, 3508, 5127
[BbvS I]	GC(AT)GC	22	340, 442, 568, 613, 616, 619, 649, 664, 667, 733, 784, 802, 1541, 1715, 1987, 2461, 2652, 3505, 3508, 3844, 4312, 5127
Bcl I	T/GATCA	1	2770

[1] Restriction enzyme names in brackets (e.g., [AflII]) are not commercially available as of September, 1985.
[2] The direction of an asymmetric cut is indicated by -> or <- in the restriction enzyme name.
[3] The restriction enzyme recognition sequence (5' - - >3'). The position of endonuclease cleavage is indicated (/). Ambiguous bases in the recognition sequence are enclosed in parentheses.
[4] Numbers represent the first base in the sequence of the recognition site for a restriction endonuclease.

Restriction Enzyme Sites in SV40 Virus DNA (*Continued*)

Restriction enzyme	Recognition pattern	Number of sites	Base number matched
Bgl I	GCCNNNN/NGGC	1	5235
[Bin I->]	GGATC(N)$_4$/	4	873, 2533, 4099, 4769
[<-Bin I]	/(N)$_5$GATCC	2	2138, 2534
Bsm I	GAATGCN/	1	2583
Bsm I	G/CATTC	3	1891, 2682, 4528
Bsp1286 I	G(GAT)GC(CAT)/C	4	776, 1288, 2258, 3385
BstN I	CC/(AT)GG	17	160, 177, 232, 358, 910, 964, 1408, 1535, 2528, 2897, 3146, 3247, 3335, 4008, 4069, 4892, 5092
BstX I	CCANNNNN/NTGG	1	4759
[Cfr10 I]	(AG)CCGG(CT)	1	345
Dde I	C/TNAG	20	287, 722, 1337, 1436, 1641, 1711, 1857, 2375, 2793, 2846, 2929, 3057, 3290, 4387, 4499, 4638, 4808, 4858, 4918, 5228
Dpn I	GmA/TC	8	874, 2138, 2534, 2771, 3716, 4100, 4710, 4770
[Dra II]	(AG)G/GNCC(CT)	3	587, 2258, 2797
EcoR I*	/AATT	39	27, 119, 191, 447, 594, 645, 672, 729, 1170, 1186, 1385, 1397, 1783, 1796, 1949, 2368, 2407, 2431, 2614, 2678, 2949, 3030, 3078, 3111, 3329, 3345, 3464, 3493, 3534, 3754, 3934, 4152, 4174, 4293, 4393, 4587, 4624, 4672, 4784
[Eco47 III]	AGCGCT	1	832
EcoR I	G/AATTC	1	1782

(*Continued*)

Restriction Enzyme Sites in SV40 Virus DNA (*Continued*)

Restriction enzyme	Recognition pattern	Number of sites	Base number matched
EcoR I'	(AG)(AG)A/T(CT)(CT)	24	628, 796, 856, 924, 1384, 1396, 1660, 1782, 1948, 2533, 2613, 3178, 3195, 3364, 3610, 3723, 3933, 4099, 4568, 4623, 4649, 4769, 4895, 5134
EcoR II	/CC(AT)GG	17	160, 177, 232, 358, 910, 964, 1408, 1535, 2528, 2897, 3146, 3247, 3335, 4008, 4069, 4892, 5092
EcoR V	GAT/ATC	1	768
[Esp I]	GC/TNAGC	1	1710
Fnu4H I	GC/NGC	24	340, 442, 568, 613, 616, 619, 649, 661, 664, 667, 733, 784, 802, 1541, 1715, 1987, 2461, 2652, 3505, 3508, 3844, 4312, 5127, 5241
Fok I->	GGATG$(N)_9$/	4	80, 1442, 2455, 5036
<-Fok I	/$(N)_{13}$CATCC	7	3311, 3395, 3654, 3702, 3945, 4804, 4926
[Gsu I]	CTGGAG	3	653, 965, 1623
[Gsu I]	CTCCAG	3	1743, 3380, 3647
[Hae I]	(AT)GG/CC(AT)	11	330, 360, 708, 741, 755, 934, 1234, 1461, 3200, 4861, 5190
Hae II	(AG)GCGC/(CT)	1	832
Hae III	GG/CC	19	6, 331, 361, 660, 709, 742, 756, 935, 1235, 1462, 1507, 2259, 2799, 3172, 3201, 4862, 5191, 5234, 5243
[HgiE II]	ACCNNNNNNGGT	1	2174
Hha I	GCG/C	2	343, 833
Hinc II	GT(CT)/(AG)AC	7	470, 499, 519, 2057, 2297, 2666, 3733

Restriction Enzyme Sites in SV40 Virus DNA (*Continued*)

Restriction enzyme	Recognition pattern	Number of sites	Base number matched
Hind III	A/AGCTT	6	1046, 1493, 1708, 3476, 4002, 5171
Hinf I	G/ANTC	10	1739, 2824, 2848, 3373, 3610, 4376, 4459, 4568, 4592, 5135
Hpa I	GTT/AAC	4	499, 519, 2666, 3733
Hpa II	C/CGG	1	346
Hph I->	GGTGA(N)$_8$/	3	1245, 1381, 3237
<-Hph I	/(N)$_7$TCACC	1	1085
Kpn I	GGTAC/C	1	294
Mae I	C/TAG	12	1079, 1459, 1610, 1779, 2160, 2580, 3434, 3748, 4105, 4483, 5112, 5188
Mae III	/GTNAC	14	352, 820, 866, 1333, 1477, 1844, 1871, 2036, 2424, 2639, 3238, 3997, 4273, 4678
Mbo I	/GATC	8	874, 2138, 2534, 2771, 3716, 4100, 4710, 4770
Mbo II->	GAAGA(N)$_8$/	4	461, 1106, 1501, 4904
<-Mbo II	/(N)$_7$TCTTC	12	2851, 2861, 2875, 2905, 3661, 4036, 4417, 4430, 4495, 4973, 5004, 5015
Mnl I->	CCTC(N)$_7$/	27	2, 8, 286, 310, 363, 719, 744, 1088, 1435, 1664, 2198, 2481, 2743, 3379, 3398, 4028, 4082, 4472, 4517, 4807, 4917, 4937, 5024, 5139, 5193, 5205, 5208
<-Mnl I	/(N)$_7$GAGG	24	319, 658, 724, 1150, 1341, 1402, 1596, 1643, 1805, 1859, 2440, 2449, 2492, 2495, 2510, 2519, 2711, 2720, 2788, 2795, 4442, 5079, 5232, 5238
Nae I	GCC/GGC	1	345
Nco I	C/CATGG	3	37, 333, 560

(*Continued*)

Restriction Enzyme Sites in SV40 Virus DNA (*Continued*)

Restriction enzyme	Recognition pattern	Number of sites	Base number matched
Nde I	CA/TATG	2	3808, 4826
Nla III	CATG/	17	38, 129, 201, 334, 384, 561, 1913, 2542, 2774, 2840, 2856, 3234, 3582, 3866, 4052, 4373, 4797
Nla IV	GGN/NCC	16	156, 228, 294, 587, 777, 1017, 1404, 1507, 1561, 1975, 2131, 2171, 2258, 2533, 2797, 4696
Nsi I	ATGCA/T	3	126, 198, 3583
[NspB II]	C(AC)G/C(GT)G	4	270, 662, 1716, 3506
[Nsp7524 I]	(AG)CATG/(CT)	2	128, 200
Pss I	(AG)GGNCC(CT)	3	587, 2258, 2797
Pst I	CTGCA/G	2	1988, 3204
Pvu II	CAG/CTG	3	270, 1716, 3506
Rsa I	GT/AC	12	295, 970, 1467, 3072, 3225, 3576, 3870, 3927, 3942, 4168, 4876, 4987
Sau96 I	G/GNCC	11	557, 588, 1018, 1507, 2013, 2258, 2259, 2798, 3171, 3538, 5118
ScrF I	CC/NGG	17	160, 177, 232, 358, 910, 964, 1408, 1535, 2528, 2897, 3146, 3247, 3335, 4008, 4069, 4892, 5092
SfaN I->	GCATC(N)$_5$/	3	688, 4883, 4925
<-SfaN I	/(N)$_9$GATGC	3	125, 197, 2621
Sfi I	GGCCNNNN/NGGCC	1	5234
Sph I	GCATG/C	2	128, 200
Ssp I	AAT/ATT	6	1303, 1823, 3256, 4127, 4280, 5083

Restriction Enzyme Sites in SV40 Virus DNA (*Continued*)

Restriction enzyme	Recognition pattern	Number of sites	Base number matched
Stu I	AGG/CCT	7	360, 708, 741, 755, 1234, 1461, 5190
Sty I	C/C(AT)(AT)GG	8	37, 333, 560, 1078, 2812, 3455, 4409, 5187
Taq I	T/CGA	1	4739
[Tce I]	GAAGA	4	461, 1106, 1501, 4904
[Tce I]	TCTTC	12	2851, 2861, 2875, 2905, 3661, 4036, 4417, 4430, 4495, 4973, 5004, 5015
Tthlll II->	CAA(AG)CA$(N)_9$/	11	3497, 3627, 3640, 3689, 3950, 4345, 4582, 4822
<-Tthlll II	/$(N)_{11}$TG(CT)TTG	3	423, 3728, 4868
Xho II	(AG)/GATC(CT)	3	2533, 4099, 4769

Restriction Enzymes That Do Not Cut SV40 DNA

Enzyme [1,2]	Recognition site[3]	Enzyme	Recognition site	Enzyme	Recognition site
Aat II	GACGT/C	[Gdi II]	CGGCC/(AG)	Sac I	GAGCT/C
[Afl III]	A/C(AG)(CT)GT	Hga I->	GACGC$(N)_5$/	Sac II	CCGC/GG
Aha II	G(AG)/CG(CT)C	<-Hga I	/$(N)_{10}$GCGTC	Sal I	G/TCGAC
Asu II	TT/CGAA	HgiA I	G(AT)GC(AT)/C	Sca I	AGT/ACT
Ava I	C/(CT)CG(AG)G	[Mae II]	A/CGT	Sma I	CCC/GGG
Bal I	TGG/CCA	Mlu I	A/CGCGT	[Sna I]	GTATAC
Bgl II	A/GATCT	Mst II	CC/TNAGG	SnaB I	TAC/GTA
BssH II	G/CGCGC	Nar I	GG/CGCC	Spe I	A/CTAGT
BstE II	G/GTNACC	Nci I	CC/(GC)GG	Tha I	CG/CG
[Cfr I]	(CT)/GGCC(AG)	Nhe I	G/CTAGC	Tth111 I	GACN/NNGTC
Cla I	AT/CGAT	Not I	GC/GGCCGC	Xba I	T/CTAGA
[Dra III]	CACNNN/GTG	Nru I	TCG/CGA	Xho I	C/TCGAG
Fsp I	TGC/GCA	Pvu I	CGAT/CG	Xma III	C/GGCCG
[Gdi II]	(CT)/GGCCG	Rsr II	CG/G(AT)CCG	Xmn I	GAANN/NNTTC

[1] Restriction enzyme names in brackets (e.g., [AflII]) are not commercially available as of September, 1985.
[2] The direction of an asymmetric cut is indicated by -> or <- in the restriction enzyme name.
[3] The restriction enzyme recognition sequence (5'-->3'). The position of endonuclease cleavage is indicated (/). Ambiguous bases in the recognition sequence are enclosed in parentheses.

```
Palindrome ORGREP  <---¦.......................ORGREP CORE..................................................¦
        10        20 ¦       30         40         50         60         70         80      -----  ---------
GGGGGCCCCT GGCCTCCGCT TACTCTGGAG AAAAAGAAGA GAGGCATTGT AGAGGCTTCC AGAGGCAACT TGTCAAAACA GGACTGGCGC CTTGGAGGCG
CCCCCGGGGA CCGGAGGCGA ATGAGACCTC TTTTTCTTCT CTCCGTAACA TCTCCGAAGG TCTCCGTTGA ACAGTTTTGT CCTGACCGCG GAACCTCCGC
                                                                                               Nar I Bgl I

                                                                           ¦---> T antigens CDS Start
       110        120        130        140        150        160        170 ¦      180        190        200
CTGTGGGGCC ACCCAAATTG ATATAATTAA GCCCCAACCG CCTCTTCCCG CCTCATTTCA GCCTCACCAC CATCATGGAT AGAGTTCTGA GCAGAGCTGA
GACACCCCGG TGGGTTTAAC TATATTAATT CGGGGTTGGC GGAGAAGGGC GGAGTAAAGT CGGAGTGGTG GTAGTACCTA TCTCAAGACT CGTCTCGACT

       210        220        230        240        250        260        270        280        290        300
CAAAGAAAGG CTGCTAGAAC TTCTAAAACT TCCCAGACAA CTATGGGGGG ATTTTGGAAG AATGCAGCAG GCATATAAGC AGCAGTCACT GCTACTGCAC
GTTTCTTTCC GACGATCTTG AAGATTTTGA AGGGTCTGTT GATACCCCCC TAAAACCTTC TTACGTCGTC CGTATATTCG TCGTCAGTGA CGATGACGTG

       310        320        330        340        350    --- ---    370        380        390        400
CCAGACAAAG GTGGAAGCCA TGCCTTAATG CAGGAATTGA ACAGTCTCTG GGGAACATTT AAAACTGAAG TATACAATCT GAGAATGAAT CTAGGAGGAA
GGTCTGTTTC CACCTTCGGT ACGGAATTAC GTCCTTAACT TGTCAGAGAC CCCTTGTAAA TTTTGACTTC ATATGTTAGA CTCTTACTTA GATCCTCCTT
                                                          Aha III

Large T
EXON 1 End
    <---¦
       410 ¦      420        430        440        450        460        470        480        490        500
CCGGCTTCCA GGTAAGAAGG CTACATGCGG ATGGGTGGAA TCTAAGTACC AAAGACACCT TTGGTGATAG ATACTACCAG CGGTTCTGCA GAATGCCTCT
GGCCGAAGGT CCATTCTTCC GATGTACGCC TACCCACCTT AGATTCATGG TTTCTGTGGA AACCACTATC TATGATGGTC GCCAAGACGT CTTACGGAGA

       510        520        530        540        550        560        570        580        590        600
TACCTGCCTA GTAAATGTTA AATACAGCTC ATGTAGTTGT ATATTATGCC TGCTTAGAAA GCAACATAGA GAGCTCAAAG ACAAATGTGA TGCCAGGTGC
ATGGACGGAT CATTTACAAT TTATGTCGAG TACATCAACA TATAATACGG ACGAATCTTT CGTTGTATCT CTCGAGTTTC TGTTTACACT ACGGTCCACG

       610        620        630        640        650        660        670        680        690        700
CTAGTACTTG GAGAATGTTT TTGTCTTGAA TGTTACATGC AATGGTTTGG AACACCAACC CGAGATGTGC TGAACCTGTA TGCAGACTTC ATTGCAAGCA
GATCATGAAC CTCTTACAAA AACAGAACTT ACAATGTACG TTACCAAACC TTGTGGTTGG GCTCTACACG ACTTGGACAT ACGTCTGAAG TAACGTTCGT
```

(Continued)

```
                                                                            Large T, Small t, EXON 2 CDS Start
               Small, Middle t EXON 1 CDS End  <---:                                                     :->
       710        720        730        740        750        760        770        780        790       :800
TGCCTATAGA CTGGCTGGAC CTGGATGTGC ACAGCGTGTA TAATCCAAGT AAGTATCAAG AGGGCGGGTG GGTATTTACG GCCTATATTC TTACAGGGCT
ACGGATATCT GACCGACCTG GACCTACACG TGTCGCACAT ATTAGGTTCA TTCATAGTTC TCCCGCCCAC CCATAAATGC CGGATATAAG AATGTCCCGA

Small t EXON 2 CDS
End <---: :---> Middle T EXON 2 Start
       810        820        830        840        850        860        870        880        890        900
CTCCCCCTAG AACGGCGGAG CGAGGAACTG AGGAGAGCGG CCACAGTCCA CTACACGATG ACTACTGGTC ATTCAGCTAT GGAAGCAAGT ACTTCACAAG
GAGGGGGATC TTGCCGCCTC GCTCCTTGAC TCCTCTCGCC GGTGTCAGGT GATGTGCTAC TGATGACCAG TAAGTCGATA CCTTCGTTCA TGAAGTGTTC

       910        920        930        940        950        960        970        980        990       1000
GGAATGGAAT GATTTCTTCA GAAAGTGGGA CCCCAGCTAC CAGTCGCCGC CTAAGACTGC CGAGTCTTCT GAGCAACCCG ACCTATTCTG TTATGAGGAG
CCTTACCTTA CTAAAGAAGT CTTTCACCCT GGGGTCGATG GTCAGCGGCG GATTCTGACG GCTCAGAAGA CTCGTTGGGC TGGATAAGAC AATACTCCTC

      1010       1020       1030       1040       1050       1060       1070       1080       1090       1100
CCACTCCTAT CCCCCAACCC GAGTTCTCCA ACAGATACAC CCGCACATAC TGCTGGAAGA AGACGAAATC CTTGTGTTGC TGAGCCCGAT GACAGCATAT
GGTGAGGATA GGGGGTTGGG CTCAAGAGGT TGTCTATGTG GGCGTGTATG ACGACCTTCT TCTGCTTTAG GAACACAACG ACTCGGGCTA CTGTCGTATA

      1110       1120       1130       1140       1150       1160       1170       1180       1190       1200
CCCCGGACCC CCCCAGAACT CCTGTATCCA GAAAGCGACC AAGACCAGCT GGAGCCACTG GAGGAGGAGG AGGAGGAGTA CATGCCAATG GAGGATCTGT
GGGGCCTGGG GGGGTCTTGA GGACATAGGT CTTTCGCTGG TTCTGGTCGA CCTCGGTGAC CTCCTCCTCC TCCTCCTCAT GTACGGTTAC CTCCTAGACA

      1210       1220       1230       1240       1250       1260       1270       1280       1290       1300
ATTTGGACAT CCTACCGGGG GAACAAGTAC CCCAGCTCAT CCCCCCCCCT ATCATTCCCA GGGCGGGTCT GAGTCCATGG GAGGGTCTGA TTCTTCGGGA
TAAACCTGTA GGATGGCCCC CTTGTTCATG GGGTCGAGTA GGGGGGGGGA TAGTAAGGGT CCCGCCCAGA CTCAGGTACC CTCCCAGACT AAGAAGCCCT

      1310       1320       1330       1340       1350       1360       1370       1380       1390       1400
TTTGCAGAGG GCTCATTTCG ATCCGATCCT AGATGCGAGT CAGAGAATGA GAGCTACTCA CAGAGCTGCT CTCAGAGCTC ATTCAATGCA ACGCCACCTA
AAACGTCTCC CGAGTAAAGC TAGGCTAGGA TCTACGCTCA GTCTCTTACT CTCGATGAGT GTCTCGACGA GAGTCTCGAG TAAGTTACGT TGCGGTGGAT

                                                                             Middle T EXON 2 CDS End  <---:
      1410       1420       1430       1440       1450       1460       1470       1480       1490       1500
AGAAGGCTAG GGAGGACCCT GCTCCTAGTG ACTTTCCTAG CAGCCTTACT GGGTATTTGT CTCATGCTAT TTATTCTAAT AAAACGTTCC CGGCATTTCT
TCTTCCGATC CCTCCTGGGA CGAGGATCAC TGAAAGGATC GTCGGAATGA CCCATAAACA GAGTACGATA AATAAGATTA TTTTGCAAGG GCCGTAAAGA

-:    1510       1520       1530       1540       1550       1560  ------          1580       1590       1600
AGTATACTCC ACCAAAGAGA AATGCAAACA ATTATATGAT ACCATAGGGA AGTTCAGGCC CGAATTCAAA TGCCTGGTCC ATTATGAGGA GGGGGGCATG
TCATATGAGG TGGTTTCTCT TTACGTTTGT TAATATACTA TGGTATCCCT TCAAGTCCGG GCTTAAGTTT ACGGACCAGG TAATACTCCT CCCCCCGTAC
                                                                   EcoR I
```

```
      1610       1620       1630       1640       1650       1660       1670       1680       1690       1700
CTGTTCTTTC TAACTATGAC TAAGCACAGG GTTTCAGCAG TTAAGAATTA TTGCTCTAAG CTTTGCCGCA GCTTCCTAAT GTGTAAGGCA GTCACCAAGC
GACAAGAAAG ATTGATACTG ATTCGTGTCC CAAAGTCGTC AATTCTTAAT AACGAGATTC GAAACGGCGT CGAAGGATTA CACATTCCGT CAGTGGTTCG

      1710       1720       1730       1740       1750       1760       1770       1780       1790       1800
CTATGGAATG CTATCAAGTT GTAACCGCAG CACCATTTCA GTTAATAACA GAAAATAAGC CAGGCCTCCA CCAATTCGAG TTTACAGACG AGCCAGAAGA
GATACCTTAC GATAGTTCAA CATTGGCGTC GTGGTAAAGT CAATTATTGT CTTTTATTCG GTCCGGAGGT GGTTAAGCTC AAATGTCTGC TCGGTCTTCT

      1810       1820       1830       1840       1850       1860       1870       1880       1890       1900
ACAGAAAGCA GTAGACTGGA TTATGGTAGC AGACTTTGCA CTAGAAAACA ACCTTGATGA TCCCCTGTTA ATTATGGGGT ATTATCTTGA TTTTGCCAAA
TGTCTTTCGT CATCTGACCT AATACCATCG TCTGAAACGT GATCTTTTGT TGGAACTACT AGGGGACAAT TAATACCCCA TAATAGAACT AAAACGGTTT

      1910     ------       1930       1940       1950       1960       1970       1980       1990       2000
GAGGTTCCTT CATGCATAAA GTGTAGCAAA GAGGAAACCC GCCTCCAAAT ACATTGGAAA AACCATAGAA AGCATGCAGA GAATGCAGAC CTCTTCCTGA
CTCCAAGGAA GTACGTATTT CACATCGTTT CTCCTTTGGG CGGAGGTTTA TGTAACCTTT TTGGTATCTT TCGTACGTCT CTTACGTCTG GAGAAGGACT
             Nsi I

      2010       2020       2030       2040       2050       2060       2070       2080       2090       2100
ATTGTAAAGC TCAAAAGACA ATCTGTCAGC AGGCAGCTGC GAGTCTGGCA TCCAGGAGAC TGAAATTAGT AGAGTGTACC CGCAGCCAGC TATTAAAGGA
TAACATTTCG AGTTTTCTGT TAGACAGTCG TCCGTCGACG CTCAGACCGT AGGTCCTCTG ACTTTAATCA TCTCACATGG GCGTCGGTCG ATAATTTCCT

      2110       2120       2130       2140       2150       2160       2170       2180       2190       2200
GAGATTGCAA CAGTCTCTCC TCAGGCTAAA AGAACTTGGC TCCTCCGATG CTCTACTCTA CCTAGCAGGT GTCGCTTGGT ACCAGTGTCT TTTAGAGGAC
CTCTAACGTT GTCAGAGAGG AGTCCGATTT TCTTGAACCG AGGAGGCTAC GAGATGAGAT GGATCGTCCA CAGCGAACCA TGGTCACAGA AAATCTCCTG

      2210       2220       2230       2240       2250       2260       2270       2280       2290       2300
TTTCCTCAAA CCCTGTTTAA GATGCTTAAG CTGCTAACAG AAAATGTGCC AAAACGACGC AACATACTTT TTAGAGGACC AGTTAATTCA GGAAAGACAG
AAAGGAGTTT GGGACAAATT CTACGAATTC GACGATTGTC TTTTACACGG TTTTGCTGCG TTGTATGAAA AATCTCCTGG TCAATTAAGT CCTTTCTGTC

      2310       2320       2330       2340       2350       2360       2370       2380       2390       2400
GCCTAGCAGC CGCGCTTATT AGCCTGTTAG GAGGCAAGTC TCTCAACATA AATTGCCCTG CAGATAAACT TGCTTTTGAG CTTGGTGTGG CACAGGACCA
CGGATCGTCG GCGCGAATAA TCGGACAATC CTCCGTTCAG AGAGTTGTAT TTAACGGGAC GTCTATTTGA ACGAAAACTC GAACCACACC GTGTCCTGGT

      2410       2420       2430       2440       2450       2460       2470       2480       2490       2500
GTTTGTGGTC TGTTTTGAAG ATGTAAAGGG TCAAATAGCC TTGAACAAAC AACTGCAGCC AGGGATGGGA GTGGCTAATC TAGATAATCT CAGGACTACC
CAAACACCAG ACAAAACTTC TACATTTCCC AGTTTATCGG AACTTGTTTG TTGACGTCGG TCCCTACCCT CACCGATTAG ATCTATTAGA GTCCTGATGG

      2510       2520       2530       2540       2550       2560       2570       2580       2590       2600
TGGAATGGGA GTGTAAAGGT CAATCTAGAA AAAAAGCACA GCAACAAAAG GTCACAACTC TTTCCACCCT GTGTGTGTAC AATGAATGAA TATCTCCTAC
ACCTTACCCT CACATTTCCA GTTAGATCTT TTTTTCGTGT CGTTGTTTTC CAGTGTTGAG AAAGGTGGGA CACACACATG TTACTTACTT ATAGAGGATG
```


(*Continued*)

```
      2610       2620       2630       2640       2650       2660       2670       2680       2690       2700
CACAAACAGT ATGGGCCCGG TTTCACATGG TGTTGGATTT CACCTGCAAA CCCCATCTGG CCCAATCTCT TGAAAAGTGT GAATTTTTGC AAAGGGAAAG
GTGTTTGTCA TACCCGGGCC AAAGTGTACC ACAACCTAAA GTGGACGTTT GGGGTAGACC GGGTTAGAGA ACTTTTCACA CTTAAAAACG TTTCCCTTTC

      2710       2720       2730       2740       2750       2760       2770       2780       2790       2800
AATTATTCAG AGTGGAGATA CCCTTGCCCT ATTACTCATA TGGAATTTCA CTTCAGATGT ATTTGATCCT GATATTCAGG GTCTTGTGAA GGAGGTTCGT
TTAATAAGTC TCACCTCTAT GGGAACGGGA TAATGAGTAT ACCTTAAAGT GAAGTCTACA TAAACTAGGA CTATAAGTCC CAGAACACTT CCTCCAAGCA

      2810       2820       2830       2840       2850       2860       2870       2880       2890       2900
GACCAGTTTG CTAGTGAGTG CTCCTACAGT TTGTTTTGTG ATATACTTTG TAATGTGCAA GAAGGCGACG ACCCCTTGAA GGACATATGT GATATAGCTG
CTGGTCAAAC GATCACTCAC GAGGATGTCA AACAAAACAC TATATGAAAC ATTACACGTT CTTCCGCTGC TGGGGAACTT CCTGTATACA CTATATCGAC

Large T EXON 2
 CDS End <---¦
      2910       2920       2930       2940       2950       2960       2970       2980       2990       3000
AATACACAGT TTATTGAATA AACATTAATT TCCAGGAAAT ACAGTCTTTG TTTTTCCAAA GCGGTCAACA TAGCGCGTCA TATCAGGGTC CCCCGGTACA
TTATGTGTCA AATAACTTAT TTGTAATTAA AGGTCCTTTA TGTCAGAAAC AAAAAGGTTT CGCCAGTTGT ATCGCGCAGT ATAGTCCCAG GGGGCCATGT
                                ¦---> VP1 CDS End

      3010       3020       3030       3040       3050       3060       3070       3080       3090       3100
GGTTCAGTCC CATCATACAC TCTAACCTCC TCTACCTGGG TGTTCTCCCC TTCCATGGGT TGGCCCTGCA CTTGGGGGAG CATGTTGTTG AAAAGGGAAC
CCAAGTCAGG GTAGTATGTG AGATTGGAGG AGATGGACCC ACAAGAGGGG AAGGTACCCA ACCGGGACGT GAACCCCCTC GTACAACAAC TTTTCCCTTG

      3110       3120       3130       3140       3150       3160       3170       3180       3190       3200
TTATGAGGGA GGCCATGGGA TAGGGATTTT TGACCCATCT TTTTCTCAGG GTGATTTTGA AATATCTGGG AAAGCCCTTC TCCAGTGATG ACACATAGTT
AATACTCCCT CCGGTACCCT ATCCCTAAAA ACTGGGTAGA AAAAGAGTCC CACTAAAACT TTATAGACCC TTTCGGGAAG AGGTCACTAC TGTGTATCAA

      3210       3220       3230       3240       3250       3260       3270       3280       3290       3300
TCTTGTAACT CTCCAGCCCA TTATATCTAC ACAGGAGAGG TATAGGCCCT CTCCTTTACA GAGGGGCCCA ACTCCATTTT CATCTAGGAG CACAGTTGTC
AGAACATTGA GAGGTCGGGT AATATAGATG TGTCCTCTCC ATATCCGGGA GAGGAAATGT CTCCCCGGGT TGAGGTAAAA GTAGATCCTC GTGTCAACAG

      3310       3320       3330       3340       3350       3360       3370       3380       3390       3400
AGGGTGTTTG TGAACTGCAG GACGGGTGGA GCTGTTGTGC CTCCAGTGTA ATTGCCAAAG TACCTTGTGT TCTCATTTTT TGCTGGATCT GGATGCCAGA
TCCCACAAAC ACTTGACGTC CTGCCCACCT CGACAACACG GAGGTCACAT TAACGGTTTC ATGGAACACA AGAGTAAAAA ACGACCTAGA CCTACGGTCT

      3410       3420       3430       3440       3450       3460       3470       3480       3490       3500
TTTCAACTGG ATACATTCCG TCCTTATCCA GCTTGGCCTT GCTAATTGGA TTCAGGACTT GGTCTTTGTT GACCATGTCC TTCTTTGTGA TTGTTTTGAT
AAAGTTGACC TATGTAAGGC AGGAATAGGT CGAACCGGAA CGATTAACCT AAGTCCTGAA CCAGAAACAA CTGGTACAGG AAGAAACACT AACAAAACTA
```

```
      3510       3520       3530       3540       3550       3560       3570       3580       3590       3600
TGTTACTACC CCTTCTTCCT TGTATTTTGT TCTGGCATCT GTCACAAGTC CCTGGAGGTC AAGCGGTTCC CCGCCACCAG CAAACACATG ATATTGGCTG
ACAATGATGG GGAAGAAGGA ACATAAAACA AGACCGTAGA CAGTGTTCAG GGACCTCCAG TTCGCCAAGG GGCGGTGGTC GTTTGTGTAC TATAACCGAC

      3610       3620       3630       3640       3650       3660       3670       3680       3690       3700
CCTTCCACTG GAGTGGAATT TCCTTTGTGT TTACTGAATC TGTGGGTTTT GTTGAACCCA TGCACATCTA ACAGTGAGCC AGAGCCCACC ACCTCGGTTT
GGAAGGTGAC CTCACCTTAA AGGAAACACA AATGACTTAG ACACCCAAAA CAACTTGGGT ACGTGTAGAT TGTCACTCGG TCTCGGGTGG TGGAGCCAAA

      3710       3720       3730       3740       3750       3760       3770       3780       3790       3800
TCACTGAGAC TGCCTCCCAC ATTTGTAGGG TGTCACAGGT GAGGTCCTCA TTGAGGCATG GGAAGCTGGA CTTTGCCATA CTCCATGTGG GAAGTGTATT
AGTGACTCTG ACGGAGGGTG TAAACATCCC ACAGTGTCCA CTCCAGGAGT AACTCCGTAC CCTTCGACCT GAAACGGTAT GAGGTACACC CTTCACATAA

      3810       3820       3830       3840       3850       3860       3870       3880       3890       3900
ATTTCGGGGA ATCCACGTAT CTGATGTAGC CAAATTAATC CCTCTGCTCC AACCATAGTA TTGCCCTCCC TCTGTTAGGC TTTCAGGGGT GGGTGGCTGC
TAAAGCCCCT TAGGTGCATA GACTACATCG GTTTAATTAG GGAGACGAGG TTGGTATCAT AACGGGAGGG AGACAATCCG AAAGTCCCCA CCCACCGACG

      3910       3920       3930       3940       3950       3960       3970       3980       3990       4000
CCCATTCTGG GGTTCAGAAA AGCTTCTATT TCTGTCACAC TGTCTGGCCC TGTCACAAGG TCCAGCACCT CCATACCCCC TTTAATAAGC AGTTTGGGAA
GGGTAAGACC CCAAGTCTTT TCGAAGATAA AGACAGTGTG ACAGACCGGG ACAGTGTTCC AGGTCGTGGA GGTATGGGGG AAATTATTCG TCAAACCCTT

      4010       4020       4030       4040       4050       4060       4070       4080       4090       4100
CGGGTGCGGG TCTTGGACAG GCCTTTGTAC ATTTTGTCTC GCATTTAGAG ACGCCGCTTT TTCTTTTGGG GGCCATCTTC CTCTATGACT GTTGCCCAAG
GCCCACGCCC AGAACCTGTC CGGAAACATG TAAAACAGAG CGTAAATCTC TGCGGCGAAA AAGAAAACCC CCGGTAGAAG GAGATACTGA CAACGGGTTC
                                                     |---> VP2, VP3 CDS End             |
                                                                 VP1 CDS Start  <---|

      --- ---    4120       4130       4140       4150       4160       4170       4180       4190       4200
TAGGTGTGAT ATCACCGTAC AGCCCTAGAA TTAAAGGAAG CATCCAGTCA GGAGTGACTC TTTGGTGGGA CACCACCTGG GCCTGATAGA ACTTTATCAC
ATCCACACTA TAGTGGCATG TCGGGATCTT AATTTCCTTC GTAGGTCAGT CCTCACTGAG AAACCACCCT GTGGTGGACC CGGACTATCT TGAAATAGTG
    EcoR V

      4210       4220       4230       4240       4250       4260       4270       4280       4290       4300
CTCACCTGAC TCATCCTGTA TATGTGCTGC AGGGGTAGGC CCACCATTTC CCATGCTCCC TTCTATGCGA TTAAAGAGTG CTCTCCTCTG TGGTGGATTA
GAGTGGACTG AGTAGGACAT ATACACGACG TCCCCATCCG GGTGGTAAAG GGTACGAGGG AAGATACGCT AATTTCTCAC GAGAGGAGAC ACCACCTAAT

      4310       4320       4330       4340       4350       4360       4370       4380       4390       4400
AGACCTAGTT GCCTATAATA GTCACTTAGT GATGAGTACC CTGAGCTCGC TGATGAGGCA CCTCTGTTGA TTGCATCTAT GGCTGACTGA GGAGCATTAG
TCTGGATCAA CGGATATTAT CAGTGAATCA CTACTCATGG GACTCGAGCG ACTACTCCGT GGAGACAACT AACGTAGATA CCGACTGACT CCTCGTAATC

      4410       4420       4430       4440       4450       4460       4470       4480       4490       4500
AAACCACCCA CCGGGTGTTT TCAAGTAGCC TAGCTAGGCC ATCCAGGAAT GTATGTGTCT GTCTTACAGT TAGTTCTCTT ACAGCTCCTT CCAGTCTGTG
TTTGGTGGGT GGCCCACAAA AGTTCATCGG ATCGATCCGG TAGGTCCTTA CATACACAGA CAGAATGTCA ATCAAGAGAA TGTCGAGGAA GGTCAGACAC
```


(Con tinued)

The Nucleotide Sequence of Polyoma Virus (A2 Strain) DNA (*Continued*)

```
      4510       4520       4530       4540       4550       4560       4570       4580       4590       4600
TTGTGTTTCC TGCACAACCA TTTGCCACAC ATATCTTCCC ACGGAATGAA GTAGGCCATG GCCCCAATCA TGTACTACAT TTAGAGCATG AGCAAACTGA
AACACAAAGG ACGTGTTGGT AAACGGTGTG TATAGAAGGG TGCCTTACTT CATCCGGTAC CGGGGTTAGT ACATGATGTA AATCTCGTAC TCGTTTGACT

      4610       4620       4630       4640       4650       4660       4670       4680       4690       4700
TTAACTCCTG GGAAGTATAT ATCGAGAAGG GCTGGATCCC GCCATGGTAT CAACGCCATA TTTCTATTTA CAGTAGGGAC CTCTTCGTTG TGTAGGTACC
AATTGAGGAC CCTTCATATA TAGCTCTTCC CGACCTAGGG CGGTACCATA GTTGCGGTAT AAAGATAAAT GTCATCCCTG GAGAAGCAAC ACATCCATGG
                                            VP3 CDS Start  <---¦

      4710       4720       4730       4740       4750       4760       4770       4780       4790       4800
GCTGTATTCC TAGGGAAATA GTAGAGGCAC CTTGAACTGT CTGCATCAGC CATATAGCCC CCGCTGTTCG ACTTACAAAC ACAGGCACAG TACTGACAAA
CGACATAAGG ATCCCTTTAT CATCTCCGTG GAACTTGACA GACGTAGTCG GTATATCGGG GGCGACAAGC TGAATGTTTG TGTCCGTGTC ATGACTGTTT

      4810       4820       4830       4840       4850       4860       4870       4880       4890       4900
CCCATACACC TCCTCTGAAA TACCCATAGT TGCTAGGGCT GTCTCCGAAC TCATTACACC CTCCAAAGTC AGAGCTGTAA TTTCGCCATC AAGGGCAGCG
GGGTATGTGG AGGAGACTTT ATGGGTATCA ACGATCCCGA CAGAGGCTTG AGTAATGTGG GAGGTTTCAG TCTCGACATT AAAGCGGTAG TTCCCGTCGC

      4910       4920       4930       4940       4950       4960       4970       4980       4990       5000
AGGGCTTCTC CAGATAAAAT AGCTTCTGCC GAGAGTCCCG TAAGGGTAGA CACTTCAGCT AATCCCTCGA TGAGGTCTAC TAGAATAGTC AGTGCGGCTC
TCCCGAAGAG GTCTATTTTA TCGAAGACGG CTCTCAGGGC ATTCCCATCT GTGAAGTCGA TTAGGGAGCT ACTCCAGATG ATCTTATCAG TCACGCCGAG

                                           ¦---> ORGREP alpha element
      5010       5020       5030       5040 ¦     5050       5060       5070       5080       5090       5100
CCATTTTGAA AATTCACTTA CTTGATCAGC TTCAGAAGAT GGCGGAGGGC CTCCAACACA GTAATTTTCC TCCCGACTCT TAAAATAGAA AATGTCAAGT
GGTAAAACTT TTAAGTGAAT GAACTAGTCG AAGTCTTCTA CCGCCTCCCG GAGGTTGTGT CATTAAAAGG AGGGCTGAGA ATTTTATCTT TTACAGTTCA
<--¦
VP2 CDS Start

                                                                                                 ORGREP
    ORGREP alpha element  <---¦                                                             ¦--->  beta element
      5110       5120       5130 ¦     5140       5150       5160       5170       5180       ¦ 5190       5200
CAGTTAAGCA GGAAGTGACT AACTGACCGC AGCTGGCCGT GCGACATCCT CTTTTAATTA GTTGCTAGGC AACTGCCCTC CAGAGGGCAG TGTGGTTTTG
GTCAATTCGT CCTTCACTGA TTGACTGGCG TCGACCGGCA CGCTGTAGGA GAAAATTAAT CAACGATCCG TTGACGGGAG GTCTCCCGTC ACACCAAAAC

                                                                   ¦........ORGREP CORE...........
                                                                   ¦                ¦-> ORGREP
                                       ORGREP beta element  <---¦¦                  ¦  Palindrome
      5210       5220       5230       5240       5250       5260       5270       5280       5290
CAAGAGGAAG CAAAAAGCCT CTCCACCCAG GCCTAGAATG TTTCCACCCA ATCATTACTA TGACAACAGC TGTTTTTTTT AGTATTAAGC AGAGGCC
GTTCTCCTTC GTTTTTCGGA GAGGTGGGTC CGGATCTTAC AAAGGTGGGT TAGTAATGAT ACTGTTGTCG ACAAAAAAAA TCATAATTCG TCTCCGG
```

Restriction Enzyme Sites in Polyoma Virus (A2 Strain) DNA

Restriction enzyme[1,2]	Recognition sequence [3]	Number of sites	Base number matched[4]
Acc I	GT/(AC)(GT)AC	5	370, 1502, 1811, 4946, 4975
[Afl II]	C/TTAAG	1	2225
Aha II	G(AG)/CG(CT)C	2	86, 4050
Aha III	TTT/AAA	1	358
Alu I	AG/CT	31	195, 526, 572, 875, 935, 1147, 1234, 1352, 1364, 1376, 1659, 1670, 2008, 2035, 2088, 2229, 2379, 2896, 3330, 3430, 3764, 3921, 4344, 4432, 4483, 4873, 4921, 4957, 5028, 5131, 5268
Apa I	GGGCC/C	3	3, 2613, 3264
Ava I	C/(CT)CG(AG)G	2	659, 1018
Ava II	G/G(AT)CC	11	717, 928, 1105, 1414, 1576, 2276, 2395, 2987, 3743, 3959, 4677
[Avr II]	C/CTAGG	1	4709
BamH I	G/GATCC	1	4634
Ban I	G/G(CT)(AG)CC	6	86, 596, 2178, 4357, 4695, 4726
Ban II	G(AG)GC(CT)/C	10	3, 571, 796, 1082, 1309, 1375, 2613, 3264, 3682, 4343
Bbv I->	GCAGC(N)$_8$/	11	264, 279, 1440, 1668, 1727, 2033, 2082, 2306, 2455, 4895, 5129
<-Bbv I	/(N)$_{12}$GCTGC	7	210, 1365, 2036, 2230, 3597, 3896, 4226
[BbvS I]	GC(AT)GC	18	210, 264, 279, 1365, 1440, 1668, 1727, 2033, 2036, 2082, 2230, 2306, 2455, 3597, 3896, 4226, 4895, 5129

[1] Restriction enzyme names in brackets (e.g., [AflII]) are not commercially available as of September 1985.
[2] The direction of an asymmetric cut is indicated by –> or <– in the restriction enzyme name.
[3] The restriction enzyme recognition sequence (5′– –>3′). The position of endonuclease cleavage is indicated (/). Ambiguous bases in the recognition sequence are enclosed in parentheses.
[4] Numbers represent the first base in the sequence of the recognition site for a restriction endonuclease.

(*Continued*)

Restriction Enzyme Sites in Polyoma Virus (A2 Strain) DNA (*Continued*)

Restriction enzyme	Recognition sequence	Number of sites	Base number matched
Bcl I	T/GATCA	1	5023
Bgl I	GCCNNNN/NGGC	1	89
[Bin I->]	GGATC$(N)_4$/	3	1193, 3385, 4634
[<-Bin I]	/$(N)_5$GATCC	5	1320, 1325, 1859, 2765, 4635
Bsm I	GAATGCN/	4	260, 491, 1706, 1981
Bsp1286 I	G(AGT)GC(ACT)/C	14	3, 571, 727, 796, 1082, 1309, 1375, 2613, 2818, 3264, 3288, 3682, 4278, 4343
BstN I	CC/(AT)GG	17	8, 408, 593, 720, 1258, 1573, 1760, 2052, 2459, 2499, 2932, 3035, 3551, 4176, 4443, 4607, 5227
BstX I	CCANNNNN/NTGG	3	167, 1695, 3605
[Cfr I]	(CT)/GGCC(AG)	2	838, 5134
[Cfr10 I]	(AG)CCGG(CT)	1	400
Dde I	C/TNAG	20	187, 379, 442, 553, 828, 951, 969, 1080, 1269, 1371, 1398, 1620, 1656, 2120, 2489, 3145, 3704, 4325, 4341, 4387
Dpn I	GmA/TC	8	1194, 1320, 1325, 1859, 2765, 3386, 4635, 5024
[Dra II]	(AG)G/GNCC(CT)	10	2, 3, 927, 1413, 2986, 3244, 3263, 3742, 4676, 5046
[Dra III]	CACNNN/GTG	5	730, 2565, 2624, 4323, 4409
EcoR I*	/AATT	25	116, 125, 335, 1530, 1563, 1646, 1773, 1870, 2000, 2064, 2285, 2351, 2682, 2701, 2744, 2927, 3350, 3444, 3617, 3833, 4129, 4879, 5011, 5063, 5156

Restriction Enzyme Sites in Polyoma Virus (A2 Strain) DNA (*Continued*)

Restriction enzyme	Recognition sequence	Number of sites	Base number matched
EcoR I	G/AATTC	1	1562
EcoR I'	(AG)(AG)A/T(CT)(CT)	19	249, 387, 438, 1066, 1193, 1298, 1562, 2635, 2681, 2743, 3124, 3385, 3398, 3448, 3616, 3636, 3809, 4634, 5010
EcoR II	/CC(AT)GG	17	8, 408, 593, 720, 1258, 1573, 1760, 2052, 2459, 2499, 2932, 3035, 3551, 4176, 4443, 4607, 5227
EcoR V	GAT/ATC	1	4108
[Esp I]	GC/TNAGC	1	1079
Fnu4H I	GC/NGC	24	210, 264, 279, 837, 946, 1365, 1440, 1665, 1668, 1727, 2033, 2036, 2082, 2230, 2306, 2309, 2455, 3597, 3896, 4053, 4226, 4895, 4994, 5129
Fok I->	GGATG(N)$_9$/	4	429, 723, 2463, 3391
<-Fok I	/(N)$_{13}$CATCC	7	1208, 1238, 2049, 4141, 4212, 4440, 5145
[Gdi II]	(CT)/GGCCG	1	5134
[Gdi II]	CGGCC/(AG)	1	838
[Gsu I]	CTGGAG	5	25, 1149, 1158, 3552, 3608
[Gsu I]	CTCCAG	5	3180, 3211, 3341, 4908, 5178
[Hae I]	(AT)GG/CC(AT)	9	10, 1762, 2299, 3110, 3434, 4019, 4436, 4553, 5229
Hae II	(AG)GCGC/(CT)	2	86, 97
Hae III	GG/CC	27	4, 11, 107, 780, 839, 1557, 1763, 2300, 2614, 2659, 3062, 3111, 3245, 3265, 3435, 3946, 4020, 4071, 4180, 4238, 4437, 4554, 4560, 5048, 5135, 5230, 5294

(*Continued*)

Restriction Enzyme Sites in Polyoma Virus (A2 Strain) DNA (*Continued*)

Restriction enzyme	Recognition sequence	Number of sites	Base number matched
Hga I->	GACGC$(N)_5$/	2	2256, 4050
<-Hga I	/$(N)_{10}$GCGTC	1	2975
HgiA I	G(AT)GC(AT)/C	7	571, 727, 1375, 2818, 3288, 4278, 4343
Hha I	GCG/C	4	87, 98, 2312, 2973
Hinc II	GT(CT)/(AG)AC	2	2964, 3468
Hind III	A/AGCTT	2	1658, 3920
Hinf I	G/ANTC	14	387, 438, 962, 1271, 1289, 1337, 2041, 3449, 3636, 3809, 4156, 4208, 4933, 5075
Hpa II	C/CGG	8	401, 1103, 1215, 1490, 2617, 2993, 4411, 5296
Hph I->	GGTGA$(N)_8$/	3	463, 3150, 3738
<-Hph I	/$(N)_7$TCACC	6	164, 1692, 2640, 4112, 4197, 4202
Kpn I	GGTAC/C	2	2178, 4695
Mae I	C/TAG	26	214, 391, 508, 601, 807, 1329, 1407, 1425, 1437, 1499, 1841, 2162, 2303, 2480, 2525, 2811, 3284, 4125, 4305, 4430, 4434, 4710, 4833, 4980, 5165, 5233
[Mae II]	A/CGT	2	1484, 3815
Mae III	/GTNAC	16	285, 632, 1428, 1691, 1721, 2551, 2799, 3205, 3502, 3541, 3732, 3934, 3952, 4154, 4321, 5115
Mbo I	/GATC	8	1194, 1320, 1325, 1859, 2765, 3386, 4635, 5024
Mbo II->	GAAGA$(N)_8$/	7	36, 257, 1056, 1059, 1796, 2417, 5035
<-Mbo II	/$(N)_7$TCTTC	9	143, 915, 965, 1292, 1992, 3514, 4076, 4534, 4682

Restriction Enzyme Sites in Polyoma Virus (A2 Strain) DNA (*Continued*)

Restriction enzyme	Recognition sequence	Number of sites	Base number matched
Mnl I->	CCTC(N)$_7$/	36	13, 141, 151, 162, 496, 1765, 1942, 1990, 2119, 2142, 2204, 3026, 3029, 3248, 3340, 3692, 3713, 3746, 3841, 3865, 3869, 3968, 4080, 4200, 4285, 4361, 4680, 4809, 4812, 4860, 4965, 5050, 5069, 5148, 5177, 5218
<-Mnl I	/(N)$_7$GAGG	42	41, 52, 62, 95, 395, 760, 822, 830, 995, 1161, 1164, 1167, 1170, 1173, 1191, 1281, 1307, 1412, 1586, 1589, 1901, 1931, 2195, 2274, 2331, 2792, 3105, 3109, 3237, 3261, 3555, 3741, 3753, 4355, 4389, 4724, 4900, 4972, 5045, 5183, 5204, 5292
Mst II	CC/TNAGG	1	2119
Nar I	GG/CGCC	1	86
Nci I	CC/(CG)GG	7	1102, 1215, 1489, 2616, 2992, 4411, 5296
Nco I	C/CATGG	5	1275, 3053, 3113, 4556, 4642
Nde I	CA/TATG	2	2737, 2884
Nla III	CATG/	26	174, 319, 424, 530, 636, 699, 1181, 1276, 1463, 1597, 1911, 1973, 2626, 3054, 3081, 3114, 3474, 3587, 3659, 3757, 3784, 4252, 4557, 4569, 4587, 4643
Nla IV	GGN/NCC	30	2, 3, 4, 86, 105, 397, 596, 927, 928, 997, 1105, 1151, 1414, 1903, 2138, 2178, 2613, 2986, 2987, 3263, 3264, 3565, 4069, 4357, 4560, 4634, 4676, 4695, 4726, 4996

(*Continued*)

Restriction Enzyme Sites in Polyoma Virus (A2 Strain) DNA (*Continued*)

Restriction enzyme	Recognition sequence	Number of sites	Base number matched
Nsi I	ATGCA/T	1	1912
[NspB II]	C(AC)G/C(GT)G	7	478, 1146, 2034, 4699, 4761, 5130, 5267
[Nsp7524 I]	(AG)CATG/(CT)	7	423, 635, 698, 1180, 1596, 1972, 3080
Pss I	(AG)GGNCC(CT)	10	2, 3, 927, 1413, 2986, 3244, 3263, 3742, 4676, 5046
Pst I	CTGCA/G	5	486, 2358, 2453, 3315, 4227
Pvu II	CAG/CTG	4	1146, 2034, 5130, 5267
Rsa I	GT/AC	16	446, 604, 889, 1178, 1227, 2076, 2179, 2577, 2996, 3360, 4027, 4117, 4336, 4572, 4696, 4790
Sac I	GAGCT/C	3	571, 1375, 4343
Sau96 I	G/GNCC	28	3, 4, 106, 717, 928, 1105, 1414, 1557, 1576, 2276, 2395, 2613, 2614, 2659, 2987, 3062, 3245, 3264, 3265, 3743, 3946, 3959, 4070, 4179, 4238, 4560, 4677, 5047
Sca I	AGT/ACT	3	603, 888, 4789
ScrF I	CC/NGG	24	8, 408, 593, 720, 1102, 1215, 1258, 1489, 1573, 1760, 2052, 2459, 2499, 2616, 2932, 2992, 3035, 3551, 4176, 4411, 4443, 4607, 5227, 5296
SfaN I->	GCATC(N)$_5$/	5	2048, 3535, 4140, 4373, 4743
<-SfaN I	/(N)$_9$GATGC	5	589, 1332, 2147, 2221, 3392
[Sna I]	GTATAC	2	370, 1502
Sph I	GCATG/C	3	698, 1596, 1972

Restriction Enzyme Sites in Polyoma Virus (A2 Strain) DNA (*Continued*)

Restriction enzyme	Recognition sequence	Number of sites	Base number matched
Stu I	AGG/CCT	4	1762, 2299, 4019, 5229
Sty I	C/C(AT)(AT)GG	7	90, 1275, 3053, 3113, 4556, 4642, 4709
Taq I	T/CGA	5	1318, 1776, 4622, 4768, 4967
[Taq II->]	CACCCA$(N)_{11}$/	5	110, 298, 4405, 5224, 5245
[<-Taq II]	/$(N)_{9}$TGGGTG	3	432, 3037, 3890
[Tce I]	GAAGA	7	36, 257, 1056, 1059, 1796, 2417, 5035
[Tce I]	TCTTC	9	143, 915, 965, 1292, 1992, 3514, 4076, 4534, 4682
Tha I	CG/CG	2	2311, 2974
Tth111 I	GACN/NNGTC	2	3456, 3471
Tth111 II->	CAA(AG)CA$(N)_{9}$/	6	695, 1525, 2446, 2603, 3581, 4776
<-Tth111I I	/$(N)_{11}$TG(CT)TTG	1	3305
Xba I	T/CTAGA	2	2479, 2524
Xho II	(AG)/GATC(CT)	3	1193, 3385, 4634
Xmn I	GAANN/NNTTC	2	907, 3917

Restriction Enzymes That Do Not Cut Polyoma (A2 Strain) DNA

Enzyme[1,2]	Recognition site[3]	Enzyme	Recognition site	Enzyme	Recognition site
Aat II	GACGT/C	[HgiE II]	ACCNNNNNNGGT	Sfi I	GGCCNNNN/NGGCC
[Afl III]	A/C(AG)(CT)GT	Hpa I	GTT/AAC	Sma I	CCC/GGG
Asu II	TT/CGAA	Mlu I	A/CGCGT	SnaB I	TAC/GTA
Bal I	TGG/CCA	Nae I	GCC/GGC	Spe I	A/CTAGT
Bgl II	A/GATCT	Nhe I	G/CTAGC	Ssp I	AAT/ATT
Bsm I	G/CATTC	Not I	GC/GGCCGC	[Taq II->]	GACCGA$(N)_{11}$/
BssH II	G/CGCGC	Nru I	TCG/CGA	[<-Taq II]	/$(N)_{9}$TCGGTC
BstE II	G/GTNACC	Pvu I	CGAT/CG	Xho I	C/TCGAG
Cla I	AT/CGAT	Rsr II	CG/G(AT)CCG	Xma III	C/GGCCG
[Eco47 III]	AGCGCT	Sac II	CCGC/GG		
Fsp I	TGC/GCA	Sal I	G/TCGAC		

[1] Restriction enzyme names in brackets (e.g., [AflIII]) are not commercially available as of September, 1985.
[2] The direction of an asymmetric cut is indicated by -> or <- in the restriction enzyme name.
[3] The restriction enzyme recognition sequence (5' - ->3'). The position of endonuclease cleavage is indicated (/). Ambiguous bases in the recognition sequence are enclosed in parentheses.

The Nucleotide Sequence of BK Virus (Dunlop Strain) DNA

```
        10         20         30         40         50         60         70         80         90        100
TTTTGCAAAA ATTGCAAAAG AATAGGGATT TCCCCAAATA GTTTTGCTAG GCCTCAGAAA AAGCCTCCAC ACCCTTACTA CTTGAGAGAA AGGGTGGAGG
AAAACGTTTT TAACGTTTTC TTATCCCTAA AGGGGTTTAT CAAAACGATC CGGAGTCTTT TTCGGAGGTG TGGGAATGAT GAACTCTCTT TCCCACCTCC

       110        120        130        140        150        160        170        180        190        200
CAGAGGCGGC CTCGGCCTCT TATATATTAT AAAAAAAAAG GCCACAGGGA GGAGCTGCTT ACCCATGGAA TGCAGCCAAA CCATGACCTC AGGAAGGAAA
GTCTCCGCCG GAGCCGGAGA ATATATAATA TTTTTTTTTC CGGTGTCCCT CCTCGACGAA TGGGTACCTT ACGTCGGTTT GGTACTGGAG TCCTTCCTTT

       210        220        230        240        250        260        270        280        290        300
GTGCATGACT CACAGGGGAA TGCAGCCAAA CCATGACCTC AGGAAGGAAA GTGCATGACT CACAGGGAGG AGCTGCTTAC CCATGGAATG CAGCCAAACC
CACGTACTGA GTGTCCCCTT ACGTCGGTTT GGTACTGGAG TCCTTCCTTT CACGTACTGA GTGTCCCTCC TCGACGAATG GGTACCTTAC GTCGGTTTGG

                                                                                     :---> Agnoprotein
                                                                                     :       CDS Start
       310        320        330        340        350        360        370        380        390   ----
ATGACCTCAG GAAGGAAAGT GCATGACAGA CATGTTTTGC GAGCCTAGGA ATCTTGGCCT TGTCCCCAGT TAAACTGGAC AAAGGCCATG GTTCTGCGCC
TACTGGAGTC CTTCCTTTCA CGTACTGTCT GTACAAAACG CTCGGATCCT TAGAACCGGA ACAGGGGTCA ATTTGACCTG TTTCCGGTAC CAAGACGCGG
                                                                                                      Hha I

       410        420        430        440        450        460        470        480        490        500
AGCTGTCACG ACAAGCTTCA GTGAAAGTTG GTAAAACCTG GACTGGAACA AAAAAAAGAG CTCAGAGGAT TTTTATTTTT ATTTTAGAGC TTTTGCTGGA
TCGACAGTGC TGTTCGAAGT CACTTTCAAC CATTTTGGAC CTGACCTTGT TTTTTTTCTC GAGTCTCCTA AAAATAAAAA TAAAATCTCG AAAACGACCT

                                                                 Agnoprotein CDS End <---:
       510        520        530        540        550        560        570        580        590        600
ATTTGTAGA  GGTGAAGACA GTGTAGACGG GAAAAACAAA AGTACCACTG CTTTACCTGC TGTAAAAGAC TCTGTAAAAG ACTCCTAGGT AAGTAATCCC
TAAAACATCT CCACTTCTGT CACATCTGCC CTTTTTGTTT TCATGGTGAC GAAATGGACG ACATTTTCTG AGACATTTTC TGAGGATCCA TTCATTAGGG

                      :---> VP2 CDS Start
       610        620 :      630        640        650        660        670        680        690        700
TTTTTTTTTG TATTTCCAGG TTCATGGGTG CTGCTCTAGC ACTTTTGGGG GACCTAGTTG CCAGTGTATC TGAGGCTGCT GCTGCCACAG GATTTTCAGT
AAAAAAAAAC ATAAAGGTCC AAGTACCCAC GACGAGATCG TGAAAACCCC CTGGATCAAC GGTCACATAG ACTCCGACGA CGACGGTGTC CTAAAAGTCA
```

(*Continued*)

The Nucleotide Sequence of BK Virus (Dunlop Strain) DNA (*Continued*)

```
       710        720        730        740        750        760        770        780        790        800
GGCTGAAATT GCTGCTGGGG AGGCTGCTGC TGCTATAGAA GTTCAAATTG CATCCCTTGC TACTGTAGAG GGCATAACAA GTACCTCAGA GGCTATAGCT
CCGACTTTAA CGACGACCCC TCCGACGACG ACGATATCTT CAAGTTTAAC GTAGGGAACG ATGACATCTC CCGTATTGTT CATGGAGTCT CCGATATCGA

       810        820        830        840        850        860        870        880        890        900
GCTATAGGCC TAACTCCTCA AACATATGCT GTAATTGCTG GTGCTCCTGG GGCTATTGCT GGGTTTGCTG CTTTAATTCA AACTGTTAGT GGTATTAGTT
CGATATCCGG ATTGAGGAGT TTGTATACGA CATTAACGAC CACGAGGACC CCGATAACGA CCCAAACGAC GAAATTAAGT TTGACAATCA CCATAATCAA

                                                                                     |---> VP3 CDS Start
       910        920        930        940        950        960        970        980 |      990       1000
CCTTAGCTCA AGTAGGGTAT AAGTTCTTTG ATGATTGGGA TCACAAAGTT TCCACTGTAG GCCTCTATCA GCAATCAGGC ATGGCTTTGG AATTGTTTAA
GGAATCGAGT TCATCCCATA TTCAAGAAAC TACTAACCCT AGTGTTTCAA AGGTGACATC CGGAGATAGT CGTTAGTCCG TACCGAAACC TTAACAAATT

      1010       1020       1030       1040       1050       1060       1070       1080       1090       1100
CCCAGATGAG TACTATGATA TTCTGTTTCC TGGTGTAAAT ACTTTTGTTA ATAATATTCA ATACCTTGAT CCTAGGCATT GGGGTCCTTC TTTGTTTGCT
GGGTCTACTC ATGATACTAT AAGACAAAGG ACCACATTTA TGAAAACAAT TATTATAAGT TATGGAACTA GGATCCGTAA CCCCAGGAAG AAACAAACGA

      1110       1120       1130       1140       1150       1160       1170       1180       1190       1200
ACTATTTCCC AGGCTTTGTG GCATGTTATT AGGGATGATA TACCTTCTAT AACCTCACAG GAATTGCAGA GAAGAACAGA AAGATTTTTT AGAGACTCCT
TGATAAAGGG TCCGAAACAC CGTACAATAA TCCCTACTAT ATGGAAGATA TTGGAGTGTC CTTAACGTCT CTTCTTGTCT TTCTAAAAAA TCTCTGAGGA

      1210       1220       1230       1240       1250       1260       1270       1280       1290       1300
TGGCTAGATT TTTGGAGGAA ACTACCTGGA CAATTGTAAA TGCCCCTATA AACTTTTATA ATTATATTCA ACAATATTAT TCTGATCTTT CCCCTATTAG
ACCGATCTAA AAACCTCCTT TGATGGACCT GTTAACATTT ACGGGGATAT TTGAAAATAT TAATATAAGT TGTTATAATA AGACTAGAAA GGGGATAATC

      1310       1320       1330       1340       1350  ------          1370       1380       1390       1400
GCCCTCAATG GTTAGACAAG TAGCTGAAAG GGAAGGTACC CGTGTACATT TTGGCCATAC TTATAGTATA GATGATGCTG ACAGTATAGA AGAAGTTACA
CGGGAGTTAC CAATCTGTTC ATCGACTTTC CCTTCCATGG GCACATGTAA AACCGGTATG AATATCATAT CTACTACGAC TGTCATATCT TCTTCAATGT
                                                         Bal I

      1410       1420       1430       1440       1450       1460       1470       1480       1490       1500
CAAAGAATGG ACTTAAGAAA TCAACAAAGT GTACATTCAG GAGAGTTTAT AGAAAAAACT ATTGCCCCAG GAGGTGCTAA TCAAAGAACT GCTCCTCAAT
GTTTCTTACC TGAATTCTTT AGTTGTTTCA CATGTAAGTC CTCTCAAATA TCTTTTTTGA TAACGGGGTC CTCCACGATT AGTTTCTTGA CGAGGAGTTA

                                                                  |---> VP1 CDS Start
      1510       1520       1530       1540       1550       1560 |     1570       1580       1590       1600
GGATGTTGCC TTTACTTCTA GGCCTGTACG GGACTGTAAC ACCTGCTCTT GAAGCATATG AAGATGGCCC CAACCAAAAG AAAAGGAGAG TGTCCAGGGG
CCTACAACGG AAATGAAGAT CCGGACATGC CCTGACATTG TGGACGAGAA CTTCGTATAC TTCTACCGGG GTTGGTTTTC TTTTCCTCTC ACAGGTCCCC
```

```
                                                                        VP2, VP3 CDS End  <---:
      1610       1620       1630       1640       1650       1660       1670       1680       1690       1700
CAGCTCCCAA AAAGCCAAAG GAACCCGTGC AAGTGCCAAA ACTACTAATA AAAGGAGGAG TAGAAGTTCT AGAAGTTAAA ACTGGGGTAG ATGCTATTAC
GTCGAGGGTT TTTCGGTTTC CTTGGGCACG TTCACGGTTT TGATGATTAT TTTCCTCCTC ATCTTCAAGA TCTTCAATTT TGACCCCATC TACGATAATG

      1710       1720       1730     ------       1750       1760       1770       1780       1790       1800
AGAGGTAGAA TGCTTCCTAA ACCCAGAAAT GGGGGATCCA GATGAAAACC TTAGGGGCTT TAGTCTAAAG CTAAGTGCTG AAAATGACTT TAGCAGTGAT
TCTCCATCTT ACGAAGGATT TGGGTCTTTA CCCCCTAGGT CTACTTTTGG AATCCCCGAA ATCAGATTTC GATTCACGAC TTTTACTGAA ATCGTCACTA
                                    BamH I

      1810       1820       1830       1840       1850       1860       1870       1880       1890       1900
AGCCCAGAGA GAAAAATGCT TCCCTGTTAC AGCACAGCAA GAATTCCCCT CCCCAATTTA AATGAGGACC TAACCTGTGG AAATCTACTG ATGTGGGAGG
TCGGGTCTCT CTTTTTACGA AGGGACAATG TCGTGTCGTT CTTAAGGGGA GGGGTTAAAT TTACTCCTGG ATTGGACACC TTTAGATGAC TACACCCTCC

      1910       1920       1930       1940       1950       1960       1970       1980       1990       2000
CTGTAACTGT ACAAACAGAG GTTATTGGAA TAACTAGCAT GCTTAACCTT CATGCAGGGT CACAAAAAGT GCATGAGCAT GGTGGAGGAA AACCTATTCA
GACATTGACA TGTTTGTCTC CAATAACCTT ATTGATCGTA CGAATTGGAA GTACGTCCCA GTGTTTTTCA CGTACTCGTA CCACCTCCTT TTGGATAAGT

      2010       2020       2030       2040       2050       2060       2070       2080       2090       2100
AGGCAGTAAT TTCCACTTCT TTGCTGTAGG TGGAGAACCC TTGGAAATGC AGGGAGTGCT AATGAATTAC AGGTCAAAGT ACCCTGATGG TACTATAACC
TCCGTCATTA AAGGTGAAGA AACGACATCC ACCTCTTGGG AACCTTTACG TCCCTCACGA TTACTTAATG TCCAGTTTCA TGGGACTACC ATGATATTGG

      2110       2120       2130       2140       2150       2160       2170       2180       2190       2200
CCTAAAAACC CAACAGCCCA GTCCCAGGTA ATGAATACTG ACCATAAGGC CTATTTGGAC AAAAACAATG CTTATCCAGT TGAGTGCTGG GTACCTGATC
GGATTTTTGG GTTGTCGGGT CAGGGTCCAT TACTTATGAC TGGTATTCCG GATAAACCTG TTTTTGTTAC GAATAGGTCA ACTCACGACC CATGGACTAG

      2210       2220       2230       2240       2250       2260       2270       2280       2290       2300
CCAGTAGAAA TGAAAATGCT AGGTATTTTG GGACTTTCAC AGGAGGGGAA AATGTTCCCC CAGTACTTCA TGTGACCAAC ACAGCTACCA CAGTGTTGCT
GGTCATCTTT ACTTTTACGA TCCATAAAAC CCTGAAAGTG TCCTCCCCTT TTACAAGGGG GTCATGAAGT ACACTGGTTG TGTCGATGGT GTCACAACGA

      2310       2320       2330       2340       2350       2360       2370       2380       2390       2400
AGATGAACAG GGTGTGGGGC CTCTTTGTAA AGCTGATAGC CTGTATGTTT CAGCTGCTGA TATTTGTGGC CTGTTTACTA ACAGCTCTGG AACACAACAG
TCTACTTGTC CCACACCCCG GAGAAACATT TCGACTATCG GACATACAAA GTCGACGACT ATAAACACCG GACAAATGAT TGTCGAGACC TTGTGTTGTC

      2410       2420       2430       2440       2450       2460       2470       2480       2490       2500
TGGAGAGGCC TTGCAAGATA TTTTAAGATC CGCCTGAGAA AAAGATCTGT AAAGAATCCT TACCCAATTT CCTTTTTGCT AAGTGACCTT ATAAACAGGA
ACCTCTCCGG AACGTTCTAT AAAATTCTAG GCGGACTCTT TTTCTAGACA TTTCTTAGGA ATGGGTTAAA GGAAAAACGA TTCACTGGAA TATTTGTCCT
```


(Continued)

```
      2510       2520       2530       2540       2550       2560       2570       2580       2590       2600
GAACCCAGAG AGTGGATGGG CAGCCTATGT ATGGTATGGA ATCCCAGGTA GAAGAGGTTA GGGTGTTTGA TGGCACAGAA AGACTTCCAG GGGACCCAGA
CTTGGGTCTC TCACCTACCC GTCGGATACA TACCATACCT TAGGGTCCAT CTTCTCCAAT CCCACAAACT ACCGTGTCTT TCTGAAGGTC CCCTGGGTCT

                                 VP1 CDS End  <---¦
      2610       2620       2630       2640       2650       2660       2670       2680       2690       2700
TATGATAAGA TATATTGACA AACAGGGACA ATTGCAAACC AAAATGCTTT AAACAGGTGC TTTTATTGTA CATATACATT TAATAAATGC TGCTTTTGTA
ATACTATTCT ATATAACTGT TTGTCCCTGT TAACGTTTGG TTTTACGAAA TTTGTCCACG AAAATAACAT GTATATGTAA ATTATTTACG ACGAAAACAT

      2710       2720       2730       2740       2750       2760       2770       2780       2790       2800
TAAGCCACTT TTAAGCTTGT GTTATTTTGG GGGTGGTGTT TTAGGCCTTT TAAAACACTG AAAGCCTTTA CACAAATGCA ACTCTTGACT ATGGGGGTCT
ATTCGGTGAA AATTCGAACA CAATAAAACC CCCACCACAA AATCCGGAAA ATTTTGTGAC TTTCGGAAAT GTGTTTACGT TGAGAACTGA TACCCCCAGA
                            ¦---> Large T EXON 2 End

      2810       2820       2830       2840       2850       2860       2870       2880       2890       2900
GACCTTTGGG AATCTTCAGC AGGGGCTGAA GTATCTGAGA CTTGGGAAGA GCATTGTGAT TGGGATTCAG TGCTTGATCC ATGTCCAGAG TCTTCAGTTT
CTGGAAACCC TTAGAAGTCG TCCCCGACTT CATAGACTCT GAACCCTTCT CGTAACACTA ACCCTAAGTC ACGAACTAGG TACAGGTCTC AGAAGTCAAA

      2910       2920       2930       2940     ------       2960       2970       2980       2990       3000
CTGAATCCTC TTCTCTTGTA ATATCAAGAA TACATTTCCC CATGCATATA TTATATTTCA TCCTTGAAAA AGTATACATA CTTATCTCAG AATCCAGCCT
GACTTAGGAG AAGAGAACAT TATAGTTCTT ATGTAAAGGG GTACGTATAT AATATAAAGT AGGAACTTTT TCATATGTAT GAATAGAGTC TTAGGTCGGA
                                             Nsi I

      3010       3020       3030       3040       3050       3060       3070       3080       3090       3100
TTCCTTCCAT TCAACAATTC TAGATTGTAT ATCAGTTGCA AAATCAGCTA CAGGCCTAAA CCAAATTAGC AGTAGCAACA AGGTCATTCC ACTTTGTAAA
AAGGAAGGTA AGTTGTTAAG ATCTAACATA TAGTCAACGT TTTAGTCGAT GTCCGGATTT GGTTTAATCG TCATCGTTGT TCCAGTAAGG TGAAACATTT

      3110       3120       3130       3140       3150       3160       3170       3180       3190       3200
ATTCTTTTTT CAAGTAAGAA CTCTGAGTTT TGTAAGGATT TTCTTAAATA TATTTTGGGC CTAAAATCTA TTTGTCTTAC AAATCTAGCT TGCAGGGTTT
TAAGAAAAAA GTTCATTCTT GAGACTCAAA ACATTCCTAA AAGAATTTAT ATAAAACCCG GATTTTAGAT AAACAGAATG TTTAGATCGA ACGTCCCAAA

      3210       3220       3230       3240       3250       3260       3270       3280       3290       3300
TAGGGACAGG ATACTCATTC ATTGTAACCA AGCCTGGTGG AAATATTTGG GTTCTTTTGT TTAAATGTTT CTTTTCTAAA TTTACCTTAA CACTTCCATC
ATCCCTGTCC TATGAGTAAG TAACATTGGT TCGGACCACC TTTATAAACC CAAGAAAACA AATTTACAAA GAAAAGATTT AAATGGAATT GTGAAGGTAG

      3310       3320       3330       3340       3350       3360       3370       3380       3390       3400
TAAATAATCT CTCAAACTGT CTAAATTGTT TATTCCATGT CCTGAAGGCA AATCCTTTGA TTCAGCTCCT GTCCCTTTTA CATCTTCAAA AACAACCATG
ATTTATTAGA GAGTTTGACA GATTTAACAA ATAAGGTACA GGACTTCCGT TTAGGAAACT AAGTCGAGGA CAGGGAAAAT GTAGAAGTTT TTGTTGGTAC
```

```
      3410       3420       3430       3440       3450       3460       3470       3480       3490    ------
TACTGATCTA TAGCTACACC TAGCTCAAAG GTTAGCCTTT CCATGGGTAG GTTTACATTT AAGGCTTTAC CACCACACAA ATCTAATAAC CCTGCAGCTA
ATGACTAGAT ATCGATGTGG ATCGAGTTTC CAATCGGAAA GGTACCCATC CAAATGTAAA TTCCGAAATG GTGGTGTGTT TAGATTATTG GGACGTCGAT
                                                                                                    Pst I
      3510       3520       3530       3540       3550       3560       3570       3580       3590       3600
GTGTTGTTTT TCCACTATCA ATGGGACCTT TAAATAACCA GTATCTTCTT TTAGGTACAT TGAAAACAAT ACAGTGCAAA AAATCAAATA TTACAGAATC
CACAACAAAA AGGTGATAGT TACCCTGGAA ATTTATTGGT CATAGAAGAA AATCCATGTA ACTTTTGTTA TGTCACGTTT TTTAGTTTAT AATGTCTTAG

      3610       3620       3630       3640       3650       3660       3670       3680       3690       3700
CATTTTAGGT AGCAAACAGT GCAGCCAAGC AACACCTGCC ATATATTGTT CTAGTACAGC ATTTCCATGA GCTCCAAATA TTAAATCCAT TTTATCTAAT
GTAAAATCCA TCGTTTGTCA CGTCGGTTCG TTGTGGACGG TATATAACAA GATCATGTCG TAAAGGTACT CGAGGTTTAT AATTTAGGTA AAATAGATTA

      3710       3720       3730       3740       3750       3760       3770       3780       3790       3800
ATATGATTGA ATCTTTCTGT TAGCATTTCT TCCCTGGTCA TATGAAGGGT ATCTACTCTT TTTTTAGCTA AAACTGTATC TACTGCTTGC TGACAAATAC
TATACTAACT TAGAAAGACA ATCGTAAAGA AGGGACCAGT ATACTTCCCA TAGATGAGAA AAAAATCGAT TTTGACATAG ATGACGAACG ACTGTTTATG

      3810       3820       3830       3840       3850       3860       3870       3880       3890       3900
TTTTTTGATT TTTACTTTCT GCAAAAATAA TAGCATTTGC AAAGTGCTTT TCATGATACT TAAAGTGATA AGGCTGGTCT TTTTTCTGAC ACTTTTTACA
AAAAAACTAA AAATGAAAGA CGTTTTTATT ATCGTAAACG TTTCACGAAA AGTACTATGA ATTTCACTAT TCCGACCAGA AAAAAGACTG TGAAAAATGT

      3910       3920       3930       3940       3950       3960       3970       3980       3990       4000
CTCCTCTACA TTGTATTGAA ATTCTAAATA CATACCTAAT AATAAAAACA CATCCTCACA CTTTGTCTCT ACTGCATACT CAGTAATTAA TTTCCAAGAC
GAGGAGATGT AACATAACTT TAAGATTTAT GTATGGATTA TTATTTTTGT GTAGGAGTGT GAAACAGAGA TGACGTATGA GTCATTAATT AAAGGTTCTG

      4010       4020       4030       4040       4050       4060       4070       4080       4090       4100
ACCTGCTTTG TTTCTTCAGG CTCTTCTGGG CTAAAATCAT GCTCCTTTAA GCCCCCTTGA ATGCTTTCTT CTATAGTATG GTATGGATCT CTAGTTAAGG
TGGACGAAAC AAAGAAGTCC GAGAAGACCC GATTTTAGTA CGAGGAAATT CGGGGGAACT TACGAAAGAA GATATCATAC CATACCTAGA GATCAATTCC

      4110       4120       4130       4140       4150       4160       4170       4180       4190       4200
CACTATATAG TAAGTATTCC TTATTAACAC CCTTACAAAT TAAAAAACTA AAGGTACACA GCTTTTGACA GAAATTATTA ATTGCAGAAA CTCTATGTCT
GTGATATATC ATTCATAAGG AATAATTGTG GGAATGTTTA ATTTTTTGAT TTCCATGTGT CGAAAACTGT CTTTAATAAT TAACGTCTTT GAGATACAGA

      4210       4220       4230       4240       4250       4260       4270       4280       4290       4300
ATGTGGAGTT AAAAAGAATA TAATATTATG CCCAGCACAC ATGTGTCTAC TAATAAAAGT TACAGAATAT TTTTCCATAA GTTTTTTATA CAGAATTTGA
TACACCTCAA TTTTTCTTAT ATTATAATAC GGGTCGTGTG TACACAGATG ATTATTTTCA ATGTCTTATA AAAAGGTATT CAAAAAATAT GTCTTAAACT

      4310       4320       4330       4340       4350       4360       4370       4380       4390       4400
GCTTTTTCTT TAGTAGTATA CACAGCAAAG CAGGCAAGGG TTCTATTACT AAATACAGCT TGACTAAGAA ACTGGTGTAG ATCAGAGGGA AAGTCTTTAG
CGAAAAAGAA ATCATCATAT GTGTCGTTTC GTCCGTTCCC AAGATAATGA TTTATGTCGA ACTGATTCTT TGACCACATC TAGTCTCCCT TTCAGAAATC
```


(Continued)

The Nucleotide Sequence of BK Virus (Dunlop Strain) DNA (*Continued*)

```
      4410       4420       4430       4440       4450       4460       4470       4480       4490       4500
GGTCTTCTAC CTTTCTTTTT TTTTTGGGTG GTGTTGAGTG TTGAGAATCT GCTGTTGCTT CTTCATCACT GGCAAACATA TCTTCATGGC AAAATAAATC
CCAGAAGATG GAAAGAAAAA AAAAACCCAC CACAACTCAC AACTCTTAGA CGACAACGAA GAAGTAGTGA CCGTTTGTAT AGAAGTACCG TTTTATTTAG

      4510       4520       4530       4540       4550       4560       4570       4580       4590       4600
TTCATCCCAT TTTTCATTAA AGGAACTCCA CCAGGACTCC CACTCTTCTG TTCCATAGGT TGGCACCTAT AAAAAAAATA ATTACTTAGG GCCTTTTAAT
AAGTAGGGTA AAAAGTAATT TCCTTGAGGT GGTCCTGAGG GTGAGAAGAC AAGGTATCCA ACCGTGGATA TTTTTTTTAT TAATGAATCC CGGAAAATTA
                                           Large T EXON 2 CDS Start  <---¦

      4610       4620       4630       4640       4650       4660       4670       4680       4690       4700
ATTTTATTAT TTATCTAAAT ATAAGTTAGT TACCTTAAAG CTTTAGATCT CTGAAGGGAG TTTCTCCAAT TATTTGGACC CACCATTGCA GAGTTTCTTC
TAAAATAATA AATAGATTTA TATTCAATCA ATGGAATTTC GAAATCTAGA GACTTCCCTC AAAGAGGTTA ATAAACCTGG GTGGTAACGT CTCAAAGAAG
                                          ¦---> Small t CDS End

      4710       4720       4730       4740       4750       4760       4770       4780       4790       4800
AGTTAGGTCT AAGCCAAACC ACTGTGTGAA GCAGTCAATG CAGTAGCAAT CTATCCAAAC CAAGGGCTCT TTTCTTAAAA ATTTTCTATT TAAATGCCTT
TCAATCCAGA TTCGGTTTGG TGACACACTT CGTCAGTTAC GTCATCGTTA GATAGGTTTG GTTCCCGAGA AAAGAATTTT TAAAAGATAA ATTTACGGAA

      4810       4820       4830       4840       4850       4860       4870       4880       4890       4900
AATCTAAGCT GACATAGCAT GCAAGGGCAG TGCACAGAAG GCTTTTTGGA ACAAATAGGC CATTCCTTGC AGTACAGGGT ATCTGGGCAA AGAGGAAAAT
TTAGATTCGA CTGTATCGTA CGTTCCCGTC ACGTGTCTTC CGAAAAACCT TGTTTATCCG GTAAGGAACG TCATGTCCCA TAGACCCGTT TCTCCTTTTA

      4910       4920       4930       4940       4950       4960       4970       4980       4990       5000
CAGCACAAAC CTCTGAGCTA CTCCAGGTTC CAAAATCAGG CTGATGAGCT ACCTTTACAT CCTGCTCCAT TTTTTTATAC AAAGTATTCA TTCTCTTCAT
GTCGTGTTTG GAGACTCGAT GAGGTCCAAG GTTTTAGTCC GACTACTCGA TGGAAATGTA GGACGAGGTA AAAAAATATG TTTCATAAGT AAGAGAAGTA
           ¦---> Large T EXON 1 CDS End

      5010       5020       5030       5040       5050       5060       5070       5080       5090       5100
TTTATCCTCG TCGCCCCCTT TGTCAGGGTG AAATTCCTTA CACTTCCTTA AATAAGCTTT TCTCATTAAG GGAAGATTTC CCCAGGCAGC TCTTTCAAGG
AAATAGGAGC AGCGGGGGAA ACAGTCCCAC TTTAAGGAAT GTGAAGGAAT TTATTCGAAA AGAGTAATTC CCTTCTAAAG GGGTCCGTCG AGAAAGTTCC

      5110       5120       5130       5140       5150
CCTAAAAGGT CCATGAGCTC CATGGATTCT TCCCTGTTAA GAACTTTATC CAT
GGATTTTCCA GGTACTCGAG GTACCTAAGA AGGGACAATT CTTGAAATAG GTA
                 Large T EXON 1, Small t CDS Start <---¦
```

Restriction Enzyme Sites in BK Virus (Dunlop Strain) DNA

Restriction enzyme[1,2]	Recognition pattern[3]	Number of sites	Base number matched[4]
Acc I	GT/(AC)(GT)AC	4	523, 2972, 4245, 4316
[Afl II]	C/TTAAG	1	1412
[Afl III]	A/C(AG)(CT)GT	2	330, 4239
Aha III	TTT/AAA	6	1857, 2648, 2749, 3260, 3529, 4789
Alu I	AG/CT	34	153, 271, 401, 414, 459, 488, 797, 905, 1322, 1602, 1769, 2283, 2331, 2352, 2383, 2714, 3046, 3187, 3364, 3412, 3422, 3496, 3670, 3766, 4160, 4300, 4357, 4639, 4807, 4916, 4947, 5055, 5088, 5116
Ava II	G/G(AT)CC	7	650, 1083, 1866, 2592, 3524, 4676, 5108
[Avr II]	C/CTAGG	3	344, 584, 1071
Bal I	TGG/CCA	1	1352
BamH I	G/GATCC	1	1734
Ban I	G/G(CT)(AG)CC	3	1335, 2190, 4562
Ban II	G(AG)GC(CT)/C	4	458, 3669, 4764, 5115
Bbv I->	GCAGC(N)8/	8	172, 222, 290, 1600, 2520, 3494, 3621, 5086
<-Bbv I	/(N)12GCTGC	14	154, 272, 630, 675, 678, 681, 711, 723, 726, 729, 798, 867, 2353, 2689
[BbvS I]	GC(AT)GC	22	154, 172, 222, 272, 290, 630, 675, 678, 681, 711, 723, 726, 729, 798, 867, 1600, 2353, 2520, 2689, 3494, 3621, 5086
Bgl II	A/GATCT	2	2443, 4645

[1] Restriction enzyme names in brackets (e.g., [AflII]) are not commercially available as of September, 1985.
[2] The direction of an asymmetric cut is indicated by -> or <- in the restriction enzyme name.
[3] The restriction enzyme recognition sequence (5'-->3'). The position of endonuclease cleavage is indicated (/). Ambiguous bases in the recognition sequence are enclosed in parentheses.
[4] Numbers represent the first base in the sequence of the recognition site for a restriction endonuclease.

(Continued)

Restriction Enzyme Sites in BK Virus (Dunlop Strain) DNA (*Continued*)

Restriction enzyme	Recognition pattern	Number of sites	Base number matched
[Bin I->]	GGATC(N)$_4$/	3	938, 1734, 4085
[<-Bin I]	/(N)$_5$GATCC	5	1068, 1735, 2197, 2427, 2876
Bsm I	GAATGCN/	5	168, 218, 286, 1708, 4059
Bsp1286 I	G(AGT)GC(ACT)/C	6	458, 841, 3669, 4764, 4830, 5115
BstN I	CC/(AT)GG	16	437, 616, 846, 1029, 1109, 1225, 1467, 1594, 2124, 2544, 2587, 3233, 3733, 4531, 4923, 5082
BstX I	CCANNNNN/NTGG	2	616, 4666
[Cfr I]	(CT)/GGCC(AG)	1	1352
Dde I	C/TNAG	21	53, 188, 238, 306, 461, 670, 785, 902, 1750, 1771, 2434, 2479, 2835, 2986, 3123, 3979, 4364, 4585, 4709, 4804, 4913
Dpn I	GmA/TC	12	939, 1068, 1284, 1735, 2197, 2427, 2444, 2876, 3405, 4086, 4380, 4646
[Dra II]	(AG)G/GNCC(CT)	8	649, 1082, 1299, 1865, 2316, 2591, 3523, 4588
[Dra III]	CACNNN/GTG	1	4720
EcoR I*	/AATT	32	10, 500, 707, 746, 833, 875, 991, 1162, 1232, 1260, 1842, 1855, 2008, 2065, 2466, 2630, 3016, 3064, 3100, 3279, 3324, 3920, 3985, 3989, 4138, 4173, 4180, 4294, 4580, 4668, 4780, 5032
EcoR I	G/AATTC	1	1841

Restriction Enzyme Sites in BK Virus (Dunlop Strain) DNA (*Continued*)

Restriction enzyme	Recognition pattern	Number of sites	Base number matched
EcoR I'	(AG)(AG)A/T(CT)(CT)	38	26, 349, 467, 499, 690, 1182, 1206, 1734, 1841, 1881, 2426, 2443, 2454, 2539, 2810, 2863, 2903, 2990, 3099, 3136, 3164, 3181, 3278, 3350, 3479, 3596, 3683, 3709, 3919, 4085, 4293, 4445, 4496, 4645, 4779, 5031, 5074, 5124
EcoR II	/CC(AT)GG	16	437, 616, 846, 1029, 1109, 1225, 1467, 1594, 2124, 2544, 2587, 3233, 3733, 4531, 4923, 5082
Fnu4H I	GC/NGC	23	106, 154, 172, 222, 272, 290, 630, 675, 678, 681, 711, 723, 726, 729, 798, 867, 1600, 2353, 2520, 2689, 3494, 3621, 5086
Fok I->	GGATG$(N)_9$/	3	1133, 1501, 2514
<-Fok I	/$(N)_{13}$CATCC	5	751, 2959, 3951, 4503, 4958
[Gsu I]	CTCCAG	1	4921
[Hae I]	(AT)GG/CC(AT)	15	49, 139, 355, 383, 806, 959, 1352, 1520, 2147, 2367, 2406, 2743, 3052, 4857, 5098
Hae III	GG/CC	22	50, 108, 114, 140, 356, 384, 807, 960, 1300, 1353, 1521, 1566, 2148, 2318, 2368, 2407, 2744, 3053, 3158, 4590, 4858, 5099
HgiA I	G(AT)GC(AT)/C	5	458, 841, 3669, 4830, 5115
[HgiE II]	ACCNNNNNNGGT	1	3227
Hha I	GCG/C	1	396
Hind III	A/AGCTT	4	413, 2713, 4638, 5054

(*Continued*)

Restriction Enzyme Sites in BK Virus (Dunlop Strain) DNA (*Continued*)

Restriction enzyme	Recognition pattern	Number of sites	Base number matched
Hinf I	G/ANTC	19	207, 257, 349, 568, 580, 1194, 2454, 2539, 2810, 2864, 2888, 2903, 2990, 3359, 3596, 3709, 4445, 4535, 5125
Hph I->	GGTGA$(N)_8$/	2	511, 5027
Kpn I	GGTAC/C	2	1335, 2190
Mae I	C/TAG	18	47, 345, 585, 636, 654, 1072, 1204, 1518, 1669, 1934, 2219, 2299, 3020, 3185, 3420, 3498, 3651, 4091
Mae III	/GTNAC	11	405, 1395, 1536, 1826, 1903, 1959, 2272, 2483, 3224, 4259, 4629
Mbo I	/GATC	12	939, 1068, 1284, 1735, 2197, 2427, 2444, 2876, 3405, 4086, 4380, 4646
Mbo II->	GAAGA$(N)_8$/	7	514, 1171, 1389, 1560, 2551, 2846, 5072
<-Mbo II	/$(N)_7$TCTTC	17	2813, 2891, 2909, 3383, 3544, 3728, 4013, 4022, 4067, 4403, 4460, 4481, 4499, 4544, 4696, 4994, 5128
Mnl I->	CCTC$(N)_7$/	20	52, 64, 110, 116, 187, 237, 305, 784, 816, 962, 1153, 1303, 1494, 1848, 2320, 2907, 3903, 3954, 4910, 5006
<-Mnl I	/$(N)_7$GAGG	23	97, 103, 149, 267, 465, 509, 672, 720, 768, 789, 1215, 1471, 1655, 1702, 1864, 1897, 1918, 1985, 2243, 2405, 2554, 4385, 4892
Mst II	CC/TNAGG	4	187, 237, 305, 1749
Nco I	C/CATGG	5	163, 281, 386, 3441, 5120
Nde I	CA/TATG	3	823, 1555, 3739

Restriction Enzyme Sites in BK Virus (Dunlop Strain) DNA (*Continued*)

Restriction enzyme	Recognition pattern	Number of sites	Base number matched
Nla III	CATG/	31	164, 182, 204, 232, 254, 282, 300, 322, 331, 387, 623, 980, 1122, 1938, 1951, 1972, 1978, 2269, 2880, 2941, 3336, 3397, 3442, 3666, 3852, 4038, 4240, 4485, 4818, 5112, 5121
Nla IV	GGN/NCC	14	649, 1082, 1335, 1566, 1620, 1734, 2190, 2316, 2591, 2592, 3523, 4562, 4676, 4926
Nsi I	ATGCA/T	1	2942
[NspB II]	C(AC)G/C(GT)G	2	400, 2351
[Nsp7524 I]	(AG)CATG/(CT)	5	330, 1121, 1937, 4239, 4817
Pss I	(AG)GGNCC(CT)	8	649, 1082, 1299, 1865, 2316, 2591, 3523, 4588
Pst I	CTGCA/G	1	3492
Pvu II	CAG/CTG	2	400, 2351
Rsa I	GT/AC	18	542, 781, 1010, 1336, 1344, 1431, 1526, 1909, 2079, 2090, 2191, 2263, 2668, 3400, 3555, 3654, 4154, 4872
Sac I	GAGCT/C	3	458, 3669, 5115
Sau96 I	G/GNCC	12	650, 1083, 1300, 1566, 1866, 2317, 2592, 3157, 3524, 4589, 4676, 5108
Sca I	AGT/ACT	2	1009, 2262
ScrF I	CC/NGG	16	437, 616, 846, 1029, 1109, 1225, 1467, 1594, 2124, 2544, 2587, 3233, 3733, 4531, 4923, 5082
SfaN I->	GCATC$(N)_5$/	1	750
<-SfaN I	/$(N)_9$GATGC	2	1374, 1690

(*Continued*)

Restriction Enzyme Sites in BK Virus (Dunlop Strain) DNA (*Continued*)

Restriction enzyme	Recognition pattern	Number of sites	Base number matched
[Sna I]	GTATAC	2	2972, 4316
Sph I	GCATG/C	2	1937, 4817
Ssp I	AAT/ATT	8	1053, 1273, 3242, 3587, 3677, 4222, 4266, 4598
Stu I	AGG/CCT	9	49, 806, 959, 1520, 2147, 2406, 2743, 3052, 5098
Sty I	C/C(AT)(AT)GG	11	163, 281, 344, 386, 584, 1071, 1198, 2039, 3441, 4760, 5120
[<-Taq II]	/$(N)_9$TGGGTG	2	625, 4425
[Tce I]	GAAGA	7	514, 1171, 1389, 1560, 2551, 2846, 5072
[Tce I]	TCTTC	17	2813, 2891, 2909, 3383, 3544, 3728, 4013, 4022, 4067, 4403, 4460, 4481, 4499, 4544, 4696, 4994, 5128
Tth111 II->	CAA(AG)CA$(N)_9$/	7	819, 1912, 2619, 3613, 3626, 4473
<-Tth111 II	/$(N)_{11}$TG(CT)TTG	4	1093, 2564, 2871, 3784
Xba I	T/CTAGA	2	1668, 3019
Xho II	(AG)/GATC(CT)	5	1734, 2426, 2443, 4085, 4645
Xmn I	GAANN/NNTTC	3	2248, 2578, 4059

Restriction Enzymes That Do Not Cut BK Virus DNA

Enzyme[1,2]	Recognition site[3]	Enzyme	Recognition site	Enzyme	Recognition site
Aat II	GACGT/C	[Gdi II]	CGGCC/(AG)	Pvu I	CGAT/CG
Aha II	G(AG)/CG(CT)C	[Gsu I]	CTGGAG	Rsr II	CG/G(AT)CCG
Apa I	GGGCC/C	Hae II	(AG)GCGC/(CT)	Sac II	CCGC/GG
Asu I	TT/CGAA	Hga I->	GACGC$(N)_5$/	Sal I	G/TCGAC
Ava I	C/(CT)CG(AG)G	<-Hga I	/$(N)_{10}$GCGTC	Sfi I	GGCCNNNN/NGGCC
Bcl I	T/GATCA	Hinc II	GT(CT)/(AG)AC	Sma I	CCC/GGG
Bgl I	GCCNNNN/NGGC	Hpa I	GTT/AAC	SnaB I	TAC/GTA
Bsm I	G/CATTC	Hpa II	C/CGG	Spe I	A/CTAGT
BssH II	G/CGCGC	<-Hph I	/$(N)_7$TCACC	Taq I	T/CGA
BstE II	G/GTNACC	[Mae II]	A/CGT	[Taq II->]	GACCGA$(N)_{11}$/
[Cfr10 I]	(AG)CCGG(CT)	Mlu I	A/CGCGT	[Taq II->]	CACCCA$(N)_{11}$/
Cla I	AT/CGAT	Nae I	GCC/GGC	[<-Taq II]	/$(N)_9$TCGGTC
[Eco47 III]	AGCGCT	Nar I	GG/CGCC	Tha I	CG/CG
EcoR V	GAT/ATC	Nci I	CC/(CG)GG	Tth111 I	GACN/NNGTC
[Esp I]	GC/TNAGC	Nhe I	G/CTAGC	Xho I	C/TCGAG
Fsp I	TGC/GCA	Not I	GC/GGCCGC	Xma III	C/GGCCG
[Gdi II]	(CT)/GGCCG	Nru I	TCG/CGA		

[1] Restriction enzyme names in brackets (e.g., [AflII]) are not commercially available as of September, 1985.
[2] The direction of an asymmetric cut is indicated by –> or <– in the restriction enzyme name.
[3] The restriction enzyme recognition sequence (5'– –>3'). The position of endonuclease cleavage is indicated (/). Ambiguous bases in the recognition sequence are enclosed in parentheses.

The Nucleotide Sequence of BK Virus (MM Strain) DNA

```
------    10         20         30         40         50         60         70         80         90      -----
 AATTCCCCTC CCCAATTTAA ATGAGGACCT AACCTGTGGA AATCTACTGA TGTGGGAGGC TGTAACTGTA CAAACAGAGG TTATTGGAAT AACTAGCATG
 TTAAGGGGAG GGGTTAAATT TACTCCTGGA TTGGACACCT TTAGATGACT ACACCCTCCG ACATTGACAT GTTTGTCTCC AATAACCTTA TTGATCGTAC
EcoR I                                                                                                  Sph I

        110        120        130        140        150        160        170        180        190        200
 CTTAACCTTC ATGCAGGGTC ACAAAAAGTG CATGAGCATG GTGGAGGAAA ACCTATTCAA GGCAGTAATT TCCACTTCTT TGCTGTTGGT GGAGACCCCT
 GAATTGGAAG TACGTCCCAG TGTTTTTCAC GTACTCGTAC CACCTCCTTT TGGATAAGTT CCGTCATTAA AGGTGAAGAA ACGACAACCA CCTCTGGGGA

        210        220        230        240        250        260        270        280        290        300
 TGGAAATGCA GGGAGTGCTA ATGAATTACA GGACAAAGTA CCCAGATGGT ACTATAACCC CTAAAAACCC AACAGCCCAG TCCCAGGTAA TGAATACTGA
 ACCTTTACGT CCCTCACGAT TACTTAATGT CCTGTTTCAT GGGTCTACCA TGATATTGGG GATTTTTGGG TTGTCGGGTC AGGGTCCATT ACTTATGACT

        310        320        330        340        350        360        370        380        390        400
 CCATAAGGCC TATTTGGACA AAAACAATGC TTATCCAGTT GAGTGCTGGG TTCCTGATCC TAGTAGAAAT GAAAATACTA GGTATTTTGG GACTTTCACA
 GGTATTCCGG ATAAACCTGT TTTTGTTACG AATAGGTCAA CTCACGACCC AAGGACTAGG ATCATCTTTA CTTTTATGAT CCATAAAACC CTGAAAGTGT

        410        420        430        440        450        460        470        480        490        500
 GGAGGGGAAA ATGTTCCCCC AGTACTTCAT GTGACCAACA CAGCTACCAC AGTGTTGCTA GATGAACAGG GTGTGGGGCC TCTTTGTAAA GCTGATAGCC
 CCTCCCCTTT TACAAGGGGG TCATGAAGTA CACTGGTTGT GTCGATGGTG TCACAACGAT CTACTTGTCC CACACCCCGG AGAAACATTT CGACTATCGG

        510        520        530        540        550        560        570        580        590        600
 TGTATGTTTC AGCTGCTGAT ATTTGTGGCC TGTTTACTAA CAGCTCTGGA ACACAACAGT GGAGAGGCCT TGCAAGATAT TTTAAGATCC GCCTGAGAAA
 ACATACAAAG TCGACGACTA TAAACACCGG ACAAATGATT GTCGAGACCT TGTGTTGTCA CCTCTCCGGA ACGTTCTATA AAATTCTAGG CGGACTCTTT

 ------610        620        630        640        650        660        670        680        690        700
 AAGATCTGTA AAGAATCCTT ACCTAATTTC CTTTTTGCTA AGTGACCTTA TAAACAGGAG AACCCAGAGA GTGGATGGGC AGCCTATGTA TGGTATGGAA
 TTCTAGACAT TTCTTAGGAA TGGATTAAAG GAAAAACGAT TCACTGGAAT ATTTGTCCTC TTGGGTCTCT CACCTACCCG TCGGATACAT ACCATACCTT
  Bgl II

        710        720        730        740        750        760        770        780        790        800
 TCCCAGGTAG AAGAGGTTAG GGTGTTTGAT GGCACAGAAA GACTTCCAGG GGACCCAGAT ATGATAAGAT ATATTGACAA ACAGGGACAA TTGCAAACCA
 AGGGTCCATC TTCTCCAATC CCACAAACTA CCGTGTCTTT CTGAAGGTCC CCTGGGTCTA TACTATTCTA TATAACTGTT TGTCCCTGTT AACGTTTGGT
```

```
VP1 CDS End
  <---:
       810        820        830        840        850        860        870        880        890        900
AAATGCTTTA AACAGGTGCT TTTATTGTAC ATATACATTT AATAAATGCT GCTTTTGTAT AAGCCACTTT TAAGCTTGTG TTATTTTGGG GGTGGTGTTT
TTTACGAAAT TTGTCCACGA AAATAACATG TATATGTAAA TTATTTACGA CGAAAACATA TTCGGTGAAA ATTCGAACAC AATAAAACCC CCACCACAAA
                                                                                            :---> Large T EXON
                                                                                                  2  CDS End

       910        920        930        940        950        960        970        980        990       1000
TAGGCCTTTT AAAACACTGA AAGCCTTTAC ACAAATGCAA CTCTTGACTA TGGGGGTCTG ACCTTTGGGA ATCTTCAGCA GGGGCTGAAG TATCTGAGAC
ATCCGGAAAA TTTTGTGACT TTCGGAAATG TGTTTACGTT GAGAACTGAT ACCCCCAGAC TGGAAACCCT TAGAAGTCGT CCCCGACTTC ATAGACTCTG

      1010       1020       1030       1040       1050       1060       1070       1080       1090       1100
TTGGGAAGAG CATTGTGATT GGGATTCAGT GCTTGATCCA TGTCCAGAGT CTTCAGTTTC TGAATCCTCT TCTCTTGTAA TATCAAGAAT ACATTTCCCC
AACCCTTCTC GTAACACTAA CCCTAAGTCA CGAACTAGGT ACAGGTCTCA GAAGTCAAAG ACTTAGGAGA AGAGAACATT ATAGTTCTTA TGTAAAGGGG

------1110       1120       1130       1140       1150       1160       1170       1180       1190       1200
ATGCATATAT TATATTTCAT CCTTGAAAAA GTATACATAC TTATCTCAGA ATCCAGCCTT TCCTTCCATT CAACAATTCT AGATTGTATA TCAGTTGCAA
TACGTATATA ATATAAAGTA GGAACTTTTT CATATGTATG AATAGAGTCT TAGGTCGGAA AGGAAGGTAA GTTGTTAAGA TCTAACATAT AGTCAACGTT
Nsi I

      1210       1220       1230       1240       1250       1260       1270       1280       1290       1300
AATCAGCTAC AGGCCTAAAC CAAATTAGCA GTAGCAACAA GGTCATTCCA CTTTGTAAAA TTCTTTTTTC AAGTAAGAAC TCTGAGTTTT GTAAGGATTT
TTAGTCGATG TCCGGATTTG GTTTAATCGT CATCGTTGTT CCAGTAAGGT GAAACATTTT AAGAAAAAAG TTCATTCTTG AGACTCAAAA CATTCCTAAA

      1310       1320       1330       1340       1350       1360       1370       1380       1390       1400
TCTTAAATAT ATTTTGGGCC TAAAATCTAT TTGTCTTACA AATCTAGCTT GCAGGGTTTT AGGGACAGGA TACTCATTCA TTGTAACCAA GCCTGGTGGA
AGAATTTATA TAAAACCCGG ATTTTAGATA AACAGAATGT TTAGATCGAA CGTCCCAAAA TCCCTGTCCT ATGAGTAAGT AACATTGGTT CGGACCACCT

      1410       1420       1430       1440       1450       1460       1470       1480       1490       1500
AATATTTGGG TTCTTTTGTT TAAATGTTTC TTTTCTAAAT TTACCTTAAC ACTTCCATCT AAATAATCTC TCAAACTGTC TAAATTGTTT ATTCCATGTC
TTATAAACCC AAGAAAACAA ATTTACAAAG AAAAGATTTA AATGGAATTG TGAAGGTAGA TTTATTAGAG AGTTTGACAG ATTTAACAAA TAAGGTACAG

      1510       1520       1530       1540       1550       1560       1570       1580       1590       1600
CTGAAGGCAA ATCCTTTGAT TCAGCTCCTG TCCCTTTTAC ATCTTCAAAA ACAACCATGT ACTGATCTAT AGCTACACCT AGCTCAAAGG TTAGCCTTTC
GACTTCCGTT TAGGAAACTA AGTCGAGGAC AGGGAAAATG TAGAAGTTTT TGTTGGTACA TGACTAGATA TCGATGTGGA TCGAGTTTCC AATCGGAAAG
```

(*Continued*)

The Nucleotide Sequence of BK Virus (MM Strain) DNA (*Continued*)

```
      1610       1620       1630       1640       1650 ------1660       1670       1680       1690       1700
CATGGGTAGG TTTACATTTA AGGCTTTACC TCTACACAAA TCTAACAACC CTGCAGCTAG TGTTGTTTTT CCACTATCAA TGGGACCTTT AAATAACCAG
GTACCCATCC AAATGTAAAT TCCGAAATGG AGATGTGTTT AGATTGTTGG GACGTCGATC ACAACAAAAA GGTGATAGTT ACCCTGGAAA TTTATTGGTC
                                                       Pst I

      1710       1720       1730       1740       1750       1760       1770       1780       1790       1800
TATCTTCTTT TAGGTACATT GAAAACAATA CAGTGCAAAA AATCAAATAT TACAGAATCC ATTTTAGGTA GCAAACAGTG CAGCCAAGCA ACACCTGCCA
ATAGAAGAAA ATCCATGTAA CTTTTGTTAT GTCACGTTTT TTAGTTTATA ATGTCTTAGG TAAAATCCAT CGTTTGTCAC GTCGGTTCGT TGTGGACGGT

      1810       1820       1830       1840       1850       1860       1870       1880       1890       1900
TATATTGTTC TAGTACAGCA TTTCCATGAG CTCCAAATAT TAAATCCATT TTATCTAATA TATGATTGAA TCTTTCTGTT AGCATTTCTT CCCTGGTCAT
ATATAACAAG ATCATGTCGT AAAGGTACTC GAGGTTTATA ATTTAGGTAA AATAGATTAT ATACTAACTT AGAAAGACAA TCGTAAAGAA GGGACCAGTA

      1910       1920       1930       1940       1950       1960       1970       1980       1990       2000
ATGAAGGGTA TCTACTCTTT TTTTAGCTAA AACTGTATCT ACTGCTTGCT GACAAATAAC TTTTTTGTTT TTACTTTCTG CAAAAATAAT AGCATTTGCA
TACTTCCCAT AGATGAGAAA AAAATCGATT TTGACATAGA TGACGAACGA CTGTTTATTG AAAAAACAAA AATGAAAGAC GTTTTTATTA TCGTAAACGT

      2010       2020       2030       2040       2050       2060       2070       2080       2090       2100
AAGTGCTTTT CATGATACTT AAAGTGATAA GGCTGGTCTT TTTTCTGACA CTTTTTACAC TCCTCTACAT TGTATTGAAA TTCTAAATAC ATACCTAATA
TTCACGAAAA GTACTATGAA TTTCACTATT CCGACCAGAA AAAAGACTGT GAAAAATGTG AGGAGATGTA ACATAACTTT AAGATTTATG TATGGATTAT

      2110       2120       2130       2140       2150       2160       2170       2180       2190       2200
ATAAAAACAC ATCCTCACAC TTTGTCTCTA CTGCATACTC AGTAATTAAT TTCCAAGACA CCTGCTTTGT TTCTTCAGGC TCTTCTGGGT TAAAATCATG
TATTTTTGTG TAGGAGTGTG AAACAGAGAT GACGTATGAG TCATTAATTA AAGGTTCTGT GGACGAAACA AAGAAGTCCG AGAAGACCCA ATTTTAGTAC

      2210       2220       2230       2240       2250       2260       2270       2280       2290       2300
CTCCTTTAAG CCCCCTTGAA TGCTTTCTTC TATAGTATGG TATGGATCTC TAGTTAAGGC ACTATATAGT AAGTATTCCT TATTAACACC CTTACAAATT
GAGGAAATTC GGGGGAACTT ACGAAAGAAG ATATCATACC ATACCTAGAG ATCAATTCCG TGATATATCA TTCATAAGGA ATAATTGTGG GAATGTTTAA

      2310       2320       2330       2340       2350       2360       2370       2380       2390       2400
AAAAAACTAA AGGTACACAG CTTTTGACAG AAATTATTAA TTGCAGAAAC TCTATGTCTA TGTGGAGTTA AAAAGAATAT AATATTATGC CCAGCACACA
TTTTTTGATT TCCATGTGTC GAAAACTGTC TTTAATAATT AACGTCTTTG AGATACAGAT ACACCTCAAT TTTTCTTATA TTATAATACG GGTCGTGTGT

      2410       2420       2430       2440       2450       2460       2470       2480       2490       2500
TGTGTCTACT AATAAAAGTT ACAGAATATT TTTCCATAAG TTTTTTATAC AGAATTTGAG CTTTTTCTTT AGTAGTATAC ACAGCAAAGC AGGCAAGGGT
ACACAGATGA TTATTTTCAA TGTCTTATAA AAAGGTATTC AAAAAATATG TCTTAAACTC GAAAAAGAAA TCATCATATG TGTCGTTTCG TCCGTTCCCA
```

```
      2510       2520       2530       2540       2550       2560       2570       2580       2590       2600
TCTATTACTA AATACAGCTT GACTAAGAAA CTGGTGTAGA TCAGAGGGAA AGTCTTTAGG GTCTTCTACC TTTCTTTTTT TTTTGGGTGG TGTTGAGTGT
AGATAATGAT TTATGTCGAA CTGATTCTTT GACCACATCT AGTCTCCCTT TCAGAAATCC CAGAAGATGG AAAGAAAAAA AAAACCCACC ACAACTCACA

      2610       2620       2630       2640       2650       2660       2670       2680       2690       2700
TGAGAATCTG CTGTTGCTTC TTCATCACTG GCAAACATAT CTTCATGGCA AAATAAATCT TCATCCCATT TTTCATTAAA GGAACTCCAC CAGGACTCCC
ACTCTTAGAC GACAACGAAG AAGTAGTGAC CGTTTGTATA GAAGTACCGT TTTATTTAGA AGTAGGGTAA AAAGTAATTT CCTTGAGGTG GTCCTGAGGG

      2710       2720       2730       2740       2750       2760       2770       2780       2790       2800
ACTCTTCTGT TCCATAGGTT GGCACCTATA AAAAAAATAA TTACTTAGGG CATAGGCCAT TCCTTGCAGT ACAGGGTATC TGGGCAAAGA GGAAAATCAG
TGAGAAGACA AGGTATCCAA CCGTGGATAT TTTTTTTATT AATGAATCCC GTATCCGGTA AGGAACGTCA TGTCCCATAG ACCCGTTTCT CCTTTTAGTC
    Large T EXON 2          ¦                                  ¦---> Small t CDS End
      CDS Start      <---¦

      2810       2820       2830       2840       2850       2860       2870       2880       2890       2900
CACAAACCTC TGAGCTACTC CAGGTTCCAA AATCAGGCTG ATGAGCTACC TTTACATCCT GCTCCATTTT TTTATACAAA GTATTCATTC TCTTCATTTT
GTGTTTGGAG ACTCGATGAG GTCCAAGGTT TTAGTCCGAC TACTCGATGG AAATGTAGGA CGAGGTAAAA AAATATGTTT CATAAGTAAG AGAAGTAAAA
     ¦--->  Large T EXON 1 CDS End

      2910       2920       2930       2940       2950       2960       2970       2980       2990       3000
ATCCTCGTCG CCCCCTTTGT CAGGGTGAAA TTCCTTACAC TTCCTTAAAG CTTTTCTCAT TAAGGGAAGA TTTCCCCAGG CAGCTCTTTC AAGGCCTAAA
TAGGAGCAGC GGGGGAAACA GTCCCACTTT AAGGAATGTG AAGGAATTTC GAAAAGAGTA ATTCCCTTCT AAAGGGGTCC GTCGAGAAAG TTCCGGATTT

      3010       3020       3030       3040       3050       3060       3070       3080       3090       3100
AGGTCCATGA GCTCCATGGA TTCTTCCCTG TTAAGAACTT TATCCATTTT TGCAAAAATT GCAAAAGAAT AGGGATTTCC CCAAATAGTT TTGCTAGGCC
TCCAGGTACT CGAGGTACCT AAGAAGGGAC AATTCTTGAA ATAGGTAAAA ACGTTTTTAA CGTTTTCTTA TCCCTAAAGG GGTTTATCAA AACGATCCGG
           Small t, Large T EXON 1 CDS Start  <---¦

      3110       3120       3130       3140       3150       3160       3170       3180       3190       3200
TCAGAAAAAG CCTCCACACC CTTACTACTT GAGAGAAAGG GTGGAGGCAG AGGCGGCCTC GGCCTCTTAT ATATTATAAA AAAAAAGGCC ACAGGGAGGA
AGTCTTTTTC GGAGGTGTGG GAATGATGAA CTCTCTTTCC CACCTCCGTC TCCGCCGGAG CCGGAGAATA TATAATATTT TTTTTTCCGG TGTCCCTCCT

      3210       3220       3230       3240       3250       3260       3270       3280       3290       3300
GCTGCTTACC CATGGAATGC AGCCAAACCA TGACCTCAGG AAGGAAAGTG CATGACTGGG CAGCCAGCCA GTGGCAGTTA ATAGTGAAAC CCCGCCCCTA
CGACGAATGG GTACCTTACG TCGGTTTGGT ACTGGAGTCC TTCCTTTCAC GTACTGACCC GTCGGTCGGT CACCGTCAAT TATCACTTTG GGGCGGGGAT
```


(Continued)

```
      3310       3320       3330       3340       3350       3360       3370       3380       3390       3400
AAATCTCTCT TACCCATGGA ATGCAGCCAA ACCATGACCT CAGGAAGGAA AGTGCATGAC TGGGCAGCCA GCCAGTGGCA GTTAATAGTG AAACCATGCC
TTTAGAGAGA ATGGGTACCT TACGTCGGTT TGGTACTGGA GTCCTTCCTT TCACGTACTG ACCCGTCGGT CGGTCACCGT CAATTATCAC TTTGGTACGG

      3410       3420       3430       3440       3450       3460       3470       3480       3490       3500
AAACCATGAC CTCAGGAAGG AAAGTGCATG ACTGGGCAGC CAGCCAGTGG CAGTTAATTT GCGAGCCTAG GAATCTTGGC CTTGTCCCCA GTTAAACTGG
TTTGGTACTG GAGTCCTTCC TTTCACGTAC TGACCCGTCG GTCGGTCACC GTCAATTAAA CGCTCGGATC CTTAGAACCG GAACAGGGGT CAATTTGACC

                ¦---> Agnoprotein CDS Start
      3510           --- -      3530       3540       3550       3560       3570       3580       3590       3600
ACAAAGGCCA TGGTTCTGCG CCAGCTGTCA CGACAAGCTT CAGTGAAAGT TGGTAAAACC TGGACTGGAA CAAAAAAAAG AGCTCAGAGG ATTTTTATTT
TGTTTCCGGT ACCAAGACGC GGTCGACAGT GCTGTTCGAA GTCACTTTCA ACCATTTTGG ACCTGACCTT GTTTTTTTTC TCGAGTCTCC TAAAAATAAA
                     Hha I

      3610       3620       3630       3640       3650       3660       3670       3680       3690       3700
TTATTTTAGA GCTTTTGCTG GAATTTTGTA GAGGTGAAGA CAGTGTAGAC GGGAAAAACA AAAGTACCAC TGCTTTACCT GCTGTAAAAG ACTCTGTAAA
AATAAAATCT CGAAAACGAC CTTAAAACAT CTCCACTTCT GTCACATCTG CCCTTTTTGT TTTCATGGTG ACGAAATGGA CGACATTTTC TGAGACATTT

Agnoprotein CDS End
 <---¦                                                  ¦---> VP2 CDS Start
      3710       3720       3730           --- ---------       3760       3770       3780       3790       3800
AGACTCCTAG GTAAGTAATC CCTTTTTTTT TGTATTTCCA GGTTCATGGG TGCTGCTCTA GCACTTTTGG GGGACCTAGT TGCCAGTGTA TCTGAGGCTG
TCTGAGGATC CATTCATTAG GGAAAAAAAA ACATAAAGGT CCAAGTACCC ACGACGAGAT CGTGAAAACC CCCTGGATCA ACGGTCACAT AGACTCCGAC
                                            BstX I

      3810       3820       3830       3840       3850       3860       3870       3880       3890       3900
CTGCTGCCAC AGGATTTTCA GTGGCTGAAA TTGCTGCTGG GGAGGCTGCT GCTGCTATAG AAGTTCAAAT TGCATCCCTT GCTACTGTAG AGGGCATAAC
GACGACGGTG TCCTAAAAGT CACCGACTTT AACGACGACC CCTCCGACGA CGACGATATC TTCAAGTTTA ACGTAGGGAA CGATGACATC TCCCGTATTG

      3910       3920       3930       3940       3950       3960       3970       3980       3990       4000
AAGTACCTCA GAGGCTATAG CTGCTATAGG CCTAACTCCT CAAACATATG CTGTAATTGC TGGTGCTCCT GGGGCTATTG CTGGGTTTGC TGCTTTAATT
TTCATGGAGT CTCCGATATC GACGATATCC GGATTGAGGA GTTTGTATAC GACATTAACG ACCACGAGGA CCCCGATAAC GACCCAAACG ACGAAATTAA

      4010       4020       4030       4040       4050       4060       4070       4080       4090       4100
CAAACTGTTA GTGGTATTAG TTCCTTGGCT CAAGTAGGGT ATAGGTTCTT TAGTGATTGG GATCACAAAG TTTCCACTGT AGGCCTCTAT CAGCAATCAG
GTTTGACAAT CACCATAATC AAGGAACCGA GTTCATCCCA TATCCAAGAA ATCACTAACC CTAGTGTTTC AAAGGTGACA TCCGGAGATA GTCGTTAGTC
```

```
 |---> VP3 CDS Start
      4110       4120       4130       4140       4150       4160       4170       4180       4190       4200
GCATGGCTTT GGAATTGTTT AACCCAGATG AGTACTATGA TATTCTGTTT CCTGGTGTAA ATACTTTTGT TAATAATATT CAATACCTTG ATCCTAGGCA
CGTACCGAAA CCTTAACAAA TTGGGTCTAC TCATGATACT ATAAGACAAA GGACCACATT TATGAAAACA ATTATTATAA GTTATGGAAC TAGGATCCGT

      4210       4220       4230       4240       4250       4260       4270       4280       4290       4300
TTGGGGTCCT TCTTTGTTTG CTACTATTTC CCAGGCTTTG TGGCATGTTA TTAGGGATGA TATACCTTCT ATAACCTCAC AGGAATTGCA GAGAAGAACA
AACCCCAGGA AGAAACAAAC GATGATAAAG GGTCCGAAAC ACCGTACAAT AATCCCTACT ATATGGAAGA TATTGGAGTG TCCTTAACGT CTCTTCTTGT

      4310       4320       4330       4340       4350       4360       4370       4380       4390       4400
GAAAGATTTT TTAGAGACTC CTTGGCTAGA TTTTTGGAGG AAACTACCTG GACAATTGTA AATGCCCCTA TAAACTTTTA TAATTATATT CAACAATATT
CTTTCTAAAA AATCTCTGAG GAACCGATCT AAAAACCTCC TTTGATGGAC CTGTTAACAT TTACGGGGAT ATTTGAAAAT ATTAATATAA GTTGTTATAA

      4410       4420       4430       4440       4450       ---- --    4470       ------        4490       4500
ATTCTGATCT TTCCCCTATT AGGCCCTCAA TGGTTAGACA AGTAGCTGAA AGGGAAGGTA CCCGTGTACA TTTTGGCCAT ACTTATAGTA TAGATGATGC
TAAGACTAGA AAGGGGATAA TCCGGGAGTT ACCAATCTGT TCATCGACTT TCCCTTCCAT GGGCACATGT AAAACCGGTA TGAATATCAT ATCTACTACG
                                                          Kpn I              Bal I

      4510       4520       4530       4540       4550       4560       4570       4580       4590       4600
TGACAGTATA GAAGAAGTTA CACAAAGAAT GGACTTAAGA AATCAACAAA GTGTACATTC AGGAGAGTTT ATAGAAAAAA CTATTGCCCC AGGAGGTGCT
ACTGTCATAT CTTCTTCAAT GTGTTTCTTA CCTGAATTCT TTAGTTGTTT CACATGTAAG TCCTCTCAAA TATCTTTTTT GATAACGGGG TCCTCCACGA

                                                                                         |---> VP1 CDS Start
      4610       4620       4630       4640       4650       4660       4670       4680      |4690       4700
AATCAAAGAA CTGCTCCTCA ATGGATGTTG CCTTTACTTC TAGGCCTGTA CGGGACTGTA ACACCTGCTC TTGAAGCATG TGAAGATGGC CCCAACCAAA
TTAGTTTCTT GACGAGGAGT TACCTACAAC GGAAATGAAG ATCCGGACAT GCCCTGACAT TGTGGACGAG AACTTCGTAC ACTTCTACCG GGGTTGGTTT

                                                                                 VP2, VP3 CDS End  <---|
      4710       4720       4730       4740       4750       4760       4770       4780       4790       4800
AGAAAAGGAG AGTGTCCAGG GGCAGCTCCC AAAAAGCCAA AGGAACCCGT GCAAGTGCCA AAACTACTAA TAAAAGGAGG AGTAGAAGTT CTAGAAGTTA
TCTTTTCCTC TCACAGGTCC CCGTCGAGGG TTTTTCGGTT TCCTTGGGCA CGTTCACGGT TTTGATGATT ATTTTCCTCC TCATCTTCAA GATCTTCAAT

|     4810       4820       4830       4840       4850       ----- -    4870       4880       4890       4900
AAACTGGGGT AGATGCTATT ACAGAGGTAG AATGCTTCCT AAACCCAGAA ATGGGGGATC CAGATGAAAA CCTTAGGGGC TTTAGTCTAA AGCTAAGTGC
TTTGACCCCA TCTACGATAA TGTCTCCATC TTACGAAGGA TTTGGGTCTT TACCCCCTAG GTCTACTTTT GGAATCCCCG AAATCAGATT TCGATTCACG
                                                            BamH I

      4910       4920       4930       4940       4950       4960  -
TGAAAATGAC TTTAGCAGTG ATAGCCCAGA GAGAAAAATG CTTCCCTGTT ACAGCACAGC AAG
ACTTTTACTG AAATCGTCAC TATCGGGTCT CTCTTTTTAC GAAGGGACAA TGTCGTGTCG TTC
```

Restriction Enzyme Sites in BK Virus (MM Strain) DNA

Restriction enzyme[1,2]	Recognition pattern[3]	Number of sites	Base number matched[4]
Acc I	GT/(AC)(GT)AC	4	1131, 2404, 2475, 3645
[Afl II]	C/TTAAG	1	4534
[Afl III]	A/C(AG)(CT)GT	1	2398
Aha III	TTT/AAA	5	16, 807, 908, 1419, 1688
Alu I	AG/CT	30	442, 490, 511, 542, 873, 1205, 1346, 1523, 1571, 1581, 1655, 1829, 1925, 2319, 2459, 2516, 2813, 2844, 2949, 2982, 3010, 3200, 3523, 3536, 3581, 3610, 3919, 4444, 4724, 4891
Ava II	G/G(AT)CC	6	25, 751, 1683, 3002, 3772, 4205
[Avr II]	C/CTAGG	3	3466, 3706, 4193
Bal I	TGG/CCA	1	4474
BamH I	G/GATCC	1	4856
Ban I	G/G(CT)(AG)CC	2	2721, 4457
Ban II	G(AG)GC(CT)/C	3	1828, 3009, 3580
Bbv I->	GCAGC$(N)_8$/	10	679, 1653, 1780, 2980, 3219, 3260, 3323, 3364, 3436, 4722
<-Bbv I	/$(N)_{12}$GCTGC	13	512, 848, 3201, 3752, 3797, 3800, 3803, 3833, 3845, 3848, 3851, 3920, 3989
[BbvS I]	GC(AT)GC	23	512, 679, 848, 1653, 1780, 2980, 3201, 3219, 3260, 3323, 3364, 3436, 3752, 3797, 3800, 3803, 3833, 3845, 3848, 3851, 3920, 3989, 4722
Bgl II	A/GATCT	1	602

[1] Restriction enzyme names in brackets (e.g., [AflII]) are not commercially available as of September, 1985.
[2] The direction of an asymmetric cut is indicated by -> or <- in the restriction enzyme name.
[3] The restriction enzyme recognition sequence (5′-->3′). The position of endonuclease cleavage is indicated (/). Ambiguous bases in the recognition sequence are enclosed in parentheses.
[4] Numbers represent the first base in the sequence of the recognition site for a restriction endonuclease.

Restriction Enzyme Sites in BK Virus (MM Strain) DNA (*Continued*)

Restriction enzyme	Recognition pattern	Number of sites	Base number matched
[Bin I->]	GGATC(N)$_4$/	3	2244, 4060, 4856
[<-Bin I]	/(N)$_5$GATCC	5	356, 586, 1035, 4190, 4857
Bsm I	GAATGCN/	4	2218, 3215, 3319, 4830
Bsp1286 I	G(AGT)GC(ACT)/C	4	1828, 3009, 3580, 3963
BstN I	CC/(AT)GG	16	283, 703, 746, 1392, 1892, 2690, 2820, 2976, 3559, 3738, 3968, 4151, 4231, 4347, 4589, 4716
BstX I	CCANNNNN/NTGG	1	3738
[Cfr I]	(CT)/GGCC(AG)	1	4474
Dde I	C/TNAG	18	593, 638, 994, 1145, 1282, 2138, 2523, 2744, 2810, 3100, 3235, 3339, 3411, 3583, 3792, 3907, 4872, 4893
Dpn I	GmA/TC	11	356, 586, 603, 1035, 1564, 2245, 2539, 4061, 4190, 4406, 4857
[Dra II]	(AG)G/GNCC(CT)	7	24, 475, 750, 1682, 3771, 4204, 4421
EcoR I*	/AATT	31	1, 14, 167, 224, 625, 789, 1175, 1223, 1259, 1438, 1483, 2079, 2144, 2148, 2297, 2332, 2339, 2453, 2739, 2929, 3057, 3456, 3622, 3829, 3868, 3955, 3997, 4113, 4284, 4354, 4382
EcoR I	G/AATTC	1	4963

(*Continued*)

Restriction Enzyme Sites in BK Virus (MM Strain) DNA (*Continued*)

Restriction enzyme	Recognition pattern	Number of sites	Base number matched
EcoR I'	(AG)(AG)A/T(CT)(CT)	37	40, 585, 602, 613, 698, 969, 1022, 1062, 1149, 1258, 1295, 1323, 1340, 1437, 1509, 1638, 1755, 1842, 1868, 2078, 2244, 2452, 2604, 2655, 2928, 2968, 3018, 3073, 3301, 3471, 3589, 3621, 3812, 4304, 4328, 4856, 4963
EcoR II	/CC(AT)GG	16	283, 703, 746, 1392, 1892, 2690, 2820, 2976, 3559, 3738, 3968, 4151, 4231, 4347, 4589, 4716
Fnu4H I	GC/NGC	24	512, 679, 848, 1653, 1780, 2980, 3153, 3201, 3219, 3260, 3323, 3364, 3436, 3752, 3797, 3800, 3803, 3833, 3845, 3848, 3851, 3920, 3989, 4722
Fok I->	GGATG(N)$_9$/	3	673, 4255, 4623
<-Fok I	/(N)$_{13}$CATCC	5	1118, 2110, 2662, 2855, 3873
[Gsu I]	CTCCAG	1	2818
[Hae I]	(AT)GG/CC(AT)	15	306, 526, 565, 902, 1211, 2754, 2992, 3096, 3186, 3477, 3505, 3928, 4081, 4474, 4642
Hae III	GG/CC	21	307, 477, 527, 566, 903, 1212, 1317, 2755, 2993, 3097, 3155, 3161, 3187, 3478, 3506, 3929, 4082, 4422, 4475, 4643, 4688
HgiA I	G(AT)GC(AT)/C	4	1828, 3009, 3580, 3963
[HgiE II]	ACCNNNNNNGGT	1	1386
Hha I	GCG/C	1	3518
Hind III	A/AGCTT	3	872, 2948, 3535

Restriction Enzyme Sites in BK Virus (MM Strain) DNA (*Continued*)

Restriction enzyme	Recognition pattern	Number of sites	Base number matched
Hinf I	G/ANTC	13	613, 698, 969, 1023, 1047, 1062, 1149, 1518, 1755, 1868, 2604, 2694, 3019, 3471, 3690, 3702, 4316
Hph I->	GGTGA$(N)_8$/	2	2924, 3633
Kpn I	GGTAC/C	1	4457
Mae I	C/TAG	15	93, 360, 378, 458, 1179, 1344, 1579, 1657, 1810, 2250, 3094, 3467, 3707, 3758, 3776, 4194, 4326, 4640, 4791
Mae III	/GTNAC	10	62, 118, 431, 642, 1383, 2418, 3527, 4517, 4658, 4948
Mbo I	/GATC	11	356, 586, 603, 1035, 1564, 2245, 2539, 4061, 4190, 4406, 4857
Mbo II->	GAAGA$(N)_8$/	7	710, 1005, 2966, 3636, 4293, 4511, 4682
<-Mbo II	/$(N)_7$TCTTC	16	972, 1050, 1068, 1542, 1703, 1887, 2172, 2181, 2226, 2562, 2619, 2640, 2658, 2703, 2891, 3022
Mnl I->	CCTC$(N)_7$/	21	7, 479, 1066, 1629, 2062, 2113, 2807, 2903, 3099, 3111, 3157, 3163, 3234, 3338, 3410, 3906, 3938, 4084, 4275, 4425, 4616
<-Mnl I	/$(N)_7$GAGG	22	23, 56, 77, 144, 402, 564, 713, 2544, 2789, 3144, 3150, 3196, 3587, 3631, 3794, 3842, 3890, 3911, 4337, 4593, 4777, 4824
Mst II	CC/TNAGG	4	3234, 3338, 3410, 4871
Nco I	C/CATGG	5	1600, 3014, 3210, 3314, 3508
Nde I	CA/TATG	2	1898, 3945

(*Continued*)

Restriction Enzyme Sites in BK Virus (MM Strain) DNA (*Continued*)

Restriction enzyme	Recognition pattern	Number of sites	Base number matched
Nla III	CATG/	31	97, 110, 131, 137, 428, 1039, 1100, 1495, 1556, 1601, 1825, 2011, 2197, 2399, 2644, 3006, 3015, 3211, 3229, 3251, 3315, 3333, 3355, 3395, 3405, 3427, 3509, 3745, 4102, 4244, 4677
Nla IV	GGN/NCC	13	349, 475, 750, 751, 1682, 2721, 2823, 3771, 4204, 4457, 4688, 4742, 4856
Nsi I	ATGCA/T	1	1101
[NspB II]	C(AC)G/C(GT)G	2	510, 3522
[Nsp7524 I]	(AG)CATG/(CT)	4	96, 2398, 4243, 4676
Pss I	(AG)GGNCC(CT)	7	24, 475, 750, 1682, 3771, 4204, 4421
Pst I	CTGCA/G	1	1651
Pvu II	CAG/CTG	2	510, 3522
Rsa I	GT/AC	17	68, 238, 249, 422, 827, 1559, 1714, 1813, 2313, 2769, 3664, 3903, 4132, 4458, 4466, 4553, 4648
Sac I	GAGCT/C	3	1828, 3009, 3580
Sau96 I	G/GNCC	10	25, 476, 751, 1316, 1683, 3002, 3772, 4205, 4422, 4688
Sca I	AGT/ACT	2	421, 4131
ScrF I	CC/NGG	16	283, 703, 746, 1392, 1892, 2690, 2820, 2976, 3559, 3738, 3968, 4151, 4231, 4347, 4589, 4716
SfaN I->	GCATC(N)$_5$/	1	3872
<-SfaN I	/(N)$_9$GATGC	2	4496, 4812
[Sna I]	GTATAC	2	1131, 2475

Restriction Enzyme Sites in BK Virus (MM Strain) DNA (*Continued*)

Restriction enzyme	Recognition pattern	Number of sites	Base number matched
Sph I	GCATG/C	1	96
Ssp I	AAT/ATT	7	1401, 1746, 1836, 2381, 2425, 4175, 4395
Stu I	AGG/CCT	9	306, 565, 902, 1211, 2992, 3096, 3928, 4081, 4642
Sty I	C/C(AT)(AT)GG	11	198, 1600, 3014, 3210, 3314, 3466, 3508, 3706, 4023, 4193, 4320
[<-Taq II]	/9(N_9TGGGTG	2	2584, 3747
[Tce I]	GAAGA	7	710, 1005, 2966, 3636, 4293, 4511, 4682
[Tce I]	TCTTC	16	972, 1050, 1068, 1542, 1703, 1887, 2172, 2181, 2226, 2562, 2619, 2640, 2658, 2703, 2891, 3022
Tth111 II->	CAA(AG)CA$(N)_9$/	6	71, 778, 1772, 1785, 2632, 3941
<-Tth111 II	/$(N)_{11}$TG(CT)TTG	4	723, 1030, 1943, 4215
Xba I	T/CTAGA	2	1178, 4790
Xho II	(AG)/GATC(CT)	4	585, 602, 2244, 4856
Xmn I	GAANN/NNTTC	3	407, 737, 2218

Restriction Enzymes That Do Not Cut BK Virus (MM Strain) DNA

Enzyme[1,2]	Recognition site[3]	Enzyme	Recognition site	Enzyme	Recognition site
Aat II	GACGT/C	[Gdi II]	(CT)/GGCCG	Nru I	TCG/CGA
Aha II	G(AG)/CG(CT)C	[Gdi II]	CGGCC/(AG)	Pvu I	CGAT/CG
Apa I	GGGCC/C	[Gsu I]	CTGGAG	Rsr II	CG/G(AT)CCG
Asu II	TT/CGAA	Hae II	(AG)GCGC/(CT)	SacI I	CCGCG/G
Ava I	C/(CT)CG(AG)G	Hga I->	GACGC$(N)_5$/	Sal I	G/TCGAC
Bcl I	T/GATCA	<-Hga I	/$(N)_{10}$GCGTC	Sfi I	GGCCNNNN/NGGCC
Bgl I	GCCNNNN/NGGC	Hinc II	GT(CT)/(AG)AC	Sma I	CCC/GGG
Bsm I	G/CATTC	Hpa I	GTT/AAC	SnaB I	TAC/GTA
BssH II	G/CGCGC	Hpa II	C/CGG	Spe I	A/CTAGT
BstE II	G/GTNACC	<-Hph I	/7$(N)_7$TCACC	Taq I	T/CGA
[Cfr10 I]	(AG)CCGG(CT)	[Mae II]	/ACGT	[Taq II->]	GACCGA$(N)_9$/
Cla I	AT/CGAT	Mlu I	A/CGCGT	[Taq II->]	CACCCA$(N)_9$/
[Dra III]	CACNNN/GTG	Nae I	GCC/GGC	[<-Taq II]	/$(N)_{11}$TCGGTC
[Eco47 III	AGCGCT	Nar I	GG/CGCC	Tha	CG/CG
EcoR V	GAT/ATC	Nci I	CC/(CG)GG	Tth111 I	GACN/NNGTC
[Esp I]	GC/TNAGC	Nhe I	G/CTAGC	Xho I	C/TCGAG
Fsp I	TGC/GCA	Not I	GC/GGCCGC	Xma III	C/GGCCG

[1] Restriction enzyme names in brackets (e.g., [AflII]) are not commercially available as of September, 1985.
[2] The direction of an asymmetric cut is indicated by –> or <– in the restriction enzyme name.
[3] The restriction enzyme recognition sequence (5′ – –>3′). The position of endonuclease cleavage is indicated (/). Ambiguous bases in the recognition sequence are enclosed in parentheses.

```
        10         20         30         40         50         60         70         80         90        100
GCCTCGGCCT CCTGTATATA TAAAAAAAAG GGAAGGGATG GCTGCCAGCC AAGCATGAGC TCATACCTAG GGAGCCAACC AGCTAACAGC CAGTAAACAA
CGGAGCCGGA GGACATATAT ATTTTTTTTC CCTTCCCTAC CGACGGTCGG TTCGTACTCG AGTATGGATC CCTCGGTTGG TCGATTGTCG GTCATTTGTT
ORGREP      :--->       98 bp tandem repeat number 1
 CORE SEQ...:

       110        120        130        140        150        160        170        180        190        200
AGCACAAGGC TGTATATATA AAAAAAAGGG AAGGGATGGC TGCCAGCCAA GCATGAGCTC ATACCTAGGG AGCCAACCAG CTAACAGCCA GTAAACAAAG
TCGTGTTCCG ACATATATAT TTTTTTTCCC TTCCCTACCG ACGGTCGGTT CGTACTCGAG TATGGATCCC TCGGTTGGTC GATTGTCGGT CATTTGTTTC
    <---::--->         98 bp tandem repeat number 2

                                                                             :--->  Agnoprotein CDS Start
       210        220        230        240        250        260        270 :280        290        300
CACAAGGGGA AGTGGAAAGC AGCCAAGGGA ACATGTTTTG CGAGCCAGAG CTGTTTTGGC TTGTCACCAG CTGGCCATGG TTCTTCGCCA GCTGGTCACGT
GTGTTCCCCT TCACCTTTCG TCGGTTCCCT TGTACAAAAC GCTCGGTCTC GACAAAACCG AACAGTGGTC GACCGGTACC AAGAAGCGGT CGACCAGTGCA
  <---:

       310        320        330        340        350        360        370        380        390        400
AAGGCTTCTG TGAAAGTTAG TAAAACCTGG AGTGGAACTA AAAAAAGAGC TCAAAGGATT TTAATTTTTT TGTTAGAATT TTTGCTGGAC TTTTGCACAG
TTCCGAAGAC ACTTTCAATC ATTTTGGACC TCACCTTGAT TTTTTTCTCG AGTTTCCTAA AATTAAAAAA ACAATCTTAA AAACGACCTG AAAACGTGTC

                                                                   Agnoprotein CDS End <---:
       410        420        430        440        450        460        470        480        490        500
GTGAAGACAG TGTAGACGGG AAAAAAAGAC AGAGACACAG TGGTTTGACT GAGCAGACAT ACAGTGCTTT GCCTGAACCA AAAGCTACAT AGGTAAGTAA
CACTTCTGTC ACATCTGCCC TTTTTTTCTG TCTCTGTGTC ACCAAACTGA CTCGTCTGTA TGTCACGAAA CGGACTTGGT TTTCGATGTA TCCATTCATT

                      :--->   VP2 CDS Start
       510        520 :      530        540        550        560        570        580        590        600
TGTTTTTTTT TGTGTTTTCA GGTTCATGGG TGCCGCACTT GCACTTTTGG GGGACCTAGT TGCTACTGTT TCTGAGGCTG CTGCTGCCAC AGGATTTTCA
ACAAAAAAAA ACACAAAAGT CCAAGTACCC ACGGCGTGAA CGTGAAAACC CCCTGGATCA ACGATGACAA AGACTCCGAC GACGACGGTG TCCTAAAAGT

       610        620        630        640        650        660        670        680        690        700
GTAGCTGAAA TTGCTGCTGG AGAGGCTGCT GCTACTATAG AAGTTGAAAT TGCATCCCTT GCTACTGTAG AGGGGATTAC AAGTACCTCT GAGGCTATAG
CATCGACTTT AACGACGACC TCTCCGACGA CGATGATATC TTCAACTTTA ACGTAGGGAA CGATGACATC TCCCCTAATG TTCATGGAGA CTCCGATATC
```


(Continued)

The Nucleotide Sequence of JC Virus (Madl Strain) DNA (*Continued*)

```
       710        720        730        740       --- -        760        770        780        790        800
CTGCTATAGG CCTTACTCCT GAAACATATG CTGTAATAAC TGGAGCTCCG GGGGCTGTAG CTGGGTTTGC TGCATTGGTT CAAACTGTAA CTGGTGGTAG
GACGATATCC GGAATGAGGA CTTTGTATAC GACATTATTG ACCTCGAGGC CCCCGACATC GACCCAAACG ACGTAACCAA GTTTGACATT GACCACCATC
                                                   Hpa II
                                                                                      |---> VP3 CDS Start
       810        820        830        840        850        860        870        880 |      890        900
TGCTATTGCT CAGTTGGGAT ATAGATTTTT TGCTGACTGG GATCATAAAG TTTCAACAGT TGGGCTTTTT CAGCAGCCAG CTATGGCTTT ACAATTATTT
ACGATAACGA GTCAACCCTA TATCTAAAAA ACGACTGACC CTAGTATTTC AAAGTTGTCA ACCCGAAAAA GTCGTCGGTC GATACCGAAA TGTTAATAAA

       910        920        930        940         -- ----   960        970        980     ------       1000
AATCCAGAAG ACTACTATGA TATTTTATTT CCTGGAGTGA ATGCCTTTGT TAACAATATT CACTATTTAG ATCCTAGACA TTGGGGCCCG TCCTTGTTCT
TTAGGTCTTC TGATGATACT ATAAAATAAA GGACCTCACT TACGGAAACA ATTGTTATAA GTGATAAATC TAGGATCTGT AACCCCGGGC AGGAACAAGA
                                                     Hpa I                                Apa I

      1010       1020       1030       1040       1050       1060       1070       1080       1090       1100
CCACAATCTC CCAGGCTTTT TGGAATCTTG TTAGAGATGA TTTGCCAGCC TTAACCTCTC AGGAAATTCA GAGAAGAACC CAAAAACTAT TTGTTGAAAG
GGTGTTAGAG GGTCCGAAAA ACCTTAGAAC AATCTCTACT AAACGGTCGG AATTGGAGAG TCCTTTAAGT CTCTTCTTGG GTTTTTGATA AACAACTTTC

      1110       1120       1130       1140       1150       1160       1170       1180       1190       1200
TTTAGCAAGG TTTTTGGAAG AAACTACTTG GGCAATAGTT AATTCACCAG CTAACTTATA TAATTATATT TCAGACTATT ATTCTAGATT GTCTCCAGTT
AAATCGTTCC AAAAACCTTC TTTGATGAAC CCGTTATCAA TTAAGTGGTC GATTGAATAT ATTAATATAA AGTCTGATAA TAAGATCTAA CAGAGGTCAA

      1210       1220       1230       1240       1250       1260       1270       1280       1290       1300
AGGCCCTCTA TGGTAAGGCA AGTTGCCCAA AGGGAGGGAA CCTATATTTC TTTTGGCCAC TCATACACCC AAAGTATAGA TGATGCAGAC AGCATTCAAG
TCCGGGAGAT ACCATTCCGT TCAACGGGTT TCCCTCCCTT GGATATAAAG AAAACCGGTG AGTATGTGGG TTTCATATCT ACTACGTCTG TCGTAAGTTC

      1310          - ----- 1330       1340       1350       1360       1370       1380       1390       1400
AAGTTACCCA AAGGCTAGAT TTAAAAACCC CAAATGTGCA ATCTGGTGAA TTTATAGAAA GAAGTATTGC ACCAGGAGGT GCAAATCAAA GATCTGCTCC
TTCAATGGGT TTCCGATCTA AATTTTTGGG GTTTACACGT TAGACCACTT AAATATCTTT CTTCATAACG TGGTCCTCCA CGTTTAGTTT CTAGACGAGG
                Aha III
                                                                    |--->  VP1 CDS Start
      1410       1420       1430       1440       1450       1460       1470       1480       1490       1500
TCAATGGATG TTGCCTTTAC TTTTAGGGTT GTACGGGACT GTAACACCTG CTCTTGAAGC ATATGAAGAT GGCCCCAACA AAAAGAAAAG GAGAAAGGAA
AGTTACCTAC AACGGAAATG AAAATCCCAA CATGCCCTGA CATTGTGGAC GAGAACTTCG TATACTTCTA CCGGGGTTGT TTTTCTTTTC CTCTTTCCTT

                                 VP2, VP3 CDS End  <---|
      1510       1520       1530       1540       1550       1560     . 1570       1580       1590       1600
GGACCCCGTG CAAGTTCCAA AACTTCTTAT AAGAGGAGGA GTAGAAGTTC TAGAAGTTAA AACTGGGGTT GACTCAATTA CAGAGGTAGA ATGCTTTTTA
CCTGGGGCAC GTTCAAGGTT TTGAAGAATA TTCTCCTCCT CATCTTCAAG ATCTTCAATT TTGACCCCAA CTGAGTTAAT GTCTCCATCT TACGAAAAAT
```

```
      1610    -------        1630        1640        1650        1660        1670        1680        1690        1700
ACTCCAGAAA  TGGGTGACCC  AGATGAGCAT  CTTAGGGGTT  TTAGTAAGTC  AATATCTATA  TCAGATACAT  TTGAAAGTGA  CTCCCCAAAT  AGGGACATGC
TGAGGTCTTT  ACCCACTGGG  TCTACTCGTA  GAATCCCCAA  AATCATTCAG  TTATAGATAT  AGTCTATGTA  AACTTTCACT  GAGGGGTTTA  TCCCTGTACG
               BstE II

      1710        1720    ------         1740        1750        1760        1770        1780        1790        1800
TTCCTTGTTA  CAGTGTGGCC  AGAATTCCAC  TACCCAATCT  AAATGAGGAT  CTAACCTGTG  GAAATATACT  CATGTGGGAG  GCTGTGACCT  TAAAAACTGA
AAGGAACAAT  GTCACACCGG  TCTTAAGGTG  ATGGGTTAGA  TTTACTCCTA  GATTGGACAC  CTTTATATGA  GTACACCCTC  CGACACTGGA  ATTTTTGACT
                          EcoR I

      1810        1820        1830        1840        1850        1860        1870        1880        1890        1900
GGTTATAGGG  GTGACAAGTT  TGATGAATGT  GCACTCTAAT  GGGCAAGCAA  CTCATGACAA  TGGTGCAGGG  AAGCCAGTGC  AGGGCACCAG  CTTTCATTTT
CCAATATCCC  CACTGTTCAA  ACTACTTACA  CGTGAGATTA  CCCGTTCGTT  GAGTACTGTT  ACCACGTCCC  TTCGGTCACG  TCCCGTGGTC  GAAAGTAAAA

      1910        1920        1930        1940        1950        1960        1970        1980        1990        2000
TTTTCTGTTG  GGGGGGAGGC  TTTAGAATTA  CAGGGGGTGC  TTTTTAATTA  CAGAACAAAG  TACCCAGATG  GAACAATTTT  TCCAAAGAAT  GCCACAGTGC
AAAAGACAAC  CCCCCCTCCG  AAATCTTAAT  GTCCCCCACG  AAAAATTAAT  GTCTTGTTTC  ATGGGTCTAC  CTTGTTAAAA  AGGTTTCTTA  CGGTGTCACG

      2010        2020        2030        2040        2050        2060        2070        2080        2090        2100
AATCTCAAGT  CATGAACACA  GAGCACAAGG  CGTACCTAGA  TAAGAACAAA  GCATATCCTG  TTGAATGTTG  GGTTCCTGAT  CCCACCAGAA  ATGAAAACAC
TTAGAGTTCA  GTACTTGTGT  CTCGTGTTCC  GCATGGATCT  ATTCTTGTTT  CGTATAGGAC  AACTTACAAC  CCAAGGACTA  GGGTGGTCTT  TACTTTTGTG

      2110        2120        2130        2140        2150        2160        2170        2180        2190        2200
AAGATATTTT  GGGACACTAA  CAGGAGGAGA  AAATGTTCCT  CCAGTTCTTC  ATATAACAAA  CACTGCCACA  ACAGTGTTGC  TTGATGAATT  TGGTGTTGGG
TTCTATAAAA  CCCTGTGATT  GTCCTCCTCT  TTTACAAGGA  GGTCAAGAAG  TATATTGTTT  GTGACGGTGT  TGTCACAACG  AACTACTTAA  ACCACAACCC

      2210        2220        2230        2240        2250        2260        2270        2280        2290        2300
CCACTTTGCA  AAGGTGACAA  CTTATACTTG  TCAGCTGTTG  ATGTCTGTGG  CATGTTTACA  AACAGGTCTG  GTTCCCAGCA  GTGGAGAGGA  CTCTCCAGAT
GGTGAAACGT  TTCCACTGTT  GAATATGAAC  AGTCGACAAC  TACAGACACC  GTACAAATGT  TTGTCCAGAC  CAAGGGTCGT  CACCTCTCCT  GAGAGGTCTA

      2310        2320        2330        2340        2350        2360        2370        2380        2390        2400
ATTTTAAGGT  GCAGCTAAGG  AAAAGGAGGG  TTAAAAACCC  CTACCCAATT  TCTTTCCTTC  TTACTGATTT  AATTAACAGA  AGGACTCCTA  GAGTTGATGG
TAAAATTCCA  CGTCGATTCC  TTTTCCTCCC  AATTTTTGGG  GATGGGTTAA  AGAAAGGAAG  AATGACTAAA  TTAATTGTCT  TCCTGAGGAT  CTCAACTACC

      2410        2420        2430        2440        2450        2460        2470        2480        2490        2500
GCAGCCTATG  TATGGCATGG  ATGCTCAAGT  AGAGGAGGTT  AGAGTTTTTG  AGGGAACAGA  GGAGCTTCCA  GGGGACCCAG  ACATGATGAG  ATACGTTGAC
CGTCGGATAC  ATACCGTACC  TACGAGTTCA  TCTCCTCCAA  TCTCAAAAAC  TCCCTTGTCT  CCTCGAAGGT  CCCCTGGGTC  TGTACTACTC  TATGCAACTG
```

(Continued)

```
                VP1 CDS End  <---¦
      2510       2520       2530       2540       2550       2560       2570       2580       2590       2600
AAATATGGAC AGTTGCAGAC AAAAATGCTG TAATCAAAAG CCTTTATTGT AATATGCAGT ACATTTTAAT AAAGTATAAC CAGCTTTACT TAACAGTTGC
TTTATACCTG TCAACGTCTG TTTTTACGAC ATTAGTTTTC GGAAATAACA TTATACGTCA TGTAAAATTA TTTCATATTG GTCGAAATGA ATTGTCAACG

      2610       2620       2630       2640       2650       2660       2670       2680       2690       2700
AGTTATTTTG GGGGAGGGGT CTTTGGTTTT TTGAAACATT GAAAGCCTTT ACAGATGTGA AAAGTGCAGT TTTCCTGTGT GTCTGCACCA GAGGCTTCTG
TCAATAAAAC CCCCTCCCCA GAAACCAAAA AACTTTGTAA CTTTCGGAAA TGTCTACACT TTTCACGTCA AAAGGACACA CAGACGTGGT CTCCGAAGAC
   ¦---> Large T EXON 2 CDS End

      2710       2720       2730       2740       2750       2760       2770       2780       2790       2800
AGACCTGGGA AAAGCATTGT GATTGTGATT CAGTGCTTGA TCCATGTCCA GAGTCTTCTG CTTCAGAATC TTCCTCTCTA GGAAAGTCAA GAATGGGTCT
TCTGGACCCT TTTCGTAACA CTAACACTAA GTCACGAACT AGGTACAGGT CTCAGAAGAC GAAGTCTTAG AAGGAGAGAT CCTTTCAGTT CTTACCCAGA

      2810       2820       2830       2840       2850       2860       2870       2880       2890       2900
CCCCATACCA ACATTAGCTT TCATAGTAGA AAATGTATAC ATGCTTATTT CTAAATCCAG CCTTTCTTTC CACTGCACAA TCCTCTCATG AATGGCAGCT
GGGGTATGGT TGTAATCGAA AGTATCATCT TTTACATATG TACGAATAAA GATTTAGGTC GGAAAGAAAG GTGACGTGTT AGGAGAGTAC TTACCGTCGA

      2910       2920       2930       2940       2950       2960       2970       2980       2990       3000
GCAAAGTCAG CAACTGGCCT AAACCAGATT AAAAGCAAAA GCAAAGTCAT ACCACTTTGC AAAATCCTTT TTTCTAGCAA ATACTCAGAG CAGCTTAGTG
CGTTTCAGTC GTTGACCGGA TTTGGTCTAA TTTTCGTTTT CGTTTCAGTA TGGTGAAACG TTTTAGGAAA AAAGATCGTT TATGAGTCTC GTCGAATCAC

      3010       3020       3030       3040       3050       3060       3070       3080       3090       3100
ATTTTCTCAG GTAGGCCTTT GGTCTAAAAT CTATCTGCCT TACAAATCTG GCCTGTAAAG TTCTAGGCAC TGAATATTCA TTCATGGTTA CAATTCCAGG
TAAAAGAGTC CATCCGGAAA CCAGATTTTA GATAGACGGA ATGTTTAGAC CGGACATTTC AAGATCCGTG ACTTATAAGT AAGTACCAAT GTTAAGGTCC

      3110       3120       3130       3140       3150       3160       3170       3180       3190       3200
TGGAAACACC TGTGTTCTTT TGTTTTGGTG TTTTCTCTCT AAATTAACTT TTACACTTCC ATCTAAGTAA TCTCTTAAGC AATCAAGGTT GCTTATGCCA
ACCTTTGTGG ACACAAGAAA ACAAAACCAC AAAAGAGAGA TTTAATTGAA AATGTGAAGG TAGATTCATT AGAGAATTCG TTAGTTCCAA CGAATACGGT

      3210       3220       3230       3240       3250       3260       3270       3280       3290       3300
TGCCCTGAAG GTAAATCCCT TGACTCTGCA CCAGTGCCTT TTACATCCTC AAATACAACC ATAAACTGAT CTATACCCAC TCCTAATTCA AAGTTTAATC
ACGGGACTTC CATTTAGGGA ACTGAGACGT GGTCACGGAA AATGTAGGAG TTTATGTTGG TATTTGACTA GATATGGGTG AGGATTAAGT TTCAAATTAG

      3310       3320       3330       3340       3350     ------       3370       3380       3390       3400
TTTCTAATGG CATATTAACA TTTAATGACT TTCCCCCACA GAGATCAAGT AAAGCTGCAG CTAAAGTAGT TTTGCCACTG TCTATTGGCC CCTTGAATAG
AAAGATTACC GTATAATTGT AAATTACTGA AAGGGGGTGT CTCTAGTTCA TTTCGACGTC GATTTCATCA AAACGGTGAC AGATAACCGG GGAACTTATC
                                                          Pst I
```

```
      3410       3420       3430       3440       3450       3460       3470       3480       3490       3500
CCAGTACCTT TTTTTTGGAA TGTTTAATAC AATGCATTTT AGAAAGTCAT AAATAACAGT GTCCATTTGA GGCAGCAAGC AATGAATCCA GGCCACCCCA
GGTCATGGAA AAAAAACCTT ACAAATTATG TTACGTAAAA TCTTTCAGTA TTTATTGTCA CAGGTAAACT CCGTCGTTCG TTACTTAGGT CCGGTGGGGT

      3510       3520       3530       3540       3550       3560       3570       3580       3590       3600
GCCATATATT GCTCTAAAAC AGCATTGCCA TGTGCCCCAA AAATTAAGTC CATTTTATCA AGCAAGAAAT TAAACCTTTC AACTAACATT TCTTCTCTGG
CGGTATATAA CGAGATTTTG TCGTAACGGT ACACGGGGTT TTTAATTCAG GTAAAATAGT TCGTTCTTTA ATTTGGAAAG TTGATTGTAA AGAAGAGACC

      3610       3620       3630       3640       3650       3660       3670       3680       3690       3700
TCATGTGGAT GCTGTCAACC CTTTGTTTGG CTGCTACAGT ATCAACAGCC TGCTGGCAAA TGCTTTTTTG ATTTTTGCTA TCTGCAAAAA TTTGGGCATT
AGTACACCTA CGACAGTTGG GAAACAAACC GACGATGTCA TAGTTGTCGG ACGACCGTTT ACGAAAAAAC TAAAAACGAT AGACGTTTTT AAACCCGTAA

      3710       3720       3730       3740       3750       3760       3770       3780       3790       3800
ATAATAGTGT TTTTCATGAT GGTTAAAGTG ATTTGGCTGA TCCTTTTTTT CACATTTTTT GCATTGCTGT GGGTTTTCCT GAAAGTCTAA GTACATGCCC
TATTATCACA AAAAGTACTA CCAATTTCAC TAAACCGACT AGGAAAAAAA GTGTAAAAAA CGTAACGACA CCCAAAAGGA CTTTCAGATT CATGTACGGG

      3810       3820       3830       3840       3850       3860       3870       3880       3890       3900
ATAAGCAAAA AAACATCCTC ACATTTGGTT TCCAAGGCAT ACTGTGTAAC TAATTTCCAT GAAACCTGCT TAGTTTCTTC TGGTTC TTCT GGGTTAAAGT
TATTCGTTTT TTTGTAGGAG TGTAAACCAA AGGTTCCGTA TGACACATTG ATTAAAGGTA CTTTGGACGA ATCAAAGAAG ACCAAG AAGA CCCAATTTCA

      3910       3920       3930       3940       3950       3960       3970       3980       3990       4000
CATGCTCCTT AAGGCCCCCC TGAATACTTT CTTCCACTAC TGCATATGGC TGTCTACACA GGGCACTATA AAACAAGTAT TCCTTA TTCA CACCTTTACA
GTACGAGGAA TTCCGGGGGG ACTTATGAAA GAAGGTGATG ACGTATACCG ACAGATGTGT CCCGTGATAT TTTGTTCATA AGGAAT AAGT GTGGAAATGT

      4010       4020       4030       4040       4050       4060       4070       4080       4090       4100
AATTAAAAAA CTAAAGGTAC ATAGTTTTTG ACAGTAGTTA TTAATTGCTG ACACTCTATG TCTATGTGGT GTTAAGAAAA ACAAAA TATT ATGACCCCCA
TTAATTTTTT GATTTCCATG TATCAAAAAC TGTCATCAAT AATTAACGAC TGTGAGATAC AGATACACCA CAATTCTTTT TGTTTT ATAA TACTGGGGGT

      4110       4120       4130       4140       4150       4160       4170       4180       4190       4200
AAACCATGTC TACTTATAAA AGTTACAGAA TATTTTTCCA TAAGTTTCTT ATATAAAATT TGAGCTTTTT CTTTAGTGGT ATACAC AGCA AAAGAAGCAA
TTTGGTACAG ATGAATATTT TCAATGTCTT ATAAAAAGGT ATTCAAAGAA TATATTTTAA ACTCGAAAAA GAAATCACCA TATGTG TCGT TTTCTTCGTT

      4210       4220       4230       4240       4250       4260       4270       4280       4290       4300
CAGTTCTATT ACTAAACACA GCTTGACTGA GGAATGCATG CAGATCTACA GGAAAGTCTT TAGGGTCTTC TACCTTTTTT TTCTTTTTAG GTGGGGTAGA
GTCAAGATAA TGATTTGTGT CGAACTGACT CCTTACGTAC GTCTAGATGT CCTTTCAGAA ATCCCAGAAG ATGGAAAAAA AAGAAAAATC CACCCCATCT

      ----  --       4320       4330       4340       4350       4360       4370       4380       4390       4400
GTGTTGGGAT CCTGTGTTTT CATCATCACT GGCAAACATT TCTTCATGGC AAAACAGGTC TTCATCCCAC TTCTCATTAA ATGTATTCCA CCAGGATTCC
CACAACCCTA GGACACAAAA GTAGTAGTGA CCGTTTGTAA AGAAGTACCG TTTTGTCCAG AAGTAGGGTG. AAGAGTAATT TACATAAGGT GGTCCTAAGG
     BamH I
```


(Continued)

```
      4410       4420       4430       4440       4450       4460       4470       4480       4490       4500
CATTCATCTG TTCCATAGGT TGGCACCTAA AAAAAAACAA TTAAGTTTAT TGTAAAAAAC AAAATGCCCT GCAAAAGAAA AATAGTGGTT TACCTTAAAG
GTAAGTAGAC AAGGTATCCA ACCGTGGATT TTTTTTTGTT AATTCAAATA ACATTTTTTG TTTTACGGGA CGTTTTCTTT TTATCACCAA ATGGAATTTC
 Large T EXON 2 Start <---¦¦--->    Large T INTRON End                                                  ¦--->

      4510       4520       4530       4540       4550       4560       4570       4580       4590       4600
CTTTAGATCC CTGTAGGGGG TGTCTCCAAG AACTTTCTCC CAGCAATGAA GAGCTTCTTG GGTTAAGTCA CACCCAAACC ATTGTCTGAA GCAATCAAAG
GAAATCTAGG GACATCCCCC ACAGAGGTTC TTGAAAGAGG GTCGTTACTT CTCGAAGAAC CCAATTCAGT GTGGGTTTGG TAACAGACTT CGTTAGTTTC
Small t CDS End

      4610     ------ ------4630       4640       4650       4660       4670       4680       4690       4700
CAATAGCAAT CTATCCACAC AAGTGGGCTG CTTCTTAAAA ATTTTCTGTT TCTATGCCTT AATTTTAGCA TGCACATTAA ACAGGGGCAA TGCACTGAAG
GTTATCGTTA GATAGGTGTG TTCACCCGAC GAAGAATTTT TAAAAGACAA AGATACGGAA TTAAAATCGT ACGTGTAATT TGTCCCCGTT ACGTGACTTC
                 BstX I

      4710       4720       4730       4740       4750       4760       4770       4780       4790       4800
GATTAGTGGC ACAGTTAGGC CATTCCTTGC AATAAAGGGT ATCAGAATTA GGAGGAAAAT CACAACCAAC CTCTGAACTA TTCCATGTAC CAAAATCAGG
CTAATCACCG TGTCAATCCG GTAAGGAACG TTATTTCCCA TAGTCTTAAT CCTCCTTTTA GTGTTGGTTG GAGACTTGAT AAGGTACATG GTTTTAGTCC
                                                 Large T INTRON     <---¦ ¦---> Large T EXON 1 CDS End

      4810       4820       4830       4840       4850       4860       4870       4880       4890       4900
CTGATGAGCA ACTTTTACAC CTTGTTCCAT TTTTTTATAT AAAAAATTCA TTCTCTTCAT CTTGTCTTCG TCCCCACCTT TATCAGGGTG GAGTTCTTTG
GACTACTCGT TGAAAATGTG GAACAAGGTA AAAAAATATA TTTTTTAAGT AAGAGAAGTA GAACAGAAGC AGGGGTGGAA ATAGTCCCAC CTCAAGAAAC

      4910       4920       4930       4940       4950       4960       4970       4980       4990       5000
CATTTTTTCA GATAAGCTTT TCTCATGACA GGAATGTTCC CCCATGCAGA CCTATCAAGG CCTAATAAAT CCATAAGCTC CATGGATTCC TCCCTATTCA
GTAAAAAAGT CTATTCGAAA AGAGTACTGT CCTTACAAGG GGGTACGTCT GGATAGTTCC GGATTATTTA GGTATTCGAG GTACCTAAGG AGGGATAAGT

      5010       5020       5030       5040       5050       5060       5070       5080       5090       5100
GCACTTTGTC CATTTTAGCT TTTTGCAGCA AAAAATTACT GCAAAAAAGG GAAAAACAAG GGAATTTCCC TGGCCTCCTA AAAAGCCTCC ACGCCCTTAC
CGTGAAACAG GTAAAATCGA AAAACGTCGT TTTTTAATGA CGTTTTTTCC CTTTTTGTTC CCTTAAAGGG ACCGGAGGAT TTTTCGGAGG TGCGGGAATG
Large T EXON 1,¦
Small t CDS<---¦

      5110       5120       5130
TACTTCTGAG TAAGCTTGGA GGCGGAGGCG
ATGAAGACTC ATTCGAACCT CCGCCTCCGC
                 ¦.....ORGREP CORE
```

Restriction Enzyme Sites in JC Virus (Madl Strain) DNA

Restriction enzyme[1,2]	Recognition pattern[3]	Number of sites	Base number matched[4]
Acc I	GT/(AC)(GT)AC	5	412, 2835, 3952, 4108, 4179
[Afl II]	C/TTAAG	2	3174, 3908
[Afl III]	A/C(AG)(CT)GT	1	231
Aha III	TTT/AAA	1	1320
Alu I	AG/CT	33	58, 81, 156, 179, 249, 269, 290, 348, 483, 603, 699, 744, 759, 879, 1149, 1889, 2233, 2313, 2463, 2582, 2816, 2897, 2992, 3353, 3359, 4163, 4220, 4499, 4552, 4915, 4976, 5017, 5113
Apa I	GGGCC/C	1	984
Ava II	G/G(AT)CC	3	552, 1501, 2473
[Avr II]	C/CTAGG	2	66, 164
Bal I	TGG/CCA	3	272, 1254, 1716
BamH I	G/GATCC	1	4307
Ban I	G/G(CT)(AG)CC	3	529, 1883, 4422
Ban II	G(AG)GC(CT)/C	5	57, 155, 347, 743, 984
Bbv I->	GCAGC(N)$_8$/	9	219, 873, 2311, 2401, 2895, 2990, 3357, 3472, 5025
<-Bbv I	/(N)$_{12}$GCTGC	14	41, 139, 577, 580, 583, 613, 625, 628, 700, 769, 2898, 3354, 3630, 4627
[BbvS I]	GC(AT)GC	23	41, 139, 219, 577, 580, 583, 613, 625, 628, 700, 769, 873, 2311, 2401, 2895, 2898, 2990, 3354, 3357, 3472, 3630, 4627, 5025

[1] Restriction enzyme names in brackets (e.g., [AflII]) are not commercially available as of September, 1985.
[2] The direction of an asymmetric cut is indicated by –> or <– in the restriction enzyme name.
[3] The restriction enzyme recognition sequence (5′– –>3′). The position of endonuclease cleavage is indicated (/). Ambiguous bases in the recognition sequence are enclosed in parentheses.
[4] Numbers represent the first base in the sequence of the recognition site for a restriction endonuclease.

(Continued)

Restriction Enzyme Sites in JC Virus (Madl Strain) DNA (*Continued*)

Restriction enzyme	Recognition pattern	Number of sites	Base number matched
Bgl II	A/GATCT	2	1390, 4242
[Bin I->]	GGATC(N)$_4$/	3	840, 1747, 4307
[<-Bin I]	/(N)$_5$GATCC	6	970, 2078, 2739, 3739, 4308, 4506
Bsm I	GAATGCN/	4	939, 1589, 1987, 4232
Bsm I	G/CATTC	1	1292
Bsp1286 I	G(AGT)GC(ACT)/C	10	57, 155, 347, 743, 984, 1829, 1882, 2021, 3532, 3961
BstE II	G/GTNACC	1	1613
BstN I	CC/(AT)GG	10	326, 931, 1011, 1372, 2468, 2704, 3096, 3488, 4391, 5069
BstX I	CCANNNNN/NTGG	1	4615
[Cfr I]	(CT)/GGCC(AG)	3	272, 1254, 1716
Dde I	C/TNAG	17	449, 572, 689, 809, 1058, 1631, 1797, 2315, 2698, 2984, 2994, 3006, 3163, 3787, 3869, 4227, 5106
Dpn I	GmA/TC	12	841, 970, 1391, 1748, 2078, 2739, 3268, 3343, 3739, 4243, 4308, 4506
[Dra II]	(AG)G/GNCC(CT)	6	551, 983, 1201, 1500, 2472, 3912
EcoR I*	/AATT	34	363, 377, 609, 648, 893, 1065, 1141, 1162, 1349, 1576, 1723, 1926, 1946, 1975, 2187, 2347, 2371, 3092, 3142, 3285, 3542, 3568, 3689, 3852, 4001, 4043, 4157, 4439, 4640, 4661, 4746, 4845, 5034, 5063
EcoR I	G/AATTC	1	1722

Restriction Enzyme Sites in JC Virus (Madl Strain) DNA (*Continued*)

Restriction enzyme	Recognition pattern	Number of sites	Base number matched
EcoR I'	(AG)(AG)A/T(CT)(CT)	31	356, 376, 592, 823, 969, 1023, 1064, 1317, 1348, 1390, 1722, 1747, 2186, 2766, 2853, 2962, 3027, 3044, 3213, 3484, 3688, 4156, 4242, 4307, 4394, 4505, 4639, 4844, 4967, 4984, 5062
EcoR II	/CC(AT)GG	10	326, 931, 1011, 1372, 2468, 2704, 3096, 3488, 4391, 5069
Fnu4H I	GC/NGC	25	41, 139, 219, 532, 577, 580, 583, 613, 625, 628, 700, 769, 873, 2311, 2401, 2895, 2898, 2990, 3354, 3357, 3472, 3630, 4627, 5025, 5128
Fok I->	GGATG(N)$_9$/	5	36, 134, 1406, 2419, 3607
<-Fok I	/(N)$_{13}$CATCC	4	653, 3244, 3814, 4363
[Gsu I]	CTGGAG	4	327, 617, 740, 932
[Gsu I]	CTCCAG	4	1193, 1602, 2139, 2293
[Hae I]	(AT)GG/CC(AT)	11	272, 708, 1254, 1716, 2915, 3013, 3049, 3490, 4717, 4958, 5071
Hae III	GG/CC	19	6, 273, 709, 985, 1202, 1255, 1471, 1717, 2199, 2916, 3014, 3050, 3387, 3491, 3913, 4718, 4959, 5072, 5130
HgiA I	G(AT)GC(AT)/C	6	57, 155, 347, 743, 1829, 2021
Hinc II	GT(CT)/(AG)AC	4	949, 1568, 2495, 3614
Hind III	A/AGCTT	3	4498, 4914, 5112
Hinf I	G/ANTC	12	1023, 1571, 1679, 2289, 2383, 2727, 2751, 2766, 3222, 3484, 4395, 4985
Hpa I	GTT/AAC	1	949

(*Continued*)

Restriction Enzyme Sites in JC Virus (Madl Strain) DNA (*Continued*)

Restriction enzyme	Recognition pattern	Number of sites	Base number matched
Hpa II	C/CGG	1	748
Hph I->	GGTGA(N)$_8$/	5	400, 1345, 1613, 1810, 2213
<-Hph I	/(N)$_7$TCACC	2	264, 1144
Mae I	C/TAG	12	67, 165, 556, 974, 1184, 1315, 1550, 2036, 2388, 2778, 2974, 3063
[Mae II]	A/CGT	2	297, 2493
Mae III	/GTNAC	15	263, 294, 787, 1303, 1441, 1614, 1677, 1707, 1784, 1811, 2214, 3087, 3846, 4122, 4567
Mbo I	/GATC	12	841, 970, 1391, 1748, 2078, 2739, 3268, 3343, 3739, 4243, 4308, 4506
Mbo II->	GAAGA(N)$_8$/	6	403, 907, 1073, 1117, 1465, 4548
<-Mbo II	/(N)$_7$TCTTC	13	282, 2146, 2754, 2769, 3591, 3876, 3885, 3930, 4266, 4341, 4359, 4854, 4865
Mnl I->	CCTC(N)$_7$/	15	2, 8, 686, 1055, 1205, 1399, 2138, 2773, 2882, 3247, 3817, 4770, 4989, 5074, 5086
<-Mnl I	/(N)$_7$GAGG	27	574, 622, 670, 691, 1234, 1376, 1533, 1536, 1583, 1745, 1778, 1799, 1916, 2124, 2286, 2326, 2432, 2435, 2450, 2459, 2614, 2691, 3469, 4229, 4752, 5119, 5125
Nci I	CC/(CG)GG	1	748
Nco I	C/CATGG	2	275, 4980
Nde I	CA/TATG	3	725, 1460, 3943

Restriction Enzyme Sites in JC Virus (Madl Strain) DNA (*Continued*)

Restriction enzyme	Recognition pattern	Number of sites	Base number matched
Nla III	CATG/	31	54, 152, 232, 276, 525, 1696, 1771, 1853, 2011, 2251, 2416, 2482, 2743, 2840, 2887, 3083, 3199, 3529, 3602, 3715, 3794, 3858, 3901, 4105, 4237, 4345, 4669, 4784, 4924, 4943, 4981
Nla IV	GGN/NCC	18	71, 169, 529, 551, 983, 984, 1237, 1471, 1501, 1883, 2071, 2270, 2472, 2473, 3387, 3913, 4307, 4422
Nsi I	ATGCA/T	2	3432, 4234
[NspB II]	C(AC)G/C(GT)G	4	268, 289, 2232, 2896
[Nsp7524 I]	(AG)CATG/(CT)	7	231, 1695, 2250, 2839, 3793, 4236, 4668
Pss I	(AG)GGNCC(CT)	6	551, 983, 1201, 1500, 2472, 3912
Pst I	CTGCA/G	1	3355
Pvu II	CAG/CTG	4	268, 289, 2232, 2896
Rsa I	GT/AC	9	683, 1431, 1960, 2032, 2559, 3404, 3791, 4017, 4787
Sac I	GAGCT/C	4	57, 155, 347, 743
Sau96 I	G/GNCC	10	552, 984, 985, 1202, 1471, 1501, 2198, 2473, 3387, 3913
ScrF I	CC/NGG	11	326, 748, 931, 1011, 1372, 2468, 2704, 3096, 3488, 4391, 5069
SfaN I->	GCATC(N)$_5$/	2	652, 1627
<-SfaN I	/(N)$_9$GATGC	3	1282, 2420, 3608
[Sna I]	GTATAC	2	2835, 4179
Sph I	GCATG/C	2	4236, 4668

(*Continued*)

Restriction Enzyme Sites in JC Virus (Madl Strain) DNA (*Continued*)

Restriction enzyme	Recognition pattern	Number of sites	Base number matched
Ssp I	AAT/ATT	4	955, 3073, 4085, 4129
Stu I	AGG/CCT	3	708, 3013, 4958
Sty I	C/C(AT)(AT)GG	6	66, 164, 223, 275, 3832, 4980
[Taq II->]	CACCCA(N)$_{11}$/	2	1266, 4571
[<-Taq II]	/(N)$_9$TGGGTG	2	527, 1611
[Tce I]	GAAGA	6	403, 907, 1073, 1117, 1465, 4548
[Tce I]	TCTTC	13	282, 2146, 2754, 2769, 3591, 3876, 3885, 3930, 4266, 4341, 4359, 4854, 4865
Tthlll II->	CAA(AG)CA(N)$_9$/	8	50, 148, 1844, 2157, 2259, 3476, 3559, 4333
<-Tthlll II	/(N)$_{11}$TG(CT)TTG	3	2178, 2734, 3624
Xba I	T/CTAGA	2	1183, 1549
Xho II	(AG)/GATC(CT)	6	969, 1390, 1747, 4242, 4307, 4505
Xmn I	GAANN/NNTTC	3	2129, 3922, 4548

Restriction Enzymes That Do Not Cut JC Virus DNA

Enzyme[1,2]	Recognition site[3]	Enzyme	Recognition site	Enzyme	Recognition site
Aat II	GACGT/C	[Gdi II]	CGGCC/(AG)	Rsr II	CG/G(AT)CCG
Aha II	G(AG)/CG(CT)C	Hae II	(AG)GCGC/(CT)	Sac II	CCGC/GG
Asu II	TT/CGAA	Hga I->	GACGC$(N)_5$/	Sal I	G/TCGAC
Ava I	C/(CT)CG(AG)G	<-Hga I	/$(N)_{10}$GCGTC	Sca I	AGT/ACT
Bcl I	TG/ATCA	[HgiE II]	ACCNNNNNNGGT	Sfi I	GGCCNNNN/NGGCC
Bgl I	GCCNNNN/NGGC	Hha I	GCG/C	Sma I	CCC/GGG
BssH II	G/CGCGC	Kpn I	GGTAC/C	SnaB I	TAC/GTA
[Cfr10 I]	(AG)CCGG(CT)	Mlu I	A/CGCGT	Spe I	A/CTAGT
Cla I	AT/CGAT	Mst II	CC/TNAGG	Taq I	T/CGA
[Dra III]	CACNNN/GTG	Nae I	GCC/GGC	[Taq II->]	GACCGA$(N)_{11}$/
[Eco47 III]	AGCGCT	Nar I	GG/CGCC	[<-Taq II]	/$(N)_9$TCGGTC
EcoR V	GAT/ATC	Nhe I	G/CTAGC	Tha I	CG/CG
[Esp I]	GC/TNAGC	Not I	GC/GGCCGC	Tth111 I	GACN/NNGTC
Fsp I	TGC/GCA	Nru I	TCG/CGA	Xho I	C/TCGAG
[Gdi II]	(CT)/GGCCG	Pvu I	CGAT/CG	Xma III	C/GGCCG

[1] Restriction enzyme names in brackets (e.g., [AflII]) are not commercially available as of September, 1985.
[2] The direction of an asymmetric cut is indicated by –> or <– in the restriction enzyme name.
[3] The restriction enzyme recognition sequence (5′ – –>3′). The position of endonuclease cleavage is indicated (/). Ambiguous bases in the recognition sequence are enclosed in parentheses.

Index